University of Plymouth Library
Subject to status this item may be renewed
via your Voyager account
http://voyager.plymouth.ac.uk
Tel: (01752) 232323

Plant Pigments and their Manipulation

Annual Plant Reviews

A series for researchers and postgraduates in the plant sciences. Each volume in this series focuses on a theme of topical importance, and emphasis is placed on rapid publication.

Editorial Board:

Titles in the series:

1. Arabidopsis
Edited by M. Anderson and J. A. Roberts

2. Biochemistry of Plant Secondary Metabolism
Edited by M. Wink

3. Functions of Plant Secondary Metabolites and Their Exploitation in Biotechnology
Edited by M. Wink

4. Molecular Plant Pathology
Edited by M. Dickinson and J. Beynon

5. Vacuolar Compartments
Edited by D. G. Robinson and J. C. Rogers

6. Plant Reproduction
Edited by S. D. O'Neill and J. A. Roberts

7. Protein–Protein Interactions in Plant Biology
Edited by M. T. McManus, W. A. Laing and A. C. Allan

8. The Plant Cell Wall
Edited by J. K. C. Rose

9. The Golgi Apparatus and the Plant Secretory Pathway
Edited by D. G. Robinson

10. The Plant Cytoskeleton in Cell Differentiation and Development
Edited by P. J. Hussey

11. Plant–Pathogen Interactions
Edited by N. J. Talbot

12. Polarity in Plants
Edited by K. Lindsey

13. Plastids
Edited by S. G. Møller

14. Plant Pigments and their Manipulation
Edited by K. M. Davies

Plant Pigments and their Manipulation

Edited by

KEVIN M. DAVIES
Crop & Food Research
Palmerston North
New Zealand

Blackwell
Publishing

CRC Press

© 2004 by Blackwell Publishing Ltd

Editorial offices:
Blackwell Publishing Ltd, 9600 Garsington Road, Oxford OX4 2DQ, UK
  Tel: +44 (0)1865 776868
Blackwell Publishing Asia Pty Ltd, 550 Swanston Street, Carlton, Victoria 3053, Australia
  Tel: +61 (0)3 8359 1011

ISBN 1–4051–1737–0
ISSN 1460–1494

Published in the USA and Canada (only) by
CRC Press LLC, 2000 Corporate Blvd., N.W., Boca Raton, FL 33431, USA
Orders from the USA and Canada (only) to
CRC Press LLC

USA and Canada only:
ISBN 0–8493–2350–9
ISSN 1097–7570

The right of the Author to be identified as the Author of this Work has been asserted in accordance with the Copyright, Designs and Patents Act 1988.

This book contains information obtained from authentic and highly regarded sources. Reprinted material is quoted with permission, and sources are indicated. Reasonable efforts have been made to publish reliable data and information, but the author and the publisher cannot assume responsibility for the validity of all materials or for the consequences of their use.

Trademark notice: Product or corporate names may be trademarks or registered trademarks, and are used only for identification and explanation, without intent to infringe.

First published 2004

Library of Congress
Cataloging-in-Publication Data:
A catalog record for this title is available from
the Library of Congress

British Library Cataloguing-in-Publication
Data:
A catalogue record for this title is available
from the British Library

Set in 10/12pt Times
by Kolam Information Services Pvt. Ltd, Pondicherry, India
Printed and bound in Great Britain
by MPG Ltd, Bodmin, Cornwall

The publisher's policy is to use permanent paper from mills that operate a sustainable forestry policy, and which has been manufactured from pulp processed using acid-free and elementary chlorine-free practices. Furthermore, the publisher ensures that the text paper and cover board used have met acceptable environmental accreditation standards.

For further information on
Blackwell Publishing, visit our website:
www.blackwellpublishing.com

Contents

5 Condensed tannins   150
GREG TANNER

List of contributors

Professor Øyvind M. Andersen Department of Chemistry, University of Bergen, Allegt. 41, N-5007 Bergen, Norway

Dr Laurent Christinet Laboratory of Plant Cell Genetics, Department of Plant Molecular Biology, Université de Lausanne, CH 1015 Lausanne, Switzerland

Ms Abby J. Cuttriss School of Biochemistry and Molecular Biology, Australian National University, Canberra, ACT 0200, Australia

Dr Kevin M. Davies New Zealand Institute for Crop & Food Research Limited, Private Bag 11-600, Palmerston North, New Zealand

Professor George W. Francis Department of Chemistry, University of Bergen, Allegt. 41, N-5007 Bergen, Norway

Professor Brian R. Jordan Soil, Plant & Ecological Sciences Division, PO Box 84, Lincoln University, Canterbury, New Zealand

Professor Mary Ann Lila Department of Natural Resources & Environmental Sciences, University of Illinois, Urbana, Illinois 61801, USA

Dr Barry J. Pogson School of Biochemistry and Molecular Biology, Australian National University, Canberra, ACT 0200, Australia

Dr Kathy E. Schwinn New Zealand Institute for Crop & Food Research Limited, Private Bag 11-600, Palmerston North, New Zealand

Dr Greg J. Tanner   CSIRO Plant Industry, GPO Box 1600, Canberra ACT 2601, Australia

Dr Robert D. Willows   Department of Biological Sciences, Macquarie University, North Ryde, NSW 2109, Australia

Professor Jean-Pierre Zrÿd Laboratory of Plant Cell Genetics, Department of Plant Molecular Biology, Université de Lausanne, CH 1015 Lausanne, Switzerland

Preface

It is difficult to overstate the importance of plant pigments in biology. Chlorophylls are arguably the most important organic compounds on earth, as they are required for photosynthesis. Carotenoids are also necessary for the survival of both plants and mammals, through their roles in photosynthesis and nutrition, respectively. The other plant pigment groups, such as flavonoids and betalains, also have important roles in both the biology of plants and the organisms with which plants interact. For example, the flavonoids (both pigmented and non-pigmented) serve biological functions in plants as diverse as signalling to micro-organisms, protecting against pathogens, ameliorating biotic and abiotic stress, influencing auxin transport, enabling plant fertility and providing visual signals to insects and animals for pollination and seed dispersal.

Human use of plant pigment extracts dates back 10 000 years, and it is thus not surprising that plant pigmentation is one of the oldest subjects in formal plant science. The conspicuousness of floral colour traits and the easy identification of mutants have led to many discoveries with impacts much wider than in the pigmentation field alone, from the studies of Mendel, Darwin, McClintock and others. The structural diversity of plant pigments has attracted chemists since the beginning of the twentieth century, and for carotenoids and anthocyanins well over 600 individual structures have now been defined for each class. Given the wealth of data on plant pigments, this volume of *Annual Plant Reviews* cannot hope to be exhaustive on the subject. Rather we have tried to provide an overview of pigment chemistry and biology, and an up-to-date account of pigment biosynthesis and the modification of their production using biotechnology.

Chapter 1, the introduction, presents an overview of pigmentation and some of the general functions of plant pigments. Following this are Chapters 2 to 6, covering the major plant pigment groups: chlorophylls, carotenoids, flavonoids (the coloured flavonoids of flowers and tannins) and betalains. Chapter 7 presents an overview of pigments outside these groups that are of particular significance in biology or commerce. In addition to these, there are three chapters focusing on research areas that are common across the pigment groups. Chapters 8 and 9 provide comprehensive reviews of two areas of particular current interest – the roles of plant pigments in human health and the amelioration of the effects of UV radiation, respectively. Chapter 10 provides an overview of modern techniques used for the extraction, separation, identification and quantification of the major pigment classes.

Together the chapters cover a wide scope of pigmentation research, from the importance of structural diversity in generating the range of colours seen in plants through to improving human health properties of crops by increasing pigment levels in transgenic plants. We hope that this volume will be of use to researchers, professionals and advanced students in both the academic and industry sectors. The contributors come from Australasia, Europe and the USA, and it is my pleasure as the Editor to thank them all for the time they have put into preparing chapters of a consistently high standard. I am also grateful to Graeme MacKintosh and David McDade at Blackwell Publishing for their assistance in preparing the manuscript. Finally, my thanks to all the present and past members of the Plant Pigments Group of Crop & Food Research for their hard work on flavonoid biosynthesis over many years, to our long-term collaborators Cathie Martin and Ken Markham for their encouragement of our group, and to Crop & Food Research for allowing me the time needed for this book.

Kevin M. Davies

1 An introduction to plant pigments in biology and commerce

Kevin M. Davies

1.1 Introduction

This introductory chapter presents a general overview of plant pigmentation, together with some general functional and economic aspects not covered in detail in the chapters on specific pigment groups.

1.2 Plant pigmentation

1.2.1 The physical basis of pigmentation

Plant pigmentation is generated by the electronic structure of the pigment interacting with sunlight to alter the wavelengths that are either transmitted or reflected by the plant tissue. The specific colour perceived will depend on the abilities of the observer. Humans without colour blindness can detect wavelengths between approximately 380 and 730 nm, representing the visible spectrum of red, orange, yellow, green, blue, indigo and violet. So chlorophyll with peak absorbancies at 430 and 680 nm will leave wavelengths forming a green colour. Of course, often the colours are the result of a mix of residual wavelengths; for example, anthocyanins absorbing yellow-green light wavelengths of 520–530 nm will generate mauve colours formed by the reflection of a mix of orange, red and blue wavelengths. Thus the pigments can be described in two ways: the wavelength of maximum absorbance (λ_{max}) and the colour perceived by humans. Further details of the generation of colours and the behaviour of light in plant tissues can be found in Hendry (1996) and Chapter 10 of this volume.

The names of many common pigments convey little information to the general reader, as they tend to reflect historical discoveries rather than a set naming system. For example, carotene was first isolated from *Daucus carota* (carrot), violaxanthin from *Viola tricolor* (pansy), and the common anthocyanidins, pelargonidin, cyanidin, peonidin, delphinidin, petunidin and malvidin from *Pelargonium, Centaurea, Paeonia, Delphinium, Petunia* and *Malva*, respectively. However, these trivial names are often well-established, familiar to workers in the field and allow easy flow of text. Thus, the trivial names are used extensively in this book. More complete names, giving details of constitution and stereochemistry, have been developed for many compounds to meet the standards of the

International Union of Pure and Applied Chemistry (IUPAC) and International Union of Biochemistry (IUB) (Weedon & Moss, 1995). The details of such nomenclature, and lists of the IUPAC semi-systematic names, are available for some plant pigment groups in more specialised publications (Pfander, 1987; Harborne, 1988, 1994; Kull & Pfander, 1995; Weedon & Moss, 1995). Bohm (1998) gives a guide to relating the trivial flavonoid names to those in the *Chemical Abstracts*. In the same book he also provides a table listing the meanings of the trivial names often used for flavonoid di- and trisaccharides (e.g. sophoroside). The trivial names in this book are often accompanied by structural diagrams of the compounds, providing much of the information that would come from the full name. These diagrams commonly include basic representations of the stereochemistry, with a solid triangle for a bond representing above the plane of the paper and a dashed triangle for below the plane of the paper.

1.2.2   Structural variation of plant pigments

Plant pigments exist in many varied forms, some with highly complex and large structures. For example, over 600 naturally occurring carotenoid structures have been identified (Britton *et al.*, 1995) and over 7000 flavonoids, including over 500 anthocyanins (Chapter 10). The complexity of some pigments is well illustrated by the anthocyanin Ternatin A1, which consists of the base 15-carbon anthocyanin modified with seven molecules of glucose, four molecules of 4-coumaric acid and one molecule of malonic acid, corresponding to $C_{96}H_{107}O_{53}^+$ (Terahara *et al.*, 1990). In this book we have grouped plant pigments on a common structural and biosynthetic basis into four major groups (Table 1.1), and Chapters 2–6 focus on these pigments. In addition to these major groups there is a great array of pigments that are of limited taxonomic occurrence, and often poorly characterised. Some of the more notable of these are covered in Chapter 7.

Table 1.1 Major pigments of plants and their occurrence in other organisms

Pigment	Common types	Occurrence
Betalains	Betacyanins	The Caryophyllales and some fungi
	Betaxanthins	
Carotenoids	Carotenes	Photosynthetic plants and bacteria
	Xanthophylls	Retained from the diet by some birds, fish, and crustaceans
Chlorophylls	Chlorophyll	All photosynthetic plants
Flavonoids	Anthocyanins	Widespread and common in plants,
	Aurones	including angiosperms,
	Chalcones	gymnosperms, ferns, fern allies and
	Flavonols	bryophytes. Retained from the diet
	Proanthocyanidins	by some insects

The most obvious and widespread pigments of plants are, of course, the chlorophylls. These are cyclic tetrapyrrole pigments chelated with magnesium, and they share structural features with the haem and bile pigments of animals. Also associated with photosynthesis, but additionally providing bright colours to flowers and fruits, are the carotenoids. Carotenoids are terpenoid pigments present in all photosynthetic plants and they also occur in photosynthetic bacteria such as *Erwinia* and *Rhodobacter*. Annual production of carotenoids by plants, algae and dinoflagellates has been estimated at 100 million tons (Britton *et al.*, 1995). The flavonoids are phenylpropanoid compounds of widespread occurrence. There are several major classes of flavonoids; however, only a few of these provide pigments to plants, in particular the anthocyanins and proanthocyanidins (condensed tannins). Reviews of the biosynthesis and function of non-pigmented flavonoids can be found in earlier volumes in this Annual Plant Reviews series (Wink, 1999a, 1999b) and in Bohm (1998). The betalains are nitrogenous pigments that are the most taxonomically restricted of the major plant pigment groups, being found only in a few families of the order Caryophyllales and some fungi. Curiously, their occurrence is mutually exclusive to that of the anthocyanins.

Within plants, the major pigment groups show wide occurrence in the different tissues. For example, flavonoids occur in almost all tissues, carotenoids in leaves, roots, tubers, seeds, fruits and flowers, and even chlorophylls occur in flowers and fruits as well as leaves. Within tissues, there is often distinct localisation of pigment types in different cell layers. For example, anthocyanins are typically found in epidermal cells in petals and sub-epidermally in leaves, and chlorophyll in the sub-epidermal photosynthetic cell layers of leaves. The subcellular localisation of the different pigment groups is also generally distinct. The chlorophylls and carotenoids are principally lipid-soluble, plastid-located pigments, although there are examples of water-soluble carotenoids, at least some of which are located in the vacuole via plastid–vacuole interactions (Bouvier *et al.*, 2003a). The betalains are water-soluble and vacuolar-located. While flavonoids occur in many subcellular locations, as well as extracellularly, the coloured flavonoids are principally found in the vacuole (Bohm, 1998). For flavonoids, the subcellular localisation is just one of several factors that determine the behaviour of the pigment molecule in the cell and the colour generated from it (Brouillard & Dangles, 1993). The mechanisms by which pigments such as the flavonoids are directed to the correct subcellular compartment are poorly defined, although some of the steps for anthocyanins have been elucidated (Winefield, 2002). Interactions of plant pigments with other cellular compounds have been well defined for flavonoids and small molecules (Brouillard & Dangles, 1993), and it is known that both flavonoids and carotenoids interact with specific proteins in the cell (Vishnevetsky *et al.*, 1999; Winefield, 2002).

Modern phytochemical techniques can enable the rapid identification of the class of pigment present in a plant tissue of interest, and with more extensive analysis the detailed structure of the compound may be elucidated. The techniques required for identification and analysis of pigments in a plant of interest can be daunting to newer researchers in the field. With this in mind, Chapter 10 provides an overview of the modern and most recent techniques used for extraction, separation, identification and quantification of the major pigment classes in plants, providing sufficient detail to support the practical application of such techniques.

1.2.3   The history of plant pigment research

Pigmentation is one of the oldest subjects in formal plant science and has lead to many discoveries with impacts much wider than those in the pigmentation field alone. Early studies included discovery of the purple pigment from *Viola* as a natural pH indicator (Boyle, 1664), comparisons of the solubility of different pigments (Nehemiah Grew in 1682, quoted extensively in Onslow, 1925 and cited in Bohm, 1998), and Mendel's studies on the genetics of flower and seed colour in *Pisum sativum* – pea (Bhattacharya & Bhattacharya, 2001). The first publications on carotenoids date from the early nineteenth century, and detail the basis of colours of common food colourants of the time, some of which are still in use today, e.g. saffron and annatto (Eugster, 1995). The term chlorophyll was first used by Pelletier and Caventou (1818), and the different chlorophyll pigments were first separated from each other by Stokes (1864), Sorby (1873) and Tswett (1906) (cited in Jackson, 1976 and Eugster, 1995). Tswett's research included the development of column chromatography, one of the basic methods of modern biochemistry, and allowed the ground-breaking work of Willstätter on the structure of carotenoids. The conspicuousness of the flower colour trait and the easy identification of mutants has established it as a favourite system for geneticists ever since Mendel, from pioneering studies such as those of Darwin (1868), Onslow/Whedale (1925/1907) through to transposon studies and recent breakthroughs such as those on transcriptional regulation in plants featured in Chapter 4. Studies on transposons in pigment genes of maize were part of the work that earned Barbara McClintock a Nobel Prize in 1983, and plant pigments have also featured in research earning Nobel prizes for Willstätter (1915), Fischer (1930, the chemistry of blood pigments and chlorophyll), Karrer (1937, work on the structures of plant pigments, in particular carotenoids) and Kuhn (1938, work on the structure of carotenoids and related vitamins). Studies on anthocyanins have been responsible for several important breakthroughs in plant science, including the isolation and identification of a plant transcription factor gene for the first time (Cone *et al.*, 1986; Paz-Ares *et al.*, 1986), the isolation of one of the first cDNAs for a plant cytochrome P450 enzyme (Holton *et al.*, 1993), the first demonstration of antisense RNA technology in a transgenic plant (van der Krol *et al.*, 1988) and the first description of transgene co-suppression (Napoli *et al.*, 1990; van der Krol *et al.*, 1990).

1.2.4   The biosynthesis of plant pigments

The biosynthetic pathways for the major pigments of plants are now well defined at the genetic and enzymatic level. Of course, some knowledge gaps do exist, but these are becoming fewer. In the case of flavonoids, much is also known on the transcriptional regulation of the pathway. However, little is known on gene regulation mechanisms for any of the other major pigmentation pathways. Changes in the transcription rate of pigment biosynthetic genes usually precede pigment production, and it is thought that transcriptional regulation is the major controlling step for the pigmentation pathways studied to date. However, translational and post-translational regulation may also be important in specific cases, particularly for the photosynthetic pigments. Indeed, the general emphasis on transcriptional regulation may in part be due to lack of appropriate studies to identify post-transcriptional control. This is illustrated by the recent use of genomics to show rhythmic expression of many *Arabidopsis thaliana* genes, including those of the phenylpropanoid and carotenoid pathways, with the potential involvement of changing RNA stability in generating the rhythms (Staiger, 2002).

Although the biosynthetic pathways of many plant pigments are now well defined, little is known of the turnover and degradation of most pigments. There is an obvious loss of chlorophylls during the autumn senescence of leaves, and some information is available on chlorophyll degradation. However, anthocyanin levels can also change rapidly, e.g. in flowers of the Yesterday, Today and Tomorrow plant (*Brunfelsia calcina*), which can turn from white to purple and back to white within three days. Simpson *et al.* (1976) provided a comprehensive review on the knowledge to that date on the metabolism of the various pigment groups. However, this earlier chemistry and biochemistry has not been followed by extensive findings on the molecular biology of the metabolism of pigments in plants. Indeed, the metabolism is perhaps better understood for the fate of pigments when absorbed from the diets of animals than it is for the plant tissues themselves. One gene has been identified that has a direct impact on colour fading of flowers, the *Fading* locus of *Petunia hybrida* (de Vlaming *et al.*, 1982). Fully coloured flowers turn rapidly to white towards the end of flower development when the *Fading* gene is dominant and the genetic background is appropriate with regard to other genes that affect the type of anthocyanin present and the flower petal pH. Active degradation of the anthocyanins is thought to occur, rather than a change to a colourless form of anthocyanin (Schram *et al.*, 1984).

1.3 The functions of pigments in plants

1.3.1 The function of pigments in vegetative tissues

Chlorophylls and carotenoids are required for photosynthesis, chlorophylls for the capture of light energy and as the primary electron donors and carotenoids as essential structural components of the photosynthetic apparatus, where they protect against photo-oxidation. The roles of the pigments in these processes are covered in Chapters 2 and 3. Plant pigments are also involved in other interactions of plants with light, in particular the response to UV radiation (described in detail in Chapter 9), which is of growing concern with regard to changes in the global environment.

Anthocyanins also frequently occur in vegetative tissues. The most spectacular example is their contribution to autumn colours in leaves of many deciduous species, which they generate in combination with the retention of carotenoids and loss of chlorophyll (Matile, 2000; Hoch *et al.*, 2001; Lee, 2002). In non-senescing tissues their occurrence is more sporadic. Some species accumulate them in significant amounts in healthy leaves, providing red or purple colours to the foliage. In other cases anthocyanin production is induced in leaves in response to stresses such as cold, high light levels, pest and pathogen attack or deficiency of nutrients such as phosphate and nitrogen. Anthocyanin colouration in leaves can vary with season, environment, between individuals of a population and between different leaves on a single plant. It is commonly thought that anthocyanins have a role in protecting the photosynthetic apparatus from damage in many of these situations, and those tissues that show more anthocyanin accumulation are often at greater photoinhibitory risk, e.g. during nutrient reabsorption in senescing leaves or in cold temperatures (Hoch *et al.*, 2001). However, the details of how anthocyanins achieve this are not determined. One hypothesis is that anthocyanins help attenuate the light levels, modifying the quantity and quality of light incident on the chloroplasts and thus reducing excitation pressure (Gould *et al.*, 2002a; Steyn *et al.*, 2002). However, this would not account for their accumulation in other stress situations. An alternative is that anthocyanins are acting as both direct light screens under high light stress and general antioxidants against harmful reactive oxygen species in the various other stress situations in which they are prevalent (Gould *et al.*, 2002a, 2002b, 2002c). Supporting data include observations that red-leafed morphs of some shade species have a significant antioxidant advantage over green morphs, that anthocyanins can enhance oxidative protection in species more directly exposed to the sun, and that anthocyanins can reduce photoinhibition and photobleaching of chlorophyll under light stress conditions (Gould *et al.*, 2002c; Neill *et al.*, 2002; Steyn *et al.*, 2002). Recently, the role of anthocyanins in improving foliar nutrient reabsorption during senescence, through the shielding of the photosynthetic apparatus from excess light, was tested using wild-type and anthocyanin-deficient mutants of three deciduous woody species under varying environmental conditions (Hoch *et al.*, 2003). Nitrogen reabsorption efficiencies of the mutants were significantly lower than the wild-type counterparts, supporting the protection hypothesis of anthocyanins in senescing leaves.

There are likely to be many other functions of pigments in plants that have not been determined as yet. A recent example of a new function for anthocyanins is that of protecting light-sensitive phototoxic plant defence compounds from degradation, which was described by Page and Towers (2002). Thiarubrines are phototoxic plant pigments that decompose to thiophenes when exposed to sunlight. In *Ambrosia chamissonis*, they occur in laticifers that are surrounded by anthocyanin-containing cells. Page and Towers (2002) were able to show that the anthocyanins around the laticifers functioned to photoprotect these defence compounds.

1.3.2   The function of pigments in reproductive tissues

The most obvious function of plant pigments, with the exception of chlorophyll, is to provide colour to flowers and fruit for attraction of pollinators and seed-dispersal agents. These colours arise predominantly from flavonoid and carotenoid pigments, and a short guide to the likely pigments producing specific colours in flowers and fruits of plants is given in Table 1.2. There is, however, also a range of less common pigments that generate colours in specific species.

For many angiosperms, colour is key to attracting pollinators, whether they are bees, butterflies, other insects or birds, although it is frequently one of a number of factors, including fragrance, floral shape and nectar reward, which combine to determine pollinator choice. Flavonoids are the most common flower colour pigments and it is on these that most research has been done. The role of flavonoids in pollination was the subject of an extensive review by Harborne and Grayer (1994), in which the authors were able to identify many general trends with regard to pollinator preference for different colours. However, they also noted the shortage of detailed studies on specific pigments and pollinators. This has changed greatly in the last decade, with many studies determining pollinator preference with regard to individual colours (Melendez-Ackerman et al., 1997; Gumbert et al., 1999; Oberrath & Bohning-Gaese, 1999; Gigord et al., 2001; Johnson & Midgley, 2001; Jones & Reithel, 2001; Landeck, 2002), fragrances (Odell et al., 1999; Raguso & Willis, 2002) and even petal epidermal cell shape (Glover & Martin, 1998; Comba et al., 2000). These studies cover a mix of approaches, including observational field studies, laboratory studies using model flowers and studies of the frequency of natural colour variants or white-flowered mutants within a population. Table 1.3 presents some of the general colour preferences identified for different pollinators. Most beetle-pollinated flowers are cream, white or green, and it was thought that beetles were generally insensitive to anthocyanin colours, such as red and blue (Harborne & Grayer, 1994). However, beetle selection of specific cyanic colours has now been demonstrated in some studies, including that of Johnson and Midgley (2001), who found that monkey beetles (Scarabaeidae: Hopliini) preferred orange-coloured model flowers to red-, yellow- or blue-coloured ones.

For pollinators such as bees that can detect light in the UV spectrum, UV-absorbing pigments also influence flower selection. The main contributors to the UV absorbance of the flower are the chalcone- and flavonol-type flavonoids. Flavonols are very common in flowers, often being in greater abundance than the coloured pigments. The flavonoids may form UV-visible patterning in petals, often in combination with UV-reflective carotenoid pigments (Harborne & Grayer, 1994; Bohm, 1998).

Table 1.2 The most common pigment types associated with flower and fruit colours in plants. (The terms pelargonidin, cyanidin and delphinidin are used to refer to pelargonidin, cyanidin or delphinidin derived anthocyanins.)

Colour	Specific pigment type	Pigment group	Examples
Cream	Flavonols or flavones	Flavonoid	Most cream flowers
Pink to red		Carotenoid	Some red flowers and fruit, e.g. *Lycopersicon esculentum* (tomato) fruit
	Pelargonidin and/or cyanidin	Flavonoid	Most pink flowers and some fruit, e.g. *Eustoma grandiflorum* (lisianthus) flowers
	Pelargonidin and/or cyanidin	Flavonoid	Most red flowers and some fruit, e.g. *Malus* (apple) fruit
	Anthocyanin and carotenoid mix	Flavonoid and carotenoid	A few examples, e.g. *Tulipa* flowers
	Betacyanin	Betalain	A few examples in the Caryophyllales, e.g. *Bougainvillea* flowers
Orange		Carotenoid	Most orange flowers and fruit, e.g. *Tagetes erecta* (marigold) flowers
	Pelargonidin alone	Flavonoid	A few examples, e.g. *Pelargonium* flowers
	Anthocyanin and aurone mix	Flavonoid	Rare occurrence, e.g. *Antirrhinum majus* (snapdragon) flowers
	Anthocyanin and chalcone mix	Flavonoid	Rare occurrence, e.g. *Dianthus* (carnation) flowers
	Betacyanin	Betalain	A few examples in the Caryophyllales, e.g. *Portulaca* (purslane) flowers
Yellow		Carotenoid	Most yellow flowers and fruit
	Aurone	Flavonoid	Rare occurrence, e.g. *Antirrhinum majus* flowers
	Chalcone	Flavonoid	Rare occurrence, e.g. *Dianthus* flowers
	Flavonol	Flavonoid	Rare occurrence, e.g. *Gossypium* (cotton) flowers
	Betaxanthin	Betalain	A few examples in the Caryophyllales, e.g. *Portulaca* flowers
Green		Chlorophyll	All green flowers and fruit
Blue	Delphinidin	Flavonoid	Most blue flowers and fruit
	Cyanidin	Flavonoid	Rare occurrence, e.g. *Ipomoea* (morning glory) flowers
Purple		Carotenoid	Rare occurrence, e.g. *Capsicum* (pepper) fruit
	Cyanidin and/or delphinidin	Flavonoid	Most mauve flowers, e.g. *Petunia* and some purple fruit, e.g. *Solanum melongena* (eggplant)
		Flavonoid and carotenoid mix	Some flowers, e.g. *Cymbidium* orchids
Black	Delphinidin	Flavonoid and carotenoid mix	Some black flowers, e.g. *Viola* (pansy)

Table 1.3 Flower colour preferences of some pollinators, presented in terms of colour perceived by humans

Pollinator	Flower colour preference
Bees	Blue, yellow and UV-absorbing pigments
Birds	Bright red and scarlet
Beetles	White, cream, green and occasionally orange and red
Butterflies	Strong pinks, reds and mauves
Flies	White, green, dark brown and purple
Moths	White, cream and occasionally red
Wasps	Purple and blue

In some cases, colour combinations and floral patterning may help attract a range of pollinators, including organisms as varied as hummingbirds and bumble bees, or provide more specific signals within the flower (discussed extensively in Harborne & Grayer, 1994 and particularly Bohm, 1998). For example, distinctive spots on the flower lip or pigment lines in the flower tube may act as a nectar guide to bees. This is well illustrated by the yellow face and two yellow throat stripes of aurone pigment in *Antirrhinum majus* (snapdragon) flowers. There are a few cases of elegant plant–animal co-evolution in which colour patterning has been shown to be part of floral mimicry. In particular, flowers of the orchid genus *Ophrys* use scent, shape and colour to mimic female bees, causing the male bee to attempt copulation, thus achieving pollination (Schiestl *et al.*, 1999; Paxton & Tengo, 2001; Ayasse *et al.*, 2003). The relationship between different orchid species and different bee species can be highly specific, and it is likely based on the pattern of scent compounds produced by the flower (Schiestl *et al.*, 1999). There have been recent breakthroughs in understanding the genetic basis of colour patterning in flowers, discussed briefly in Chapter 4.

Change in flower colour during the later stages of flower development or in response to pollination has been recognised for many years. Indeed, as early as the late nineteenth century, Müller described the colour change of *Lantana* flowers from yellow to purple over an ageing period of three days, and the preference of pollinators for the younger, yellow flowers (see Weiss, 1991). That the yellow flowers were fertile, offered nectar and pollen that older flowers did not, and were preferred by the butterfly pollinators was later confirmed for *Lantana* species by Weiss (1991). The same study also offered a possible reason for maintenance of the older flowers – larger inflorescences were more successful in attracting butterflies. In most cases, a change in colour is likely to be associated with a change in nectar and pollen availability (Weiss, 1991, 1995; Harborne & Grayer, 1994; Bohm, 1998; Oberrath & Bohning-Gaese, 1999). At least 200 other plant genera contain species that show colour change during flower development and interact with a wide range of pollinator species, making it a common occurrence in plant reproduction (Weiss, 1991).

The specific studies on individual plant and pollinator species are supported at a higher level by general evolutionary and ecological trends (Harborne & Grayer, 1994; Gumbert et al., 1999). In particular, blue flower colours are more common in the temperate ecosystems, in which bees are key pollinators, while bright red colours are more prevalent in tropical ecosystems, in which other insects and birds are more important.

Carotenoids and flavonoids also commonly colour pollen, although their functions in pollen are not well elucidated. They have been shown to have a role in signalling to pollinators (Lunau, 2000), and it is possible they also have protective activities against various stresses. Colourless flavonoids are known to be involved in plant fertility in some species (Taylor & Jorgensen, 1992; Jorgensen et al., 2002), but this has not been shown for coloured flavonoids. The role of pigments in fruit is an obvious one, of signalling the ripeness of the fruit to seed-dispersal agents. Both carotenoids and flavonoids commonly provide fruit colours.

Flavonoids and carotenoids can also colour seeds, e.g. the yellow carotenoids and purple flavonoids of maize kernels. Such pigmentation may be related to timing of seed germination or plant defence. In some cases the pigments may reach very high levels. In the resinous seed coating of *Bixa orellana* (Fig. 1.1), which is the source of the commercial food colourant annatto, levels of the apocarotenoid bixin can reach 10% dry weight (Britton, 1996).

1.3.3   The roles of plant pigments in non-plant organisms

Many organisms absorb plant pigments from their diet, and may sequester them until they reach high levels. Flavonoid uptake has been demonstrated for a range of insects. In particular, butterflies and grasshoppers have been shown to take up the colourless or weakly pigmented flavonols and flavones, reaching levels of 2% dry weight of the wings of some butterflies. The sequestered compounds possibly act as visual attractants to mates (Harborne & Grayer, 1994; Bohm, 1998). Uptake of the more strongly pigmented flavonoids, such as anthocyanins, to the level of providing pigmentation to the insect has been implied by observation of insect colour on different plant food sources but it has not been characterised in detail. Flavonoids are taken up into the bloodstream in much smaller amounts by mammals, including humans, and their role in human health is a subject of much current research, which is covered in detail in Chapter 8. The polyphenolic tannins, reviewed in Chapter 5, also impact on animal health through the amelioration of bloat in ruminant animals.

Plant-derived carotenoids feature as key pigments in such familiar animal tissues as flamingo feathers (principally ketocarotenoids), shrimp and lobster shells, beetle shells, egg yolks and fish flesh (such as astaxanthin in goldfish) (Britton et al., 1995; Britton, 1996). In marine invertebrates, the carotenoids can occur as 'carotenoprotein' complexes. These can generate vivid colours, including blues, greens, reds and purples. Two examples of varying carotenoid–protein interactions in lobster are the blue astaxanthin-protein compound crustacyanin of

Fig. 1.1 The red-coloured flower, seed pods and seed (see inset) of the tropical bush *Bixa orellana*, which is the source of the natural food colourant annatto. (Photograph by the author.)

the carapace and the green astaxanthin-lipoglycoprotein compound ovoverdin of the ovaries (Zagalsky, 1995; Britton *et al.*, 1997). The effect of the protein on the carotenoid colour in invertebrates can often be observed upon cooking of the animal, as the heat can lead to the breakdown of the complex and a dramatic colour change (Britton *et al.*, 1995). Carotenoid occurrence in bird plumage has also been extensively studied in relation to the specific chemical structures that occur, which can be modified by the bird's metabolism (McGraw *et al.*, 2002), and the impact of the colours on mating behaviour (Olson & Owens, 1998). Like flavonoids, carotenoids are also of much interest with regard to their health-promoting effects in the human diet, a subject also discussed in Chapter 8.

An extreme example of the use of ingested pigments by animals, in this case from algae in the diet, is shown by photosynthetic molluscan sea slugs, in particular *Elysia chlortica* (Rumpho *et al.*, 2000). In these organisms, chloroplasts are taken up intact from the food source and are maintained intracellularly in specific cells that line the highly branched digestive system. The chloroplasts remain functional for at least nine months, providing both camouflage for the mollusc and a supply of carbon. The mollusc has been shown in the laboratory to be able to sustain itself in the absence of food by photoautotrophic CO_2 fixation. The uptake of chloroplasts is widespread in the molluscan order Ascoglossa, but in some cases this may represent use only for camouflage and not the remarkable interaction described for *E. chlortica*.

1.4   Economic aspects of plant pigments

Human use of plant pigment extracts dates back as long as recorded history. The most obvious use has been the application of pigments such as henna for tattooing and carthamin, indigo and other pigments to generate bright colours in clothing. For example, anthraquinone, indigoid and flavonoid pigments have been identified in the fourth century AD Egyptian textiles (Orska-Gawrys *et al.*, 2003). Further back than this the use of carthamin extract (from safflower, *Carthamus tinctorius*) to dye the wrappings of mummies has been reported, and there is written evidence from 4600 years ago documenting human use of indigo (Gilbert & Cooke, 2001). Furthermore, a key part of garment manufacture since the time of early human societies has been the tanning of animal leathers with polymeric phenols, with record use dating back 10 000 years (Bohm, 1998). Today there are still tanneries carrying out leather treatment and dyeing with plant pigments in much the same manner as they have been doing for centuries (Fig. 1.2).

From an economic perspective, putting to one side the vital role of chlorophylls and carotenoids in photosynthesis, the most obvious contribution of plant pigments to agriculture is with regard to consumer choice of fresh fruit, vegetables and floriculture products. However, the impact of the non-photosynthetic pigments is often much wider. For example, they are also of economic importance as

Fig. 1.2 Traditional tanning and dyeing of leather, including the use of many plant-derived compounds, in the Moroccan city of Fez. (Photograph by the author, taken in 2002.)

flavour and colour components of teas, wine and other beverages, as natural food colourants, for the health of ruminant animals, as plant defence agents and for amelioration of damaging UV light. The great range of non-coloured flavonoids and alkaloids produced by plants is often key to plant defence, and the biosynthesis of these compounds, and their importance to plant defence, agriculture and the pharmaceutical industries, have been extensively reviewed in previous Annual Plant Reviews (Wink, 1999a, 1999b).

1.4.1 Natural food colourants

Food quality is first assessed by its visual characteristics such as colour. Fresh food is often highly coloured by the major plant pigment groups, e.g. carotenoids and anthocyanins in fruit and chlorophylls in green vegetables. However, for processed foods the pigmentation is often lost during manufacturing, and the visual appeal of the final product is enhanced using added colourants. Until the discovery of synthetic dyes in the mid-nineteenth century, the food industry was solely reliant on natural food colourants. Although their use in many applications was superseded by synthetic dyes, in recent years there has been a return to the use of natural colourants, and an increased interest in new sources and improving their performance in food applications. Four plant pigment types are widely used as food colourants: annatto, anthocyanins, betalains (beetroot pigment) and curcumin (the main pigment of turmeric spice). Together with the insect-derived pigment cochineal, they account for over 90% of the market for natural food colourants (Hendry, 1996). Use of chlorophyll as a food colourant is very limited in comparison to these pigments, principally because of its poor stability during food processing or in response to light or acid conditions in the final food product. Plant pigments used as colourants in smaller amounts include the carotenoids xanthophyll and lutein, carthamus yellow from *C. tinctorius* petals, iridoids derived from *Gardenia* fruit and the carotenoid derivatives crocin and crocetin from *Crocus sativus* (saffron). One of the most important natural colourants is caramel, which is used extensively in both the food and beverage industries, particularly in soft drinks. However, as it is not a true plant pigment, being derived directly from sugar by processing, it is not considered here. Plant pigments are also key components of spices sold in large amounts, in particular paprika (containing a mix of carotenoids), turmeric and saffron.

In addition to their use as food colourants, there is extensive use of pigment extracts as animal feed supplements. For example, carotenoids such as lutein are used in poultry/egg production and aquaculture. Furthermore, large amounts of nature-identical synthetic pigments are used, in particular β-carotene. The types of pigments used in different food applications are determined by their solubility, and for the water-soluble pigments, their behaviour in response to pH. The anthocyanins and betalains are water-soluble, and the chlorophylls, curcumin and carotenoids typically oil-soluble.

Annatto is one of the oldest known dyes used for foods, textiles and cosmetics. It is extracted from the resinous coating on the seeds of the tropical bush *B. orellana*. The species occurs in the wild in tropical North America and was used by Native Americans in pre-Columbian times as a source of pigment. Today, around 7000 tonnes of seed are processed annually by the pigment industry. The main pigment of annatto is *cis*-bixin, a monomethyl ester of the diapocarotenoic acid norbixin (Fig. 1.3), and supplies yellow to orange colours. It is sparingly soluble in oil and is principally used in dairy and fat-based foods. Present in smaller quantities in the pigment extract is a water-soluble carotenoid, *cis*-norbixin, which can also be generated by alkaline treatment of bixin (Britton, 1996).

Curcumin (Fig. 1.3) is the principal pigment in the spice turmeric, which is extracted from the rhizomes of *Curcuma longa*, a perennial member of the ginger family (Zingiberaceae) that has been cultivated in Asia for many centuries. It supplies strong yellow colours and is generally oil-soluble. Turmeric has traditionally been used for colouring and flavouring meals; it is still used in large quantities for this purpose as well as extensively in a wide range of processed foods. Around 300 000 tonnes are produced annually in India, mostly for spice with a small amount for preparation of pure curcumin (Francis, 1996).

Anthocyanins are widely used as food colourants (Jackman & Smith, 1996). First described as pH indicators, the colours of anthocyanin vary greatly on the pH of the food, but generally are used only in acidic foods and provide red to blue

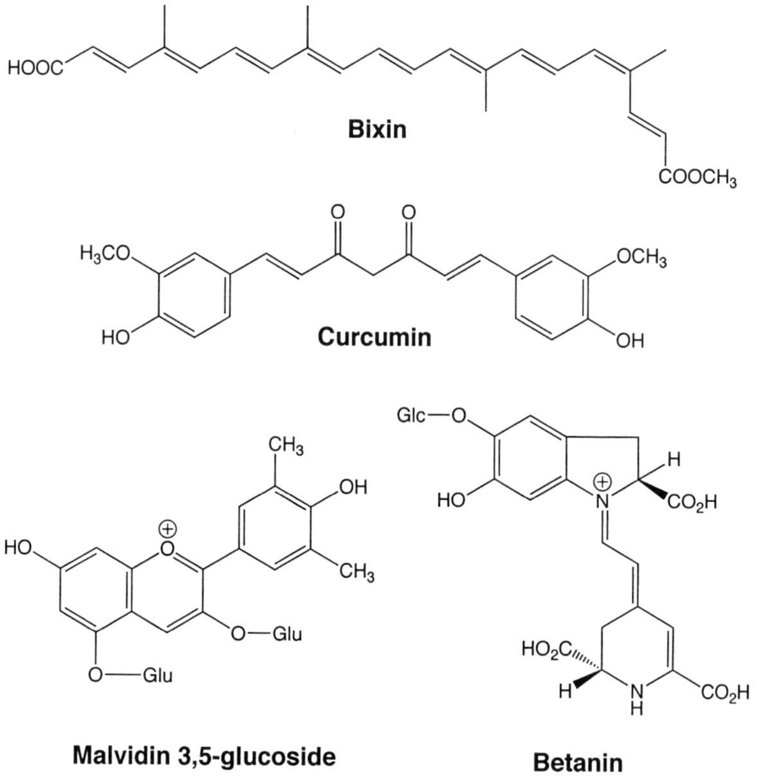

Fig. 1.3 Diagrammatic representation of the major coloured constituents of common plant-based food colourants. Bixin is the main pigment of annatto, curcumin the main pigment in turmeric, malvidin 3,5-O-glucoside a representative anthocyanin from grape extract and betanin the major betalain of beetroot extract.

colours. Although they are present in many sources, commercial anthocyanin extracts are predominantly prepared from *Vitis* species (grape)/berry skin or leas (barrel sediment) from the wine industry, which is available cheaply and in large amounts. Around 10 000 tonnes of grape skin are processed annually in Europe yielding about 50 tonnes of anthocyanin. However, although around 20 different anthocyanins have been reported for grapes, the common commercial preparations contain principally only relatively simple 3- and 3,5-diglucosides of cyanidin, delphinidin and malvidin (Fig. 1.3). These have limited colour stability with regard to pH, and are therefore limited in their food applications. This has led to research into sources of more complex, and thus more stable, anthocyanins to widen their food applications (Jackman & Smith, 1996; Giusti & Wrolstad, 2003; Schwarz *et al.*, 2003). Commercial extracts are now available from species such as *Ribes nigrum* (black currant), *Sambucus nigra* (elderberry) and *Brassica oleracea* (red cabbage). Furthermore, extracts from *Raphanus sativus* (red radish) and *Solanum tuberosum* (red potatoes) are potential sources of acylated anthocyanins to replace the synthetic dye Allura Red (Giusti & Wrolstad, 2003). This interest in new anthocyanin sources has been intensified by the growing body of evidence on the health benefits of anthocyanins.

Beetroot is the main source of betalain food colourants, principally the red betanin (Fig. 1.3) and lesser quantities of yellow betaxanthins. In Europe 20 000 tonnes of beetroot are processed annually for juice and pigment extraction, and betalains can account for 2% of beet-soluble solids (Jackman & Smith, 1996). Betalains are water-soluble pigments, and as they are unstable to heat and light, are used principally in foods with short shelf lives that do not need high heat treatment.

At present, most plant-derived food colourants are sourced directly from plant material generated through traditional plant breeding. However, the amount of most plant pigment types, and the structural variation within each type, is affected by seasonal and environmental factors, making availability of defined source material variable. In addition, current source material is limited in the type of pigment produced, particularly for anthocyanins. Thus, there is much interest in applying plant biotechnology to colourant production, both for modifying the type of pigment produced and for developing alternative production routes. To date, most research has been on tissue culture methodologies, and genetic modification (GM) has not been applied to the same extent as it has for food crops and floriculture.

Microbial fermentation has been extensively applied to food colourant production, including GM-developed strains. It is mostly based on endogenous microbial pigments, e.g. production of β-carotene in the algae *Dunaliella salina*, the blue pigment phycocyanin in the filamentous blue-green algae *Spirulina platensis*, and astaxanthin in the algae *Haematococcus pluvialis* and the yeast *Phaffia rhodozyma* (O'Callaghan, 1996). Recently, Bouvier *et al.* (2003b) have reported the production of the apocarotenoid bixin in *Escherichia coli*. Production of simple flavonoids in microbial systems has also been achieved (Hwang *et al.*, 2003), but to date this has not been extended to the coloured flavonoids. The first successful biotechnology system for plant-derived pigments was plant cell culture production of the naphthoquinone pigment shikonin. Plant cell culture systems have also been developed for betalains (Strack *et al.*, 2003) and anthocyanins (Dörenburg & Knorr, 1996). However, commercial production has not occurred for either of these pigment groups, principally due to the economics of biotechnology production compared to extraction of field crops. Application of GM techniques may improve the economics by allowing production of higher-value pigments with structures designed for specific food applications. Negative aspects of use of GM techniques include the potential need for additional regulatory approval and the uncertainty of public acceptance of GM products.

1.4.2   Modification of pigment biosynthesis in transgenic plants

The availability of the biosynthetic genes for several plant pigment pathways, and regulatory genes for flavonoids, has allowed genetic modification approaches for altering pigment production in transgenic plants. This book addresses both the biosynthesis and function of the plant pigments and the prospects for directed modification of their biosynthetic pathways. Modification of plant pigment biosynthesis using genetic technologies can be broken down into a few major approaches: abolishing or reducing flux into sections of a pathway by targeting biosynthetic genes with anti-sense/sense-suppression constructs; increasing levels of specific pathway compounds by overproducing endogenous biosynthetic activities; redirecting a pathway to produce novel compounds by the introduction of new enzyme activities; and changing regulation of a pathway using gene constructs for regulatory factors. All these approaches have been successfully used for modifying flavonoid biosynthesis, and some have been successful in the modification of other pigment pathways, in particular carotenoids and shikonin. To date, one transgenic crop with modified pigment production is commercially available – carnations containing transgenes that allow accumulation of delphinidin-derived anthocyanins for novel mauve flower colours (Lu *et al.*, 2003). This is likely to be the first of many such ornamental products, as suggested by at least a dozen field trials of GM ornamentals with modified pigmentation that have been conducted recently by various biotechnology companies, universities and research institutes. The targets are cut flower crops, pot plants, bedding plants and garden species.

There are also opportunities for altering plant colour using non-plant genes that code for biosynthetic activities for novel pigments, some of which are protein-based. There are only a few published examples of these approaches to date, and as the source pigments are not of plant origin, they are only mentioned briefly here rather than in the main chapters of this book. Green fluorescent protein (GFP) from the jellyfish *Aequorea victoria* has been used extensively as a molecular marker (Tsien, 1998; Stewart, 2001), and there has been a study on producing GFP flowers as ornamental products (Mercuri *et al.*, 2001). Mutagenesis studies have generated GFP variants with new colours and improved fluorescence in plants (Stewart, 2001), and searches for related sequences have led to a red fluorescent protein (Campbell *et al.*, 2002). Indeed, there is a wide range of protein-based GFP-like chromophores in aquatic organisms of the phylum Cnidaria, which includes the class Hydrozoa containing *A. victoria*. Perhaps the most notable group of protein chromophores of the phylum is that providing the brilliant colours in coral, in particular the pocilloporin pigments (Dove *et al.*, 2001). Brugliera *et al.* (2002) have identified GFP-like proteins in a range of coral reef organisms as possible sources for genes that could impart colours to plants, although reports on their performance in transgenics have yet to be published. Preferably, these pigments would readily provide colour in visible light without the need for special light sources.

Currently there is widespread interest in applying GM techniques to improving health attributes of crops by changing plant pigment content. Several approaches have been taken to modifying carotenoid biosynthesis in food plants, and these are reviewed in Chapter 3. Of special note to date is the 'Golden Rice' line developed to have higher levels of β-carotene, the precursor of vitamin A (Hoa *et al.*, 2003). This research featured widely in the international media, with the project leader,

Dr Ingo Potrykus, appearing on the cover of *Time* magazine. It is planned that full-scale production of the GM rice will occur within five years, and it is likely to be followed by related products, such as high β-carotene mustard seed oil (Fraley, 2003; Potrykus, 2003) and zeaxanthin-rich potatoes (Römer *et al.*, 2002). With respect to flavonoids, although there have been many successes altering their biosynthesis in ornamental crops, there are few examples in crop plants. Bovy *et al.* (2002) introduced into tomato transgenes for two transcription factors (LC and C1) that are involved in the upregulation of flavonoid biosynthesis in maize. Normally, tomato fruit contain only small amounts of flavonoids in their peel, and none in their flesh. However, the transgenic plants accumulated colourless flavonoids in the flesh of the fruit and increased levels of anthocyanins in the leaves. Nine major flavonoids could be detected in the flesh, principally flavonol and dihydroflavonol glycosides (Le Gall *et al.*, 2003). The total flavonoid content of the whole fruit (flesh and peel together) was tenfold higher for some transgenic lines compared to non-transgenic controls. This experiment is a good illustration of the promise that genes for transcription factors offer for increasing flavonoid levels in food crops. Chapter 4 reviews the knowledge of flavonoid transcription factors from studies of anthocyanin biosynthesis in flowers.

Flavonoid biosynthetic genes have also been used successfully to increase the flavonoid levels of tomato fruit (Verhoeyen *et al.*, 2002). A 78-fold increase in total flavonoid levels was achieved in some transgenic lines overexpressing a gene for the flavonoid biosynthetic enzyme chalcone isomerase. Furthermore, transgenes for two other flavonoid biosynthetic enzymes, chalcone synthase and flavonol synthase, were able to increase further the flavonoid levels in the transgenic lines overexpressing chalcone isomerase.

Given the key roles of plant pigments in so many important plant processes, it is likely that they will remain the focus of major biotechnology programmes. The recent successes in elucidating the biosynthesis of many plant pigment types, as featured in this book, will provide strong underpinnings for such studies.

References

Ayasse, M., Schiestl, F. P., Paulus, H. F., Ibarra, F. & Francke, W. (2003) Pollinator attraction in a sexually deceptive orchid by means of unconventional chemicals. *Proc. Royal Soc. London Series B – Biol. Sci.*, **270**, 517–522.

Bhattacharya, C. & Bhattacharya, N. (2001) Mendel's experiment regarding anthocyanin inheritance in the light of 100 years after its rediscovery. *Crop Res.*, **21**, 181–187.

Bohm, B. A. (1998) *Introduction to Flavonoids*. Harwood Academic Publishers, Amsterdam, The Netherlands.

Bouvier, F., Suire, C., Mutterer, J. & Camara, B. (2003a) Oxidative remodeling of chromoplast carotenoids: identification of the carotenoid dioxygenase CsCCD and CsZCD genes involved in crocus secondary metabolite biogenesis. *Plant Cell*, **15**, 47–62.

Bouvier, F., Dogbo, O. & Camara, B. (2003b) Biosynthesis of the food and cosmetic plant pigment bixin (Annatto). *Science*, **300**, 2089–2092.

Bovy, A., de Vos, R., Kemper, M., Schijlen, E., Pertejo, M.A., Muir, S., Collins, G., Robinson, S., Verhoeyen, M., Hughes, S., Santos-Buelga, C. & van Tunen, A. (2002) High-flavonol tomatoes resulting from the heterologous expression of the maize transcription factor genes LC and C1. *Plant Cell*, **14**, 2509–2526.

Boyle, R. (1664) *Experiments and Considerations Touching Colours: First Occasionally Written Among Some Other Essays to a Friend and Now Suffered to Come Abroad as the Beginning of an Experimental History of Colours.* Henry Herringman, London, UK. Available on microfilm at *Early English Books Online*, http://www.lib.umich.edu/eebo/

Britton, G. (1996) Carotenoids, in *Natural Food Colorants* (eds G. A. F. Hendry & J. D. Houghton), Chapman & Hall, London, UK, pp. 197–243.

Britton, G., Liaaen-Jensen, S. & Pfander, H. (1995) Carotenoids today and challenges for the future, in *Carotenoids, Vol. 1A: Isolation and Analysis* (eds G. Britton, S. Liaaen-Jensen, & H. Pfander), Birkhäuser Verlag, Basel, Switzerland, pp.13–26.

Britton, G., Weesie, R. J., Askin, D., Warburton, J. D., GallardoGuerrero, L., Jansen, F. J., de Groot, H. J. M., Lugtenburg, J., Cornard, J. P. & Merlin, J. C. (1997) Carotenoid blues: structural studies on caroteno-proteins. *Pure Appl. Chem.*, **69**, 2075–2084.

Brouillard, R. & Dangles, O. (1993) Flavonoids and flower colour, in *The Flavonoids: Advances in Research Since 1986* (ed. J. B. Harborne), Chapman & Hall, London, UK, pp. 565–587.

Brugliera, F., Karan, M., Prescott, M., Mason, J., Dove, S. G., Hoegh-Guldberg, I. O. & Jones, E. L. (2002) Cell visual characteristic-modifying sequences. Patent application, No. WO 02/070703 A2.

Campbell, R. E., Tour, O., Palmer, A. E., Steinbach, P. A., Baird, G. S., Zacharias, D. A. & Tsien, R. Y. (2002) A monomeric red fluorescent protein. *Proc. Natl. Acad. Sci. USA*, **99**, 7877–7882.

Comba, L., Corbet, S. A., Hunt, H., Outram, S., Parker, J. S. & Glover, B. J. (2000) The role of genes influencing the corolla in pollination of *Antirrhinum majus. Plant Cell Environ.*, **23**, 639–647.

Cone, K. C., Burr, F. A. & Burr, B. (1986) Molecular analysis of the maize anthocyanin regulatory locus *C1. Proc. Nat. Acad. Sci. USA*, **83**, 9631–9635.

Darwin, C. R. (1868) The variation of animals and plants under domestication. John Murray, London, UK.

de Valming, P., van Eekeres, J. E. M. & Wiering, H. (1982) A gene for flower colour fading in *Petunia hybrida. Theor. Appl. Genet.*, **61**, 41–46.

Dörenburg, H. & Knorr, D. (1996) Generation of colors and flavors in plant cell and tissue cultures. *Crit. Rev. Plant Sci.*, **15**, 141–168.

Dove, S. G., Hoegh-Guldberg & O., Ranganathan, S. (2001) Major colour patterns of reef-building corals are due to a family of GFP-like proteins. *Coral Reefs*, **19**, 197–204.

Eugster, C. H. (1995) History: 175 years of carotenoid chemistry, in *Carotenoids, Vol. 1A: Isolation and Analysis* (eds G. Britton, S. Liaaen-Jensen, & H. Pfander), Birkhäuser Verlag, Basel, Switzerland, pp. 1–12.

Fraley, R. T. (2003) Improving the nutritional quality of plants, in *Plant Biotechnology 2002 and Beyond* (ed. I. K. Vasil), Kluwer Academic Publishers, Dordrecht, The Netherlands, pp. 61–67.

Francis, F. J. (1996) Less common natural colourants, in *Natural Food Colorants* (eds G. A. F. Hendry & J. D. Houghton), Chapman & Hall, London, UK, pp. 112–132.

Gigord, L. D. B., Macnair, M. R. & Smithson, A. (2001) Negative frequency-dependent selection maintains a dramatic flower color polymorphism in the rewardless orchid *Dactylorhiza sambucina* (L.) Soo. *Proc. Natl. Acad. Sci. USA*, **98**, 6253–6255.

Gilbert, K. G. & Cooke, D. T. (2001) Dyes from plants: past usage, present understanding and potential. *Plant Growth Reg.*, **34**, 57–69.

Giusti, M. M. & Wrolstad, R. E. (2003) Acylated anthocyanins from edible sources and their application in food systems. *Biochem. Eng. J.*, **14**, 217–225.

Glover, B. J. & Martin, C. R. (1998) The role of petal cell shape and pigmentation in pollination success in *Antirrhinum majus. Heredity*, **80**, 778–784.

Gould, K. S., Vogelmann, T. C., Han, T. & Clearwater, M. J. (2002a) Profiles of photosynthesis within red and green leaves of *Quintinia serrata* A. Cunn. *Physiol. Plant.*, **116**, 127–133.

Gould, K. S., Neill, S. O. & Vogelmann, T. C. (2002b) A unified explanation for anthocyanins in leaves? *Adv. Bot. Res.*, **37**, 167–192.

Gould, K. S., Mckelvie, J. & Markham, K. R. (2002c) Do anthocyanins function as antioxidants in leaves? Imaging of H_2O_2 in red and green leaves after mechanical injury. *Plant Cell Environ.*, **25**, 1261–1269.

Gumbert, A., Kunze, J. & Chittka, L. (1999) Floral diversity in plant communities, bee colour space and a null model. *Proc. Royal Soc. Lon. Series B – Biol. Sci.*, **266**, 1711–1716.

Harborne, J. B. (ed.) (1988) *The Flavonoids, Advances in Research Since 1980*, Chapman & Hall, London, UK.

Harborne, J. B. (ed.) (1994) *The Flavonoids, Advances in Research Since 1986*, Chapman & Hall, London, UK.

Harborne, J. B. & Grayer, R. J. (1994) Flavonoids and insects, in *The Flavonoids: Advances in Research Since 1986* (ed. J. B. Harborne), Chapman & Hall, London, UK, pp. 589–618.

Hendry, B. S. (1996) Natural food colours, in *Natural Food Colorants* (eds G. A. F. Hendry & J. D. Houghton), Chapman & Hall, London, UK, pp. 40–79.

Hoa, T. T. C., Al-Babili, S., Schaub, P., Potrykus, I. & Beyer, P. (2003) Golden indica and japonica rice lines amenable to deregulation. *Plant Physiol.*, **133**, 161–169.

Hoch, W. A., Zeldin, E. L. & McCown, B. H. (2001) Physiological significance of anthocyanins during autumnal leaf senescence. *Tree Physiol.*, **21**, 1–8.

Hoch, W. A., Singsaas, E. L. & McCown, B. H. (2003) Resorption protection: anthocyanins facilitate nutrient recovery in autumn by shielding leaves from potentially damaging light levels. *Plant Physiol.*, **133**, 1296–1305.

Holton, T. A., Brugliera, F., Lester, D. R., Tanaka, Y., Hyland, C. D., Menting, J. G. T., Lu, C-Y., Farcy, E., Stevenson, T. W. & Cornish, E. C. (1993) Cloning and expression of cytochrome P450 genes controlling flower colour. *Nature*, **366**, 276–279.

Hwang, E. I., Kaneko, M., Ohnishi, Y. & Horinouchi, S. (2003) Production of plant-specific flavanones by *Escherichia coli* containing an artificial gene cluster. *Appl. Environ. Microbiol.*, **69**, 2699–2706.

Jackman, R. L. & Smith, J. L. (1996) Anthocyanins and betalains, in *Natural Food Colorants* (eds G. A. F. Hendry & J. D. Houghton), Chapman & Hall, London, UK, pp. 244–309.

Jackson, A. H. (1976) Structure, properties and distribution of chlorophylls, in *Chemistry and Biochemistry of Plant Pigments* (ed. T. W. Goodwin), Academic Press, London, UK, pp. 1–63.

Johnson, S. D. & Midgley, J. F. (2001) Pollination by monkey beetles (Scarabaeidae: Hopliini): do color and dark centers of flowers influence alighting behavior? *Environ. Ent.*, **30** 861–868.

Jones, K. N. & Reithel, J. S. (2001) Pollinator-mediated selection on a flower color polymorphism in experimental populations of *Antirrhinum* (Scrophulariaceae). *Am. J. Bot.*, **88**, 447–454.

Jorgensen, R. A., Que, Q. D. & Napoli, C. A. (2002) Maternally controlled ovule abortion results from cosuppression of dihydroflavonol-4-reductase or flavonoid-3′,5′-hydroxylase genes in *Petunia hybrida. Funct. Plant Biol.*, **29**, 1501–1506.

Kull, D. & Pfander, H. (1995) Appendix: list of new carotenoids, in *Carotenoids*, Vol. 1A: *Isolation and Analysis* (eds G. Britton, S. Liaaen-Jensen, & H. Pfander), Birkhäuser Verlag, Basel, Switzerland, pp. 295–317.

Landeck, I. (2002) Feeding spectrum of the hairy flower wasp *Scolia hirta* in Lusatia (Central Europe) with special focus on flower colour, morphology of flowers and inflorescences (Hymenoptra: Scoliidae). *Ent. Generalis*, **26**, 107–120.

Le Gall, G., DuPont, M. S., Mellon, F. A., Davis, A. L., Collins, G. J., Verhoeyen, M. E. & Colquhoun, I. J. (2003) Characterization and content of flavonoid glycosides in genetically modified tomato (*Lycopersicon esculentum*) fruits. *J. Ag. Food Chem.*, **51**, 2438–2446.

Lee, D. W. (2002) Anthocyanins in autumn leaf senescence. *Adv. Bot. Res.*, **37**, 147–165.

Lu, C., Chandler, S. F., Mason, J. G. & Brugliera, F. (2003) Florigene flowers: from laboratory to market, in *Plant Biotechnology 2002 and Beyond* (ed. I. K. Vasil), Kluwer Academic Publishers, Dordrecht, The Netherlands, pp. 333–336.

Lunau, K. (2000) The ecology and evolution of visual pollen signals. *Plant Systematics Evol.*, **222**, 89–111.

Matile, P. (2000) Biochemistry of an Indian summer: physiology of autumnal leaf coloration. *Exp. Gerontology*, **35**, 145–158.

McGraw, K. J., Adkins-Regan, E. & Parker, R. S. (2002) Anhydrolutein in the zebra finch: a new, metabolic-ally-derived carotenoid in birds. *Comp. Biochem. Physiol.* Part B – *Biochem. Mol. Biol.*, **132**, 811–818.

Melendez-Ackerman, E., Campbell, D. R. & Waser, N. M. (1997) Hummingbird behavior and mechanisms of selection on flower color in *Ipomopsis. Ecology*, **78**, 2532–2541.

Mercuri, A., Sacchetti, A., De Benedetti, L., Schiva, T. & Alberti, S. (2001). Green fluorescent flowers. *Plant Sci.*, **161**, 961–968.

Napoli, C., Lemieux, C. & Jorgensen, R. (1990) Introduction of a chimeric chalcone synthase gene into Petunia results in reversible co-suppression of homologous genes in trans. *Plant Cell*, **2**, 279–289.

Neill, S. O., Gould, K. S., Kilmartin, P. A., Mitchell, K. A. & Markham, K. R. (2002) Antioxidant capacities of green and cyanic leaves in the sun species, *Quintinia serrata. Funct. Plant Biol.*, **29**, 1437–1443.

Oberrath, R. & Bohning-Gaese, K. (1999) Floral color change and the attraction of insect pollinators in lungwort (*Pulmonaria collina*). *Oecologia*, **121**, 383–391.

O'Callaghan, M. C. (1996) Biotechnology in natural food colours. The role of bioprocessing, in *Natural Food Colorants* (eds G. A. F. Hendry & J. D. Houghton), Chapman & Hall, London, UK, pp. 80–111.

Odell, E., Raguso, R. A. & Jones, K. N. (1999) Bumblebee foraging responses to variation in floral scent and color in snapdragons (*Antirrhinum*: Scrophulariaceae). *Am. Mid. Nat.*, **142**, 257–265.

Olson, V. A. & Owens, I. P. F. (1998) Costly sexual signals: are carotenoids rare, risky or required? *Trends Ecol. Evol.*, **13**, 510–514.

Onslow, M. A. (1925) *The Anthocyanin Pigments of Plants*, 2nd edn, Cambridge University Press, Cambridge, UK.

Orska-Gawrys, J., Surowiec, I., Kehl, J., Rejniak, H., Urbaniak-Walczak, K. & Trojanowitcz, M. (2003) Identification of natural dyes in archeological Coptic textiles by liquid chromatography with diode array detection. *J. Chromatogr. A.*, **898**, 239–248.

Page, J. E. & Towers, G. H. N. (2002) Anthocyanins protect light-sensitive thiarubrine phototoxins. *Planta*, **215**, 478–484.

Paxton, R. J. & Tengo, J. (2001) Doubly duped males: the sweet and sour of the orchid's bouquet. *Trends Ecol. Evol.*, **16**, 167–169.

Paz–Ares, J., Wienand, U., Peterson, P. A. & Saedler, H. (1986) Molecular cloning of the *c* locus of *Zea mays*: a locus regulating the anthocyanin pathway. *EMBO J.*, **5**, 829–833.

Pfander, H. (1987) *The Key to Carotenoids*, 2nd edn, Birkhäuser Verlag, Basel, Switzerland.

Potrykus, I. (2003) Nutritional improvement of rice to reduce malnutrition in developing countries, in *Plant Biotechnology 2002 and Beyond* (ed. I. K. Vasil), Kluwer Academic Publishers, Dordrecht, The Netherlands, pp. 401–406.

Raguso, R. A. & Willis, M. A. (2002) Synergy between visual and olfactory cues in nectar feeding by naïve hawkmoths, *Manduca sexta. Animal Behav.*, **64**, 685–695.

Römer, S., Lubeck, J., Kauder, F., Steiger, S., Adomat, C. & Sandmann, G. (2002) Genetic engineering of a zeaxanthin-rich potato by antisense inactivation and co-suppression of carotenoid epoxidation. *Metab. Eng.*, **4**, 263–272.

Rumpho, M. E., Summer, E. J. & Manhart, J. R. (2000) Solar-powered seaslugs. Mollusc/Algal chloroplast symbiosis. *Plant Physiol.*, **123**, 29–38.

Schiestl, F. P., Ayasse, M., Paulus, H. F., Lofstedt, C., Hansson, B. S., Ibarra, F. & Francke, W. (1999) Orchid pollination by sexual swindle. *Nature*, **399**, 421–422.

Schram, A. W., Jonsson, L. M. V. & Bennink, G. J. H. (1984) Biochemistry of flavonoid synthesis in *Petunia hybrida*, in *Monogrpahs on Theoretical and Applied Genetics 9: Petunia* (ed. K. C. Sink), Springer-Verlag, Berlin, pp. 68–76.

Schwarz, M., Hillebrand, S., Habben, S., Degenhardt, A. & Winterhalter, P. (2003) Application of high-speed countercurrent chromatography to the large-scale isolation of anthocyanins. *Biochem. Eng. J.*, **14**, 179–189.

Simpson, K. L., Lee, T-C., Rodriguez, D. B. & Chichester, C. O. (1976) Metabolism in senescent and stored tissues, in *Chemistry and Biochemistry of Plant Pigments* Vol. 1 (ed. T. W. Goodwin), Academic Press, London, UK, pp. 780–842.

Staiger D. (2002) Circadian rhythms in *Arabidopsis*: time for nuclear proteins. *Planta*, **214**, 334–344.

Stewart, C. N. (2001) The utility of green fluorescent protein in transgenic plants. *Plant Cell Rep.*, **20**, 376–382.

Steyn, W. J., Wand, S. J. E., Holcroft, D. M. & Jacobs, G. (2002) Anthocyanins in vegetative tissues: a proposed unified function in photoprotection. *New Phytologist*, **155**, 349–361.

Strack, D., Vogt, T. & Schliemann, W. (2003) Recent advances in betalain research. *Phytochemistry*, **62**, 247–269.

Taylor, L. P. & Jorgensen, R. (1992) Conditional male fertility in chalcone synthase-deficient petunia. *J. Hered.*, **83**, 11–17.

Terahara, N., Saito, N., Honda, T., Toki, K. & Osajima, Y. (1990) Structure of ternatin A1, the largest ternatin in the major blue anthocyanins from *Clitoria ternatea* flowers. *Tetrahedron Lett.*, **31**, 2921–2924.

Tsien, R. Y. (1998) The green fluorescent protein. *Annu. Rev. Biochem.*, **67**, 509–544.

van der Krol, A. R., Lenting, P. E., Veenstra, J. G., van der Meer, I. M., Koes, R. E., Gerats, A. G. M., Mol, J. N. M. & Stuitje, A. R. (1988) An antisense chalcone synthase gene in transgenic plants inhibits flower pigmentation. *Nature*, **333**, 866–869.

van der Krol, A. R., Mur, L. A., Beld, M., Mol, J. N. M. & Stuitje, A. R. (1990) Flavonoid genes in petunia: addition of a limited number of additional copies may lead to a suppression of gene activity. *Plant Cell*, **2**, 291–299.

Verhoeyen, M. E., Bovy, A., Collins, G., Muir, S., Robinson, S., de Vos, C. H. R. & Colliver, S. (2002) Increasing antioxidant levels in tomatoes through modification of the flavonoid biosynthetic pathway. *J. Exp. Bot.*, **53**, 2099–2106.

Vishnevetsky, M., Ovadis, M. & Vainstein, A. (1999) Carotenoid sequestration in plants: the role of carotenoid-associated proteins. *Trends Plant Sci.*, **4**, 232–235.

Weedon, B. C. L. & Moss, G. P. (1995) Structure and nomenclature, in *Carotenoids*, Vol. 1A: *Isolation and Analysis* (eds G. Britton, S. Liaaen-Jensen, & H. Pfander), Birkhäuser Verlag, Basel, Switzerland, pp. 27–71.

Whedale, M. W. (1907) The inheritance of flower colour in *Antirrhinum majus. Proc. Roy. Soc. Series B – Biol. Sci.*, **79**, 288–305.

Winefield, C. (2002) The final steps in anthocyanin formation: a story of modification and sequestration. *Adv. Bot. Res.*, **37**, 55–74.

Wink, M. (ed.) (1999a) *Biochemistry of Plant Secondary Metabolism*. Sheffield Academic Press, Sheffield, UK.

Wink, M. (ed.) (1999b) *Functions of Plant Secondary Metabolites and their Exploitation in Biotechnology.* Sheffield Academic Press, Sheffield, UK.

Weiss, M. R. (1991) Floral colour changes as cues for pollinators. *Nature*, **354**, 227–229.

Weiss, M. R. (1995) Floral colour change: a widespread functional convergence. *Am. J. Bot.*, **82**, 167–185.

Zagalsky, P. F. (1995) Carotenoproteins, in *Carotenoids*, Vol. 1A: *Isolation and Analysis* (eds G. Britton, S. Liaaen-Jensen, & H. Pfander), Birkhäuser Verlag, Basel, Switzerland, pp. 287–294.

2   Chlorophylls

Robert D. Willows

2.1   Introduction

Chlorophylls are the pigments that make plants green and are arguably the most important compounds on earth as they are required for the harvesting and transduction of light energy in photosynthesis. The largest proportion of light energy absorbed and transduced in photosynthesis is by direct absorption of light by chlorophylls. Manipulation of chlorophylls will therefore have a major impact on the ability of plants to carry out photosynthesis. Thus, unlike many of the other plant pigments discussed in this volume, mutations in the chlorophyll biosynthetic pathway are often lethal to plants, and manipulation of the amount of chlorophyll and/or the type of chlorophyll found within plant tissues is limited. However, analysis of mutants within the chlorophyll biosynthetic pathway has been of enormous value in elucidating the pathway as well as in providing information on the regulation and integration of chlorophyll biosynthesis with plant growth and development.

This chapter is not intended to be an extensive review of the literature on chlorophyll biosynthesis and degradation, as many reviews on these topics have appeared over the last few years (Reinbothe *et al*., 1996; Reinbothe & Reinbothe, 1996b; Porra, 1997; Rudiger, 1997; Schoefs & Bertrand, 1997; Armstrong & Apel, 1998; Averina, 1998; Beale, 1999; Matile *et al*., 1999; Litvin *et al*., 2000; von Wettstein, 2000; Papenbrock & Grimm, 2001; Schoefs, 2002; Willows, 2003). Instead, this chapter will concentrate on the effects that manipulation of the individual steps within these pathways – by either mutagenesis or by other methods such as enzyme inhibition – has on pigment levels and plant growth and development. Regulatory mutants and other regulatory mechanisms will then be covered before closing with a brief look at uses of chlorophylls and chemically modified chlorophylls as photodynamic and imaging agents.

2.1.1   *What are chlorophylls and where are they found in plants?*

Chlorophylls belong to a class of compounds known as tetrapyrroles. Examples of the various types of chlorophylls found in plants and some other naturally occurring tetrapyrroles are shown in Fig. 2.1. Chlorophylls are distinguished from other tetrapyrroles, such as haem and vitamin B12, by the presence of a magnesium ion that is coordinated within the tetrapyrrole ring and a fifth so-called isocyclic ring. Chlorophyll *a* and chlorophyll *b* are the major types of chlorophylls found in

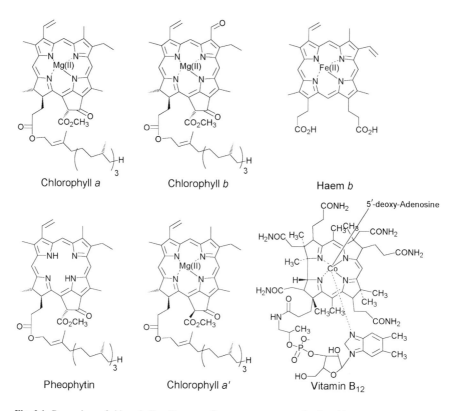

Fig. 2.1 Comparison of chlorophylls with some other common tetrapyrroles found in nature.

plants. They have a characteristic green colour due to strong absorbance of blue and red light, with chlorophyll *b* having a noticeable blue tinge as its absorbance spectrum is red-shifted compared to chlorophyll *a*. Chlorophylls also have a characteristic red fluorescence which is readily observed when plant tissue or chlorophyll-containing extracts are irradiated with blue light. The spectral properties of chlorophylls are essential for the function of chlorophyll in harvesting light energy and in the transduction of that light energy for photosynthesis.

The chlorophyll *a/b* ratio varies from 2.0–2.8 for shade-adapted plants to 3.5–4.9 for plants adapted to full-sun conditions. This variation in chlorophyll *a/b* ratios is due to differences in the ratio of photosystem I (PSI) to photosystem II (PSII) and the size and composition of the light-harvesting complexes (LHCs) associated with each photosystem. The photosystems contain chlorophyll *a* but not chlorophyll *b*, whereas the LHCs contain significant amounts of chlorophyll *b*. Shade-adapted plants tend to have more LHCs associated with their photosystems than sun-adapted plants and thus have lower chlorophyll *a/b* ratios than sun-adapted plants (Anderson, 1986; Porra, 2002).

A number of spectral variants of chlorophyll *a* are also detected *in vivo*. These spectral differences result from chlorophyll molecules in different environments within LHCs and photosystems. Other minor chlorophylls and chlorophyll derivatives have also been reported, with the most important ones being chlorophyll *a'* (Fig. 2.1), found in the PSI reaction centre, and pheophytin (Fig. 2.1), found in the PSII reaction centre. Other organisms such as algae, cyanobacteria and photosynthetic bacteria have other types of chlorophylls or bacteriochlorophylls within their LHCs and photosystems (Willows, 2003).

2.1.2   *Photochemical properties of chlorophylls*

The photochemical properties that allow chlorophylls to carry out their function in photosynthesis also present potential problems for plants. If the light energy absorbed by chlorophylls is not used in photosynthesis, the excess light energy must be dissipated in some way. This dissipation of excess light energy can occur by a number of mechanisms, including fluorescence and reaction with other compounds. The potential of excited-state chlorophylls to react with oxygen to form singlet oxygen is one of the most damaging for the plant (Reinbothe *et al.*, 1996; Matile *et al.*, 1999; von Wettstein, 2000). The generation of singlet oxygen can result in a cascade of free radical mediated reactions forming a variety of reactive oxygen species that can damage proteins and nucleic acids and possibly lead to cell death. In order to prevent these damaging effects, plants have developed mechanisms to limit the formation of singlet oxygen and free radicals. Some of these protective mechanisms include the use of antioxidants such as ascorbic acid and α-tocopherol; expression of superoxide dismutase and peroxidases; a reduction or increase in the total amount of chlorophyll by modifying the size of the chlorophyll antennae when grown under high or low light conditions, respectively; modification and use of accessory pigments, such as carotenoids via the xanthophyll cycle, to dissipate excess light energy from chlorophylls before reaction with oxygen can occur.

The phototoxic effects described above for chlorophylls can also be induced by accumulation of coloured chlorophyll biosynthetic intermediates. Thus, the chlorophyll biosynthetic pathway is tightly regulated to keep the concentration of the biosynthetic intermediates below a phototoxic level.

2.2   Chlorophyll biosynthesis

2.2.1   *Overview of chlorophyll biosynthesis*

An overview of the chlorophyll *a* biosynthetic pathway found in plants is shown in Figs 2.2 and 2.3. All tetrapyrroles such as chlorophyll are synthesised from eight molecules of the five-carbon compound, aminolevulinic acid (ALA). Plants synthesise their ALA from glutamic acid via a pathway known as the Beale or C_5 pathway, shown in Fig. 2.2. The reactions leading from ALA to protoporphyrin IX are highly conserved between plants, animals and bacteria, and many of the plant enzymes were initially identified based on sequence identity with bacterial and/or animal enzymes (Beale, 1999).

The enzymes and intermediates up to and including protoporphyrin IX are shared by the chlorophyll and haem biosynthetic pathways. Haem is required in the cytosol and mitochondria as well as within the chloroplast of plants. The chloroplast is the site of synthesis of protoporphyrinogen within plant tissues. The final two steps for mitochondrial haem synthesis take place within the mitochondria using protoporphyrinogen that is exported from the chloroplast (Jacobs & Jacobs, 1993). It is believed that the cytoplasmic haem is derived via export from the chloroplast (Grimm, 2003). Mutations in this section of the pathway are well characterised in bacteria and animals, where they are known collectively as porphyrias (de Rooij et al., 2003). A maize mutant has been described in this section of the pathway that has light-induced necrotic lesions reminiscent of some of the human porphyrias (Hu et al., 1998). Antisense ribonucleic acid (RNA) technology has also been used to partially inactivate enzymes in this section of the pathway, and these plants have similar necrotic lesions due to the accumulation of phototoxic intermediates (Mock & Grimm, 1997; von Wettstein, 2000).

The reactions that convert protoporphyrin IX to chlorophyll are unique to the chlorophyll biosynthetic pathway and the enzyme activities are found only within plant chloroplasts. Plant mutants in this section of the pathway have been known for some time. These mutants are usually pale green to pale yellow due to a lack of chlorophyll, and importantly they do not usually accumulate intermediates to high levels (Henningsen et al., 1993; von Wettstein, 2000).

Other mutants have also been characterised which are due to defects in the regulation of the biosynthetic pathway. These mutants include the light-sensitive barley (Hordeum vulgare) tigrina mutants and the Arabidopsis thaliana flu mutant, which are characterised by an overproduction of protochlorophyllide (von Wettstein, 2000).

2.2.2 ALA biosynthesis

ALA can be formed by two different pathways, and the literature on both pathways has been recently reviewed (Kannangara et al., 1994; Shoolingin-Jordan, 2003).

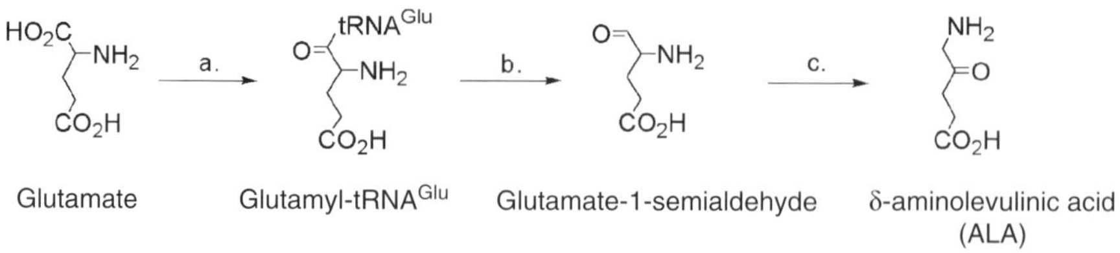

| Glutamate | Glutamyl-tRNA^{Glu} | Glutamate-1-semialdehyde | δ-aminolevulinic acid (ALA) |

Fig. 2.2 C_5 pathway for ALA biosynthesis: (a) glutamyl-tRNA synthetase; (b) glutamyl-tRNA reductase; (c) glutamate-1-semialdehyde aminotransferase.

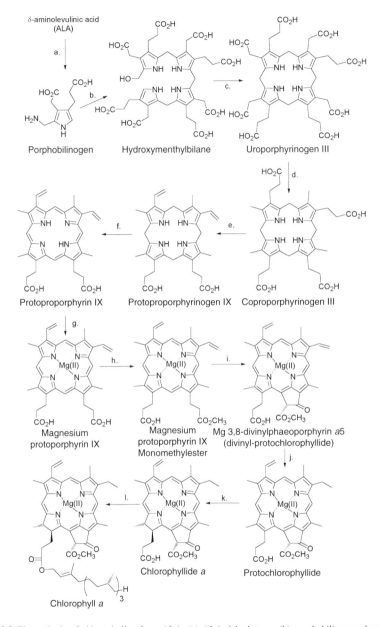

Fig. 2.3 Biosynthesis of chlorophyll *a* from ALA: (a) ALA dehydratase; (b) porphobilinogen deaminase; (c) uroporphyrinogen III synthase; (d) uroporphyrinogen decarboxylase; (e) coproporphyrinogen oxidase; (f) protoporphyrinogen oxidase; (g) magnesium chelatase; (h) S-adenosylmethionine:magnesium protoporphyrin IX O-methyltransferase; (i) magnesium protoporphyrin IX monomethylester oxidative cyclase; (j) 8-vinyl reductase; (k) protochlorophyllide oxidoreductase; (l) chlorophyll synthase.

The C_5 pathway is found in all photosynthetic eukaryotes such as plants and algae, in some species of trypanosomes, in all archaebacteria and in most bacteria (with the exception of the α-proteobacteria). The second pathway for ALA biosynthesis is known as the Shemin pathway in which ALA is synthesised from succinyl-CoA and glycine by ALA-synthase, and this pathway is found in animals and the α-proteobacteria.

Glutamyl-tRNA synthetase is the first step in the C_5 pathway and catalyses the adenosine triphosphate (ATP)-dependent condensation of glutamate and tRNA to form glutamyl-tRNA^{Glu}. The glutamyl-tRNA^{Glu} produced in this step is used both for protein biosynthesis and for ALA biosynthesis (Kumar et al., 1996). The barley chloroplast enzyme has been purified (Bruyant & Kannangara, 1987) and the sequences for the barley and tobacco enzymes are available in the GenBank database (accessions X83523 and X83524 respectively). No plant mutants have been described for this enzyme, probably because it is essential for both chlorophyll and protein biosynthesis within the chloroplast.

The second enzyme in the pathway is glutamyl-tRNA reductase (GTR), which catalyses the NADPH-dependent reduction of glutamyl-tRNA to glutamate-1-semialdehyde (GSA). The sequences of this enzyme have a high degree of identity to the corresponding bacterial and archaebacterial proteins and the structure of an archael GTR has been determined by X-ray crystallography (Moser et al., 2001). The plant enzyme has a 31–34 amino acid N-terminal extension compared to the bacterial enzyme (Vothknecht et al., 1998), and all plant species studied to date have at least two differentially expressed gtr genes (Beale, 1999; Grimm, 2003). Haem was reported to be a potent feedback inhibitor of this enzyme and the N-terminal extension found on the plant GTR has been suggested to allow haem binding to effect this feedback inhibition (Vothknecht et al., 1998).

Although no plant GTR mutants have been described, a Euglena gracilis mutant was discovered which has a point mutation in the tRNA^{Glu}. This mutation prevents chlorophyll synthesis as the mutated tRNA is not a substrate for the GTR but the tRNA is still able to participate in chloroplastic protein synthesis (Stange-Thomann et al., 1994). A specific inhibitor of GTR is a glutamyl-tRNA substrate analogue called glutamycin (Kannangara et al., 1994; Moser et al., 1999). This inhibitor has been only used in in vitro studies and its effect in whole plants has not been determined. Other inhibitors of bacterial GTR have been described but it is not known if these inhibit the plant enzymes (Loida et al., 1999).

Glutamate-1-semialdehyde aminotransferase (GSA-AT) is the final enzyme in the C_5 ALA biosynthetic pathway. This enzyme has a pyridoxal-phosphate cofactor and the gene for the enzyme has been cloned and sequenced from a number of organisms (Kannangara et al., 1994). The structure of the cyanobacterial enzyme has been determined by X-ray crystallography (Hennig et al., 1997). It has been suggested – based on this structure and that of the archael GTR structure – that a GTR/GSA-AT complex forms, which allows channelling of the unstable GSA directly from GTR to GSA-AT (Moser et al., 2001).

Gabaculine is an effective and specific inhibitor of GSA-AT that works both *in vivo* and *in vitro* (Hill *et al.*, 1985; Hoober *et al.*, 1988; Nair *et al.*, 1990). Feeding gabaculine to barley and *Pisum sativum* (pea) has been shown to effectively inhibit chlorophyll synthesis. The inhibitory effects of gabaculine can be overcome by feeding ALA, and this has been an effective experimental tool in a variety of studies on chlorophyll and haem synthesis in plants.

Antisense RNA technology has been used to probe the effect of reducing the activity of GTR and GSA-AT in *A. thaliana* and *Nicotiana tabacum* (tobacco) respectively (Höefgen *et al.*, 1994; Kumar & Soll, 2000). This technology works by introducing all or part of the protein-coding region of a gene in the opposite orientation together with an appropriate promoter. The promoter produces RNA which is complimentary to the normal mRNA and which results in double-stranded RNA formation *in vivo*. This double-stranded RNA is rapidly degraded and causes a net reduction in the quantity of the specific mRNA available for protein synthesis, which results in reduction in the amount of enzyme produced. A number of tobacco lines expressing the GSA-AT gene in the antisense orientation produced chlorophyll-deficient phenotypes. These lines varied considerably in their phenotype; some lines were uniformly pale green to white, others had white veins and green mesophyll while some had green veins and white mesophyll tissue, suggesting that positional effects of the inserted antisense gene were important (Höefgen *et al.*, 1994).

Similarly the *A. thaliana* GTR antisense plants also exhibited varying degrees of chlorophyll deficiency, ranging from patchy yellow leaves to totally yellow plants. These plants had decreased levels of chlorophyll, haem and ALA that were proportional to the amount of GTR (Kumar & Soll, 2000).

2.2.3 ALA to protoporphyrin IX

2.2.3.1 ALA-dehydratase
ALA-dehydratase (ALA-D), also known as porphobilinogen (PBG) synthase, catalyses the condensation of two molecules of ALA to form the monopyrrole PBG. These two ALA molecules bind to the enzyme at what are referred to as the 'A' and 'P' substrate-binding sites. The enzyme has been purified from a large number of sources including bacteria, yeast, animals and higher plants. There is a high degree of amino acid sequence similarity between the enzymes from various sources, and the X-ray structures from yeast, *Pseudomonas aeruginosa* and *Escherichia coli* are similar (Frankenberg *et al.*, 1999a, 1999b; Shoolingin-Jordan, 2003). The enzymes from different sources can be grouped into two classes based on their divalent metal ion requirement. The enzymes from plants and some bacteria require magnesium ions, which are not absolutely required for catalysis (at least in *P. aeruginosa)* (Frankenberg *et al.*, 1999b). The other class of enzymes, found in animals, some bacteria and fungi, has a bound zinc, which is absolutely required for catalysis. A number of the zinc-requiring enzymes can also be activated by magnesium ions although

magnesium cannot replace the catalytic requirement for zinc (Shoolingin-Jordan, 2003).

The majority of ALA-dehydratase inhibitor studies have been with the bacterial or animal enzymes. The animal enzyme and other zinc-containing enzymes are exceptionally sensitive to heavy metals such as lead. This inhibition can be reduced by addition of exogenous zinc, indicating that the heavy metals probably bind to the zinc-binding site. Most other types of inhibitors described are ALA substrate analogues and bind either covalently or non-covalently to the 'A' or 'P' binding sites on the enzyme (Shoolingin-Jordan, 2003). Levulinic acid is a substrate analogue that forms a Schiff base with a lysine residue at the 'P' binding site on the enzyme; it inhibits the tobacco (Shieh *et al.*, 1973) and pea enzymes (Senior *et al.*, 1996). Succinyl acetone is a substrate analogue which is a potent inhibitor and is ~ 2000 times more effective at inhibiting the pea ALA-dehydratase than the yeast or *E. coli* enzymes (Senior *et al.*, 1996).

2.2.3.2 *Hydroxymethyl bilane synthase*
Hydroxymethyl bilane (HMB) is the first tetrapyrrole intermediate in the synthesis of all naturally occurring tetrapyrroles. HMB is synthesised by the deamination of four molecules of PBG by the enzyme HMB synthase, which is also known as PBG deaminase. HMB synthase was originally called uroporphyrinogen I synthase because HMB is unstable and rapidly cyclises to the cyclic tetrapyrrole uroporphyrinogen I (Shoolingin-Jordan, 2003). HMB synthases are characterised from several plant sources including *A. thaliana*, pea and spinach (Liedgens *et al.*, 1983; Spano & Timko, 1991; Jones & Jordan, 1994). The amino acid sequences of HMB synthases from diverse sources are highly conserved with 60% similarity between those of *E. coli* and humans (Shoolingin-Jordan, 2003). The enzyme from all known sources is a heat-stable monomer of 34–44 kDa with a dipyrromethane cofactor covalently bound. It was the first enzyme in the tetrapyrrole biosynthetic pathway to have its structure solved (Shoolingin-Jordan, 2003).

2.2.3.3 *Uroporphyrinogen III synthase*
Uroporphyrinogen III synthase catalyses the rearrangement and cyclisation of HMB to uroporphyrinogen III. This enzyme has been particularly difficult to study as the substrate is unstable and the enzyme itself is particularly sensitive to heat (Shoolingin-Jordan, 2003). The structure of the human enzyme has recently been determined by X-ray crystallography and has been shown to have an unusual dumbbell-like conformation with no obvious catalytic residues (Schubert *et al.*, 2002). In contrast to HMB synthase, there is very little cross-species sequence similarity between the known uroporphyrinogen III synthases. Activity of wheat germ uroporphyrinogen III synthase was described in the 1960s (Stevens & Frydman, 1968), but since then no further details of the plant enzyme have emerged. Although no plant uroporphyrinogen III synthase sequences have been described in the literature, two *A. thaliana* sequences have been recently annotated

as uroporphyrinogen III synthases in the GenBank database (accession numbers AAM64811 and CAC85287).

2.2.3.4 *Uroporphyrinogen III decarboxylase*

Uroporphyrinogen III decarboxylase catalyses the stepwise decarboxylation of uroporphyrinogen III to coproporphyrinogen III. This enzyme has been studied in detail in animals but it is only relatively recently that the details of the plant enzyme have emerged (Shoolingin-Jordan, 2003). The X-ray structure of the enzyme from tobacco has been recently determined and a comparison of this structure with that of the human enzyme reveals a high degree of structural conservation (Martins *et al.*, 2001a, 2001b). The degree of structural similarity suggests that mechanisms for the plant and human enzymes will be similar. For the human enzyme decarboxylations occur in a random order at high uroporphyrinogen III concentrations, while at physiological concentrations decarboxylation begins at ring D and proceeds in a clockwise direction around the macrocycle (Luo & Lim, 1991, 1993).

A naturally occurring maize uroporphyrinogen III decarboxylase mutant has been described. This mutant is characterised by necrotic spots on its leaves that are correlated with light exposure and accumulated uroporphyrinogen. This is the first known case of natural porphyria in a plant (Hu *et al.*, 1998). Tobacco plants expressing antisense mRNA to uroporphyrinogen III decarboxylase have a similar phenotype with light-induced necrotic lesions due to accumulation of uroporphyrin within the leaves (Mock & Grimm, 1997; Mock *et al.*, 1999). These plants are particularly interesting as although they have this porphyric phenotype, they are not pale green like the GSA-AT mutants and they have only slightly reduced chlorophyll levels. This suggests that this enzymatic step is not rate limiting in chlorophyll synthesis.

2.2.3.5 *Coproporphyrinogen III oxidase*

The oxidative decarboxylation of coproporphyrinogen III to protoporphyrinogen IX is catalysed by coproporphyrinogen III oxidase. Two forms of coproporphyrinogen oxidases are known within living cells. The aerobic form found in eukaryotes, bacteria and cyanobacteria uses molecular oxygen as the final electron acceptor. A second form known as the anaerobic coproporphyrinogen oxidase has been identified by mutagenesis studies in bacteria, but apart from its sequence very little is known about this enzyme (Akhtar, 2003). The plant enzyme is a member of the aerobic class of coproporphyrinogen III oxidases and has been cloned and sequenced from a number of sources. All of the plant sequences have a predicted chloroplast transit sequence indicating chloroplast localisation (Beale, 1999), and the tobacco enzyme is imported into chloroplasts and is located in the stroma (Kruse *et al.*, 1995b).

A chlorophyll-deficient mutant of sweet clover has been suggested to be a coproporphyrinogen oxidase mutant as coproporphyrin accumulates in this mutant upon feeding ALA (Bevins *et al.*, 1992). Transgenic tobacco plants expressing

coproporphyrinogen III oxidase in the antisense orientation have a similar phenotype to the uroporphyrinogen III decarboxylase mutants described previously
(Kruse *et al.*, 1995a; Mock *et al.*, 1999).

2.2.3.6   *Protoporphyrinogen IX oxidase*

The oxidation of protoporphyrinogen to protoporphyrin IX is the last common step
in the haem and chlorophyll biosynthetic pathways. The enzyme protoporphyrinogen IX oxidase catalyses this six-electron oxidation and both anaerobic and
aerobic forms of protoporphyrinogen oxidase have been identified. The plant
enzyme is of the aerobic type and requires molecular oxygen.

All of the previous enzymes within the pathway are localised exclusively within
the chloroplast of higher plants, but protoporphyrinogen IX oxidase activity is
found in both chloroplasts and mitochondria (Jacobs *et al.*, 1982; Jacobs & Jacobs,
1987). Recently the sequences of two protoporphyrinogen IX oxidase genes have
been identified in tobacco. These genes were found to code for the chloroplast and
mitochondrial isoforms of the enzyme (Lermontova *et al.*, 1997).

Protoporphyrinogen oxidase is the target of the diphenylether class of herbicides (Matringe *et al.*, 1989; Duke *et al.*, 1991). These herbicides cause a massive
build-up of protoporphyrinogen IX within the cell, which then moves to the cell
membrane where it is oxidised to protoporphyrin IX. The protoporphyrin IX then
acts as a photodynamic sensitiser producing reactive oxygen species in the
presence of light, which in turn causes membrane and cellular damage and cell
death (Matringe *et al.*, 1989; Duke *et al.*, 1991). The movement of the protoporphyrinogen IX out of the chloroplast appears to be an active transport process,
and an ABC-type transporter in the chloroplast membrane is thought to be responsible (Jacobs & Jacobs, 1993; Moller *et al.*, 2001). A number of mechanisms are
possible for herbicide resistance. A *Chlamydomonas* mutant was described that
has a mutation in the protoporphyrinogen oxidase, which renders it insensitive to
the diphenylether herbicides (Randolph-Anderson *et al.*, 1998). The sensitivity to
diphenylether herbicides was overcome in transgenic tobacco, rice and soybean
plants that were engineered to overexpress a protoporphyrinogen IX oxidase
within the plastid or mitochondria (Lee *et al.*, 2000; Lermontova & Grimm,
2000; Warabi *et al.*, 2001).

2.2.4   *Protoporphyrin IX to chlorophyll*

2.2.4.1   *Magnesium chelatase*

Magnesium chelation is the first committed step in chlorophyll biosynthesis as the
earlier enzymatic steps are shared with the haem biosynthetic pathway. Superfi-
cially, Mg^{2+} insertion into protoporphyrin resembles Fe^{2+} insertion that is catalysed by ferrochelatase. Ferrochelatase is a single subunit enzyme of ca 40 kDa
and is encoded by a single gene. The enzyme catalyses Fe^{2+} insertion into
protoporphyrin without the involvement of any additional cofactors (Ferreira,
1999). In contrast, magnesium chelatase has a requirement for ATP and consists

of three different protein subunits, known as BchI, BchD and BchH in organisms that synthesise bacteriochlorophylls and ChlI, ChlD and ChlH in organisms that synthesise chlorophylls (see Willows & Hansson, 2003).

The magnesium chelatase reaction has been dissected into two phases. The first phase involves formation of an activation complex between the BchI/ChlI and BchD/ChlD proteins that is dependent on both the protein concentration and ATP (Walker & Weinstein, 1991b; Jensen *et al.*, 1996a; Willows *et al.*, 1996; Guo *et al.*, 1998; Willows & Beale, 1998). This activation complex catalyses magnesium insertion into protoporphyrin only when combined with the BchH/ChlH protein, Mg-ATP, protoporphyrin IX and Mg^{2+}. The BchH/ChlH protein behaves as a substrate in the magnesium chelatase reaction and has a K_m in the low micromolar range (Jensen *et al.*, 1998; Willows & Beale, 1998; Gibson *et al.*, 1999). The structure of the BchI protein from *Rhodobacter capsulatus* was determined by X-ray crystallography (Fodje *et al.*, 2001). Both the structure and primary sequence of BchI show that it belongs to the extended *A*TPases *A*ssociated with a variety of cellular *A*ctivities (AAA+). This is one of the largest and most diverse classes of proteins known, and AAA+ proteins are represented in all organisms from all kingdoms (Confalonieri & Duguet, 1995; Vale, 2000). The AAA+ proteins are known to form nucleotide-dependent ring structures, which are usually hexameric, and many types can form double hexameric rings. In the double-ring structure the second ring of AAA+ modules often have an inactive ATPase and this ring presumably has a structural role. AAA+ proteins have also been called mechanoenzymes due to observed large conformational changes that occur upon ATP hydrolysis and also the mechanical nature of the processes in which many of these proteins are involved (Vale, 2000). It has been proposed that the BchI/ChlI protein forms an ATP-dependent hexameric ring which interacts with a similar BchD/ChlD hexameric ring to form a double-ring complex (Fodje *et al.*, 2001; Willows & Hansson, 2003). This structure then presumably catalyses an ATP-dependent conformation change in the BchH/ChlH to effect magnesium insertion into protoporphyrin IX which is bound to BchH/ChlH (Hansson *et al.*, 2002; Willows & Hansson, 2003).

Twenty barley mutants at three genetic loci, termed *xantha-f, xantha-g* and *xantha-h,* have been characterised as magnesium chelatase mutants due to their chlorophyll-deficient yellow phenotype and their accumulation of protoporphyrin IX (Henningsen *et al.*, 1993). Analysis of the mRNA transcript levels of *chlI, chlH* and *chlD* genes in a number of the *xantha-f, xantha-g* and *xantha-h* mutants indicated that *chlI* is the *xantha-h* locus, *chlH* is the *xantha-f* locus, (Jensen *et al.*, 1996b) and *chlD* is the *xantha-g* locus (Petersen *et al.*, 1999). Each of the three semidominant alleles of the *xantha-h* locus, originally isolated as pale green *chlorina* mutants, has a single missense mutation leading to either Asp to Asn, Arg to Lys or Leu to Phe amino acid substitutions in the 42 kDa magnesium chelatase ChlI protein (Hansson *et al.*, 1999).

A. thaliana mutants have been identified that have defective magnesium chelatase *chlI* genes. The two allelic mutants *cs* and *ch-42* have defects in a gene with strong identity to *bchI* (Koncz *et al.*, 1990). These mutants can still make small amounts of chlorophyll using a second *chlI* gene (Rissler *et al.*, 2002). Two *A. thaliana* mutants that have mutations in the *chlH* gene have also been described and characterised (Mochizuki *et al.*, 2001). These two mutants are called *cch* (conditional chlorina) and *gun5* (genomes uncoupled) (Susek *et al.*, 1993). The *gun* mutants were selected for their ability to express the chlorophyll *a/b*-binding proteins of PSII, *Lhcb1*, under conditions where it is normally not expressed. The *gun5* and *cch* alleles were found to have missense mutations resulting in an Ala to Val substitution in *gun5* and a Pro to Leu substitution in *cch*. The ChlH protein thus appears to have a dual function in chlorophyll biosynthesis and chloroplast-to-nucleus signal transduction.

Other *chlH* gene mutants have been described in *Antirrhinum majus* (snap-dragon) and *Chlamydomonas reinhardtii*. The snapdragon mutant *oli*-605 (Luo *et al.*, 1991) was found to have a transposon insertion in the *chlH* gene. The gene encodes a large protein of 1359 amino acids that had strong sequence identity to cobaltochelatase CobN subunit and to the BchH protein, and this was the first indication that the plant and bacterial magnesium chelatase genes were similar (Hudson *et al.*, 1993). The *br$_s$-1* mutant in *C. reinhardtii* as well as another chlorophyll-deficient mutant called *chl1* also have mutations in the *chlH* gene (Chekounova *et al.*, 2001). These mutations in *chlH* are insertions that cause a frame shift resulting in a truncated protein product, which does not accumulate (Chekounova *et al.*, 2001).

Transgenic tobacco plants that express the *chlI* and *chlH* genes in the antisense orientation have been produced. These plants have a uniformly pale green pheno-type typical of some of the barley mutants in these two genes. Perhaps unsurpris-ingly, protoporphyrin IX did not accumulate in these mutants, which tends to support the theory that feedback inhibition of ALA biosynthesis by haem is a major controlling factor in the pathway. However, the transcript levels of the *gtr* and *alad* genes were also reduced in these lines, suggesting that expression may be synchronised with magnesium chelatase transcripts in some way (Papenbrock *et al.*, 2000a, 2000b).

Studies reporting compounds that specifically inhibit magnesium chelatase have been somewhat confusing as inhibition of activity has been examined in a variety of ways ranging from *in vivo* studies, *in organello* studies to true *in vitro* inhibition experiments. The *in vivo* and *in organello* studies suffer from problems of access of the inhibitor to the enzyme, which was highlighted in a number of papers. *In organello* experiments using intact cucumber chloroplasts showed that N-methylprotoporphyrin and N-methylmesoporphyrin resulted in 50% inhibition of magnesium chelatase at $3 \mu M$ but that protochlorophyllide, chlorophyllide, haem or magnesium-protoporphyrin had very little effect on activity (Walker & Weinstein, 1991a). Similarly, intact barley chloroplast experiments showed pheo-phorbide inhibition at $0.92 \mu M$ while chlorophyllide and zinc pheophorbide were only slightly inhibitory, and chlorophyll and pheophytin were not inhibitory (Pöpperl *et al.*, 1997). These results are in contrast to experiments where *in situ* generation of magnesium protoporphyrin or its monomethylester in pea seedlings caused inhibition of magnesium chelatase, which was inversely dependent on the total concentration of magnesium protoporphyrin and its monomethylester (Averina et al., 1995, 1996). It seems that the metal containing porphyrins in earlier studies may not have effectively penetrated the chloroplast and thus their effect on magnesium chelatase is unclear. The inhibition observed by pheophorbide may have a physiological function as it forms during leaf senescence when chlorophyll is being degraded; thus it may act to prevent further chlorophyll biosynthesis by inhibition of magnesium chelatase (Pöpperl et al., 1997).

Protein-modifying reagents also inhibit magnesium chelatase activity (Walker & Weinstein, 1991a). Reagents that modify cysteine residues in proteins were potent inhibitors of magnesium chelatase with 50% inhibition achieved using 50 µM N-ethylmaleimide (NEM), 100 µM p-chloromercuribenzoyl sulphonate (PCMBS) or 50 µM p-chloromercuribenzoate (PCMB) (Walker & Weinstein, 1991a). This suggests that a cysteine residue is involved in catalysis. In addition to protein-modifying agents and tetrapyrrole substrate analogues, magnesium chelatase can be inhibited by a variety of other compounds. These inhibitors can be classified according to their probable mode of inhibition as ATPase inhibitors, metal ion chelators, and other non-classifiable inhibitors. The cucumber magnesium chelatase was inhibited by the non-hydrolysable ATP analogues, β,γ-methylene ATP and β,γ-imino ATP, and it was also inhibited by the metal ion chelators 2,2-dipyridyl and 1,10-phenanthroline (Walker & Weinstein, 1991a). The metal ion chelators inhibit both the ATPase activity of individual subunits as well as magnesium chelatase activity (Hansson & Kannangara, 1997). In addition, Co(III)-ATP-1,10-phenanthroline, a reagent that labels ATPases, inhibited magnesium chelatase activity and bound to all three subunits (Hansson & Kannangara, 1997).

A number of other inhibitors of magnesium chelatase activity have also been described but their mode of action is unclear. All three hydroxybenzoic acid methylester isomers were found to inhibit magnesium chelatase of cress and barley seedlings (Wubert et al., 1997). Magnesium chelatase from barley and R. sphaeroides was inhibited by chloramphenicol and p-aminosalicylic acid (Kannangara et al., 1997). Light has also been shown to inhibit the magnesium chelatase of barley (Pöpperl et al., 1997) and Rhodobacter (Willows & Beale, 1998), and this mode of inhibition probably occurs via photo-oxidative damage of the BchH/ChlH subunit (Willows & Beale, 1998; Willows et al., 2003). The inhibition by light of barley magnesium chelatase was demonstrated with isolated chloroplasts, in contrast with the situation in planta where isolated chloroplasts of barley from etiolated barley seedlings exposed to 4 h of light have considerably higher activity than chloroplasts from plants not exposed to light (Jensen et al., 1996b). This increase in activity in planta can be attributed to the increased synthesis of the ChlI and ChlH subunits (Jensen et al., 1996b) and is supported by data showing that magnesium protoporphyrin and magnesium protoporphyrin monomethylester levels increase dramatically in leaves from barley or tobacco when transferred from dark to light (Pöpperl et al., 1998).

2.2.4.2   S-adenosylmethionine:magnesium protoporphyrin IX-O-methyltransferase (E.C. 2.1.1.11)

S-adenosylmethionine:magnesium protoporphyrin IX-O-methyltransferase cata-lyses the S-adenosylmethionine-dependent methylation of the carboxyl group of the 13-propionate on magnesium protoporphyrin IX. This enzyme is membrane-associated and the activity has been characterised for a number of plant species (reviewed in Bollivar, 2003). The gene for S-adenosylmethionine:magnesium protoporphyrin IX-O-methyltransferase has also been cloned and sequenced from tobacco and *A. thaliana* (Block *et al.*, 2002), and antisense transgenic tobacco plants have been produced and are the subject of a patent (Reindl *et al.*, 2001). Two barley mutants *xantha-n* and *albina-e* have no detectable S-adenosylmethionine:magnesium protoporphyrin IX-O-methyltransferase activity. These mutants also have defective membrane structure and it is suggested that these mutations may be a pleiotropic effect of the defective membrane structure (Moller *et al.*, 1997).

Sinefungin, an inhibitor of S-adenosylmethionine-dependent methylation reac-tions, was found to be an effective inhibitor of the barley enzyme when delivered via the transpiration stream (Vothknecht *et al.*, 1995). The S-adenosylmethionine substrate analogue, S-adenosyl-ethionine, was an effective inhibitor of the barley enzyme (Shieh *et al.*, 1978). The two products of the enzyme, S-adenosyl-homocysteine and magnesium protoporphyrin IX monomethylester are also inhibitors of the O-methyltransferase (Shieh *et al.*, 1978; Yaronskaya *et al.*, 1993).

2.2.4.3   Magnesium-protoporphyrin IX monomethylester oxidative cyclase

An oxidative cyclisation is required to create the fifth ring of chlorophyll; this reaction is catalysed by magnesium protoporphyrin IX monomethylester oxidative cyclase. The origin of the oxygen atom in the fifth ring was studied by ^{18}O labelling using $^{18}O_2$ and/or $H_2^{18}O$. Cucumber cotyledons incubated in a nitrogen atmosphere containing 20% $^{18}O_2$ resulted in ^{18}O incorporation into the oxo group of the isocyclic ring (Walker *et al.*, 1989). This is in contrast to photosynthetic bacteria where the oxo group is usually obtained from water (Porra *et al.*, 1995, 1998).

Oxidative cyclase activity has been demonstrated with chloroplasts of *C. reinhardtii* (Bollivar & Beale, 1995), developing chloroplasts from cucumber cotyledons (Vijayan *et al.*, 1992), lysed cucumber and *C. reinhardtii* chloroplasts (Walker *et al.*, 1991; Whyte *et al.*, 1992; Whyte & Castelfranco, 1993; Bollivar & Beale, 1996) and with cell-free extracts from cyanobacteria (Bollivar & Beale, 1996). The oxidative cyclase from cucumber chloroplasts was resolved into membrane and soluble components. As the oxo group on the fifth ring is derived from molecular oxygen in plants (Walker *et al.*, 1989), it was considered possible that a cytochrome P450-type enzyme was involved in the reaction. However, inhibition of the cucumber cyclase by inhibitors of P450 enzymes was not consistent. The reconstituted system was inhibited by haemoprotein inhibitors such as azide and KCN, but very little inhibition was observed in intact chloroplasts using these. Benzoquinone and quinol were also strong inhibitors of the cyclase (Whyte & Castelfranco, 1993). In contrast to the cyclases from cucumber, *C. reinhardtii* cyclase activity was found associated with the membranes and did not require a soluble component. The *C. reinhardtii* cyclase was not inhibited by the flavoprotein inhibitor quinacrine or by the haemoprotein inhibitors CO, KCN, or NaN$_3$ (Bollivar & Beale, 1996).

A metal ion appears to be essential for the reaction as divalent metal ion chelators are potent inhibitors of the enzyme. For example, pretreatment of the cucumber pellet fraction with either 8-hydroxyquinoline or desferal mesylate inhibited cyclase activity, indicating a metal ion requirement in this fraction (Walker *et al.*, 1991). Similarly, plants treated with β-thujaplicin, an effective chelator of Fe^{2+} (Tanaka *et al.*, 1995), and ALA accumulate magnesium protoporphyrin monomethylester and do not make protochlorophyllide (Oster *et al.*, 1996). The most common inhibitor of cyclase is the herbicide 2,2′-dipyridyl which is also a Fe^{2+} chelator and results in reduced chlorophyll synthesis (Mostowska *et al.*, 1996). One common feature of all cyclases studied is that they are all inhibited by chelators of Fe^{2+}, suggesting that nonhaem iron is involved in the reaction. As only hydrophobic Fe^{2+} chelators appear to be effective inhibitors, it was suggested that the Fe^{2+} requirement is associated with the cyclase membrane fraction (Bollivar & Beale, 1996).

Although no plant genes have been positively identified, hints at the identity of the plant oxidative cyclase genes come from *Chlamydomonas* mutants and the purple bacterium *Rubrivivax gelatinosus*. Unlike many purple bacteria *Rx. gelatinosus* is able to synthesise bacteriochlorophyll *a* under both aerobic and anaerobic conditions. Disruption of the *acsF* gene of *Rx. gelatinosus* prevents bacteriochlorophyll *a* synthesis and causes accumulation of magnesium protoporphyrin IX monomethylester under aerobic conditions but not under conditions of low aeration. The designation *acsF* stands for aerobic cyclisation system Fe-containing subunit, as AcsF and its homologues have a conserved putative binuclear-iron-cluster-binding motif (Pinta *et al.*, 2002). The AcsF protein is homologous to previously identified genes in *C. reinhardtii* called *Crd1* (Moseley *et al.*, 2000) and *Cth1* (Moseley *et al.*, 2002) and homologues of *AcsF* were also identified in *A. thaliana* and *Synechocystis* (Pinta *et al.*, 2002).

Crd1 and *Cth1* expression in *C. reinhardtii* is reciprocal and is regulated by copper and/or oxygenation conditions. *Crd1* is expressed under low aeration and/or low copper conditions, and *Cth1* is expressed under oxygenated and copper-sufficient conditions. Mutation of either of these genes and growth under conditions where the alternative protein is not expressed result in a chlorotic phenotype with reduced PSI and light harvesting 1 accumulation (Moseley *et al.*, 2000, 2002). These results suggest that the Crd1 and Cth1 proteins probably encode two isoforms of the oxidative cyclase. Two mutant loci in barley called *xantha-l*[35] and *viridis-k*[23] also have defective cyclase activity (Walker *et al.*, 1997), raising the possibility of two isoforms of the enzyme in barley.

2.2.4.4 Reduction of the 8-vinyl group

The reduction of the 8-vinyl group can probably occur at any stage from proto-porphyrin IX to chlorophyllide *a*. This finding is supported by numerous studies in which 8-ethyl and 8-vinyl derivatives of these intermediates have been detected (Rebeiz *et al.*, 1994; Kim & Rebeiz, 1995; Parham & Rebeiz, 1995; Kim *et al.*, 1997). The relative amounts of mono- and 8-vinyl intermediates and the stage at which reduction occurs is complex and depends on numerous factors such as species, developmental stage, time in the dark or light, the age of the tissue, and light intensity (Rebeiz *et al.*, 1994). Virtually all photosynthetic organisms require reduction of the 8-vinyl group of chlorophyll to an ethyl group. However, certain marine *Prochlorococcus* species accumulate 8-vinyl chlorophylls *a* and *b* in addition to or instead of the 8-ethyl pigments (Goericke & Repeta, 1992, 1993).

A method was described for the separation of 8-vinyl-protochlorophyllide and 8-ethyl-protochlorophyllide using a solid phase polyethylene column. This was used to analyse the biosynthesis of both these intermediates in wheat and cucumber cotyledons. The activity in wheat was higher than in cucumber and it was suggested that the reaction is reversible (Whyte and Griffiths, 1993). An 8-vinyl reductase activity was detected in plastid membranes from cucumber that converts 8-vinyl-chlorophyllide *a* to chlorophyllide *a* but is unable to convert 8-vinyl-protochlorophyllide to 8-ethyl-protochlorophyllide (Parham & Rebeiz, 1992, 1995). It has been suggested that a soluble component may mediate the substrate specificity of the 8-vinyl reductase allowing other 8-vinyl intermediates to be converted to 8-ethyl forms, which would explain the diversity of 8-ethyl intermediates that have been observed (Kim *et al.*, 1997). No plant genes have yet been identified that are required for reduction of the 8-vinyl group.

2.2.4.5 Protochlorophyllide oxidoreductases

Two types of enzymes have been identified that reduce the D pyrrole ring of protochlorophyllide to form chlorophyllide. Of these two enzymes the NADPH-protochlorophyllide oxidoreductase (EC 1.3.1.33 or EC 1.6.99.1, abbreviated POR) has been the subject of a large number of reviews (Hendrich & Bereza, 1993; Fujita, 1996; Reinbothe *et al.*, 1996; Reinbothe & Reinbothe, 1996a, 1996b; Adamson *et al.*, 1997; Lebedev & Timko, 1998; Schoefs, 1999, 2001; Rudiger, 2003).

POR is a single subunit enzyme that requires light as a substrate and it appears to be present in all organisms that synthesise chlorophyll but has not been found in bacteriochlorophyll-synthesising organisms. The second type of enzyme, known as the light-independent protochlorophyllide oxidoreductase or DPOR, consists of three subunits (Armstrong, 1998; Fujita & Bauer, 2003). The multi-subunit DPOR has not been found in flowering plants (angiosperms) but appears to be present in most other chlorophyll and bacteriochlorophyll-synthesising organisms and allows these organisms to make chlorophyll in the dark. When angiosperms are germinated in the dark, they accumulate small amounts of protochlorophyllide bound together with NADPH and POR in a ternary complex. These plants are unable to synthesise chlorophyll until the bound protochlorophyllide is converted to chlorophyllide when exposed to light. This has lead to the widespread belief that angiosperms are unable to synthesise chlorophyll in the dark. However, there are numerous reports that mature green leaves of some angiosperms can synthesise chlorophyll in the dark (reviewed in Adamson *et al.*, 1997). This suggests that DPOR may be present in mature leaves of some angiosperms or that another as yet uncharacterised mechanism exists for chlorophyll synthesis in the dark in these plants.

2.2.4.5.1 Light dependent POR

The first POR-encoding gene, *por*, was isolated from barley (Schulz *et al.*, 1989). Since then, *por* genes have been isolated and sequenced from many plants and algae including *A. thaliana* (Armstrong *et al.*, 1995), *Triticum aestivum* (Teakle & Griffiths, 1993), *N. tabacum* (Masuda *et al.*, 2002), cucumber (Fusada *et al.*, 2000), pea (Spano *et al.*, 1992a; He *et al.*, 1994), *Pinus taeda* (Spano *et al.*, 1992b), *Pinus mugo* (Forreiter & Apel, 1993), *C. reinhardtii* (Li & Timko, 1996) as well as from the cyanobacteria *Synechocystis* PCC6803 (Suzuki & Bauer, 1995). Multiple isoforms of POR have been found in *A. thaliana* (Armstrong *et al.*, 1995; Oosawa *et al.*, 2000; Su *et al.*, 2001), *P. taeda* (Skinner & Timko, 1998), *P. mugo* (Forreiter & Apel, 1993), barley (Holtorf *et al.*, 1995) and tobacco (Masuda *et al.*, 2002). In barley and *A. thaliana* the isoforms are differentially expressed and the isoform called PORA appears to have a role only in the de-etiolation process (Armstrong *et al.*, 1995; Holtorf *et al.*, 1995). However, this type of differential regulation of isoforms does not appear to be universal as the two tobacco *por* genes are similarly regulated (Masuda *et al.*, 2002). Some plants such as cucumber have only a single *por* gene indicating that multiple isoforms are not essential for plant growth and development (Fusada *et al.*, 2000).

When plants are placed in the dark, protochlorophyllide, NADPH and POR form a ternary complex within the chloroplast which is poised waiting for the final substrate, light, to allow photoconversion of the protochlorophyllide to chlorophyllide. This ternary complex is membrane-associated and forms crystalline-like structures, which are visible by electron microscopy in the etioplasts of angiosperms. These crystalline structures are called prolamellae bodies (PLBs) and their formation is dependent on the ternary complex arranged in aggregates on membranes. One of the main spectral forms of protochlorophyllide observed *in vivo* is due to these NADPH–POR–protochlorophyllide ternary complex aggregates that make up the PLBs (Wiktorsson *et al.*, 1992, 1993, 1996). Pigment binding to POR is essential for the formation of PLBs, as mutants that are unable to make protochlorophyllide do not make PLBs (Henningsen *et al.*, 1993). PLB formation can also be inhibited by treatment of plants with gabaculine, which inhibits protochlorophyllide formation (Younis *et al.*, 1995). Lipids are also essential for the formation of the PLBs (Klement *et al.*, 2000). Flavins (Belyaeva *et al.*, 2000), violaxanthin and zeaxanthin (Chahdi *et al.*, 1998) have also been detected in PLBs and may be involved in their formation. PLBs have been detected in organisms that are normally able to synthesise chlorophyll in the dark such as the yellow in the dark mutants of *C. reinhardtii*. This suggests that most PORs are capable of forming PLBs, and the demonstration that both PORA and PORB of *A. thaliana* are able to form PLBs supports this suggestion (Sperling *et al.*, 1998; Franck *et al.*, 2000). On exposure to light, protochlorophyllide is converted to chlorophyllide and then rapidly to chlorophyll. The PLBs then rapidly disperse or disaggregate as the photosystems are assembled. Protein phosphorylation appears to be involved in this disaggregation process and in the formation of the PLBs (Wiktorsson *et al.*, 1996; Kovacheva *et al.*, 2000).

2.2.4.5.2   Light-independent (dark) POR (DPOR)

Light-independent protochlorophyllide reduction has been extensively reviewed by Armstrong (1998). Cyanobacteria, green algae and most non-flowering plants have both a POR and a DPOR. Green algae and most non-flowering land plants are able to make chlorophyll in the dark using the protein products of the chloroplast-encoded genes *chlL*, *chlN* and *chlB*. Mutation or deletion of these chloroplast-encoded *chlL*, *chlN* and *chlB* genes in the green algae *C. reinhardtii* prevents chlorophyll synthesis in the dark (Roitgrund & Mets, 1990; Suzuki & Bauer, 1992; Li *et al.*, 1993; Liu *et al.*, 1993). Seven *C. reinhardtii* nuclear mutants have a similar lack of chlorophyll in the dark and in all cases these mutations prevent the translation of the chloroplast-encoded *chlL* gene (Cahoon & Timko, 2000). The *chlL* gene of *C. reinhardtii* hybridises to DNA from distantly related bacteria and non-flowering land plants but not to DNA from the representative angiosperms, *Zea mays*, *A. thaliana*, *N. tabacum* and *Bougainvillea glabra* (Suzuki & Bauer, 1992). When homologues of the *chlL*, *chlN* and *chlB* genes are present, they are invariably found in the chloroplast genomes (Lidholm & Gustafsson, 1991; Burke *et al.*, 1993; Suzuki *et al.*, 1997; see also Armstrong, 1998).

2.2.4.6   *Chlorophyll a synthase*

Chlorophyll *a* synthesis is completed with the esterification of chlorophyllide *a* with a phytol. This reaction is catalysed by chlorophyll synthase. Oat (*Avena sativa*) and *A. thaliana* chlorophyll synthase genes, *chlG*, have been cloned and the enzymes heterologously expressed in *E. coli*. These enzymes are nuclear-encoded and have a chloroplast transit sequence for translocation into the chloroplast. Phytyl-pyrophosphate and geranylgeranyl-pyrophosphate are both substrates for chlorophyll synthases. *A. thaliana* chlorophyll synthase preferred geranylgeranyl-pyrophosphate as the substrate (Oster *et al.*, 1997; Oster & Rudiger, 1997; Schmid *et al.*, 2001).

Reduction of geranylgeraniol to phytol can occur either before of after ester-ification to chlorophyllide *a*. *ChlP* is the gene encoding the enzyme involved in the reduction of geranylgeraniol to phytol. The *chlP* genes in *N. tabacum* and *A. thaliana* are found in the nuclear genome and encode a 52-kDa precursor protein. Transgenic tobacco plants expressing antisense ChlP RNA have both reduced tocopherol and chlorophyll synthesis, indicating that this enzyme provides phytol and/or phytyl-pyrophosphate for both of these pathways (Tanaka *et al.*, 1999).

One reductase activity in the chloroplast envelope converts geranylgeranyl-pyrophosphate to phytyl-pyrophosphate and a second in the thylakoids converts geranylgeraniol esterified to chlorophyllide *a* into chlorophyll *a*. A second gene may be present which encodes the second reductase.

2.2.4.7 Interconversion of chlorophyll a and b

Chlorophyll *a* oxygenase (CAO) is the enzyme that catalyses the conversion of chlorophyll *a* to chlorophyll *b*, and a reductase is found in chloroplasts which can catalyse the reverse reaction. This interconversion is thought to operate as a cycle and has been reviewed recently by Rudiger (2002).

CAO genes have been identified in *A. thaliana, Oryza sativa, Marchantia polymorpha, Dunaliella salina, Prochlorothrix hollandica* and *Prochloron didemni* (Tomitani *et al.*, 1999; Espineda *et al.*, 1999). The *A. thaliana* CAO was heterologously expressed in *E. coli* and required oxygen and reduced ferri-doxin to convert chlorophyllide *a* to chlorophyllide *b*. Traces of the 7^1-hydroxy intermediate were detected and the enzyme could also use Zn-chlorophyllide *a* as a substrate but not pheophorbide or chlorophyll *a* (Oster *et al.*, 2000). CAO has been introduced into cyanobacteria where it results in chlorophyll *b* synthesis resulting in aberrant PSI trimer assembly (Satoh & Tanaka, 2002).

2.3 Chlorophyll degradation

The half-life of chlorophyll within a normal plant leaf is estimated to be between 6 and 50 h. In addition, during the highly visible colour change which occurs in autumn in deciduous plants and also in other senescing processes, chlorophyll is also degraded. The main reason for chlorophyll degradation during senescence is so that the plant can recover nutrients, such as the nitrogen tied up in the proteins of the photosynthetic proteins, without having to risk potential photo-oxidation caused by the chlorophyll that would be liberated by this process. However, in contrast to the chlorophyll biosynthetic pathway, the pathway for chlorophyll degradation is not as well elucidated. So what happens to this chlorophyll? One of the reasons for the relative lack of information is that most of the chlorophyll degradation products are colourless and so were difficult to identify, and it was only in the 1990s that significant advances were made in the field of chlorophyll degradation. These advances have been reviewed quite extensively and the key steps are shown in Fig. 2.4 and are summarised below (Rudiger, 1997; Matile *et al.*, 1999, 2000; Kräutler, 2003).

The first two steps of chlorophyll degradation are removal of phytol, catalysed by chlorophyllase, and removal of magnesium, by a magnesium dechelatase. Chlor-ophyllase is a well-characterised esterase which has been localised to the chloroplast envelope membrane and will remove phytol from chlorophyll *a, b* and pheophytin *a* (Rudiger, 1997). The removal of magnesium from chlorophyll occurs by enzymatic sequestration of the magnesium with the assistance of a low–

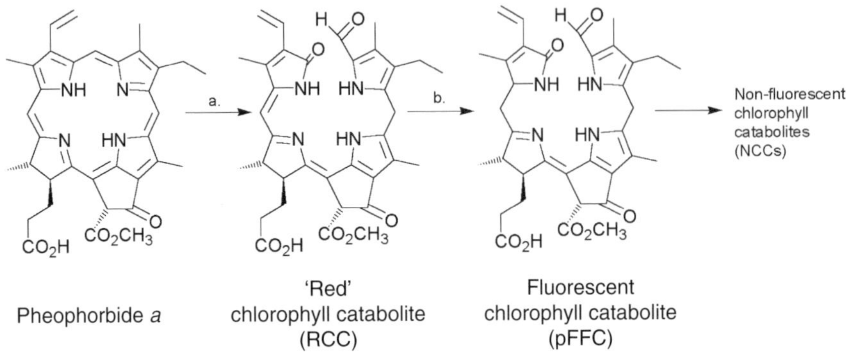

Fig. 2.4 Key steps in chlorophyll degradation: (a) pheophorbide *a* oxygenase; (b) RCC reductase.

molecular weight heat-stable substance. The magnesium is then transported out of senescing leaves and stored in the remaining part of the plant (Kräutler, 2003).

The macrocycle of pheophorbide *a* is then oxidatively opened between pyrrole rings A and B by pheophorbide *a* oxygenase to produce the 'red' chlorophyll catabolite (RCC). Pheophorbide *a* oxygenase requires molecular oxygen and ferredoxin and is located in the chloroplast envelope membrane. This RCC is then reduced by the ferredoxin-dependent RCC reductase to the slightly yellow fluorescent chlorophyll catabolite (pFCC). pFCC is unstable and is rapidly and possibly non-enzymatically converted to colourless and non-fluorescent compounds which are stored in the vacuole of the senescing leaf (Kräutler, 2003).

2.4   Regulation

The three main regulatory points in chlorophyll biosynthesis appear to be the steps involved in ALA biosynthesis, magnesium chelatase and protochlorophyllide reductase. Figure 2.5 shows an overview of the regulatory mechanisms that affect these steps as well as some other regulatory mechanisms that have not been fully elucidated. The regulation of these key steps in chlorophyll biosynthesis includes mechanisms to control (1) quantities of individual enzymes using transcriptional or translational controls; and (2) activities of various enzymatic steps within the pathway using feedback inhibitions or other modifiers of enzymatic activity. An additional regulatory feature is that both the product of the magnesium chelatase reaction and the magnesium chelatase itself are implicated in control of nuclear gene expression.

2.4.1   Regulation of ALA synthesis

Glutamyl-tRNA reductase is the rate-determining step of the entire tetrapyrrole biosynthetic pathway in plants (Grimm, 2003). Feedback inhibition and

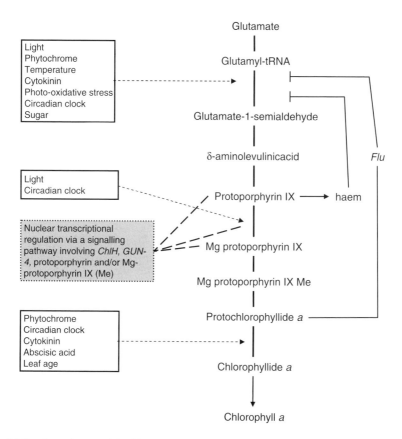

Fig. 2.5 Regulatory features of the chlorophyll biosynthetic pathway. Only key intermediates in the pathway are shown. Feedback inhibition is shown by thin lines. Boxes show effectors of gene regulation of key enzymes within the pathway. The shaded box indicates the link between chlorophyll intermediates and their effects on nuclear gene expression.

transcriptional regulation are both used to regulate the activity of this enzyme within plant cells. Feedback regulation by haem is difficult to demonstrate conclusively because of the detergent-like properties of haem and its low solubility. However, recombinant and natural barley GTR have been shown to be inhibited by haem and also appear to have a bound haem (Pontoppidan & Kannangara, 1994; Vothknecht *et al.*, 1996). The inhibition and haem binding were both abolished when the N-terminus of the barley enzyme was truncated by 30 amino acids, which tends to confirm that the haem inhibition is not an artefact (Vothknecht *et al.*, 1998). Protochlorophyllide is known to limit its own synthesis in dark-grown plants and the feedback regulation occurs at the level of ALA synthesis. This inhibition is likely to be via the *FLU* protein, as mutations in this protein result in deregulation of protochlorophyllide synthesis and it has been shown to interact with GTR (Meskauskiene *et al.*, 2001; Meskauskiene & Apel, 2002). All plants studied to date have multiple *GTR* genes that are differentially expressed in various plant organs and under a variety of conditions. Transcriptional regulators have included light, phytochrome, temperature, cytokinin, photo-oxidative stress, circadian clocks and sugar (Ujwal *et al.*, 2002; Grimm, 2003).

Glutamate 1-semialdehyde aminotransferase gene (*gsa*) expression has been studied in detail in the alga *C. reinhardtii*. Expression from *gsa* is induced by blue light and is mediated by Ca^{2+} and calmodulin. Detailed expression studies of the plant *gsa* genes have not been reported.

2.4.2   *Magnesium chelatase*

Regulation of magnesium chelatase gene expression has been examined in a number of plant species. In etiolated barley, the *chlI* and *chlH* genes are upregulated by light (Jensen *et al.*, 1996b). In green barley seedlings grown in normal daylight cycles the *chlH* gene then follows a circadian rhythm with maximal expression in the light phase (Jensen *et al.*, 1996b). The tobacco *chlH* and *chlI* genes follow a similar circadian pattern of expression but the *chlD* gene has an inverse expression pattern with maximal mRNA levels in the dark phase. In *A. thaliana* (Gibson *et al.*, 1996) and *A. majus* (Hudson *et al.*, 1993) the *chlH* gene has maximal expression in the dark and is downregulated in the light. In addition, the *chlI* gene from barley and *A. thaliana* is constitutively expressed except during the initial phases of greening.

Magnesium chelatase proteins and magnesium protoporphyrin IX have all been implicated in chloroplast–nuclear signalling. As mentioned previously, the *A. thaliana GUN-5* mutant is a result of a point mutation in the *chlH* gene and a *chlD* mutation gives a similar phenotype (Mochizuki *et al.*, 2001; Strand *et al.*, 2003). The recently identified *GUN-4* protein binds protoporphyrin IX and also interacts with the magnesium chelatase porphyrin-binding *ChlH* protein. This implicates *GUN-4* as one of the downstream signalling components in chloroplast nuclear signalling (Larkin *et al.*, 2003).

2.4.3   *Protochlorophyllide reductase*

Phytochrome, circadian clocks, cytokinin, abscisic acid and leaf age have all been implicated in control of *por* gene expression. The amount of POR protein and POR mRNA decreases rapidly in many species when etiolated plants are exposed to light (Forreiter *et al.*, 1990), suggesting phytochrome involvement in this process. Experiments using *A. thaliana* with red and far-red light treatments have confirmed phytochrome A regulates *PORA* mRNA levels (Barnes *et al.*, 1996; Sperling *et al.*, 1997, 1998). Phytochrome has also been shown to regulate the expression of the *por* gene from a lower plant (*M. paleacea*) (Suzuki *et al.*, 2001). In barley the phytochrome- and/or light-dependent regulation of *PORA* mRNA levels is dependent on a $3'$ untranslated region in the *PORA* mRNA

(Holtorf & Apel, 1996). In addition to the reduction in message, a light-dependent degradation of the PORA bound to chlorophyllide, but not protochlorophyllide, occurs and a light-induced protease has been shown to be responsible (Reinbothe *et al.*, 1995). In contrast, cucumber, which has only a single *por* gene, shows an increase in *por* message levels during the de-etiolation process (Kuroda *et al.*, 1995), and a decrease in message levels occur when plants are transferred from dark to light (Kuroda *et al.*, 2000).

The effect of plant age and leaf age on *por* gene expression has been studied in pea (He *et al.*, 1994), barley (Holtorf *et al.*, 1995; Schunmann & Ougham, 1996), wheat (Marrison *et al.*, 1996) and *A. thaliana* (Armstrong *et al.*, 1995). In barley and *A. thaliana* the *PORA* mRNA is only expressed in young etiolated tissue while the *PORB* mRNA is expressed throughout development. In light-grown seedlings of pea and wheat the youngest leaves contained the highest *POR* message levels.

The plant hormones cytokinin and abscisic acid also appear to have a role in regulation of *por* gene expression. The involvement of cytokinin in *por* regulation was inferred from the finding that cytokinins overcame the inhibition of greening caused by treatment with cadmium and mercury (Thomas & Singh, 1995, 1996), although cadmium and mercury also have a direct effect on POR enzyme activity (Boddi *et al.*, 1995; Lenti *et al.*, 2002). It was subsequently found that cytokinins directly activated *por* gene expression in cucumber (Kuroda *et al.*, 2001) and *Lupinus luteus* (Kusnetsov *et al.*, 1998). It was also found that abscisic acid inhibits *por* gene expression in *L. luteus* (Kusnetsov *et al.*, 1998).

2.5   Chlorophyll and chemically modified chlorophylls as photodynamic agents

As has been mentioned throughout this chapter, chlorophyll and its intermediates are able to act as photosensitising agents that can generate reactive oxygen species to potentially cause significant cellular damage leading to cell death. This is how the diphenylether herbicides work, and this effect has also been harnessed in a medical treatment known as photodynamic therapy. Photodynamic therapy is usually used for the treatment of certain types of cancer. The cancer cells are treated so that a photodynamic agent accumulates preferentially within the cancer cells. Irradiation with an intense laser light then kills the cancer cells without affecting the surrounding tissue. Treatments often involve feeding δ-aminolevu-linic acid so that protoporphyrin IX accumulates and the cells are then irradiated with a blue laser. However, human tissue absorbs blue light and thus this specific technique is not effective on large or deep tumours. Human tissue is partially transparent to red light and thus chlorophyll and its derivatives, which can act as photosensitisers when irradiated with red light, have been developed as photo-dynamic agents with wider application. A few examples include pheophorbide *a* methylester used as a photodynamic agent to kill lung cancer carcinoma (Sun & Leung, 2002); chlorophyll *a*-based conjugates as diagnostic and therapeutic agents (Iwai *et al.*, 1989a); pheophorbide *a* derivatives as photodynamic agents (Iwai *et al.*, 1989b); chlorophyllide *a* and a chlorine derivative which have the positron emitter ^{48}V inserted in the place of magnesium and can thus be used as a tumour-imaging tool in conjunction with photodynamic therapy (Iwai *et al.*, 1990).

References

Adamson, H. Y., Hiller, R. G. & Walmsley, J. (1997) Protochlorophyllide reduction and greening in angiosperms – an evolutionary perspective. *J. Photochem. Photobiol. B* , **41**, 201–221.

Akhtar, M. (2003) Coproporphyrinogen III and protoporphyrinogen IX oxidases, in *The Porphyrin Handbook II*, Vol. 12 (eds K. M. Kadish, K. Smith & R. Guilard), Academic Press, San Diego, USA, pp. 75–92.

Anderson, J. M. (1986) Photoregulation of the composition, function, and structure of thylakoid membranes. *Ann. Rev. Plant Physiol.*, **37**, 93–136.

Armstrong, G. & Apel, K. (1998) Molecular and genetic analysis of light-dependent chlorophyll biosynthesis. *Methods Enzymol.*, **297**, 237–244.

Armstrong, G. A. (1998) Greening in the dark: light-independent chlorophyll biosynthesis from anoxygenic photosynthetic bacteria to gymnosperms. *J. Photochem. Photobiol.* B, **43**, 87.

Armstrong, G. A., Runge, S., Frick, G., Sperling, U. & Apel, K. (1995) Identification of NADPH: proto-chlorophyllide oxidoreductases A and B: a branched pathway for light-dependent chlorophyll biosynthesis in *Arabidopsis thaliana*. *Plant Physiol.*, **108**, 1505–1517.

Averina, N. G. (1998) Mechanisms of regulation and interplastid localization of chlorophyll biosynthesis. *Membr. Cell Biol.*, **12**, 627–643.

Averina, N. G., Yaronskaya, E. B., Rassadina, V. V., Shalygo, N. V. & Walter, G. (1995) Influence of light, 5-aminolevulinic acid, homocysteine and dipyridyl on magnesium-chelatase activity and porphyrin accumulation in plants. *Photosynth.: Light Biosphere, Proc. Int. Photosynth. Congr. 10th*, **3**, 925–928.

Averina, N. G., Yaronskaya, E. B., Rassadina, V. V. & Walter, G. (1996) Response of magnesium chelatase activity in green pea (*Pisum sativum* L.) leaves to light, 5-aminolevulinic acid and dipyridyl supply. *J. Photochem. Photobiol.* B, **36**, 17–22.

Barnes, S. A., Nishizawa, N. K., Quaggio, R. B., Whitelam, G. C. & Chua, N. H. (1996) Far-red light blocks greening of Arabidopsis seedlings via a phytochrome A-mediated change in plastid development. *Plant Cell*, **8**, 601–615.

Beale, S. I. (1999) Enzymes of chlorophyll biosynthesis. *Photosynth. Res.*, **60**, 43–73.

Belyaeva, O. B., Sundqvist, C. & Litvin, F. F. (2000) Nonpigment components of the photochlorophyllide photoactive complex: studies of low-temperature blue-green fluorescence spectra. *Membr. Cell Biol.*, **13**, 337–345.

Bevins, M., Yang, C. M. & Markwell, J. (1992) Characterization of a chlorophyll-deficient mutant of sweet clover (*Melilotus alba*). *Plant Physiol. Biochem. (Paris)*, **30**, 327–31.

Block, M. A., Tewari, A. K., Albrieux, C., Marechal, E. & Joyard, J. (2002) The plant S-adenosyl-L-methionine:Mg-protoporphyrin IX methyltransferase is located in both envelope and thylakoid chloroplast membranes. *Eur. J. Biochem.*, **269**, 240–248.

Boddi, B., Oravecz, A. R. & Lehoczki, E. (1995) Effect of cadmium on organization and photoreduction of protochlorophyllide in dark-grown leaves and etioplast inner membrane preparations of wheat. *Photosynthetica*, **31**, 411–20.

Bollivar, D. W. (2003) Regulatory mechanisms of eukaryotic tetrapyrrole biosynthesis, in *The Porphyrin Handbook II*, Vol. 13 (eds K. M. Kadish, K. Smith & R. Guilard), Academic Press, San Diego, USA, pp. 49–70.

Bollivar, D. W. & Beale, S. I. (1995) Formation of the isocyclic ring of chlorophyll by isolated *Chlamydomonas reinhardtii* chloroplasts. *Photosynth. Res.*, **43**, 113–124.

Bollivar, D. W. & Beale, S. I. (1996) The chlorophyll biosynthetic enzyme Mg-protoporphyrin IX mono-methyl ester (oxidative) cyclase. Characterization and partial purification from *Chlamydomonas rein-hardtii* and *Synechocystis* sp. PCC 6803. *Plant Physiol.*, **112**, 105–114.

Bruyant, P. & Kannangara, C. G. (1987) Biosynthesis of δ-aminolevulinate in greening barley leaves. VII: purification and characterization of the glutamate-tRNA ligase. *Carlsberg Res. Commun.*, **52**, 99–109.

Burke, D. H., Alberti, M. & Hearst, J. E. (1993) *bchFNBH* bacteriochlorophyll synthesis genes of *Rhodobacter capsulatus* and identification of the third subunit of light-independent protochlorophyllide reductase in bacteria and plants. *J. Bacteriol.*, **175**, 2414–2422.

Cahoon, A. B. & Timko, M. P. (2000) *Yellow-in-the-dark* mutants of Chlamydomonas lack the CHLL subunit of light-independent protochlorophyllide reductase. *Plant Cell*, **12**, 559–568.

Chahdi, M. A. O., Schoefs, B. & Franck, F. (1998) Isolation and characterization of photoactive complexes of NADPH: protochlorophyllide oxidoreductase from wheat. *Planta*, **206**, 673–680.

Chekounova, E., Voronetskaja, V., Papenbrock, J., Grimm, B. & Beck, C. F. (2001) Characterization of *Chlamydomonas* mutants defective in the H-subunit of Mg-chelatase. *Mol. Gen. Genet.*, **266**, 363–373.

Confalonieri, F. & Duguet, M. (1995) A 200-amino acid ATPase module in search of a basic function. *Bioessays*, **17**, 639–650.

De Rooij, F. W. M., Edixhoven, A. & Wilson, J. H. P. (2003) Porphyria: a diagnostic approach, in *The Porphyrin Handbook II*, Vol. 14 (eds K. M. Kadish, K. Smith & R. Guilard), Academic Press, San Diego, USA, pp. 211–246.

Duke, S. O., Lydon, J., Becerril, J. M., Sherman, T. D., Lehnen, L. P. J. & Matsumoto, H. (1991) Protoporphyrinogen oxidase-inhibiting herbicides. *Weed Science*, **39**, 465–473.

Espineda, C. E., Linford, A. S., Devine, D. & Brusslan, J. A. (1999) The *AtCAO* gene, encoding chlorophyll *a* oxygenase, is required for chlorophyll *b* synthesis in *Arabidopsis thaliana*. *Proc. Natl. Acad. Sci. USA*, **96**, 10507–10511.

Ferreira, G. C. (1999) Ferrochelatase. *Int. J. Biochem. Cell Biol.*, **31**, 995–1000.

Fodje, M. N., Hansson, A., Hansson, M., Olsen, J. G., Gough, S., Willows, R. D. & Al-Karadaghi, S. (2001) Interplay between an AAA module and an integrin I domain may regulate the function of magnesium chelatase. *J. Mol. Biol.*, **311**, 111–122.

Forreiter, C. & Apel, K. (1993) Light-independent and light-dependent protochlorophyllide-reducing activities and two distinct NADPH-protochlorophyllide oxidoreductase polypeptides in mountain pine (*Pinus mugo*). *Planta*, **190**, 536–545.

Forreiter, C., van Cleve, B., Schmidt, A. & Apel, K. (1990) Evidence for a general light-dependent negative control of NADPH-protochlorophyllide oxidoreductase in angiosperms. *Planta*, **183**, 126–132.

Franck, F., Sperling, U., Frick, G., Pochert, B., van Cleve, B., Apel, K. & Armstrong, G.A. (2000) Regulation of etioplast pigment-protein complexes, inner membrane architecture, and protochlorophyllide a chemical heterogeneity by light-dependent NADPH:protochlorophyllide oxidoreductases A and B. *Plant Physiol.*, **124**, 1678–1696.

Frankenberg, N., Erskine, P. T., Cooper, J. B., Shoolingin-Jordan, P. M., Jahn, D. & Heinz, D. W. (1999a) High resolution crystal structure of a Mg^{2+}-dependent porphobilinogen synthase. *J. Mol. Biol.*, **289**, 591–602.

Frankenberg, N., Jahn, D. & Jaffe, E. K. (1999b) *Pseudomonas aeruginosa* contains a novel type V porphobilinogen synthase with no required catalytic metal ions. *Biochem.*, **38**, 13976–13982.

Fujita, Y. (1996) Protochlorophyllide reduction: a key step in the greening of plants. *Plant Cell Physiol.*, **37**, 411–421.

Fujita, Y. & Bauer, C. (2003) The light-independent protochlorophyllide reductase: a nitrogenase-like enzyme catalyzing a key reaction for greening in the dark, in *The Porphyrin Handbook II*, Vol. 12 (eds K. M. Kadish, K. Smith & R. Guilard), Academic Press, San Diego, USA, pp. 109–156.

Fusada, N., Masuda, T., Kuroda, H., Shiraishi, T., Shimada, H., Ohta, H. & Takamiya, K. (2000) NADPH-protochlorophyllide oxidoreductase in cucumber is encoded by a single gene and its expression is transcriptionally enhanced by illumination. *Photosynth. Res.*, **64**, 147–154.

Gibson, L. C., Marrison, J. L., Leech, R. M., Jensen, P. E., Bassham, D. C., Gibson, M. & Hunter, C. N. (1996) A putative Mg chelatase subunit from *Arabidopsis thaliana* cv C24. Sequence and transcript analysis of the gene, import of the protein into chloroplasts, and *in situ* localization of the transcript and protein. *Plant Physiol.*, **111**, 61–71.

Gibson, L. C. D., Jensen, P. E. & Hunter, C. N. (1999) Magnesium chelatase from *Rhodobacter sphaeroides*: initial characterization of the enzyme using purified subunits and evidence for a BCHI-BCHD complex. *Biochem. J.*, **337**, 243–251.

Goericke, R. & Repeta, D. J. (1992) The pigments of *Prochlorococcus marinus*: the presence of divinyl chlorophyll *a* and *b* in a marine procaryote. *Limnol. Oceanogr.*, **37**, 425–433.

Goericke, R. & Repeta, D. J. (1993) Chlorophylls *a* and *b* and divinyl chlorophylls *a* and *b* in the open subtropical North Atlantic Ocean. *Marine Ecol.: Prog. Ser.*, **101**, 307–313.

Grimm, B. (2003) Regulatory mechanisms of eukaryotic tetrapyrrole biosynthesis, in *The Porphyrin Handbook II*, Vol. 12 (eds K. M. Kadish, K. Smith & R. Guilard), Academic Press, San Diego, USA, pp. 1–32.

Guo, R., Luo, M. & Weinstein, J. D. (1998) Magnesium chelatase from developing pea leaves. *Plant Physiol.*, **116**, 605–615.

Hansson, A., Kannangara, C. G., von Wettstein, D. & Hansson, M. (1999) Molecular basis for semidominance of missense mutations in the XANTHA-H (42-kDa) subunit of magnesium chelatase. *Proc. Natl. Acad. Sci. USA*, **96**, 1744–1749.

Hansson, A., Willows, R. D., Roberts, T. H. & Hansson, M. (2002) Three semidominant barley mutants with single amino acid substitutions in the smallest magnesium chelatase subunit form defective *AAA+* hexamers. *Proc. Natl. Acad. Sci. USA*, **99**, 13944–13949.

Hansson, M. & Kannangara, C. G. (1997) ATPases and phosphate exchange activities in magnesium chelatase subunits of *Rhodobacter sphaeroides*. *Proc. Natl. Acad. Sci. USA*, **94**, 13351–13356.

He, Z.H., Li, J., Sundqvist, C. & Timko, M.P. (1994) Leaf developmental age controls expression of genes encoding enzymes of chlorophyll and heme biosynthesis in pea (*Pisum sativum* L.). *Plant Physiol.*, **106**, 537–546.

Hendrich, W. & Bereza, B. (1993) Spectroscopic characterization of protochlorophyllide and its transformation. *Photosynthetica*, **28**, 1–16.

Hennig, M., Grimm, B., Contestabile, R., John, R. A. & Jansonius, J. N. (1997) Crystal structure of glutamate-1-semialdehyde aminomutase: an alpha2-dimeric vitamin B6-dependent enzyme with asymmetry in structure and active site reactivity. *Proc. Natl. Acad. Sci. USA*, **94**, 4866–4871.

Henningsen, K. W., Boynton, J. E. & Wettstein, D. V. (1993) Mutants at *xantha* and *albina* loci in relation to chloroplast biogenesis in barley (*Hordeum vulgare* L.), Biologiske Skrifter 42, The Royal Danish Academy of Sciences and Letters, Copenhagen, Denmark.

Hill, C. M., Pearson, S. A., Smith, A. J. & Rogers, L. J. (1985) Inhibition of chlorophyll synthesis in *Hordeum vulgare* by 3-amino 2,3-dihydrobenzoic acid (gabaculin). *Biosci. Rep.*, **5**, 775–781.

Höefgen, R., Axelsen, K. B., Kannangara, C. G., Schüettke, I., Pohlenz, H. D., Willmitzer, L., Grimm, B. & von Wettstein, D. (1994) A visible marker for antisense mRNA expression in plants: inhibition of chlorophyll synthesis with a glutamate-1-semialdehyde aminotransferase antisense gene. *Proc. Natl. Acad. Sci. USA*, **91**, 1726–1730.

Holtorf, H. & Apel, K. (1996) Transcripts of the two NADPH protochlorophyllide oxidereductase genes *PorA* and *PorB* are differentially degraded in etiolated barley seedlings. *Plant Mol. Biol.*, **31**, 387–392.

Holtorf, H., Reinbothe, S., Reinbothe, C., Bereza, B. & Apel, K. (1995) Two routes of chlorophyllide synthesis that are differentially regulated by light in barley (*Hordeum vulgare* L.). *Proc. Natl. Acad. Sci. USA*, **92**, 3254–3258.

Hoober, J. K., Kahn, A., Ash, D. E., Gough, S. & Kannangara, C. G. (1988) Biosynthesis of delta-aminolevulinate in greening barley leaves. IX. Structure of the substrate, mode of gabaculine inhibition, and the catalytic mechanism of glutamate 1-semialdehyde aminotransferase. *Carlsberg Res. Commun.*, **53**, 11–25.

Hu, G., Yalpani, N., Briggs, S. P. & Johal, G. S. (1998) A porphyrin pathway impairment is responsible for the phenotype of a dominant disease lesion mimic mutant of maize. *Plant Cell*, **10**, 1095–1105.

Hudson, A., Carpenter, R., Doyle, S. & Coen, E. S. (1993) *Olive*: a key gene required for chlorophyll biosynthesis in *Antirrhinum majus*. *EMBO J.*, **12**, 3711–3719.

CHLOROPHYLLS 49

Iwai, K., Ichihara, Y., Kimura, S., Rai, H., Akatsuka, Y. & Suzuki, K. (1989a) Therapeutic effect of chlorin E-6na as a new photosensitizing agent in photodynamic therapy of mouse tumor. *J. Clin. Biochem. Nutr.*, **6**, 117–126.

Iwai, K., Ido, T., Iwata, R., Kawamura, M. & Kimura, S. (1989b) Localizing efficiency of Vanadium-48-labeled vanadylpheophorbide in tumor as a new tumor imaging agent. *Nuc. Med. Biol.*, **16**, 783–790.

Iwai, K., Kimura, S., Ido, T. & Iwata, R. (1990) Tumor uptake of Vanadium-48 vanadyl chlorine E-6 sodium as a tumor-imaging agent in tumor-bearing mice. *Nuc. Med. Biol.*, **17**, 775–780.

Jacobs, J. M. & Jacobs, N. J. (1987) Oxidation of protoporphyrinogen to protoporphyrin, a step in chlorophyll and haem biosynthesis. Purification and partial characterization of the enzyme from barley organelles. *Biochem. J.*, **244**, 219–224.

Jacobs, J. M. & Jacobs, N. J. (1993) Porphyrin accumulation and export by isolated barley (*Hordeum vulgare*) plastids: effect of diphenyl ether herbicides. *Plant Physiol.*, **101**, 1181–1188.

Jacobs, J. M., Jacobs, N. J. & De Maggio, A. E. (1982) Protoporphyrinogen oxidation in chloroplasts and plant mitochondria, a step in heme and chlorophyll synthesis. *Arch. Biochem. Biophys.*, **218**, 233–239.

Jensen, P. E., Gibson, L. C. D., Henningsen, K. W. & Hunter, C. N. (1996a) Expression of the *chlI*, *chlD*, and *chlH* genes from the *Cyanobacterium synechocystis* PCC6803 in *Escherichia coli* and demonstration that the three cognate proteins are required for magnesium-protoporphyrin chelatase activity. *J. Biol. Chem.*, **271**, 16662–16667.

Jensen, P. E., Gibson, L. C. D. & Hunter, C. N. (1998) Determinants of catalytic activity with the use of purified I, D & H subunits of the magnesium protoporphyrin IX chelatase from *Synechocystis* PCC6803. *Biochem. J.*, **334**, 335–344.

Jensen, P. E., Willows, R. D., Petersen, B. L., Vothknecht, U. C., Stummann, B. M., Kannangara, C.G., von Wettstein, D. & Henningsen, K.W. (1996b) Structural genes for Mg-chelatase subunits in barley: Xantha-f, -g and -h. *Mol. Gen. Genet.*, **250**, 383–394.

Jones, R. M. & Jordan, P. M. (1994) Purification and properties of porphobilinogen deaminase from *Arabidopsis thaliana*. *Biochem. J.*, **299**, 895–902.

Kannangara, C. G., Andersen, R. V., Pontoppidan, B., Willows, R. & von Wettstein, D. (1994) Enzymic and mechanistic studies on the conversion of glutamate to 5- aminolaevulinate, in *CIBA Foundation Symposium*, Vol. 180, pp. 3–20; discussion 21–25.

Kannangara, C. G., Vothknecht, U. C., Hansson, M. & von Wettstein, D. (1997) Magnesium chelatase: association with ribosomes and mutant complementation studies identify barley subunit xantha-G as a functional counterpart of *Rhodobacter* subunit BCHD. *Mol. Gen. Genet.*, **254**, 85–92.

Kim, J. S., Kolossov, V. & Rebeiz, C. A. (1997) Chloroplast biogenesis 76. Regulation of 4-vinyl reduction during conversion of divinyl Mg-protoporphyrin IX to monovinyl protochlorophyllide *a* is controlled by plastid membrane and stromal factors. *Photosynthetica*, **34**, 569–581.

Kim, J.-S. & Rebeiz, C. A. (1995) An improved analysis for determination of monovinyl and divinyl protoporphyrin IX. *J. Photosci.*, **2**, 103–106.

Klement, H., Oster, U. & Rudiger, W. (2000) The influence of glycerol and chloroplast lipids on the spectral shifts of pigments associated with NADPH: protochlorophyllide oxidoreductase from *Avena sativa* L. *FEBS Lett.*, **480**, 306–310.

Koncz, C., Mayerhofer, R., Koncz-Kalman, Z., Nawrath, C., Redei, G. P. & Schell, J. (1990) Isolation of a gene encoding a novel chloroplast protein by T-DNA tagging in *Arabidopsis thaliana*. *EMBO J.*, **9**, 1337–1346.

Kovacheva, S., Ryberg, M. & Sundqvist, C. (2000) ADP/ATP and protein phosphorylation dependence of phototransformable protochlorophyllide in isolated etioplast membranes. *Photosynth. Res.*, **64**, 127–136.

Kräutler, B. (2003) Chlorophyll breakdown and chlorophyll catabolites, in *The Porphyrin Handbook II*, Vol. 13 (eds K. M. Kadish, K. Smith & R. Guilard), Academic Press, San Diego, USA, pp. 183–209.

Kruse, E., Mock, H. P. & Grimm, B. (1995a) Reduction of coproporphyrinogen oxidase level by antisense RNA synthesis leads to deregulated gene expression of plastid proteins and affects the oxidative defense system. *EMBO J.*, **14**, 3712–3720.

Kruse, E., Mock, H.-P. & Grimm, B. (1995b) Coproporphyrinogen III oxidase from barley and tobacco-sequence analysis and initial expression studies. *Planta*, **196**, 796–803.

Kumar, A. M., Schaub, U., Soll, D. & Ujwal, M. L. (1996) Glutamyl-transfer RNA – at the crossroad between chlorophyll and protein biosynthesis. *Trends Plant Sci.*, **1**, 371–376.

Kumar, A. M. & Soll, D. (2000) Antisense HEMA1 RNA expression inhibits heme and chlorophyll biosynthesis in arabidopsis. *Plant Physiol.*, **122**, 49–56.

Kuroda, H., Masuda, T., Fusada, N., Ohta, H. & Takamiya, K. (2000) Expression of NADPH-Protochlorophyllide oxidoreductase gene in fully green leaves of cucumber. *Plant Cell Physiol.*, **41**, 226–229.

Kuroda, H., Masuda, T., Fusada, N., Ohta, H. & Takamiya, K. (2001) Cytokinin-induced transcriptional activation of NADPH-protochlorophyllide oxidoreductase gene in cucumber. *J. Plant Res.*, **114**, 1–7.

Kuroda, H., Masuda, T., Ohta, H., Shioi, Y. & Takamiya, K. (1995) Light-enhanced gene expression of NADPH-protochlorophyllide oxidoreductase in cucumber. *Biochem. Biophys. Res. Commun.*, **210**, 310–316.

Kusnetsov, V., Herrmann, R. G., Kulaeva, O. N. & Oelmuller, R. (1998) Cytokinin stimulates and abscisic acid inhibits greening of etiolated *Lupinus luteus* cotyledons by affecting the expression of the light-sensitive protochlorophyllide oxidoreductase. *Mol. Gen. Genet.*, **259**, 21–28.

Larkin, R. M., Alonso, J. M., Ecker, J. R. & Chory, J. (2003) Gun4, A regulator of chlorophyll synthesis and intracellular signalling. *Science*, **299**, 902–906.

Lebedev, N. & Timko, M. P. (1998) Protochlorophyllide photoreduction. *Photosynth. Res.*, **58**, 5–23.

Lee, H. J., Lee, S. B., Chung, J. S., Han, S. U., Han, O., Guh, J. O., Jeon, J. S., An, G. & Back, K. (2000) Transgenic rice plants expressing a *Bacillus subtilis* protoporphyrinogen oxidase gene are resistant to diphenyl ether herbicide oxyfluorfen. *Plant Cell Physiol.*, **41**, 743–749.

Lenti, K., Fodor, F. & Boddi, B. (2002) Mercury inhibits the activity of the NADPH:protochlorophyllide oxidoreductase (POR). *Photosynthetica*, **40**, 145–51.

Lermontova, I. & Grimm, B. (2000) Overexpression of plastidic protoporphyrinogen IX oxidase leads to resistance to the diphenyl-ether herbicide acifluorfen. *Plant Physiol.*, **122**, 75–84.

Lermontova, I., Kruse, E., Mock, H. P. & Grimm, B. (1997) Cloning and characterization of a plastidal and a mitochondrial isoform of tobacco protoporphyrinogen IX oxidase. *Proc. Natl. Acad. Sci. USA*, **94**, 8895–8900.

Li, J., Goldschmidt-Clermont, M. & Timko, M. P. (1993) Chloroplast-encoded *chlB* is required for light-independent protochlorophyllide reductase activity in *Chlamydomonas reinhardtii*. *Plant Cell*, **5**, 1817–1829.

Li, J. & Timko, M. P. (1996) The *pc-1* phenotype of *Chlamydomonas reinhardtii* results from a deletion mutation in the nuclear gene for NADPH:protochlorophyllide oxidoreductase. *Plant Mol. Biol.*, **30**, 15–37.

Lidholm, J. & Gustafsson, P. (1991) Homologues of the green algal *gidA* gene and the liverwort *frxC* gene are present on the chloroplast genomes of conifers. *Plant Mol. Biol.*, **17**, 787–798.

Liedgens, W., Lutz, C. & Schneider, H. A. (1983) Molecular properties of 5-aminolevulinic acid dehydratase from *Spinacia oleracea*. *Eur. J. Biochem.*, **135**, 75–79.

Litvin, F. F., Belyaeva, O. B. & Ignatov, N. V. (2000) Chlorophyll biosynthesis and formation of reaction centers of photochemical systems of photosynthesis. *Usp. Biol. Khim.*, **40**, 3–42.

Liu, X. Q., Xu, H. & Huang, C. (1993) Chloroplast *chlB* gene is required for light-independent chlorophyll accumulation in *Chlamydomonas reinhardtii*. *Plant Mol. Biol.*, **23**, 297–308.

Loida, P. J., Thompson, R. L., Walker, D. M. & Cajacob, C. A. (1999) Novel inhibitors of glutamyl-tRNA (Glu) reductase identified through cell-based screening of the heme/chlorophyll biosynthetic pathway. *Arch. Biochem. Biophys.*, **372**, 230–237.

Luo, D., Coen, E. S., Doyle, S. & Carpenter, R. (1991) Pigmentation mutants produced by transposon mutagenesis in *Antirrhinum majus*. *Plant J.*, **1**, 59–69.

Luo, J. L. & Lim, C. K. (1991) Random decarboxylation of uroporphyrinogen III by human hepatic uroporphyrinogen decarboxylase. *J. Chromatogr.*, **566**, 409–13.

Luo, J. & Lim, C. K. (1993) Order of uroporphyrinogen III decarboxylation on incubation of porphobilinogen and uroporphyrinogen III with erythrocyte uroporphyrinogen decarboxylase. *Biochem. J.*, **289**, 529–532.

Marrison, J. L., Schunmann, P. H. D., Ougham, H. J. & Leech, R. M. (1996) Subcellular visualization of gene transcripts encoding key proteins of the chlorophyll accumulation process in developing chloroplasts. *Plant Physiol.*, **110**, 1089–1096.

Martins, B. M., Grimm, B., Mock, H. P., Huber, R. & Messerschmidt, A. (2001a) Crystal structure and substrate binding modeling of the uroporphyrinogen III decarboxylase from *Nicotiana tabacum*. Implications for the catalytic mechanism. *J. Biol. Chem.*, **276**, 44108–44116.

Martins, B. M., Grimm, B., Mock, H. P., Richter, G., Huber, R. & Messerschmidt, A. (2001b) Tobacco uroporphyrinogen III decarboxylase: characterization, crystallization and preliminary X-ray analysis. *Acta. Crystallogr. D Biol. Crystallogr.*, **57**, 1709–1711.

Masuda, T., Fusada, N., Shiraishi, T., Kuroda, H., Awai, K., Shimada, H., Ohta, H. & Takamiya, K. (2002) Identification of two differentially regulated isoforms of protochlorophyllide oxidoreductase (POR) from tobacco revealed a wide variety of light- and development-dependent regulations of POR gene expression among angiosperms. *Photosynth. Res.*, **74**, 165–172.

Matile, P. (2000) Biochemistry of an Indian summer: physiology of autumnal leaf coloration. *Exp. Gerontol.*, **35**, 145–158.

Matile, P., Hortensteiner, S. & Thomas, H. (1999) Chlorophyll degradation. *Ann. Rev. Plant Physiol. Plant Mol. Biol.*, **50**, 67–95.

Matringe, M., Camadro, J. M., Labbe, P. & Scalla, R. (1989) Protoporphyrinogen oxidase as a molecular target for diphenyl ether herbicides. *Biochem. J.*, **260**, 231–235.

Meskauskiene, R. & Apel, K. (2002) Interaction of FLU, a negative regulator of tetrapyrrole biosynthesis, with the glutamyl-tRNA reductase requires the tetratricopeptide repeat domain of FLU. *FEBS Lett.*, **532**, 27–30.

Meskauskiene, R., Nater, M., Goslings, D., Kessler, F., op den Camp, R. & Apel, K. (2001) FLU: a negative regulator of chlorophyll biosynthesis in *Arabidopsis thaliana*. *Proc. Natl. Acad. Sci. USA*, **98**, 12826–12831.

Mochizuki, N., Brusslan, J. A., Larkin, R., Nagatani, A. & Chory, J. (2001) *Arabidopsis genomes uncoupled 5* (*GUN5*) mutant reveals the involvement of Mg-chelatase H subunit in plastid-to-nucleus signal transduction. *Proc. Natl. Acad. Sci. USA*, **98**, 2053–2058.

Mock, H. P. & Grimm, B. (1997) Reduction of uroporphyrinogen decarboxylase by antisense RNA expression affects activities of other enzymes involved in tetrapyrrole biosynthesis and leads to light-dependent necrosis. *Plant Physiol.*, **113**, 1101–1112.

Mock, H. P., Heller, W., Molina, A., Neubohn, B., Sandermann, H., Jr. & Grimm, B. (1999) Expression of uroporphyrinogen decarboxylase or coproporphyrinogen oxidase antisense RNA in tobacco induces pathogen defense responses conferring increased resistance to tobacco mosaic virus. *J. Biol. Chem.*, **274**, 4231–4238.

Moller, M. G., Petersen, B. L., Kannangara, C. G., Stummann, B. M. & Henningsen, K. W. (1997) Chlorophyll biosynthetic enzymes and plastid membrane structures in mutants of barley (*Hordeum vulgare* L). *Hereditas*, **127**, 181–191.

Moller, S. G., Kunkel, T. & Chua, N. H. (2001) A plastidic ABC protein involved in intercompartmental communication of light signaling. *Genes Dev.*, **15**, 90–103.

Moseley, J., Quinn, J., Eriksson, M. & Merchant, S. (2000) The *Crd1* gene encodes a putative di-iron enzyme required for photosystem I accumulation in copper deficiency and hypoxia in *Chlamydomonas reinhardtii*. *EMBO J.*, **19**, 2139–2151.

Moseley, J. L., Page, M. D., Alder, N. P., Eriksson, M., Quinn, J., Soto, F., Theg, S. M., Hippler, M. & Merchant, S. (2002) Reciprocal expression of two candidate di-iron enzymes affecting photosystem I and light-harvesting complex accumulation. *Plant Cell*, **14**, 673–688.

Moser, J., Lorenz, S., Hubschwerlen, C., Rompf, A. & Jahn, D. (1999) *Methanopyrus kandleri* glutamyl-tRNA reductase. *J. Biol. Chem.*, **274**, 30679–30685.

Moser, J., Schubert, W. D., Beier, V., Bringemeier, I., Jahn, D. & Heinz, D. W. (2001) V-shaped structure of glutamyl-tRNA reductase, the first enzyme of tRNA-dependent tetrapyrrole biosynthesis. *EMBO J.*, **20**, 6583–6590.

Mostowska, A., Siedlecka, M. & Parys, E. (1996) Effect of 2.2′-bipyridyl, a photodynamic herbicide, on chloroplast ultrastructure, pigment content and photosynthesis rate in pea seedlings. *Acta Physiol. Plant.*, **18**, 153–164.

Nair, S. P., Kannangara, C. G., Harwood, J. L. & John, R. A. (1990) Inhibition studies on 5-aminolevulinate biosynthesis in *Pisum sativum* L. (pea). *Biochem. Soc. Trans.*, **18**, 656–657.

Oosawa, N., Masuda, T., Awai, K., Fusada, N., Shimada, H., Ohta, H. & Takamiya, K. (2000) Identification and light-induced expression of a novel gene of NADPH-protochlorophyllide oxidoreductase isoform in *Arabidopsis thaliana*. *FEBS Lett.*, **474**, 133–136.

Oster, U., Bauer, C. E. & Rudiger, W. (1997) Characterization of chlorophyll *a* and bacteriochlorophyll *a* synthases by heterologous expression in *Escherichia coli*. *J. Biol. Chem.*, **272**, 9671–9676.

Oster, U., Brunner, H. & Rudiger, W. (1996) The greening process in cress seedlings. 5. Possible interference of chlorophyll precursors, accumulated after thujaplicin treatment, with light-regulated expression of *Lhc* genes *J. Photochem. Photobiol. B - Biol.*, **36**, 255–261.

Oster, U. & Rudiger, W. (1997) The G4 gene of *Arabidopsis thaliana* encodes a chlorophyll synthase of etiolated plants. *Botanica Acta*, **110**, 420–423.

Oster, U., Tanaka, R., Tanaka, A. & Rudiger, W. (2000) Cloning and functional expression of the gene encoding the key enzyme for chlorophyll *b* biosynthesis (CAO) from *Arabidopsis thaliana*. *Plant J.*, **21**, 305–310.

Papenbrock, J. & Grimm, B. (2001) Regulatory network of tetrapyrrole biosynthesis – studies of intracellular signalling involved in metabolic and developmental control of plastids. *Planta*, **213**, 667–681.

Papenbrock, J., Mock, H. P., Tanaka, R., Kruse, E. & Grimm, B. (2000a) Role of magnesium chelatase activity in the early steps of the tetrapyrrole biosynthetic pathway. *Plant Physiol.*, **122**, 1161–1169.

Papenbrock, J., Pfündel, E., Mock, H.-P. & Grimm, B. (2000b) Decreased and increased expression of the subunit CHL I diminishes Mg chelatase activity and reduces chlorophyll synthesis in transgenic tobacco plants. *Plant J.*, **22**, 155–164.

Parham, R. & Rebeiz, C. A. (1992) Chloroplast biogenesis: [4-vinyl] chlorophyllide a reductase is a divinyl chlorophyllide *a*-specific, NADPH-dependent enzyme. *Biochem.*, **31**, 8460–8464.

Parham, R. & Rebeiz, C. A. (1995) Chloroplast biogenesis 72: a [4-vinyl] chlorophyllide a reductase assay using divinyl chlorophyllide *a* as an exogenous substrate. *Anal. Biochem.*, **231**, 164–169.

Petersen, B. L., Moller, M. G., Jensen, P. E. & Henningsen, K. W. (1999) Identification of the Xan-g gene and expression of the Mg-chelatase encoding genes Xan-f, -g and -h in mutant and wild type barley (*Hordeum vulgare* L.). *Hereditas*, **131**, 165–170.

Pinta, V., Picaud, M., Reiss-Husson, F. & Astier, C. (2002) *Rubrivivax gelatinosus acsF* (previously *orf358*) codes for a conserved, putative binuclear-iron-cluster-containing protein involved in aerobic oxidative cyclization of Mg-protoporphyrin IX monomethylester. *J. Bacteriol.*, **184**, 746–753.

Pontoppidan, B. & Kannangara, C. G. (1994) Purification and partial characterisation of barley glutamyl-tRNA(Glu) reductase, the enzyme that directs glutamate to chlorophyll biosynthesis. *Eur. J. Biochem.*, **225**, 529–537.

Pöpperl, G., Oster, U., Blos, I. & Rudiger, W. (1997) Magnesium chelatase of *Hordeum vulgare* L. is not activated by light but inhibited by pheophorbide. *Z. Naturforsch. Teil C*, **52**, 144–152.

Pöpperl, G., Oster, U. & Ruediger, W. (1998) Light-dependent increase in chlorophyll precursors during the day-night cycle in tobacco and barley seedlings. *J. Plant Physiol.*, **153**, 40–45.

Porra, R. J. (1997) Recent progress in porphyrin and chlorophyll biosynthesis. *Photochem. Photobiol.*, **65**, 492–516.

Porra, R. J. (2002) The checkered history of the development and use of simultaneous equations for the accurate determination of chlorophylls *a* and *b*. *Photosynth. Res.*, **73**, 149–156.

Porra, R. J., Schafer, W., Katheder, I. & Scheer, H. (1995) The derivation of the oxygen atoms of the 13(1)-oxo and 3-acetyl groups of bacteriochlorophyll *a* from water in *Rhodobacter sphaeroides* cells adapting from respiratory to photosynthetic conditions: evidence for an anaerobic pathway for the formation of isocyclic ring E. *FEBS Lett.*, **371**, 21–24.

Porra, R. J., Urzinger, M., Winkler, J., Bubenzer, C. & Scheer, H. (1998) Biosynthesis of the 3-acetyl and 13′-oxo groups of bacteriochlorophyll *a* in the facultative aerobic bacterium, *Rhodovulum sulphidophilum*: the presence of both oxygenase and hydratase pathways for isocyclic ring formation. *Eur. J. Biochem.*, **257**, 185–191.

Randolph-Anderson, B. L., Sato, R., Johnson, A. M., Harris, E. H., Hauser, C. R., Oeda, K., Ishige, F., Nishio, S., Gillham, N. W. & Boynton, J. E. (1998) Isolation and characterization of a mutant protoporphyrinogen oxidase gene from *Chlamydomonas reinhardtii* conferring resistance to porphyric herbicides. *Plant Mol. Biol.*, **38**, 839–859.

Rebeiz, C. A., Parham, R., Fasoula, D. A. & Ioannides, I. M. (1994) Chlorophyll *a* biosynthetic heterogeneity. *CIBA Foundation Symposium*, **180**, 177–189; discussion 190–193.

Reinbothe, C., Apel, K. & Reinbothe, S. (1995) A light-induced protease from barley plastids degrades NADPH: protochlorophyllide oxidoreductase complexed with chlorophyllide. *Mol. Cell. Biol.*, **15**, 6206–6212.

Reinbothe, S. & Reinbothe, C. (1996a) Regulation of chlorophyll biosynthesis in angiosperms. *Plant Physiol.*, **111**, 1–7.

Reinbothe, S. & Reinbothe, C. (1996b) The regulation of enzymes involved in chlorophyll biosynthesis. *Eur. J. Biochem.*, **237**, 323–343.

Reinbothe, S., Reinbothe, C., Apel, K. & Lebedev, N. (1996) Evolution of chlorophyll biosynthesis – the challenge to survive photooxidation. *Cell*, **86**, 703–705.

Reindl, A., Reski, R., Lerchl, J., Grimm, B. & Al-Awadi, A. (2001) In *PCT Int. Appl.*, BASF Aktiengesellschaft, Germany. WO #2001009355, pp. 70.

Rissler, H. M., Collakova, E., Dellapenna, D., Whelan, J. & Pogson, B. J. (2002) Chlorophyll biosynthesis. Expression of a second *chlI* gene of magnesium chelatase in Arabidopsis supports only limited chlorophyll synthesis. *Plant Physiol.*, **128**, 770–779.

Roitgrund, C. & Mets, L. J. (1990) Localization of two novel chloroplast genome functions: trans-splicing of RNA and protochlorophyllide reduction. *Curr. Genet.*, **17**, 147–153.

Rudiger, W. (1997) Chlorophyll metabolism – from outer space down to the molecular level. *Phytochemistry*, **46**, 1151–1167.

Rudiger, W. (2002) Biosynthesis of chlorophyll *b* and the chlorophyll cycle. *Photosynth. Res.*, **74**, 184–193.

Rudiger, W. (2003) The last steps of chlorophyll biosynthesis, in *The Porphyrin Handbook II*, Vol. 12 (eds K. M. Kadish, K. Smith & R. Guilard), Academic Press, San Diego, USA, pp. 71–108.

Satoh, S. & Tanaka, A. (2002) Chlorophyll *b* inhibits the formation of photosystem I trimer in *Synechocystis* sp. PCC6803. *FEBS Lett.*, **528**, 235–240.

Schmid, H. C., Oster, U., Kogel, J., Lenz, S. & Rudiger, W. (2001) Cloning and characterisation of chlorophyll synthase from *Avena sativa*. *Biol. Chem.*, **382**, 903–911.

Schoefs, B. (1999) The light-dependent and light-independent reduction of protochlorophyllide *a* to chlorophyllide *a*. *Photosynthetica*, **36**, 481–496.

Schoefs, B. (2001) The protochlorophyllide-chlorophyllide cycle. *Photosynth. Res.*, **70**, 257–271.

Schoefs, B. (2002) Chlorophyll biosynthesis during plant greening, in *Handbook of Plant and Crop Physiology*, 2nd edn (ed. M. Pessarakli), Marcel Dekker, New York, USA, pp. 265–273.

Schoefs, B. & Bertrand, M. (1997) Chlorophyll biosynthesis, in *Handbook of Photosynthesis* (ed. M. Pessarakli), Marcel Dekker, New York, USA, pp. 49–69.

Schubert, H. L., Raux, E., Matthews, M. A., Phillips, J. D., Wilson, K. S., Hill, C. P. & Warren, M. J. (2002) Structural diversity in metal ion chelation and the structure of uroporphyrinogen III synthase. *Biochem. Soc. Trans.*, **30**, 595–600.

Schulz, R., Steinmuller, K., Klaas, M., Forreiter, C., Rasmussen, S., Hiller, C. & Apel, K. (1989) Nucleotide sequence of a cDNA coding for the NADPH-protochlorophyllide oxidoreductase (PCR) of barley (*Hordeum vulgare* L.) and its expression in *Escherichia coli*. *Mol. Gen. Genet.*, **217**, 355–361.

Schunmann, P. H. & Ougham, H. J. (1996) Identification of three cDNA clones expressed in the leaf extension zone and with altered patterns of expression in the slender mutant of barley: a tonoplast intrinsic protein, a putative structural protein and protochlorophyllide oxidoreductase. *Plant Mol. Biol.*, **31**, 529–537.

Senior, N. M., Brocklehurst, K., Cooper, J. B., Wood, S. P., Erskine, P., Shoolingin-Jordan, P. M., Thomas, P. G. & Warren, M. J. (1996) Comparative studies on the 5-aminolaevulinic acid dehydratases from *Pisum sativum*, *Escherichia coli* and *Saccharomyces cerevisiae*. *Biochem. J.*, **320**, 401–412.

Shieh, J., Miller, G. W. & Psenak, M. (1978) Properties of S-adenosyl-L-methionine-magnesium-protopor-phyrin IX methyltransferase from barley. *Plant Cell Physiol.*, **19**, 1051–1059.

Shieh, J. J., Psenak, M. & Miller, G. W. (1973) Effect of levulinic acid on the biosynthesis of porphyrins in *Nicotiana tabacum*. *Plant Sci. Lett.*, **1**, 207–211.

Shoolingin-Jordan, P. M. (2003) The biosynthesis of coproporphyrinogen, in *The Porphyrin Handbook II*, Vol. 12 (eds K. M. Kadish, K. Smith & R. Guilard), Academic Press, San Diego, USA, pp. 33–74.

Skinner, J. S. & Timko, M. P. (1998) Loblolly pine (*Pinus taeda* L.) contains multiple expressed genes encoding light-dependent NADPH: protochlorophyllide oxidoreductase (POR). *Plant Cell Physiol.*, **39**, 795–806.

Spano, A. J., He, Z., Michel, H., Hunt, D. F. & Timko, M. P. (1992a) Molecular cloning, nuclear gene structure, and developmental expression of NADPH: protochlorophyllide oxidoreductase in pea (*Pisum sativum* L.). *Plant Mol. Biol.*, **18**, 967–972.

Spano, A. J., He, Z. & Timko, M. P. (1992b) NADPH: protochlorophyllide oxidoreductases in white pine (*Pinus strobus*) and loblolly pine (*P. taeda*). Evidence for light and developmental regulation of expression and conservation in gene organization and protein structure between angiosperms and gymnosperms. *Mol. Gen. Genet.*, **236**, 86–95.

Spano, A. J. & Timko, M. P. (1991) Isolation, characterization and partial amino acid sequence of a chloroplast-localized porphobilinogen deaminase from pea (*Pisum sativum* L.). *Biochim. Biophys. Acta*, **1076**, 29–36.

Sperling, U., Franck, F., van Cleve, B., Frick, G., Apel, K. & Armstrong, G. A. (1998) Etioplast differentiation in arabidopsis: both PORA and PORB restore the prolamellar body and photoactive protochlorophyl-lide-F655 to the *cop1* photomorphogenic mutant. *Plant Cell*, **10**, 283–296.

Sperling, U., van Cleve, B., Frick, G., Apel, K. & Armstrong, G. A. (1997) Overexpression of light-dependent PORA or PORB in plants depleted of endogenous POR by far-red light enhances seedling survival in white light and protects against photooxidative damage. *Plant J.*, **12**, 649–658.

Stange-Thomann, N., Thomann, H. U., Lloyd, A. J., Lyman, H. & Soll, D. (1994) A point mutation in *Euglena gracilis* chloroplast tRNA(Glu) uncouples protein and chlorophyll biosynthesis. *Proc. Natl. Acad. Sci. USA*, **91**, 7947–7951.

Stevens, E. & Frydman, B. (1968) Isolation and properties of wheat germ uroporphyrinogen 3 cosynthetase. *Biochim. Biophys. Acta*, **151**, 429–437.

Strand, A., Asami, T., Alonso, J., Ecker, J. R. & Chory, J. (2003) Chloroplast to nucleus communication triggered by accumulation of Mg-protoporphyrin IX. *Nature*, **421**, 79–83.

Su, Q., Frick, G., Armstrong, G. & Apel, K. (2001) POR C of *Arabidopsis thaliana*: a third light- and NADPH-dependent protochlorophyllide oxidoreductase that is differentially regulated by light. *Plant Mol. Biol.*, **47**, 805–813.

Sun, X. & Leung, W. N. (2002) Photodynamic therapy with pyropheophorbide-a methyl ester in human lung carcinoma cancer cell: efficacy, localization and apoptosis. *Photochem. Photobiol.*, **75**, 644–651.

Susek, R. E., Ausubel, F. M. & Chory, J. (1993) Signal transduction mutants of Arabidopsis uncouple nuclear CAB and RBCS gene expression from chloroplast development. *Cell*, **74**, 787–799.

Suzuki, J. Y. & Bauer, C. E. (1992) Light-independent chlorophyll biosynthesis: involvement of the chloroplast gene *chlL* (*frxC*). *Plant Cell*, **4**, 929–940.

Suzuki, J. Y. & Bauer, C. E. (1995) A prokaryotic origin for light-dependent chlorophyll biosynthesis of plants. *Proc. Natl. Acad. Sci. USA*, **92**, 3749–3753.

Suzuki, J. Y., Bollivar, D. W. & Bauer, C. E. (1997) Genetic analysis of chlorophyll biosynthesis. *Annu. Rev. Genet.*, **31**, 61–89.

Suzuki, T., Takio, S., Yamamoto, I. & Satoh, T. (2001) Characterization of cDNA of the liverwort phyto-chrome gene, and phytochrome involvement in the light-dependent and light-independent protochlor-ophyllide oxidoreductase gene expression in *Marchantia paleacea* var. *diptera*. *Plant Cell Physiol.*, **42**, 576–582.

Tanaka, R., Oster, U., Kruse, E., Rudiger, W. & Grimm, B. (1999) Reduced activity of geranylgeranyl reductase leads to loss of chlorophyll and tocopherol and to partially geranylgeranylated chlorophyll in transgenic tobacco plants expressing antisense RNA for geranylgeranyl reductase. *Plant Physiol.*, **120**, 695–704.

Tanaka, T., Muto, N., Itoh, N., Dota, A., Nishina, Y., Inada, A. & Tanaka, K. (1995) Induction of differentiation of embryonal carcinoma F9 cells by iron chelators. *Res. Comm. Mol. Path. Pharm.*, **90**, 211–220.

Teakle, G. R. & Griffiths, W. T. (1993) Cloning, characterization and import studies on protochlorophyllide reductase from wheat (*Triticum aestivum*). *Biochem. J.*, **296**, 225–230.

Thomas, R. M. & Singh, V. P. (1995) Effects of three triazole derivatives on mercury induced inhibition of chlorophyll and carotenoid accumulation in cucumber cotyledons. *Indian J. Plant Physiol.*, **38**, 313–316.

Thomas, R. M. & Singh, V. P. (1996) Reduction of cadmium-induced inhibition of chlorophyll and carotenoid accumulation in *Cucumis sativus* L. by uniconazole (S. 3307). *Photosynthetica*, **32**, 145–148.

Tomitani, A., Okada, K., Miyashita, H., Matthijs, H. C., Ohno, T. & Tanaka, A. (1999) Chlorophyll *b* and phycobilins in the common ancestor of cyanobacteria and chloroplasts. *Nature*, **400**, 159–162.

Ujwal, M. L., McCormac, A. C., Goulding, A., Kumar, A. M., Söll, D. & Terry, M. J. (2002) Divergent regulation of the *HEMA* gene family encoding glutamyl-tRNA reductase in *Arabidopsis thaliana*: expression of *HEMA2* is regulated by sugars, but is independent of light and plastid signalling. *Plant Mol. Biol.*, **50**, 83–91.

Vale, R. D. (2000) AAA Proteins: Lords of the Ring. *J. Cell Biol.*, **150**, F13–F19.

Vijayan, P., Whyte, B. J. & Castelfranco, P. A. (1992) A spectrophotometric analysis of the magnesium protoporphyrin IX monomethyl ester (oxidative) cyclase. *Plant Physiol. Biochem. (Paris)*, **30**, 271–278.

von Wettstein, D. (2000) Chlorophyll biosynthesis I: from analysis of mutants to genetic engineering of the pathway. *Discoveries Plant Biol.*, **3**, 75–93.

Vothknecht, U. C., Kannangara, C. G. & von Wettstein, D. (1996) Expression of catalytically active barley glutamyl tRNA^{Glu} reductase in *Escherichia coli* as a fusion protein with glutathione S-transferase. *Proc. Natl. Acad. Sci. USA*, **93**, 9287–9291.

Vothknecht, U. C., Kannangara, C. G. & von Wettstein, D. (1998) Barley glutamyl tRNA$_{Glu}$ reductase: mutations affecting haem inhibition and enzyme activity. *Phytochemistry*, **47**, 513–519.

Vothknecht, U. C., Willows, R. D. & Kannangara, C. G. (1995) Sinefungin inhibits chlorophyll synthesis by blocking the S-adenosyl-methionine: Mg-protoporphyrin IX O-methyltransferase in greening barley leaves. *Plant Physiol. Biochem.*, **33**, 759–763.

Walker, C. J., Castelfranco, P. A. & Whyte, B. J. (1991) Synthesis of divinyl protochlorophyllide. Enzymological properties of the magnesium-protoporphyrin IX monomethyl ester oxidative cyclase system. *Biochem. J.*, **276**, 691–697.

Walker, C. J., Kannangara, C. G. & von Wettstein, D. (1997) Identification of *xantha l-35* and *viridis k-23* as mutants of the Mg-protoporphyrin monomethyl ester cyclase of chlorophyll synthesis in barley (*Hordeum vulgare*). *Plant Physiol.*, **114S**, 708.

Walker, C. J., Mansfield, K. E., Smith, K. M. & Castelfranco, P. A. (1989) Incorporation of atmospheric oxygen into the carbonyl functionality of the protochlorophyllide isocyclic ring. *Biochem. J.*, **257**, 599–602.

Walker, C. J. & Weinstein, J. D. (1991a) Further characterization of the magnesium chelatase in isolated developing cucumber chloroplasts: substrate specificity, regulation, intactness and ATP requirements. *Plant Physiol.*, **95**, 1189–1196.

Walker, C. J. & Weinstein, J. D. (1991b) *In vitro* assay of the chlorophyll biosynthetic enzyme magnesium chelatase: resolution of the activity into soluble and membrane bound fractions. *Proc. Natl. Acad. Sci. USA*, **88**, 5789–5793.

Warabi, E., Usui, K., Tanaka, Y. & Matsumoto, H. (2001) Resistance of a soybean cell line to oxyfluorfen by overproduction of mitochondrial protoporphyrinogen oxidase. *Pest Manag. Sci.*, **57**, 743–748.

Whyte, B. J. & Castelfranco, P. A. (1993) Further observations on the magnesium-protoporphyrin IX monomethyl ester (oxidative) cyclase system. *Biochem. J.*, **290**, 355–359.

Whyte, B. J., Fijayan, P. & Castelfranco, P. A. (1992) *In vitro* synthesis of protochlorophyllide: effects of magnesium and other cations on the reconstituted (oxidative) cyclase. *Plant Physiol. Biochem. (Paris)*, **30**, 279–284.

Whyte, B. J. & Griffiths, W. T. (1993) 8-Vinyl reduction and chlorophyll *a* biosynthesis in higher plants. *Biochem. J.*, **291**, 939–944.

Wiktorsson, B., Engdahl, S., Zhong, L. B., Boddi, B., Ryberg, M. & Sundqvist, C. (1993) The effect of crosslinking of the subunits of NADPH-protochlorophyllide oxidoreductase on the aggregational state of protochlorophyllide. *Photosynthetica*, **29**, 205–218.

Wiktorsson, B., Ryberg, M., Gough, S. & Sundqvist, C. (1992) Isoelectric focusing of pigment-protein complexes solubilized from non-irradiated and irradiated prolamellar bodies. *Physiol. Plant.*, **85**, 659–669.

Wiktorsson, B., Ryberg, M. & Sundqvist, C. (1996) Aggregation of NADPH-protochlorophyllide oxidore-ductase-pigment complexes is favored by protein phosphorylation. *Plant Physiol. Biochem. (Paris)*, **34**, 23–34.

Willows, R. D. (2003) Biosynthesis of chlorophylls from protoporphyrin IX. *Nat. Prod. Rep.*, **20**, 327–241.

Willows, R. D. & Beale, S. I. (1998) Heterologous expression of the Rhodobacter capsulatus BCHI, -D, and -H genes that encode magnesium chelatase subunits and characterization of the reconstituted enzyme. *J. Biol. Chem.*, **273**, 34206–34213.

Willows, R. D., Gibson, L. C. D., Kanangara, C. G., Hunter, C. N. & von Wettstein, D. (1996) Three separate proteins constitute the magnesium chelatase of *Rhodobacter sphaeroides*. *Eur. J. Biochem.*, **235**, 438–443.

Willows, R. D. & Hansson, M. (2003) Mechanism, structure and regulation of magnesium chelatase, in *The Porphyrin Handbook II*, Vol. 13 (eds K. M. Kadish, K. Smith & R. Guilard), Academic Press, San Diego, USA, pp. 1–48.

Willows, R. D., Lake, V., Roberts, T. H. & Beale, S. I. (2003) Inactivation of Mg chelatase during transition from anaerobic to aerobic growth in *Rhodobacter capsulatus*. *J. Bacteriol.*, **185**, 3249–3258.

Wubert, J., Oster, U., Blos, I. & Rudiger, W. (1997) Hydroxybenzoic acid methylesters inhibit magnesium chelatase activity in cress and barley seedlings. *Plant Physiol. Biochem.*, **35**, 581–587.

Yaronskaya, E. B., Shalygo, N. V., Rassadina, V. V. & Averina, N. G. (1993) Changes of S-adenosyl-L-methionine: magnesium protoporphyrin IX methyltransferase activity in wheat after treatment with 5-aminolevulinic acid. *Photosynthetica*, **29**, 243–247.

Younis, S., Ryberg, M. & Sundqvist, C. (1995) Plastid development in germinating wheat (*Triticum aestivum*) is enhanced by gibberellic acid and delayed by gabaculine. *Physiol. Plant.*, **95**, 336–346.

3   Carotenoids

Abby Cuttriss and Barry Pogson

3.1   Introduction

Carotenoid pigments provide many fruits and flowers with distinctive red, orange and yellow colours and a number of carotenoid-derived aromas, making them commercially important in agriculture, food manufacturing and the cosmetic industry (Fig. 3.1). However, it is their roles in photosynthesis and nutrition that account for the absolute requirement for carotenoids in the survival of plants and mammals alike. Specifically, carotenoids are a ubiquitous component of all photosynthetic organisms as they are required for assembly and function of the photosynthetic apparatus. Carotenoids are also a vital part of our diet as antioxidants and precursors to vitamin A.

Carotenoid research has a long and distinguished history (see Govindjee, 1999 for a review). Insights into carotenoid chemistry earned Nobel prizes for Paul Karrer and Richard Kuhn in 1937 and 1938, respectively. In 1943, excitation energy transfer between carotenoids and chlorophyll was first shown by Dutton *et al.* (1943). The essential role of coloured carotenoids in protection from lethal photodamage was observed in the photosynthetic bacterium, *Rhodobacter sphaeroides*, by Stanier and colleagues (Griffiths *et al.*, 1955). The xanthophyll cycle, a fundamental aspect of plant photoprotection, was later identified by Yamamoto (1962). More recent studies have begun to elucidate the mechanisms of energy transfer and photoprotection, and to identify the genes and enzymes responsible for the biosynthesis of this important class of pigments.

3.2   Structure, function and manipulation

Carotenoids are a large family (over 600 members) of isoprenoids (Table 3.1). Most are composed of a C_{40} hydrocarbon backbone with varying structural and oxygenic modifications that impart the different roles on distinct carotenoids. Carotenoids provide fruit and flowers with bright colours and their dietary uptake can pigment organisms as diverse as fish, crustaceans and birds. Their distinctive colours, typically in the yellow to red spectrum, are due to a series of conjugated double bonds. The range of colours is expanded by diverse modifications to the simple polyene chain structure. Interactions of certain carotenoids with apoproteins, such as astaxanthin with the crustacyanin protein from shellfish, can result in blue pigmentation that shifts back to red during cooking as the proteins are denatured.

Fig. 3.1 Carotenoid diversity. Carotenoids are highly diverse in terms of colour and structure. Likewise, they pigment a wide range of bacteria, fungi and plant tissues. Various yellow pigments, including lutein and zeaxanthin, become apparent in autumn leaves once the green chlorophyll has been degraded. Lutein is the major pigment in marigold flowers, which can range from white to dark orange because of differences in lutein content. Astaxanthin, complexed with crustacyanin, is responsible for the blue colour of lobsters, which shifts to red when the protein is denatured. Capsanthin and capsorubin have unusual cyclopentane rings and are the predominant pigments in red capsicum. Carrot is a source of β- and α-carotene, both of which can be cleaved to form vitamin A (11-*cis*-retinal), and the distinctive red colour of ripe tomatoes is due to accumulation of lycopene.

The carotenoid backbone is either linear or contains cyclic end groups. The most abundant end group is the β-ionone ring of β-carotene and its derivatives. Carotenoids with β-ionone groups serve as precursors for vitamin A and are

Table 3.1 Semi-systematic names of carotenoids

Trivial name	Semi-systematic name	CDBs	λ_{max} (nm)	Colour
phytoene	7,8,11,12,7′,8′,11′,12′-octahydro-ψ,ψ-carotene	3	276 286 297[a]	colourless
phytofluene	7,8,11,12,7′,8′-hexahydro-ψ,ψ-carotene	5	331 348 367[a]	colourless
ζ-carotene	7,8,7′,8′-tetrahydro-ψ,ψ-carotene	7	378 400 425[a]	pale yellow
neurosporene	7,8-dihydro-ψ,ψ-carotene	9	415 440 468[a]	yellow
pro-lycopene	7Z,9Z,7′Z,9′Z-tetra-*cis*-ψ,ψ-carotene	11	424 442 464	orange
lycopene	ψ,ψ-carotene	11	447 473 505	pink/red
γ-carotene	β,ψ-carotene	11	435 461 490[a]	pink
δ-carotene	ε,ψ-carotene	11	421 456 489[a]	yellow/orange
α-carotene	β,ε-carotene	10	421 445 473[a]	yellow
zeinoxanthin	β,ε-carotene-3,-ol	10	424 446 476	yellow
lutein	β,ε-carotene-3,3′-diol	10	426 447 474	yellow
β-carotene	7,8-dihydro-β,β-carotene	11	432 454 480	orange
β-cryptoxanthin	β,β-caroten-3-ol	11	425 449 476	yellow/orange
zeaxanthin	β,β-carotene-3,3′-diol	11	432 454 480	orange
antheraxanthin	5,6-epoxy-5,6-dihydro-β,β-carotene-3,3′-diol	10	424 448 475	yellow
violaxanthin	5,6:5′6′-diepoxy-5,6,5′6′-tetrahydro-β,β-carotene-3,3′-diol	9	417 440 470	yellow
neoxanthin	5′,6′-epoxy-6,7-didehydro-5,6,5′,6′-tetrahydro-β,β-carotene-3,5,3′-triol	8	414 437 465	yellow
astaxanthin	3,3′-dihydroxy-β,β-carotene-4,4′-dione	13	478[a]	red
capsanthin	3,3′-dihydroxy-β,κ-caroten-6′-one	11	460 483 518[b]	red
capsorubin	3,3-dihydroxy-β,κ-carotene-6,6′-dione	11	460 489 523[b]	red
torulene	3′,4′-didehydro-β,ψ-carotene	13	451 48 517[a]	red
tetradehydrolycopene	3,4,3′,4′-tetrahydro-ψ,ψ-carotene	15	480 510 540[c]	red

Trivial and semi-systematic names of carotenoids that are discussed in this chapter. Number of conjugated double bonds (CDBs) is compared to absorption maxima (λ_{max}) and colour. Absorption maxima are given in an acetonitrile/ethyl acetate solvent mix measured in line by a diode array detector (Pogson *et al.*, 1996) unless otherwise indicated.

Note: [a]hexane, [b]benzene (Britton, 1995), [c]acetonitrile (Schmidt-Dannert *et al.*, 2000).

therefore essential dietary components. Other cyclic end groups are the ε-ring, which is common in plants, and the unusual cyclopentane ring of capsanthin and capsorubin that impart the distinct red colour in peppers (Fig. 3.1) (Cunningham & Gantt, 1998).

The two major classes of carotenoids are the carotenes and their oxygenated derivatives, the xanthophylls. The most abundant xanthophylls, lutein and violaxanthin, are key components of the light-harvesting complex (LHC) of leaves, and are responsible for the yellow colour of autumn leaves that is normally masked by the green chlorophylls (xanthos = yellow and phyll = leaf). The biosynthetic pathway involves a series of desaturations, cyclisations, hydroxylations, and epoxidations, commencing with the formation of phytoene and typically terminating in lutein and neoxanthin accumulation (Fig. 3.2). The pathway is discussed in depth in Section 3.3. In brief, phytoene is formed by the condensation of geranylgeranyl diphosphate (GGPP) by phytoene synthase (PSY). Phytoene is subjected to four desaturation reactions by phytoene desaturase (PDS) and ζ-carotene desaturase (ZDS) to produce tetra-*cis*-lycopene, which is isomerised by the carotenoid isomerase (CRTISO) to produce all-*trans*-lycopene. Lycopene is cyclised twice by the β-cyclase (βLCY) to produce β-carotene, or once by βLCY and once by the ε-cyclase (εLCY) to produce α-carotene. The two carotenes are hydroxylated by the β- and ε-hydroxylases (βOH, εOH) to produce zeaxanthin and lutein, respectively. Zeaxanthin is epoxidated by the zeaxanthin epoxidase (ZE) to form violaxanthin, which is further modified by the neoxanthin synthase (NXS) to produce neoxanthin. Photosynthetic tissue will typically accumulate β-carotene and the xanthophylls lutein, violaxanthin, and neoxanthin.

3.2.1 Nutrition

Both the United Nations Children's Fund (UNICEF) and the World Health Organization (WHO) list staggering statistics on the number of child deaths and blindness attributed to vitamin A deficiency. Vitamin A deficiency is responsible for a number of disorders, ranging from impaired iron mobilisation, growth retardation and blindness to depressed immune response and increased susceptibility to infectious disease (Sommer & Davidson, 2002). More than 100 million children are vitamin A deficient; between 250 000 and 500 000 become blind every year, with half of these dying within 12 months of losing their sight http:// www.who.int/nut/vad.htm. Simply improving vitamin A status of children reduces overall child mortality by 25% http://www.unicef.org/immunization/facts_vitamina.htm. Dietary carotenoids, such as α-carotene and β-carotene, can be cleaved by mammals to form 1 or 2 molecules of 11-*cis*-retinal (vitamin A), respectively, and are referred to as provitamin A.

Provitamin A levels have been targeted for increase by both plant breeding and genetic modification. Orange-fleshed sweet potato has been adapted to local conditions by traditional breeding, and was introduced to Kenya as a new source of β-carotene (Hagenimana *et al.*, 1999). There are many examples of naturally occurring carotenoid variation (Table 3.2), but perhaps the best-known example of carotenoid enhancement is 'Golden Rice', which was originally designed to address the problem of vitamin A deficiency in developing countries. *PSY* from daffodil (*Narcissus pseudonarcissus*) and the bacterial *PDS*, *crtI*, from *Erwinia uredorva*, were targeted to the rice endosperm, because the outer aleurone is generally removed in the milling process. Even in the absence of a non-rice β-cyclase gene, transformed lines produced β-carotene (Ye *et al.*, 2000). Since then, T3 golden rice lines in elite Indica and Japonica variety backgrounds have

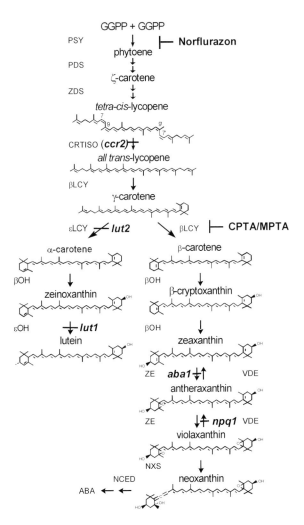

Fig. 3.2 Carotenoid biosynthetic pathway in higher plants. The pathway shows the major steps found in nearly all plant species. Additional compounds, such as lutein epoxide and capsanthin accumulate in certain species. *A. thaliana* mutations are shown in italics. Herbicide points of action are shown in bold type. PSY, phytoene synthase; PDS, phytoene desaturase; ZDS, ζ-carotene desaturase; CRTISO, carotenoid isomerase; βLCY, β-cyclase; εLCY, ε-cyclase; βOH, β-hydroxylase; εOH, ε-hydroxylase; ZE, zeaxanthin epoxidase; VDE, violaxanthin de-epoxidase; NXS, neoxanthin synthase; NCED, 9-*cis*-epoxycarotenoid dioxygenase; ABA, abscisic acid; CPTA, *N,N*-diethyl-*N*-[2-(4-chlorophenylthio)ethyl]amine; MPTA, *N,N*-diethyl-*N*-[2-(4-methylphenoxy)ethyl]amine.

been produced. Phosphomannose isomerase was used as a selectable marker to avoid the complication of antibiotic resistance selection. The elite lines are now ready for nutritional and risk assessment and breeding for local acclimation (Cuc Hoa *et al.*, 2003).

Table 3.2 Carotenoid mutants

Gene/locus	Species	Phenotype	References
phytoene destauration			
ghost	Lycopersicon esculentum	Mutation in plastid terminal oxidase. White leaves and pale yellow fruit; phytoene in leaves and fruit vs. lycopene-accumulation in wild type	(Josse et al., 2000)
pds1	Arabidopsis thaliana	Mutation in 4-hydroxyphenylpyruvate dioxygenase; phytoene accumulation	(Norris et al., 1995)
carotenoid isomerase			
ccr2	Arabidopsis thaliana	Tetra-cis-lycopene in the dark; reduced lutein in the light	(Park et al., 2002)
tangerine	Lycopersicon esculentum	Orange tetra-cis-lycopene accumulating fruit vs. red all-trans-lycopene in wild type	(Isaacson et al., 2002)
ε-cyclase			
lut2	Arabidopsis thaliana	Reduced lutein, increased xanthophyll cycle pigments	(Pogson et al., 1996)
lor1	Chlamydomonas reinhardtii	Reduced lutein and loraxanthin	(Niyogi et al., 1997a)
C-2A'-34	Scenedesmus obliquus	Reduced lutein and loraxanthin	(Bishop, 1996)
delta	Lycopersicon esculentum	Orange fruit; δ-carotene and reduced lycopene (overexpression)	(Ronen et al., 1999)
β-cyclase			
beta	Lycopersicon esculentum	Orange fruit; increased β-carotene, reduced lycopene (overexpression)	(Ronen et al., 2000)
old-gold	Lycopersicon esculentum	Deep red fruit and orange flowers; increased lycopene and reduced β-carotene (null)	(Ronen et al., 2000)
ε-hydroxylase			
lut1	Arabidopsis thaliana	Reduced lutein, increased xanthophyll cycle pigments	(Pogson et al., 1996)

zeaxanthin epoxidase

aba1	*Arabidopsis thaliana*	Increased zeaxanthin, decreased violaxanthin and neoxanthin, reduced ABA	(Koornneef *et al.*, 1982)
aba2	*Nicotiana plumbaginifolia*	Increased zeaxanthin, decreased violaxanthin and neoxanthin, reduced ABA	(Marin *et al.*, 1996)

violaxanthin de-epoxidase

npq1	*Arabidopsis thaliana*	Reduced zeaxanthin and antheraxanthin	(Niyogi *et al.*, 1998)
npq1	*Chlamydomonas reinhardtii*	Reduced zeaxanthin and antheraxanthin	(Niyogi *et al.*, 1997a)

carotenoid cleavage dioxygenases

viviparous14 (*vp14*)	*Zea mays*	Reduced ABA	(Schwartz *et al.*, 1997; Tan, 1997)
ramosus1 (*rms1*)	*Pisum sativum*	Increased branching	(Sorefan *et al.*, 2003)
max4	*Arabidopsis thaliana*	Increased branching	(Sorefan *et al.*, 2003)

regulatory or not defined

ccr1	*Arabidopsis thaliana*	Tetra-*cis*-lycopene in the dark; reduced lutein in the light	(Park *et al.*, 2002)
Or	*Brassica oleracea* var. botrytis	Orange curd; β-carotene accumulation vs. only trace carotenoids in the white wild-type curd	(Li *et al.*, 2001)
Navelate	*Citrus sinensis*	Yellow fruit; phytoene, phytofluene, β-carotene and reduced ABA vs. orange xanthophylls in wild-type	(Rodrigo *et al.*, 2003)
Cara cara	*Citrus sinensis*	Red fruit; β-carotene and lycopene vs. orange xanthophylls in wild type	(Lee, 2001)
reduced pigment (*rp*)	*Daucus carota* L.	White root; trace carotenoids vs. orange β- and α-carotene in wild type	(Goldman and Breitbach, 1996)
high pigment 1 (*hp1*)	*Lycopersicon esculentum*	Deep red fruit; increased pigmentation	(Cookson *et al.*, 2003)

Mutations affecting carotenoid biosynthesis discussed in this chapter are shown for several different plant and algal species.

Possibly the most dramatic example of carotenoid manipulation was performed in canola (*Brassica napus*). Overexpression of the bacterial *PSY* gene *crtB* in a seed-specific, plastid-targeted manner produced a remarkable 50-fold increase in carotenoid accumulation (Shewmaker *et al.*, 1999). Manipulation of carotenoid accumulation often has unexpected effects on other cellular processes. This instance is no different; concomitant with the increase in α- and β-carotene, tocopherol and chlorophyll levels decreased and fatty acid composition was altered (Shewmaker *et al.*, 1999). In order for carotenoids to accumulate to high levels, there must be an adequate sink. In this case the hydrophobic molecules can be sequestered in cellular vesicles or lipid globules within the plastid, which is perhaps why such high levels were accumulated in seeds of an oil crop. The introduction of a bacterial gene, as opposed to overexpression of the endogenous equivalent, may also have bypassed the usual regulatory controls that prevent such high levels of carotenoid accumulation.

Two carotenoids, lutein and zeaxanthin, are important for photoprotection in plants and have been implicated in protecting against age-related blindness due to macular degeneration in humans. Macular degeneration is the leading cause of blindness in the developed world and results in the loss of central and detailed vision. This is due to functional deterioration of the macula, a small portion of the retina that is exposed to direct light. The macula is known as the yellow spot because it accumulates high levels of both zeaxanthin and lutein, which have been positively correlated with lower incidence of macular degeneration (Seddon *et al.*, 1994). Consequently, a number of biotechnological approaches are underway to produce food groups with higher levels of zeaxanthin and lutein. For example, mutation of a salt-tolerant unicellular green alga, *Dunaliella salina*, has enabled accumulation of zeaxanthin as the primary xanthophyll (Jin *et al.*, 2003). In this context it should be noted that *Dunaliella* is already cultivated as a source of β-carotene by the natural products industry in Australia and Israel. The staple crop potato, which usually accumulates lutein and violaxanthin, was genetically modified to accumulate zeaxanthin (Römer *et al.*, 2002). Antisense and co-suppression of the potato zeaxanthin epoxidase increased tuber zeaxanthin levels 4–130-fold. Serendipitously, manipulation of this one enzyme also resulted in elevated transcript levels of the first step in the carotenoid pathway and a concomitant two- to threefold increase in α-tocopherol (vitamin E) (Römer *et al.*, 2002).

Astaxanthin is a powerful antioxidant, and hence a beneficial human dietary component (Jyonouchi *et al.*, 1995). However, astaxanthin's main value is as an additive to aquaculture feed as it gives salmon their characteristic pink colour (Lorenz & Cysewski, 2000). Astaxanthin is produced by a limited number of organisms, including certain marine bacteria, yeast, and some green algae (Johnson & Schroeder, 1995). For commercial purposes, it is usually chemically synthesised at some cost. This prompted Mann *et al.* (2000) to engineer tobacco flowers that produce astaxanthin by the introduction of the *Haematococcus pluvialis* β-carotene ketolase (*CrtO*) gene. Additional biotechnological approaches using marine algae and other crops are underway in a number of institutions.

The production of novel carotenoids has been made possible by recent innovations. The novel carotenoids were produced in *Escherichia coli* by molecular 'breeding' of the bacterial PDS (*crtI*) and β-cyclase (*crtY*) genes (Schmidt-Dannert *et al.*, 2000). Random shuffling of the two genes allowed the authors to produce a variety of coloured compounds, including the highly desaturated compound 3,4,3′,4′-tetra-dehydro-lycopene, despite no three-dimensional structure and limited knowledge of the enzymes' catalytic mechanism. They also produced the monocyclic carotenoid, torulene, for the first time in *E. coli*. Torulene was not only produced in a new organism, but also by an entirely new metabolic route, different from any mechanism found in nature (Schmidt-Dannert *et al.*, 2000). A similar principle has been used in several instances creating new bioactive compounds and producing carotenoids that would otherwise be inefficient to synthesise or extract (Albrecht *et al.*, 2000; Wang & Liao, 2001; Lee *et al.*, 2003).

3.2.2 Protein and membrane scaffolds

Carotenoids are found in all photosynthetic organisms, where they play a variety of crucial roles. They are involved in photosystem assembly and contribute to light harvesting by absorbing light energy in a region of the visible spectrum where chlorophyll absorption is lower and by transferring the energy to chlorophyll. Finally, carotenoids provide protection from excess light via energy dissipation, free radical detoxification and limiting damage to membranes.

Photosystem II (PSII) consists of an outer antenna surrounding a core complex comprising D1, D2 and two inner antenna proteins, CP43 and CP47. The xanthophylls are bound to the antenna and are essential for both folding and stability of the antenna LHC proteins. Thus pigment biosynthesis is tightly coordinated with LHC synthesis (Plumley & Schmidt, 1987). Several studies have suggested that the chloroplast envelope is the site of LHC assembly (Plumley & Schmidt, 1987). Free carotenoids in the chloroplast envelope have been suggested to facilitate folding of LHC apoproteins by modifying properties of the lipid bilayer (Paulsen, 1999). Blocking carotenoid biosynthesis with the herbicide norflurazon reduced membrane LHC II complex accumulation and increased recovery of it in the vacuole of *Chlamydomonas* (Hoober & Eggink, 1999).

The light-harvesting antenna is composed of major and minor LHC apoproteins that are encoded by a large family of homologous genes (Bassi *et al.*, 1993; Jansson, 1994, 1999; Green & Durnford, 1996). The structure for LHCIIb, the major LHC of PSII, was resolved to 3.4 Å and it revealed seven chlorophyll *a* molecules, five chlorophyll *b* molecules and two xanthophylls embedded in three transmembrane-spanning helices (Kuhlbrandt *et al.*, 1994). The carotenoid-binding sites have been elucidated by an elegant and thorough combination of structural studies, *in vitro* reconstitution of recombinant proteins, and analyses of pigment binding to native LHCs (Bassi & Caffarri, 2000). *In vitro* reconstitution studies have identified up to four distinct carotenoid-binding sites in LHCs (L1, L2, N1, V1), although only three of these binding sites are occupied in plants grown at moderate light intensities (Bassi & Caffarri, 2000). The L1 and L2 sites bind lutein, although either violaxanthin or zeaxanthin can be accommodated, whereas the N1 site is specific for neoxanthin. The V1 site is proposed to loosely bind violaxanthin under conditions of high light stress, aiding in accessibility for de-epoxidation to zeaxanthin, which subsequently fills the V1 site (Verhoeven *et al.*, 1999; Bassi & Caffarri, 2000). Furthermore, pigments bound to the V1 site do not contribute to light harvesting and are presumably involved only in xantho-phyll cycle activity, which is induced by high light stress (Yamamoto, 1979; Caffarri *et al.*, 2001). Finally, specific carotenoids are required for correct assem-bly of the LHC holocomplex, with the absence of lutein resulting in a loss of LHC trimers (Rissler & Pogson, 2001; Lokstein *et al.*, 2002).

The minor LHC antennae proteins, Lhcb 4, 5, and 6 (also known as CP29, CP26 and CP24, respectively), typically bind eight chlorophyll *a* molecules, two chloro-phyll *b* molecules and two xanthophyll molecules (Bassi & Caffarri, 2000). In Lhcb 4, the L1 site is occupied by lutein and L2 by violaxanthin or neoxanthin (Bassi *et al.*, 1999; Ruban *et al.*, 1999). The core complex of PSII binds β-carotene and chlorophyll *a*, and possibly small amounts of lutein (Bassi *et al.*, 1993; Satoh, 1993; Seibert, 1993; Alfonso *et al.*, 1994).

Carotenoids are important for plastid development. The *Arabidopsis thaliana crtISO* mutant accumulates tetra-*cis*-lycopene and lacks a prolamellar body (PLB) in the dark-grown plastid, the etioplast. The PLB is a uniformly curved lattice of tubular membranes, which contains several of the biochemical building blocks required for the chloroplast (Gunning & Jagoe, 1967), including the xanthophylls lutein and violaxanthin (Joyard *et al.*, 1998). Thus, a mutation in carotenoid biosynthesis disrupts membrane curvature and stabilisation, which suggests that different carotenoids either directly or indirectly impede formation of the mem-brane lattices (Park *et al.*, 2002). The absence of the PLB in *crtISO* mutants delays plastid development and greening on exposure to light, demonstrating a role for the PLB in rapid formation of a functional chloroplast, and by inference, a role for carotenoids in plastid differentiation (Park *et al.*, 2002).

In addition to directly binding LHC apoproteins, xanthophylls are also localised in the chloroplast envelope and thylakoid membranes, potentially spanning the lipid bilayer (Siefermann-Harms *et al.*, 1978; Havaux, 1998). In fact, certain xanthophylls have also been implicated in photoprotection outside of the photo-synthetic apparatus, including protection against lipid peroxidation and mainten-ance of membrane fluidity and thermostability (Tardy & Havaux, 1997; Gruszecki *et al.*, 1999). Zeaxanthin-deficient mutants of *A. thaliana*, *npq1*, exhibit enhanced lipid peroxidation and leaf necrosis upon exposure to light or chilling stress (Havaux & Niyogi, 1999). This effect cannot be attributed to the lack of non-photochemical quenching (NPQ), since the *npq4* mutant that exhibits a loss of NPQ without a change in xanthophyll cycle pool size is more tolerant to lipid peroxidation than *npq1* (Havaux & Niyogi, 1999). Therefore, zeaxanthin appears to have an additional specific photoprotective role outside of the light-harvesting antennae.

3.2.3 Photoprotection and plant fitness

Plants must maintain a balance between absorbing sufficient light for photosynthesis, whilst avoiding oxidative damage caused by too much light. In the absence of cyclic carotenoids, plastids photobleach due to the generation of reactive oxygen species that attack lipids, proteins and nucleic acids (Knox & Dodge, 1985). The bleached-white phenotype demonstrates the necessity for carotenoids in plant viability. Preventative and acclimatory measures to reduce incident light include moving leaves and chloroplasts to reducing light-harvesting antennae size. Responses to absorbed excess light include a number of complementary photoprotective mechanisms: (1) the harmless dissipation of excess energy via NPQ that is mediated by certain xanthophylls; (2) quenching of triplet chlorophylls by carotenoids; (3) antioxidants (ascorbate, tocopherols and carotenoids) and antioxidant enzymes, such as ascorbate peroxidase that detoxify free radicals; (4) repair of damaged proteins (Anderson et al., 1995; Niyogi, 1999).

NPQ-mediated dissipation of excess absorbed energy functions by non-radiative quenching of singlet chlorophyll. This may limit the formation of chlorophyll triplets, and prevent reactive oxygen species from being generated (Demmig-Adams & Adams, 1992) but requires a functional xanthophyll cycle. The xanthophyll cycle, identified by Harry Yamamoto in 1962, is a key component of the plant's high light defence strategy (Yamamoto, 1962). Under excess light, violaxanthin is de-epoxidised (removal of the two oxygen functions, or epoxy groups) to zeaxanthin via antheraxanthin. In the reverse reaction, under low or no light, zeaxanthin is epoxidised to form violaxanthin. The process is pH dependent: reduced lumenal pH activates the de-epoxidase enzyme and facilitates energy dissipation within the antennae via protonation-induced conformational changes. Another essential component of the energy dissipation mechanism is the PsbS protein of PSII (Li et al., 2000).

Mutants with altered carotenoid profiles (Table 3.2) are often affected in terms of NPQ (Niyogi et al., 1998; Pogson et al., 1998; Pogson & Rissler, 2000; Lokstein et al., 2002). The npq1 mutations of A. thaliana and Chlamydomonas are lesions in the violaxanthin de-epoxidase (VDE). These mutations and antisense VDE lines inhibit the high light–induced accumulation of zeaxanthin, which results in markedly reduced levels of NPQ (Niyogi et al., 1997a, 1998; Sun et al., 2001; Verhoeven et al., 2001). The residual NPQ in zeaxanthin-deficient lines can be essentially eradicated by eliminating lutein (Niyogi et al., 1997b, 2001); conversely, an increase in lutein resulted in faster induction of NPQ (Pogson & Rissler, 2000). Regardless of the genetic background, altered lutein levels invariably affect NPQ (Pogson et al., 1998), whether directly, or indirectly via changes to the antenna conformation, remains to be determined.

Alterations in the xanthophyll cycle pool size also affect NPQ. Inhibition of the β-carotene hydroxylase (βOH) resulted in reduced xanthophyll pool size. There was no obvious alteration to antennae composition and only a 16% decline in NPQ. A stable pool of violaxanthin remained unconverted, even under high light stress, suggesting that a portion of violaxanthin is sequestered to fill structural sites and is therefore unavailable for xanthophyll cycle conversions (Pogson & Rissler, 2000).

The physiological relevance of NPQ was demonstrated in an elegant study where the authors measured an indicator of plant fitness in *A. thaliana*: the amount of seed formed under field conditions with varying light intensities (Külheim *et al.*, 2002). Seed set in *npq1* and *npq4* lines was reduced by 30–50% in the field experiments, which was due to both a reduction in the number of fruit (siliques) and the number of seed per fruit. In the converse experiment, Horton and colleagues found that increasing the xanthophyll cycle pool (without the usual compensatory reduction in lutein) by overexpressing the bacterial βOH gene (*chyB*) enhanced stress tolerance in *A. thaliana* (Davison *et al.*, 2002). Plants could withstand extreme light and heat, and displayed less leaf necrosis, anthocyanin accumulation and lipid peroxidation. Zeaxanthin was therefore suggested to prevent oxidative damage of membranes (Davison *et al.*, 2002). The physiological relevance of xanthophylls is further exemplified by the bleaching, delayed greening and semi-lethal phenotypes observed in several carotenoid-deficient mutants (Niyogi *et al.*, 1997a; Pogson *et al.*, 1998). In the absence of zeaxanthin and lutein, *Chlamydomonas* cultures photobleach and mature *A. thaliana* leaves senesce and photobleach in high-intensity light (Niyogi *et al.*, 1997a, 2001). Zeaxanthin appears to have a specific photoprotective role outside of the light-harvesting antennae (as discussed in Section 3.2.2). Thus both NPQ and xanthophylls do indeed enhance plant fitness.

Carotenes are essential for quenching of chlorophyll triplet states in purple bacterial reaction centres and their loss results in cell death (Griffiths *et al.*, 1955; Frank & Cogdell, 1993). In plants, these fundamental roles must be undertaken by the xanthophylls bound to CP43 and CP47 or the antenna as the electron exchange mechanism of triplet quenching requires the carotenoid to be in van der Waals contact (approximately 4 Å) with chlorophyll triplets (Krinsky, 1971). Any β-carotene molecule in close proximity to the chlorophyll special pair, P680, would be oxidised due to the oxidative potential of P680 (Telfer, 2002). Instead, β-carotene may afford photoprotection by other mechanisms. A cationic β-carotene may donate an electron to the oxidised special pair, $P680^+$ (Hanley *et al.*, 1999; Vrettos *et al.*, 1999), or more likely, β-carotene could quench singlet oxygen and contribute to the signalling of D1 degradation, which triggers PSII repair (Anderson & Chow, 2002; Telfer, 2002).

3.3   Biosynthesis and regulation

3.3.1   Carotenoid biosynthesis and the plastid

Carotenoid enzymes are notoriously labile *in vitro*, confounding many classical biochemical approaches to understanding carotenoid biosynthesis. Thus, elucidation of the biosynthetic pathway has relied heavily on molecular genetics, and it is only since the 1990s that carotenoid biosynthesis has been described at the molecular level (Fig. 3.2) (Armstrong, 1997; Britton, 1998; Cunningham & Gantt, 1998; Hirschberg, 1998). Carotenoid pigments and their corresponding enzymes are located in the plastid, although all known carotenoid biosynthetic genes are encoded by the nuclear genome. The pathway is at least in part regulated via changes in transcription, especially during morphogenic changes from etioplast to chloroplast or chloroplast to chromoplast (von Lintig *et al.*, 1997; Welsch *et al.*, 2000; Bramley, 2002). As a consequence, carotenoid transcriptional regulation and therefore chloroplast–nuclear signalling must be responsive to environmental stimuli, oxidative stress, redox poise and metabolite feedback regulation. The mechanism and nature of such signals are still being elucidated.

In general, increases in carotenoid accumulation, whether during fruit ripening, flower development or production of stress-induced carotenoids in algae, have coincided with increased transcript abundance of some (but not all) key steps in the pathway (von Lintig *et al.*, 1997; Cunningham & Gantt, 1998; Grunewald *et al.*, 2000; Welsch *et al.*, 2000; Hirschberg, 2001). For example, in marigold flowers differences in transcript levels of almost all carotenoid genes suggested that mRNA transcription or stability was responsible for the dramatic differences (up to 100-fold) in carotenoid accumulation between cultivars (Moehs *et al.*, 2001).

There is, however, a paucity of information about the regulation of carotenoid accumulation in photosynthetic tissues and germinating seedlings (Cunningham & Gantt, 1998; Cunningham, 2002). Not a single component of a signalling pathway has been identified; yet it is known that the pathway is tightly regulated during development and in response to environmental stimuli (Young, 1993; Pogson *et al.*, 1996; Welsch *et al.*, 2000), and the proportion of different xanthophylls greatly affects plant viability and photoprotection (Pogson *et al.*, 1998). In order to avoid extensive photo-oxidative damage, the synthesis of carotenoids and chlorophylls and their subsequent binding to pigment-binding proteins must be precisely balanced to meet the appropriate photosynthetic demands of the various growth conditions that plants are exposed to on a daily and seasonal basis (Herrin *et al.*, 1992; Anderson *et al.*, 1995).

Carotenoid biosynthetic genes were seen to be redox-sensitive in the green alga *H. pluvialis* (Steinbrenner & Linden, 2003), which produces the high-value carotenoid astaxanthin. Transcript abundance of β*LCY*, *PSY*, *PDS* and β*OH* increased in response to increased light intensities, a trend that was eradicated by DCMU application (3-(3,4-dichlorophenyl)-1,1-dimethylurea; oxidised plastoquinone pool) and exacerbated by DBMIB treatment (2,5-dibromo-6-isopropyl-3-methyl-1,4-benzoquinone; reduced plastoquinone pool) (Steinbrenner & Linden, 2003).

Plastid–nuclear signalling has been the subject of recent interest, primarily focused on the genome-*uncoupled* (*gun*) *A. thaliana* mutants (Susek *et al.*, 1993). Magnesium protoporphyrin IX (MP), a chlorophyll precursor, has been implicated in certain instances of plastid–nucleus signalling (Larkin *et al.*, 2003; Strand *et al.*, 2003). The accumulation of MP is sufficient to regulate the expression of many photosynthetic nuclear genes in response to norflurazon-induced photo-oxidative stress. The extent to which carotenoid gene expression is regulated by the MP signal is unknown. Furthermore, the MP signal is thought to communicate overall viability of the plastid and does not modulate phytochrome-mediated changes or other environmental and developmental stimuli that regulate carotenoid gene expression (Cunningham, 2002). Nevertheless, there are compelling reasons why a chlorophyll precursor could be controlling carotenoid gene transcription; both pathways are tightly coordinated, which serves to limit damage from toxic chlorophylls using nascent carotenoids prior to and during photosystem assembly (Hoober & Eggink, 1999). Disrupting one pathway leads to a parallel reduction in the other pathway. For example, levulinic acid treatment of greening barley seedlings prevents chlorophyll production and as a result, also reduces the formation of LHCs and carotenoid accumulation (Jilani *et al.*, 1996). The treatment is reversible and demonstrates the coordinate regulation of carotenoid and chlorophyll biosynthesis. But the level of control (such as transcriptional or post-transcriptional) regulating particular biosynthetic steps is not understood. Evidence is accumulating for metabolite feedback regulation (Corona *et al.*, 1996; Fraser *et al.*, 2000; Welsch *et al.*, 2003) and alternative signals could include a signalling protein bound to different pigments or isomers (Larkin *et al.*, 2003), or it is possible to speculate that more soluble carotenoid derivatives could participate in signalling. For now, the nature of the different forms of regulation remains the subject of current research.

3.3.2   *Isoprenoids – carotenoid substrates*

In order to understand and manipulate plant pigments, one must first understand their biosynthesis. The carotenoid pathway competes with other isoprenoids such as chlorophylls, phytochrome, phylloquinone, gibberellins, abscisic acid (ABA), monoterpenes, plastoquinone, and tocopherols for the substrate, isopentenyl diphosphate (IPP) (Fig. 3.3). IPP and its isomer, dimethylallyl diphosphate (DMAPP), are the so-called 'building blocks' for carotenoids and the numerous other isoprenoids required for plant growth and survival. The biosynthesis of isoprenoid precursors is covered in great detail elsewhere (Lichtenthaler, 1999; Cunningham, 2002). A survey of genes involved in isoprenoid biosynthesis (Lange & Ghassemian, 2003) comprehensively listed all known genes involved in substrate supply, carotenoid biosynthesis and all other known isoprenoid-derived compounds, giving an indication of the diversity of this class of compounds.

There are two distinct pathways for IPP production: the cytosolic mevalonic acid (MVA) pathway and the plastidic mevalonate-independent methylerythritol 4-phosphate (MEP) pathway (Fig. 3.3) (Lange *et al.*, 2000). The cytosolic MVA pathway produces IPP from acetyl-CoA, which can then be converted reversibly

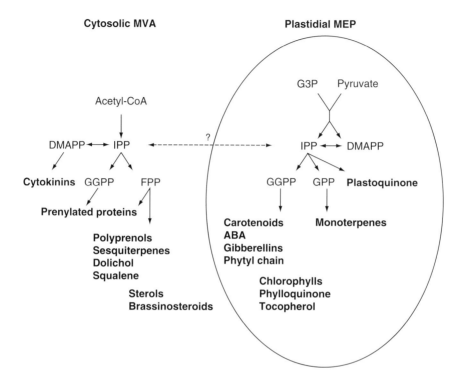

Fig. 3.3 MVA and MEP isoprenoid biosynthetic pathways. The plastidic (MEP) and cytosolic (MVA) pathways of isoprenoid biosynthesis in plants are shown. End products, including hormones and essential metabolites of each pathway are shown in bold type and the dotted arrow indicates a potential point for metabolite exchange. ABA, abscisic acid; DMAPP dimethylallyl diphosphate; FPP, farnesyl diphosphate; G3P, glyceraldehyde 3-phosphate; GGPP, geranylgeranyl diphosphate; GPP, geranyl diphosphate; IPP isopentenyl diphosphate; MEP, methylerythritol 4-phosphate; MVA, mevalonic acid.

to DMAPP by the IPP isomerase. In contrast, the MEP pathway takes place in the plastid and combines glyceraldehyde-3-phosphate (G3P) and pyruvate to eventually form IPP and DMAPP via separate branches (Fig. 3.3). The IPP isomerase is again capable of converting IPP and DMAPP (Lichtenthaler, 1999).

As stated previously, carotenoids and their respective enzymes accumulate exclusively in the plastid. Are the substrates also produced in the plastid via the MEP pathway? Labelling studies have shown that the plastidic MEP pathway supports carotenoid biosynthesis (Lichtenthaler *et al.*, 1997). Inhibitor studies reached a similar conclusion after treatment with the MEP inhibitor, fosmidomycin, reduced lycopene content in tomatoes (Zeidler *et al.*, 1998). In contrast, the MVA inhibitor, lovastatin, did not affect carotenoid accumulation (Rodriguez-Concepcion and Gruissem, 1999). Likewise the *A. thaliana Cla1* mutant, which is disrupted in the 1-deoxyxylulose-5-phosphate synthase (*DXS*) gene of the MEP pathway, is photobleached due to the absence of protective carotenoids, unless the substrate 1-deoxy-D-xylulose (DOX) is applied (Araki *et al.*, 2000; Estevez *et al.*, 2000).

The cytosolic MVA and plastidic MEP pathways are usually strictly compart-mentalised. An inhibitor study in *A. thaliana*, however, noted differences in isoprenoid metabolites (sterol, chlorophyll and carotenoids) from both pathways, despite blocking only one pathway at a time, indicating an element of crosstalk (Laule *et al.*, 2003). Despite this, gene expression levels assayed by GeneChip (Affymetrix) microarrays failed to note concomitant changes in the biosynthetic genes of these pathways (Laule *et al.*, 2003). Laule and colleagues suggested that post-transcriptional processes were probably responsible for regulating flux through isoprenoid metabolic pathways. It is also possible that microarray analysis was not sensitive enough to discern subtle changes in genes that are already expressed at low levels, as is the case with genes involved in carotenoid biosyn-thesis. Conversely, chemical complementation in tobacco cell culture lines using labelled precursors from either pathway indicated a transcriptional element to crosstalk between the MEP and MVA IPP pathways. Inhibitor treatments induced upstream genes, but this was reversed by application of precursors from the opposing IPP pathway and vice versa (Hemmerlin *et al.*, 2003). To date, only small amounts of exchange have been observed under extreme circumstances; the physiological relevance of this phenomenon is yet to be discerned.

3.3.3 Phytoene synthase

The first committed step of carotenoid biosynthesis is the condensation of two molecules of GGPP to produce the colourless carotenoid phytoene (Fig. 3.2). This reaction is catalysed in a two-step process by the PSY enzyme in higher plants and the bacterial PSY (CrtB) in prokaryotes (Armstrong, 1994). Functional studies of PSY from daffodil (*Narcissus pseudonarcissus*) established that cleavage of an N-terminal transit sequence and plastid–lipid binding were required for optimal activity. PSY was inactive in its soluble, stromal form (Schledz *et al.*, 1996), although there was no evidence of a lipid requirement in a study of tomato PSY (Fraser *et al.*, 2000). This, together with reports of a soluble PSY, GGPS and IPI complex suggests that post-translational regulation may be just as important as transcriptional regulation (Camara, 1993; Fraser *et al.*, 2000).

Tomato has two forms of PSY (Bartley & Scolnik, 1993). Knocking out the fruit-specific *PSY-1* gene results in tomato fruit with virtually no carotenoids, demonstrating that the green tissue form, *PSY-2*, cannot compensate for the disrupted *PSY-1*. Partial purification of the chloroplast PSY identified Mn^{2+} and ATP as essential cofactors for enzyme activity. The chromoplast version is less dependent on Mn^{2+}, has a lower affinity for GGPP and a more alkaline optimal pH (7.5 vs 6.5) (Fraser *et al.*, 2000) and, in its active form, is associated with the chromoplast membranes (Schledz *et al.*, 1996). Partially purified tomato chloroplast PSY was assayed in the presence of various metabolites (Fraser *et al.*, 2000). β-carotene and chlorophyll reduced *in vitro* activity by twofold whereas phytoene, ζ-carotene and lycopene did not alter activity (Fraser *et al.*, 2000).

PSY is known to be an important regulatory and rate-limiting step of carotenoid biosynthesis. Upregulation in response to light and various other stimuli invariably occurs at the transcriptional level. Regulation of *PSY* expression is mediated by phytochrome: both red and far-red light treatments enhance accumulation of *PSY* mRNA, and this enhancement is abolished in the *phyA* mutant of *A. thaliana*, showing that light regulation of *PSY* transcription is mediated by phytochrome activities (von Lintig *et al.*, 1997; Welsch *et al.*, 2000). However, the effect of light on PSY activity extends beyond transcriptional regulation. It was recently shown that PSY is associated with the PLB of etioplasts and is relatively inactive in dark-grown seedlings (Welsch *et al.*, 2000). White-light induction of photo-morphogenesis results in induction of PSY activity and relocation of PSY to the newly developing thylakoid membranes (Welsch *et al.*, 2000).

Structure/function studies of the *PSY* promoter fused to a luciferase reporter gene showed the expected basal transcriptional activity in the dark-germinated seedlings. Different light qualities increased expression; treatment with norflur-azon and gabaculine failed to have an effect; while the cyclase-blocking herbicide, CPTA, appeared to abolish light induction. The promoter study also identified a *cis*-acting motif, which was present in the promoter region of other photosyn-thesis-related genes, believed to be important in mediating the transcriptional regulation of *PSY* and other genes (Welsch *et al.*, 2003). As yet, no regulatory or signalling elements have been identified for any step of the biosynthetic pathway.

Seed-specific overexpression of the endogenous *A. thaliana PSY* gene had pleiotropic effects. Increased seed carotenoid content correlated with an increase in chlorophyll and ABA but germination was delayed, probably due to higher levels of the carotenoid-derived dormancy hormone, ABA (Lindgren *et al.*, 2003). Seeds were darker and had, on average, a 43% increase in β-carotene. Lutein, violaxanthin, lycopene, α-carotene and chlorophyll were also increased in seeds, while zeaxanthin levels increased only slightly (Lindgren *et al.*, 2003).

Overexpression of *PSY* genes in tobacco (*Nicotiana tabacum* Petit Havana SR1) resulted in dwarfism, altered leaf morphology and pigmentation. Antisense of either of tobacco's two *PSY* genes was lethal (Busch *et al.*, 2002). Likewise, constitutive expression of *PSY-1* in tomatoes caused pleiotropic effects, the most deleterious being dwarfism, which was possibly due to the diversion of substrate (GGPP) from the formation of gibberellins (Fray *et al.*, 1995). However, overexpressing the *crtB PSY* gene from *E. uredovora* in a fruit-specific manner resulted in a two- to fourfold increase in tomato fruit carotenoids. Other enzymes were not significantly altered and it appears that introduction of a bacterial gene uncoupled normal regulatory restrictions on carotenoid accumulation (Fraser *et al.*, 2002).

3.3.4 Desaturases

Phytoene is produced as a 15-*cis* isomer (Beyer, 1989; Beyer *et al.*, 1989; Beyer & Kleinig, 1991), which is desaturated to produce lycopene via a series of intermediates. Two desaturases, phytoene desaturase (PDS) and ζ-carotene desa-turase (ZDS), catalyse very similar dehydrogenation reactions to introduce four double bonds. Although associated with the membrane, detergent studies have shown that PDS is not an integral membrane protein (Schledz *et al.*, 1996). All desaturases, both bacterial and plant, contain a flavin-binding site near the N-terminus. Desaturation is linked to the respiratory redox chain (Nievelstein *et al.*, 1995) and evidence for a quinone requirement has been demonstrated in daffodil (Beyer *et al.*, 1989; Mayer *et al.*, 1992) and in phytoene-accumulating *A. thaliana* plants (Norris *et al.*, 1995). The molecular basis for this phenomenon was eluci-dated with the cloning of the gene responsible for the variegated *immutans* mutant (Carol *et al.*, 1999; Wu *et al.*, 1999). IMMUTANS is a plastid terminal oxidase (PTOX) required for PDS function, as it links desaturation to chloroplast respiratory activity. A lesion in the *PTOX* gene is also responsible for the white 'ghost' tomato phenotype, which accumulates phytoene in its leaves and fruit (Josse *et al.*, 2000).

In an early study on the regulation of carotenoid biosynthesis, a tomato *PDS* promoter was fused to a GUS reporter gene and transformed into both tomato and tobacco. Expression levels in different tissues were assessed by GUS staining (Corona *et al.*, 1996). The construct showed high levels of expression in tomato petals, anthers and ripening fruit (chromoplast-containing tissue) but low expres-sion in the equivalent tobacco tissue, which does not contain chromoplasts. The construct was active in etiolated tissue and could be unregulated in response to herbicides that block carotenoid and chlorophyll biosynthesis (Corona *et al.*, 1996). Tobacco antisense *PDS* mutants accumulated elevated levels of phytoene, suggesting increased metabolic flux (Busch *et al.*, 2002), as is observed for norflurazon treatment in the light (Simkin *et al.*, 2003), but these lines were invariably lethal. While *PSY* is clearly rate-limiting and phytochrome-mediated, the evidence for transcriptional regulation of *PDS* is more variable, depending upon the species, tissue and developmental process (Ronen *et al.*, 1999; Wetzel & Rodermel, 1998).

Constitutive expression of the bacterial phytoene desaturase from *E. uredovora*, *CrtI*, in tobacco (Misawa *et al.*, 1994) produced mutant lines that were resistant to bleaching herbicides such as norflurazon. Total carotenoid pool size was not altered, but β-carotene and its xanthophyll derivatives were increased at the expense of lutein accumulation. Another example of manipulation of carotenoids for nutri-tional purposes involves the introduction of the bacterial desaturase, *crtI*, into tomato resulting in β-carotene accumulation, and an overall decrease in carotenoid content (Römer *et al.*, 2000; Fraser *et al.*, 2001). In contrast, overexpression of β*LCY* increased β-carotene and total carotenoid content (Rosati *et al.*, 2000).

3.3.5 *Isomerase*

The two higher plant desaturases produce 7,9,7',9'-tetra-*cis*-lycopene when they are expressed in phytoene-accumulating *E. coli* (Bartley *et al.*, 1999). In contrast, bacteria only need one desaturase (CRTI) to introduce four double bonds and produce a different lycopene isomer; linear all-*trans*-lycopene (Bartley *et al.*, 1999). Evidence from bacterial studies suggests that cyclases cannot generally accept tetra-*cis*-lycopene as a substrate, requiring linear all-*trans*-lycopene to proceed (Schnurr *et al.*, 1996), although this is contradicted in isolated daffodil membranes (Beyer *et al.*, 1994). This isomer anomaly suggested a higher plant requirement for a specific isomerase enzyme (Beyer *et al.*, 1991), which had been postulated since the 1950s with the discovery of the tetra-*cis*-lycopene-accumu-lating *tangerine* tomato mutant (Tomes *et al.*, 1953; Isaacson *et al.*, 2002). Similarly, algal mutants accumulating *cis*-carotenoids have been reported (Powls & Britton, 1977; Cunningham & Schiff, 1985; Ernst & Sandmann, 1988). But the isomerase was not identified in plants until recently (Isaacson *et al.*, 2002; Park *et al.*, 2002).

The carotenoid isomerase gene, *CRTISO*, shows 20–30% identity to the bacter-ial carotenoid desaturases, *crtN* and *crtI*, particularly at conserved motifs, such as the di-nucleotide flavin-binding domain (characteristic of desaturases and cyclases). Despite this high level of identity, CRTISO demonstrated no desaturase or cyclase activity in an *E. coli* expression system (Park *et al.*, 2002). A similar level of identity in the tomato gene led Isaacson *et al.* (2002) to suggest that although ZDS could use *cis*-isomers as substrates, it could not alter the isomer state and therefore CRTISO may operate *in vivo* as part of a complex with ZDS (Isaacson *et al.*, 2002). The *CRTISO* gene was upregulated tenfold during the lycopene accumulation stages of fruit ripening: at the same time the lycopene cyclases were downregulated (Isaacson *et al.*, 2002).

In the *A. thaliana crtISO* mutant, tetra-*cis*-lycopene is photoisomerised in the light, and the biosynthetic pathway can carry on as usual, albeit with reduced lutein (Park *et al.*, 2002). The *tangerine* mutant, however, accumulates tetra-*cis*-lycopene in the fruit regardless of the light regime (Isaacson *et al.*, 2002). This highlights differences between chloroplasts and chromoplasts: carotenoids are deposited in a crystalline form in tomato chromoplasts and these may be more resistant to photoisomerisation than membrane-associated tetra-*cis*-lycopene in developing chloroplasts of *A. thaliana*.

Studies on the *A. thaliana crtISO* mutant suggested that light and the isomerase may contribute to regulating flux of metabolites through the two branches in the carotenoid biosynthetic pathway (Park *et al.*, 2002). Most carotenoids accumulated to wild-type levels in chloroplasts of CRTISO-deficient mutants, although lutein levels were markedly reduced (Park *et al.*, 2002). Thus the absence of CRTISO or specific carotenoid isomers influences partitioning of lycopene through the ε,β and β,β branches (Fig. 3.2) (Isaacson *et al.*, 2002; Park *et al.*, 2002). In accordance with this observation, expression of the bacterial *crtI* in tobacco plants, which bypasses the isomerisation steps, also resulted in a reduction in lutein levels (Misawa *et al.*, 1994). Whether this reflects an isomer substrate preference by the cyclase enzymes or regulation of the enzymes remains an interesting question to be answered, although recent experiments indicate an effect on ε-cyclase transcript levels in the *crtISO* mutant (Cuttriss & Pogson, unpublished data).

Herbicide treatments that force accumulation of lycopene in daffodil (*Narcissus pseudonarcissus*) increase *PSY* transcript abundance (Al-Babili *et al.*, 1999). This suggests a role for lycopene isomers in mediating multiple aspects of carotenoid biosynthesis, at least in some developmental contexts. Furthermore, we have identified another mutation, *ccr1*, which affects CRTISO function, in addition to other developmental processes (Park *et al.*, 2002). Its elucidation may provide insight into the function of isomerases and lycopene isomers.

3.3.6 Cyclases

After the formation of lycopene, the carotenoid biosynthetic pathway splits into two branches, distinguished by different cyclic end groups, namely β or ε. Two β-rings form the β,β branch (β-carotene and its derivatives) with one β and one ε forming the β,ε branch (α-carotene and its derivatives).

βLCY introduces a β-ionone ring to either end of lycopene to produce β-carotene. The βLCY gene was originally identified in the cyanobacterium *Synechococcus* species strain PCC7942 (Cunningham *et al.*, 1993, 1994). Hetero-dimeric β-cyclases have been observed in gram-positive bacteria (Krubasik & Sandmann, 2000), fuelling speculation that higher plant carotenoid enzymes, such as the cyclases, may act as part of a complex (Hirschberg, 2001; Cunningham, 2002). Homozygous βLCY mutants in *A. thaliana* are lethal. However, a second βLCY was identified in tomato, which was shown to be 53% identical to the classical βLCY and 86% identical to the capsanthin–capsorubin synthase (*CCS*) from pepper (*Capsicum annuum*). The alleles of this gene, *beta* and *old-gold* showed that developmentally regulated transcription is important in controlling pigment accumulation (Ronen *et al.*, 2000).

The β-cyclase and ε-cyclase enzymes each introduce a ring to form α-carotene. The *A. thaliana* εLCY can only catalyse cyclisation of one end group, demonstrated by a mutation in the βLCY that accumulated monocyclic ε,ψ-carotene (B. J. Pogson & D. DellaPenna, unpublished data). Lettuce (*Lactuca sativa*) appears to be unique among higher plants in that its εLCY can catalyse formation of a bicyclic ε,ε-carotene and its hydroxylated derivative lactucaxanthin (Phillip & Young, 1995; Cunningham & Gantt, 2001). A detailed recombinant DNA study demonstrated that a single amino acid was found to be responsible for this anomaly: the lettuce εLCY H457L mutant lost the ability to add two ε-rings, whereas the *A. thaliana* εLCY L448H was able to produce lactucaxanthin (Cunningham & Gantt, 2001). Another interesting example is from the marine cyanobacterium, *Prochlorococcus marinus* MED4, which contains a standard βLCY gene and an additional novel cyclase that is capable of forming both β and ε end groups (Stickforth *et al.*, 2003).

The ε,β-cyclic branch typically terminates in the formation of the oxygenated α-carotene derivative, lutein, although increasing numbers of species in which a lutein-5,6-epoxide cycle operates have been identified (Bungard *et al.*, 1999;

Matsubara *et al.*, 2003). To further understand the role of the terminal ε,β-carotenoid, lutein, and the partitioning of the pathway, lutein-deficient mutants, *lut1* and *lut2*, were identified (Pogson *et al.*, 1996). Increased β,β-xanthophylls accumulated in the reduced lutein lines, implying a degree of functional redundancy between lutein, violaxanthin and antheraxanthin (Pogson *et al.*, 1996). *lut1* was a recessive mutation that accumulated significant quantities of zeinoxanthin, indicating a disruption to the yet-to-be-identified ε-hydroxylase (*εOH*) gene. In contrast, *lut2* was semi-dominant and did not accumulate zeinoxanthin or any other ε-ring carotenoid; nor was β-cyclisation interrupted. It was also linked to the ε-cyclase locus (Pogson *et al.*, 1996). Complementation confirmed that *lut2* was a lesion in the *εLCY* (Pogson & Rissler, 2000).

Altered light intensity affects flux down both branches of the pathway (Ruban *et al.*, 1994) and transcript levels (Woitsch & Römer, 2003). Semi-dominance of *lut2* suggests that *εLCY* is a rate-limiting step in lutein production, and *εLCY* overexpression and co-suppression lines demonstrate that flux down the two branches can be controlled at the level of mRNA (Pogson *et al.*, 1996; Pogson & Rissler, 2000).

In addition to genetic lesions, herbicides can be used to disrupt the cyclases and study the effects on the perturbed pathway. CPTA (*N,N*-diethyl-*N*-[2-(4-chlorophenylthio)ethyl]amine) and MPTA (*N,N*-diethyl-*N*-[2-(4-methylphenoxy)ethyl]-amine) both cause lycopene accumulation although with slightly different cyclase specificities (Fedtke *et al.*, 2001). Treating tomato tissues with CPTA causes lycopene and various carotenoid-derived volatiles to accumulate (Ishida *et al.*, 1998). CPTA treatment of daffodil flowers yielded an overall increase in carotenoid content (twofold), and an increase in *PSY*, *PDS* and *βLCY* transcript and protein abundance (Al-Babili *et al.*, 1999). Lycopene was sequestered as crystals, as is seen in tomato fruit, which affected chromoplast structure. End-product regulation via an effector, possibly originating from a β-carotene derivative, was suggested as a possible mechanism.

It is important to note that control of pigment accumulation is under very different pressures in chloroplasts compared to chromoplasts, where the carotenoids are playing quite different roles. This is exemplified in tobacco, where overexpression of *CrtO*, a bacterial β-carotene ketolase gene, under the control of a tomato *PDS* promoter produces only trace amounts of astaxanthin accumulation in chloroplast-containing green tissue, but a 170% increase in total carotenoids in the chromoplast-containing nectary tissue, the most predominant of which was the algal/bacterial pigment astaxanthin (Mann *et al.*, 2000).

Amitrole (3-amino-1,2,4-triazole), like CPTA, has been shown to cause lycopene accumulation (Ridley, 1981). In addition, the authors saw a marked increase in the porphyrin chlorophyll precursors, (namely 5-aminolevulinate, Mg-protoporphyrin, Mg-protoporphyrin monomethylester and protochlorophyllide) on amitrole treatment (La Rocca *et al.*, 2001). Reduced induction of *Lhc* and *RbcS* (ribulose bisphosphate carboxylase small subunit) gene expression (La Rocca *et al.*, 2001) was in keeping with recent confirmation that Mg-protoporphyrin and its monomethylester act as a signal from the plastid to the nucleus (Strand et al., 2003). However, amitrole is also known to alter lipid composition (Di Baccio et al., 2002), inhibit catalase activity (Heim & Larrinua, 1989; Havir, 1992) and on occasion, completely fails to block pigment biosynthesis (Bouvier et al., 1997). Catalase inhibition in itself can alter carotenoid gene expression (Bouvier et al., 1998a). The variable effects of amitrole suggest that other inhibitors, such as CPTA, could provide a more targeted approach to understanding carotenoid cyclisation.

3.3.7   Hydroxylases

Xanthophylls are oxygenated carotenoids that have distinct and complementary roles in photosystem assembly, light harvesting and photoprotection. There are three hydroxylase genes in *A. thaliana* and tomato; two β-hydroxylases (β*OH*) and one ε-hydroxylase (ε*OH*). Their protein products catalyse hydroxylation of the 3-carbon of the β- and ε-rings, respectively, to form zeaxanthin and lutein (Hirschberg, 2001; Tian & DellaPenna, 2001). βOH enzymes are ferredoxin-dependent, with an iron-coordinating histidine cluster that is required for activity (Bouvier et al., 1998b). The authors proposed that the βOH enzymes break the C–H bond using iron-activated oxygen, which facilitates oxygen insertion (Bouvier et al., 1998b).

There are two distinct hydroxylation reactions of the ε- and β-rings, confirmed by the identification of the εOH locus, *lut1* (Pogson et al., 1996), and the β*OH* genes in higher plants (Sun et al., 1996). The βOH of plants shared significant identity with those from various bacterial systems with non-haem di-iron βOH activity (Perry et al., 1986; Misawa et al., 1990, 1995a, 1995b). In contrast, the recently identified εOH is a plastid-targeted cytochrome P450-type mono-oxygenase, which reveals the εOH as a member of a different gene family with a distinct enzymatic mechanism from the βOHs (Tian et al., 2004).

A truncated form of the *A. thaliana* βOH resulted in the accumulation of a monohydroxy intermediate (β-cryptoxanthin), prompting the authors to suggest that the βOH enzyme is likely to act as a homodimer (Sun et al., 1996). A second *A. thaliana* β*OH* gene was characterised recently and is 70% identical at the amino acid level (Tian & DellaPenna, 2001). Tomato also has two β*OH* genes, one of which is expressed in photosynthetic tissue, the other in flowers (Hirschberg, 2001). In contrast, both *A. thaliana* genes are expressed in all tissues analysed, albeit at different levels, and are genetically distinct from the ε*OH* (*lut1*) locus (Tian & DellaPenna, 2001).

A study of hydroxylase mutant combinations (single through to triple mutants) revealed a high level of functional redundancy in terms of β-carotene hydroxylation, with triple mutants maintaining at least 50% of the wild-type hydroxylated β-rings (mainly in the form of the monohydroxy α-carotene derivative, zeinoxanthin) despite the absence of both βOHs (Tian et al., 2003). The possibility of functional redundancy is consistent with results from βOH antisense and bacterial expression studies (Sun *et al.*, 1996; Rissler & Pogson, 2001). There were a number of explanations, including LUT1 having βOH activity, although this has since been ruled out with the identification of null *lut1* mutants that do not alter βOH activity (Tian *et al.*, 2004). A possible candidate for the βOH activity is CYP97A3, a putative P450 with 50% identity to the εOH (Tian *et al.*, 2003, 2004).

Expression analysis of β*OH*, zeaxanthin epoxidase (*ZE*) and violaxanthin de-epoxidase (*VDE*) during tobacco de-etiolation showed similar patterns of expression between β*OH* and *ZE*, but a different pattern for *VDE* expression. In addition, β*OH* expression levels were modulated by different intensities of white light (Woitsch & Römer, 2003). The authors also inferred a redox component to transcriptional regulation, as β*OH* and *ZE* transcript levels were found to be sensitive to DCMU and DBMIB treatment (Woitsch & Römer, 2003).

3.3.8   *Zeaxanthin epoxidase and violaxanthin de-epoxidase*

Light plays a critical role in modulating xanthophyll biosynthesis and accumulation (see Section 3.2.3). Under normal light conditions, when the incident light can be safely utilised for photosynthetic electron transport, ZE converts zeaxanthin to violaxanthin and the lumenal VDE remains soluble. However, under conditions of high light stress, when the incident light is in excess of the light utilised, zeaxanthin biosynthesis is triggered. VDE becomes activated by an acidification of the thylakoid lumen and docks to the thylakoid membrane, resulting in the de-epoxidation of violaxanthin to zeaxanthin (Pfundel *et al.*, 1994). The docking mechanism is poorly characterised, but is known to require the thylakoid membrane lipid monogalactosyldiacylglycerol (MGDG) (Siefermann & Yamamoto, 1975).

ZE mutants have been identified in *A. thaliana* (*aba1*) (Koornneef *et al.*, 1982; Niyogi *et al.*, 1998), tobacco (*aba2*) (Marin *et al.*, 1996) and pepper (Bouvier *et al.*, 1996). Biochemical characterisation revealed that the catalysis of the zeaxanthin to violaxanthin reaction, introducing 5,6-epoxy groups to the 3-hydroxy-β-rings, is a redox reaction that requires reduced ferredoxin (Bouvier *et al.*, 1996).

There are also numerous *VDE* mutants, termed *npq1* in both *A. thaliana* (Niyogi *et al.*, 1998) and *Chlamydomonas reinhardtii* (Niyogi *et al.*, 1997a), although the gene was originally cloned from lettuce (Bugos & Yamamoto, 1996). VDE and ZE were the first identified plant lipocalins, a class of β-barrel proteins that bind small hydrophobic molecules but are not usually catalytic (Bugos *et al.*, 1998). VDE is active below pH 6.5 and requires ascorbate as a reductant (Hager, 1969; Hager & Holocher, 1994).

3.3.9   *Neoxanthin synthase*

Conversion of violaxanthin to neoxanthin is performed by a neoxanthin synthase (NXS) enzyme. Genes that encode enzymes with limited NXS activity were originally identified in tomato, based on its resemblance to the CCS (Bouvier *et al.*, 2000) and potato – a serendipitous discovery – again based on similarity to CCS (Al-Babili *et al.*, 2000). Whether this gene encodes the primary NXS *in vivo* is a matter of debate, especially considering *A. thaliana* lacks this gene, but has NXS activity. Recently, however, putative NXS mutants that lack neoxanthin have been identified in both *A. thaliana* (*Ataba4;* A. Marion-Poll, Institut National de la Recherche Agronomique, Versailles, France, personal communication) and tomato (J. Hirschberg, The Hebrew University of Jerusalem, Jerusalem, Israel, personal communication). Thus, one of the last biosynthetic steps in the production of plant carotenoids awaits molecular characterisation.

3.3.10   Cleavage products

Vitamin A is a C_{20} carotenoid cleavage product, which, in addition to its retinoid derivatives, is essential for animal survival. Although β-carotene cleavage has long been postulated as an important step in the formation of vitamin A, it was not until 2000 that a β-carotene 15,15′-dioxygenase was cloned from *Drosophila melanogaster* (von Lintig & Vogt, 2000) and chicken (Wyss *et al.*, 2000). The deduced amino acid sequence showed homology to the plant carotenoid dioxygenase, VP14, involved in the synthesis of ABA. Any carotenoid containing an unmodified β-ionone ring has provitamin A activity, although β-carotene is one of the most active because a single β-carotene molecule is cleaved to form two retinals (vitamin A aldehyde). Retinal and its derivatives act as chromophores of the various visual pigments in animals (Wald, 1968). Vision aside, retinoic acid exerts most of the effects of vitamin A (Dowling & Wald, 1960).

The plant hormone, ABA, is involved primarily in plant stress responses, seed development and dormancy (Seo & Koshiba, 2002). ABA is a cleavage product of 9′-*cis*-neoxanthin and/or 9-*cis*-violaxanthin, an idea that was first proposed by Taylor and Smith (1967). Cleavage of 9′-cis-neoxanthin by the 9-*cis*-epoxycarotenoid dioxygenase (NCED) produces xanthoxin and was first identified in maize (*viviparous14* mutant; *vp14*) (Schwartz *et al.*, 1997; Tan, 1997). It is followed by a number of modifications that are required to produce ABA (Seo & Koshiba, 2002). *A. thaliana* encodes a family of nine NCEDs, five of which were cloned and studied in terms of expression and subcellular localisation (Tan *et al.*, 2003). All were targeted to the plastid, where their location diverged: AtNCED5 bound the thylakoid exclusively, AtNCED2, 3 and 6 were located in the thylakoid and stroma, while AtNCED9 in the stroma (Tan *et al.*, 2003). The NCED family also showed differences in tissue distribution (Tan *et al.*, 2003) and carotenoid substrate specificity (Schwartz *et al.*, 2001, 2003), thus the alternative nomenclature, carotenoid cleavage dioxygenase or CCD, may be more accurate.

The elucidation of the CCD family, and the presence of other carotenoid cleavage products that are not by-products of ABA synthesis (Milborrow *et al.*, 1988) raises interesting questions about the function of cleavage products in plant development. Intriguing examples are the *A. thaliana MAX4* and pea *RMS1* mutants. These mutants show increased branching and are both disrupted in one of the CCD genes, which was shown to be auxin-inducible (Sorefan *et al.*, 2003). The putative involvement of MAX4 in carotenoid cleavage lends itself to the possibility of a novel carotenoid-derived hormone.

As demonstrated above, carotenoid cleavage products are vital for the survival of plants and animals. They are also highly prized in the food and cosmetic industries. Bixin (annatto) is a red-coloured dicarboxylic monomethylester apocarotenoid, traditionally derived from the plant *Bixa orellana*. Bouvier *et al.* (2003a) identified the lycopene cleavage dioxygenase, bixin aldehyde dehydrogenase and norbixin carboxyl methyltransferase required to produce bixin from lycopene. Co-transforming the appropriate constructs into *E. coli* engineered to produce lycopene resulted in bixin production at a level of 5 mg/g dry weight (Bouvier *et al.*, 2003a).

Saffron, another commercially important coloured compound, can attribute the majority of its characteristic colour, flavour and aroma to the accumulation of carotenoid derivatives. A crocus zeaxanthin 7,8,7′,8′-cleavage dioxygenase was cloned and found to be targeted to the chromoplast where it initiated the production of cleavage products (Bouvier *et al.*, 2003b). Another less specific cleavage dioxygenase was also identified (Bouvier *et al.*, 2003b). Carotenoid cleavage pigments also feature in Chapter 7.

Carotenoid products are also important in the fragrance industry; enzymatic and photo-oxidative derivatives of various carotenoids are commonly used. While other aroma constituents such as esters, terpenes, and pyrazines are usually also present, these C_9 to C_{13} compounds are often essential to the odour profile (Wahlberg & Eklund, 1998).

3.4 Conclusions and future directions

The role of carotenoids in human health and plant photoprotection makes them obvious candidates for enhancement and manipulation (Sandmann, 2001). To this end, molecular genetics, in concert with classical biochemistry, has facilitated an advanced understanding of the biosynthetic pathway. We are learning to create new bioactive compounds by novel mechanisms and produce familiar carotenoids in crops that would otherwise be of low nutritional value. However, there are many aspects of carotenoid metabolism that remain to be elucidated. There is limited structural information for any of the carotenoid enzymes, with the possible exception of VDE and ZE models (Bugos *et al.*, 1998). Despite numerous transcriptional studies, no regulatory elements have been identified, and the hypothesis that carotenoid enzymes are active as multienzyme complexes is yet to be conclusively proven. We are only beginning to realise the role that carotenoid cleavage products play in plant development.

Acknowledgements

Our thanks to the many colleagues, students and staff who provided the results and insights that form the basis of this review. Thanks to Catherine Eadie for help in the preparation of Fig. 3.1.

References

Al-Babili, S., Hartung, W., Kleinig, H. & Beyer, P. (1999) CPTA modulates levels of carotenogenic proteins and their mRNAs and affects carotenoid and ABA content as well as chromoplast structure in *Narcissus pseudonarcissus* flowers. *Plant Biol.*, **1**, 607–612.

Al-Babili, S., Hugueney, P., Schledz, M., Welsch, R., Frohnmeyer, H., Laule, O. & Beyer, P. (2000) Identification of a novel gene coding for neoxanthin synthase from *Solanum tuberosum. FEBS Lett.*, **485**, 168–172.

Albrecht, M., Takaichi, S., Steiger, S., Wang, Z. Y. & Sandmann, G. (2000) Novel hydroxycarotenoids with improved antioxidative properties produced by gene combination in *Escherichia coli. Nat. Biotechnol.*, **18**, 843–846.

Alfonso, M., Montoya, G., Cases, R., Rodriguez, R. & Picorel, R. (1994) Core antenna complexes, CP43 and CP47, of higher plant photosystem II: spectral properties, pigment stoichiometry, and amino acid composition. *Biochemistry*, **33**, 10494–10500.

Anderson, J. M. & Chow, W. S. (2002) Structural and functional dynamics of plant photosystem II. *Philos. Trans. R. Soc. Lond. Ser. B-Biol. Sci.*, **357**, 1421–1430.

Anderson, J. M., Chow, W. S. & Park, Y. I. (1995) The grand design of photosynthesis: acclimation of the photosynthetic apparatus to environmental cues. *Photosynth. Res.*, **46**, 129–139.

Araki, N., Kusumi, K., Masamoto, K., Niwa, Y. & Iba, K. (2000) Temperature-sensitive Arabidopsis mutant defective in 1-deoxy-D-xylulose-5-phosphate synthase within the plastid non-mevalonate pathway of isoprenoid biosynthesis. *Physiol. Plant.*, **108**, 19–24.

Armstrong, G. A. (1994) Eubacteria show their true colors: genetics of carotenoid pigment biosynthesis from microbes to plants. *J. Bacteriol.*, **176**, 4795–4802.

Armstrong, G. A. (1997) Genetics of eubacterial carotenoid biosynthesis: a colorful tale. *Annu. Rev. Microbiol.*, **51**, 629–659.

Bartley, G. E., Scolnik, P. A. & Beyer, P. (1999) Two *Arabidopsis thaliana* carotene desaturases, phytoene desaturase and zeta-carotene desaturase, expressed in *Escherichia coli*, catalyze a poly-*cis* pathway to yield pro-lycopene. *Eur. J. Biochem.*, **259**, 396–403.

Bartley, G. E. and Scolnik, P. A. (1993) cDNA cloning, expression during development, and genome mapping of *psy2*, a second tomato gene encoding phytoene synthase. *J. Biol. Chem.*, **268**, 25718–25721.

Bassi, R. & Caffarri, S. (2000) Lhc proteins and the regulation of photosynthetic light harvesting function by xanthophylls. *Photosynth. Res.*, **64**, 243–256.

Bassi, R., Croce, R., Cugini, D. & Sandona, D. (1999) Mutational analysis of a higher plant antenna protein provides identification of chromophores bound into multiple sites. *Proc. Natl. Acad. Sci. USA*, **96**, 10056–10061.

Bassi, R., Pineau, B., Dainese, P. & Marquardt, J. (1993) Carotenoid-binding proteins of photosystem II. *Eur. J. Biochem.*, **212**, 297–303.

Beyer, P. (1989) Carotene biosynthesis in daffodil chromoplasts: on the membrane-integral desaturation and cyclization reactions, in *Physiology, Biochemistry, and Genetics of Nongreen Plastids* (eds C. D. Boyer, J. C. Shannon & R. C. Hardison), The American Society of Plant Physiologists, Rockville, MD, pp. 157–170.

Beyer, P., Mayer, M. P. & Kleinig, H. (1989) Molecular oxygen and the state of geometric isomerism of intermediates are essential in the carotene desaturation and cyclisation reactions in daffodil chromoplasts. *Eur. J. Biochem.*, **184**, 141–150.

Beyer, P. & Kleinig, H. (1991) Carotenoid biosynthesis in higher plants membrane-bound desaturation and cyclization reactions in chromoplast membranes from *Narcissus pseudonarcissus*. *Biol. Chem.*, **372**, 527.

Beyer, P., Kroncke, U. & Nievelstein, V. (1991) On the mechanism of the lycopene isomerase cyclase reaction in *Narcissus pseudonarcissus* chromoplasts. *J. Biol. Chem.*, **266**, 17072–17078.

Beyer, P., Nievelstein, V., Al-Babili, S., Bonk, M. & Kleinig, H. (1994) Biochemical aspects of carotene destauration and cyclization in chromoplast membranes from *Narcissus pseudonarcissus*. *Pure Appl. Chem.*, **66**, 1047–1056.

Bishop, N. I. (1996) The β,ε-carotenoid, lutein, is specifically required for the formation of the oligomeric forms of the light harvesting complexes in the green alga, *Scenedesmus obliquus*. *J. Photochem. Photobiol. B-Biol.*, **36**, 279–283.

Bouvier, F., Backhaus, R. A. & Camara, B. (1998a) Induction and control of chromoplast-specific carotenoid genes by oxidative stress. *J. Biol. Chem.*, **273**, 30651–30659.

Bouvier, F., D'Harlingue, A., Backhaus, R. A., Kumagai, M. H. & Camara, B. (2000) Identification of neoxanthin synthase as a carotenoid cyclase paralog. *Eur. J. Biochem.*, **267**, 6346–6352.

Bouvier, F., D'Harlingue, A. & Camara, B. (1997) Molecular analysis of carotenoid cyclase inhibition. *Arch. Biochem. Biophys.*, **346**, 53–64.

Bouvier, F., D'Harlingue, A., Hugueney, P., Marin, E., MarionPoll, A. & Camara, B. (1996) Xanthophyll biosynthesis – cloning, expression, functional reconstitution, and regulation of beta-cyclohexenyl carotenoid epoxidase from pepper (*Capsicum annuum*). *J. Biol. Chem.*, **271**, 28861–28867.

Bouvier, F., Dogbo, O. & Camara, B. (2003a) Biosynthesis of the food and cosmetic plant pigment bixin (annatto). *Science*, **300**, 2089–2091.

Bouvier, F., Keller, Y., D'Harlingue, A. & Camara, B. (1998b) Xanthophyll biosynthesis: molecular and functional characterization of carotenoid hydroxylases from pepper fruits (*Capsicum annuum* L.). *Biochim. Biophys. Acta-Lipids Lipid Metab.*, **1391**, 320–328.

Bouvier, F., Suire, C., Mutterer, J. & Camara, B. (2003b) Oxidative remodeling of chromoplast carotenoids: identification of the carotenoid dioxygenase *CsCCD* and *CsZCD* genes involved in Crocus secondary metabolite biogenesis. *Plant Cell*, **15**, 47–62.

Bramley, P. M. (2002) Regulation of carotenoid formation during tomato fruit ripening and development. *J. Exp. Bot.*, **53**, 2107–2113.

Britton, G. (1998) Overview of carotenoid biosynthesis, in *Biosynthesis and Metabolism*, Vol. 3 (eds G. Britton, S. Liaaen Jensen & H. Pfander), Birkhäuser Verlag, Basel, Switzerland, pp. 13–148.

Britton, G., Liaaen-Jensen, S. & Pfander, H. (eds) (1995) *Spectroscopy*, Birkhäuser Verlag, Basel, Switzerland.

Bugos, R. C. & Yamamoto, H. Y. (1996) Molecular cloning of violaxanthin de-epoxidase from romaine lettuce and expression in *Esherichia coli*. *Proc. Natl. Acad. Sci. USA*, **93**, 6320–6325.

Bugos, R. C., Hieber, A. D. & Yamamoto, H. Y. (1998) Xanthophyll cycle enzymes are members of the lipocalin family, the first identified from plants. *J. Biol. Chem.*, **273**, 15321–15324.

Bungard, R. A., Ruban, A. V., Hibberd, J. M., Press, M. C., Horton, P. & Scholes, J. D. (1999) Unusual carotenoid composition and a new type of xanthophyll cycle in plants. *Proc. Natl. Acad. Sci. USA*, **96**, 1135–1139.

Busch, M., Seuter, A. & Hain, R. (2002) Functional analysis of the early steps of carotenoid biosynthesis in tobacco. *Plant Physiol.*, **128**, 439–453.

Caffarri, S., Croce, R., Breton, J. & Bassi, R. (2001) The major antenna complex of photosystem II has a xanthophyll binding site not involved in light harvesting. *J. Biol. Chem.*, **276**, 35924–35933.

Camara, B. (1993) Plant phytoene synthase complex – component enzymes, immunology, and biogenesis. *Method Enzymol.*, **214**, 352–365.

Carol, P., Stevenson, D., Bisanz, C., Breitenbach, J., Sandmann, G., Mache, R., Coupland, G. & Kuntz, M. (1999) Mutations in the Arabidopsis gene *immutans* cause a variegated phenotype by inactivating a chloroplast terminal oxidase associated with phytoene desaturation. *Plant Cell*, **11**, 57–68.

Cookson, P. J., Kiano, J. W., Shipton, C. A., Fraser, P. D., Römer, S., Schuch, W., Bramley, P. M. & Pyke, K. A. (2003) Increases in cell elongation, plastid compartment size and phytoene synthase activity underlie the phenotype of the high pigment-1 mutant of tomato. *Planta*, **217**, 896–903.

Corona, V., Aracri, B., Kosturkova, G., Bartley, G. E., Pitto, L., Giorgetti, L., Scolnik, P. A. & Giuliano, G. (1996) Regulation of a carotenoid biosynthesis gene promoter during plant development. *Plant J.*, **9**, 505–512.

Cuc Hoa, T. T., Al-Babili, S., Schaub, P., Potrykus, I. & Beyer, P. (2003) Golden Indica and Japonica rice lines amenable to deregulation. *Plant Physiol.*, **133**, 161–169.

Cunningham, F. & Schiff, J. (1985) Photoisomerization of delta-carotene stereoisomers in cells of *Euglena gracillis* mutant W₃BUL and in solution. *Photochem. Photobiol. Sci.*, **42**, 295–307.

Cunningham, F. X. (2002) Regulation of carotenoid synthesis and accumulation in plants. *Pure Appl. Chem.*, **74**, 1409–1417.

Cunningham, F. X. & Gantt, E. (1998) Genes and enzymes of carotenoid biosynthesis in plants. *Annu. Rev. Plant Physiol. Plant Mol. Biol.*, **49**, 557–583.

Cunningham, F. X. & Gantt, E. (2001) One ring or two? Determination of ring number in carotenoids by lycopene epsilon-cyclases. *Proc. Natl. Acad. Sci. USA*, **98**, 2905–2910.

Cunningham, F. X., Chamovitz, D., Misawa, N., Gantt, E. & Hirschberg, J. (1993) Cloning and functional expression in *Escherichia coli* of a cyanobacterial gene for lycopene cyclase, the enzyme that catalyzes the biosynthesis of β-carotene. *FEBS Lett.*, **328**, 130–138.

Cunningham, F. X., Sun, Z. R., Chamovitz, D., Hirschberg, J. & Gantt, E. (1994) Molecular structure and enzymatic function of lycopene cyclase from the cyanobacterium *Synechococcus* sp strain PCC7942. *Plant Cell*, **6**, 1107–1121.

Davison, P. A., Hunter, C. N. & Horton, P. (2002) Overexpression of beta-carotene hydroxylase enhances stress tolerance in Arabidopsis. *Nature*, **418**, 203–206.

Demmig-Adams, B. & Adams, W. W., III (1992) Photoprotection and other responses of plants to high light stress. *Annu. Rev. Plant Physiol. Plant Mol. Biol.*, **43**, 599–626.

Di Baccio, D., Quartacci, M. F., Dalla Vecchia, F., La Rocca, N., Rascio, N. & Navari-Izzo, F. (2002) Bleaching herbicide effects on plastids of dark-grown plants: lipid composition of etioplasts in amitrole and norflurazon-treated barley leaves. *J. Exp. Bot.*, **53**, 1857–1865.

Dowling, J. E. & Wald, G. (1960) The biological function of vitamin A acid. *Proc. Natl. Acad. Sci. USA*, **46**, 587.

Dutton, H. J., Manning, W. M. & Duggar, B. M. (1943) Chl fluorescence and energy transfer in the diatom *Nitzscia closterium*. *J. Phys. Chem*, **47**, 308–317.

Ernst, S. & Sandmann, G. (1988) Poly-*cis*-carotene pathway in the *Scenedesmus* mutant C-6D. *Arch. Microbiol.*, **150**, 590–594.

Estevez, J. M., Cantero, A., Romero, C., Kawaide, H., Jimenez, L. F., Kuzuyama, T., Seto, H., Kamiya, Y. & Leon, P. (2000) Analysis of the expression of *CLA1*, a gene that encodes the 1-deoxyxylulose 5-phosphate synthase of the 2-C-methyl-D-erythritol-4-phosphate pathway in Arabidopsis. *Plant Physiol.*, **124**, 95–103.

Fedtke, C., Depka, B., Schallner, O., Tietjen, K., Trebst, A., Wollweber, D. & Wroblowsky, H.-J. (2001) Mode of action of new diethylamines in lycopene cyclase inhibition and in photosystem II turnover. *Pest Manag. Sci.*, **57**, 278–282.

Frank, H. & Cogdell, R. J. (1993) Photochemistry and function of carotenoids in photosynthesis, in *Carotenoids in Photosynthesis* (eds A. J. Young & G. Britton), Chapman & Hall, London, UK, pp. 253–326.

Fraser, P. D., Schuch, W. & Bramley, P. M. (2000) Phytoene synthase from tomato (*Lycopersicon esculentum*) chloroplasts – partial purification and biochemical properties. *Planta*, **211**, 361–369.

Fraser, P. D., Römer, S., Kiano, J. W., Shipton, C. A., Mills, P. B., Drake, R., Schuch, W. & Bramley, P. M. (2001) Elevation of carotenoids in tomato by genetic manipulation. *J. Sci. Food Agric.*, **81**, 822–827.

Fraser, P. D., Römer, S., Shipton, C. A., Mills, P. B., Kiano, J. W., Misawa, N., Drake, R. G., Schuch, W. & Bramley, P. M. (2002) Evaluation of transgenic tomato plants expressing an additional phytoene synthase in a fruit-specific manner. *Proc. Natl. Acad. Sci. USA*, **99**, 1092–1097.

Fray, R. G., Wallace, A., Fraser, P. D., Valero, D., Hedden, P., Bramley, P. M. & Grierson, D. (1995) Constitutive expression of a fruit phytoene synthase gene in transgenic tomatoes causes dwarfism by redirecting metabolites from the gibberellin pathway. *Plant J.*, **8**, 693–701.

Goldman, I. L. & Breitbach, D. N. (1996) Inheritance of a recessive character controlling reduced carotenoid pigmentation in carrot (*Daucus carota* L). *J. Hered.*, **87**, 380–382.

Govindjee (1999) Carotenoids in photosynthesis: an historical perspective, in *The Photochemistry of Carotenoids*, Vol. 8 (eds H. Frank, A. J. Young, G. Britton & R. J. Cogdell), Kluwer Academic Publishers, Dordrecht, The Netherlands, pp. 1–19.

Green, B. R. & Durnford, D. G. (1996) The chlorophyll-carotenoid proteins of oxygenic photosynthesis. *Annu. Rev. Plant Physiol. Plant Mol. Biol.*, **47**, 685–714.

Griffiths, M., Sistrom, W. R., Cohen-Bazire, G. & Stanier, R. Y. (1955) Function of carotenoids in photosynthesis. *Nature*, **1776**, 1211–1214.

Grunewald, K., Eckert, M., Hirschberg, J. & Hagen, C. (2000) Phytoene desaturase is localized exclusively in the chloroplast and up-regulated at the mRNA level during accumulation of secondary carotenoids in *Haematococcus pluvialis* (Volvocales, Chlorophyceae). *Plant Physiol.*, **122**, 1261–1268.

Gruszecki, W. I., Grudzinski, W., Banaszek-Glos, A., Matula, M., Kernen, P., Krupa, Z. & Sielewiesiuk, J. (1999) Xanthophyll pigments in light-harvesting complex II in monomolecular layers: localisation, energy transfer and orientation. *Biochim. Biophys. Acta-Bioenerg.*, **1412**, 173–183.

Gunning, B. & Jagoe, M. (1967) The prolamellar body, in *Biochemistry of Chloroplasts*, Vol. II (ed. T. Goodwin), Academic Press, London, UK, pp. 655–676.

Hagenimana, V., Anyango-Oyunga, M., Low, J., Njdroge, S. M., Gichuki, S. T. & Kabira, J. (1999) The effects of women farmers' adoption of orange-fleshed sweet potatoes: raising vitamin A intake in Kenya, in *Research Report Series*, International Center for Research on Women, Washington, DC, pp. 1–28.

Hager, A. (1969) Lichtbedingte pH-Erniedrigung in einem Chloroplasten-Kompartiment als Ursache der enzymatischen Violaxanthin-Zeaxanthin-Umwandlung: Beziehung zur Photophosphorylierung. *Planta*, **89**, 224–243.

Hager, A. & Holocher, K. (1994) Localization of the xanthophyll-cycle enzyme violaxanthin de-epoxidase within the thylakoid lumen and abolition of its mobility by a (light-dependent) pH decrease. *Planta*, **192**, 581–589.

Hanley, J., Deligiannakis, Y., Pascal, A., Faller, P. & Rutherford, A. W. (1999) Carotenoid oxidation in photosystem II. *Biochemistry*, **38**, 8189–8195.

Havaux, M. (1998) Carotenoids as membrane stabilizers in chloroplasts. *Trends Plant Sci.*, **3**, 147–151.

Havaux, M. & Niyogi, K. K. (1999) The violaxanthin cycle protects plants from photooxidative damage by more than one mechanism. *Proc. Natl. Acad. Sci. USA*, **96**, 8762–8767.

Havir, E. A. (1992) The *in vivo* and *in vitro* inhibition of catalase from leaves of *Nicotiana sylvestris* by 3-amino-1,2,4-triazole. *Plant Physiol.*, **99**, 533–537.

Heim, D. R. & Larrinua, I. M. (1989) Primary site of action of amitrole in *Arabidopsis thaliana* involves inhibition of root elongation but not of histidine or pigment biosynthesis. *Plant Physiol.*, **91**, 1226–1231.

Hemmerlin, A., Hoeffler, J. F., Meyer, O., Tritsch, D., Kagan, I. A., Grosdemange-Billiard, C., Rohmer, M. & Bach, T. J. (2003) Cross-talk between the cytosolic mevalonate and the plastidial methylerythritol phosphate pathways in tobacco bright yellow-2 cells. *J. Biol. Chem.*, **278**, 26666–26676.

Herrin, D. L., Battey, J. F., Greer, K. & Schmidt, G. W. (1992) Regulation of chlorophyll apoprotein expression and accumlation. Requirements for carotenoids and chlorophyll. *J. Biol. Chem.*, **267**, 8260–8269.

Hirschberg, J. (1998) Molecular biology of carotenoid biosynthesis, in *Biosynthesis and Metabolism*, Vol. 3 (eds G. Britton, S. Liaaen Jensen & H. Pfander), Birkhäuser Verlag, Basel, Switzerland, pp. 149–94.

Hirschberg, J. (2001) Carotenoid biosynthesis in flowering plants. *Curr. Opin. Plant Biol.*, **4**, 210–218.

Hoober, J. K. & Eggink, L. L. (1999) Assembly of light-harvesting complex II and biogenesis of thylakoid membranes in chloroplasts. *Photosynth. Res.,* **61**, 197–215.

Isaacson, T., Ronen, G., Zamir, D. & Hirschberg, J. (2002) Cloning of *tangerine* from tomato reveals a carotenoid isomerase essential for the production of beta-carotene and xanthophylls in plants. *Plant Cell,* **14**, 333–342.

Ishida, B. K., Mahoney, N. E. & Ling, L. C. (1998) Increased lycopene and flavor volatile production in tomato calyces and fruit cultured in vitro and the effect of 2-(4-chlorophenylthio)triethylamine. *J. Agric. Food Chem.,* **46**, 4577–4582.

Jansson, S. (1994) The light-harvesting chlorophyll *a/b*-binding proteins. *Biochim. Biophys. Acta,* **1184**, 1–19.

Jansson, S. (1999) A guide to the Lhc genes and their relatives in Arabidopsis. *Trends Plant Sci.,* **4**, 236–240.

Jilani, A., Kar, S., Bose, S. & Tripathy, B. C. (1996) Regulation of the carotenoid content and chloroplast development by levulinic acid. *Physiol. Plant.,* **96**, 139–145.

Jin, E. S., Feth, B. & Melis, A. (2003) A mutant of the green alga *Dunaliella salina* constitutively accumulates zeaxanthin under all growth conditions. *Biotechnol. Bioeng.,* **81**, 115–124.

Johnson, E. A. & Schroeder, W. A. (1995) Microbial carotenoids. *Adv. Biochem. Eng. Biotechnol.,* **53**, 119–178.

Josse, E. M., Simkin, A. J., Gaffe, J., Laboure, A.-M., Kuntz, M. & Carol, P. (2000) A plastid terminal oxidase associated with carotenoid desaturation during chromoplast differentiation. *Plant Physiol.,* **123**, 1427–1436.

Joyard, J., Teyssier, E., Miege, C., Berny-Seigneurin, D., Marechal, E., Block, M. A., Dorne, A. J., Rolland, N., Ajlani, G. & Douce, R. (1998) The biochemical machinery of plastid envelope membranes. *Plant Physiol.,* **118**, 715–723.

Jyonouchi, H., Sun, S. I. & Gross, M. (1995) Effect of carotenoids on in vitro immunoglobin production by human peripheral blood mononuclear cells: astaxanthin, a carotenoid without vitamin A activity, enhances in vitro immunoglobin production in response to a T-dependent stimulant and antigen. *Nutr. Cancer,* **23**, 171–183.

Kloosterziel, L. K. M., Gil, M. A., Ruijs, G. J., Jacobsen, S. E., Olszewski, N. E., Schwartz, S. H., Zeevaart, J. A. D. & Koornneef, M. (1996) Isolation and characterization of abscisic acid-deficient Arabidopsis mutants at two new loci. *Plant J.,* **10**, 655–661.

Knox, J. & Dodge, A. (1985) Singlet oxygen and plants. *Phytochem.,* **24**, 889–896.

Koornneef, M., Jorner, M. L., Brinkhorst van der Swan, D. L. C. & Karssen, C. M. (1982) The isolation of abscisic acid (ABA) deficient mutants by selection of induced revertants in non-germinating gibberellin sensitive lines of *Arabidopsis thaliana* (L.) Heynh. *Theor. Appl. Genet.,* **61**, 385–393.

Krinsky, N. I. (1971) Function of carotenoids, in *Carotenoids* (ed. O. Isler, Guttman, G., & Solms, U.) Birkhäuser Verlag, Basel, Switzerland, pp. 669–716.

Krubasik, P. & Sandmann, G. (2000) A cartenogenic gene cluster from *Brevibacterium linens* with novel lycopene cyclase genes involved in the synthesis of aromatic carotenoids. *Mol. Gen. Genet.,* **263**, 423–432.

Kuhlbrandt, W., Wang, D. N. & Fujiyoshi, Y. (1994) Atomic model of plant light-harvesting complex by electron crystallography. *Nature,* **367**, 614–621.

Külheim, C., Ågren, J. & Jansson, S. (2002) Rapid regulation of light harvesting and plant fitness in the field. *Science,* **297**, 91–93.

La Rocca, N., Rascio, N., Oster, U. & Rudiger, W. (2001) Amitrole treatment of etiolated barley seedlings leads to deregulation of tetrapyrrole synthesis and to reduced expression of Lhc and RbcS genes. *Planta,* **213**, 101–108.

Lange, B. M., Rujan, T., Martin, W. & Croteau, R. (2000) Isoprenoid biosynthesis: the evolution of two ancient and distinct pathways across genomes. *Proc. Natl. Acad. Sci. USA,* **97**, 13172–13177.

Lange, B. M. & Ghassemian, M. (2003) Genome organization in *Arabidopsis thaliana*: a survey for genes involved in isoprenoid and chlorophyll metabolism. *Plant Mol. Biol.,* **51**, 925–948.

Larkin, R. M., Alonso, J., Ecker, J. & Chory, J. (2003) GUN4, a regulator of chlorophyll synthesis and intracellular signaling. *Science,* **299**, 902–906.

Laule, O., Furholz, A., Chang, H.-S., Zhu, T., Wang, X., Heifetz, P. B., Gruissem, W. & Lange, B. M. (2003) Crosstalk between cytosolic and plastidial pathways of isoprenoid biosynthesis in *Arabidopsis thaliana*. *Proc. Natl. Acad. Sci. USA,* **100**, 6866–6871.

Lee, H. S. (2001) Characterisation of carotenoids in juice of red Navel orange (Cara Cara). *J. Agric. Food Chem.*, **49**, 2563–2568.

Lee, P. C., Momen, A. Z. R., Mijts, B. N. & Schmidt-Dannert, C. (2003) Biosynthesis of structurally novel carotenoids in *Escherichia coli*. *Chem. Biol.*, **10**, 453–462.

Li, L., Paolillo, D. J., Parthasarathy, M. V., DiMuzio, E. M. & Garvin, D. F. (2001) A novel gene mutation that confers abnormal patterns of β-carotene accumulation in cauliflower (*Brassica oleracea* var. botrytis). *Plant J.*, **26**, 59–67.

Li, X. P., Bjorkman, O., Shih, C., Grossman, A. R., Rosenquist, M., Jansson, S. & Niyogi, K. K. (2000) A pigment-binding protein essential for regulation of photosynthetic light harvesting. *Nature*, **403**, 391–395.

Lichtenthaler, H. K. (1999) The 1-deoxy-D-xylulose-5-phosphate pathway of isoprenoid biosynthesis in plants. *Annu. Rev. Plant Physiol. Plant Mol. Biol.*, **50**, 47–65.

Lichtenthaler, H. K., Schwender, J., Disch, A. & Rohmer, M. (1997) Biosynthesis of isoprenoids in higher plant chloroplasts proceeds via a mevalonate independent pathway. *FEBS Lett.*, **400**, 271–274.

Lindgren, L., Stahlberg, K. G. & Hoglund, A.-S. (2003) Seed-specific overexpression of an endogenous Arabidopsis phytoene synthase gene results in delayed germination and increased levels of carotenoids, chlorophyll, and abscisic acid. *Plant Physiol.*, **132**, 779–785.

Lokstein, H., Tian, L., Polle, J. E. W. & DellaPenna, D. (2002) Xanthophyll biosynthetic mutants of Arabidopsis thaliana: altered nonphotochemical quenching of chlorophyll fluorescence is due to changes in Photosystem II antenna size and stability. *Biochim. Biophys. Acta-Bioenerg.*, **1553**, 309–319.

Lorenz, R. T. & Cysewski, G. R. (2000) Commercial potential for *Haematococcus* microalgae as a natural source of astaxanthin. *Trends Biotechnol.*, **18**, 160–167.

Mann, V., Harker, M., Pecker, I. & Hirschberg, J. (2000) Metabolic engineering of astaxanthin production in tobacco flowers. *Nat. Biotechnol.*, **18**, 888–892.

Marin, E., Nussaume, L., Quesada, A., Gonneau, M., Sotta, B., Hugueney, P., Frey, A. & Marion-Poll, A. (1996) Molecular identification of zeaxanthin epoxidase of *Nicotiana plumbaginifolia*, a gene involved in abscisic acid and corresponding to the ABA locus of *Arabidopsis thaliana*. *EMBO J.*, **15**, 2331–2342.

Matsubara, S., Morosinotto, T., Bassi, R., Christian, A.-L., Fischer-Schliebs, E., Luttge, U., Orthen, B., Franco, A. C., Scarano, F. R., Forster, B., Pogson, B. J. & Osmond, C. B. (2003) Occurence of the lutein-epoxide cycle in mistletoes of the Loranthaceae and Viscaceae. *Planta*, **217**, 868–879.

Mayer, M. P., Nievelstein, V. & Beyer, P. (1992) Purification and characterization of a NADPH dependent oxidoreductase from chromoplasts of *Narcissus pseudonarcissus*: a redox-mediator possibly involved in carotene desaturation. *Plant Physiol. Biochem.*, **30**, 389–398.

Milborrow, B. V., Nonhebel, H. M. & Willows, R. D. (1988) 2,7-Dimethylocta-2,4-dienedioic acid is not a by-product of abscisic acid biosynthesis. *Plant Sci.*, **56**, 49–53.

Misawa, N., Kajiwara, S., Kondo, K., Yokoyama, A., Satomi, Y., Saito, T., Miki, W. & Ohtani, T. (1995a) Canthaxanthin biosynthesis by the conversion of methylene to keto groups in a hydrocarbon beta-carotene by a single gene. *Biochem. Biophys. Res. Commun.*, **209**, 867–876.

Misawa, N., Masamoto, K., Hori, T., Ohtani, T., Boger, P. & Sandmann, G. (1994) Expression of an Erwinia phytoene desaturase gene not only confers multiple resistance to herbicides interfering with carotenoid biosynthesis but also alters xanthophyll metabolism in transgenic plants. *Plant J.*, **6**, 481–489.

Misawa, N., Nakagawa, M., Kobayashi, K., Yamano, S., Izawa, Y., Nakamura, K. & Harashima, K. (1990) Elucidation of the *Erwinia uredovora* carotenoid biosynthetic pathway by functional analysis of gene products expressed in *Escherichia coli*. *J. Bacteriol.*, **172**, 6704–6712.

Misawa, N., Satomi, Y., Kondo, K., Yokoyama, A., Kajiwara, S., Saito, T., Ohtani, T. & Miki, W. (1995b) Structure and functional analysis of a marine bacterial carotenoid biosynthesis gene cluster and astaxanthin biosynthetic pathway proposed at the gene level. *J. Bacteriol.*, **177**, 6575–6584.

Moehs, C. P., Tian, L., Osteryoung, K. W. & DellaPenna, D. (2001) Analysis of carotenoid biosynthetic gene expression during marigold petal development. *Plant Mol. Biol.*, **45**, 281–293.

Nievelstein, V., Vanderkerckhove, J., Tadros, M. H., Lintig, J. V., Nitschke, W. & Beyer, P. (1995) Carotene desaturation is linked to a respiratory redox pathway in *Narcissus pseudonarcissus* chromoplast membranes involvement of a 23-kDa oxygen-evolving-complex-like protein. *Eur. J. Biochem.*, **233**, 864–872.

Niyogi, K. K. (1999) Photoprotection revisited: genetic and molecular approaches. *Annu. Rev. Plant Physiol. Plant Mol. Biol.*, **50**, 333–359.

Niyogi, K. K., Björkman, O. & Grossman, A. R. (1997a) Chlamydomonas xanthophyll cycle mutants identified by video imaging of chlorophyll fluorescence quenching. *Plant Cell*, **9**, 1369–1380.

Niyogi, K. K., Björkman, O. & Grossman, A. R. (1997b) The roles of specific xanthophylls in photoprotection. *Proc. Natl. Acad. Sci. USA*, **94**, 14162–14167.

Niyogi, K. K., Grossman, A. R. & Björkman, O. (1998) Arabidopsis mutants define a central role for the xanthophyll cycle in the regulation of photosynthetic energy conversion. *Plant Cell*, **10**, 1121–1134.

Niyogi, K. K., Shih, C., Chow, W. S., Pogson, B. J., DellaPenna, D. & Björkman, O. (2001) Photoprotection in a zeaxanthin- and lutein-deficient double mutant of Arabidopsis. *Photosynth. Res.*, **67**, 139–145.

Norris, S. R., Barrette, T. R. & DellaPenna, D. (1995) Genetic dissection of carotenoid synthesis in Arabidopsis defines plastoquinone as an essential component of phytoene desaturation. *Plant Cell*, **7**, 2139–2149.

Park, H., Kreunen, S. S., Cuttriss, A. J., DellaPenna, D. & Pogson, B. J. (2002) Identification of the carotenoid isomerase provides insight into carotenoid biosynthesis, prolamellar body formation, and photomorphogenesis. *Plant Cell*, **14**, 321–332.

Paulsen, H. (1999) Carotenoids and the assembly of light-harvesting complexes., in *The Photochemistry of Carotenoids*, Vol. 8 (eds H. Frank, A. Young, G. Britton & R. Cogdell), Kluwer Academic Publishers, Amsterdam, The Netherlands, pp. 123–135.

Perry, K. L., Simonitch, T. A., Harrisonlavoie, K. J. & Liu, S. T. (1986) Cloning and regulation of *Erwinia herbicola* pigment genes. *J. Bacteriol.*, **168**, 607–612.

Pfundel, E. E., Renganathan, M., Gilmore, A. M., Yamamoto, H. Y. & Dilley, R. A., (1994) Intrathylakoid pH in isolated pea chloroplasts as probed by violaxanthin deepoxidation. *Plant Physiol.*, **106**, 1647–1658.

Phillip, D. & Young, A. J. (1995) Occurrence of the carotenoid *lactucaxantin* in higher plant LHC II. *Photosynth. Res.*, **43**, 273–282.

Plumley, F. G. & Schmidt, G. W. (1987) Reconstitution of chlorophyll *a/b* light-harvesting complexes: xanthophyll-dependent assembly and energy transfer. *Proc. Natl. Acad. Sci. USA*, **84**, 146–150.

Pogson, B., McDonald, K. A., Truong, M., Britton, G. & DellaPenna, D. (1996) Arabidopsis carotenoid mutants demonstrate that lutein is not essential for photosynthesis in higher plants. *Plant Cell*, **8**, 1627–1639.

Pogson, B. J., Niyogi, K. K., Björkman, O. & DellaPenna, D. (1998) Altered xanthophyll compositions adversely affect chlorophyll accumulation and nonphotochemical quenching in Arabidopsis mutants. *Proc. Natl. Acad. Sci. USA*, **95**, 13324–13329.

Pogson, B. J. & Rissler, H. M. (2000) Genetic manipulation of carotenoid biosynthesis and photoprotection. *Philos. Trans. R. Soc. Lond. Ser. B-Biol. Sci.*, **355**, 1395–1403.

Powls, R. & Britton, G. (1977) The roles of isomers of phytoene, phytofluene and zeta-carotene in carotenoid biosynthesis by a mutant strain of *Scenedesmus obliquus*. *Arch. Microbiol.*, **115**, 175–179.

Ridley, S. M. (1981) Carotenoids and herbicide action, in *Carotenoid Chemistry and Biochemistry* (eds G. Britton & T. W. Goodwin), Pergamon, Oxford, UK, pp. 353–369.

Rissler, H. M. & Pogson, B. J. (2001) Antisense inhibition of the beta-carotene hydroxylase enzyme in Arabidopsis and the implications for carotenoid accumulation, photoprotection and antenna assembly. *Photosynth. Res.*, **67**, 127–137.

Rodrigo, M.-J., Marcos, J. F., Alférez, F., Mallent, M. D. & Zacarías, L. (2003) Characterization of Pinalate, a novel *Citrus sinensis* mutant with a fruit specific alteration that results in yellow pigmentation and decreased ABA content. *J. Exp. Bot.*, **54**, 727–738.

Rodriguez-Concepcion, M. & Gruissem, W. (1999) Arachidonic acid alters tomato HMG expression and fruit growth and induces 3-hydroxy-3-methylglutaryl coenzyme A reductase-independent lycopene accumulation. *Plant Physiol.*, **119**, 41–48.

Römer, S., Fraser, P. D., Kiano, J. W., Shipton, C. A., Misawa, N., Schuch, W. & Bramley, P. M. (2000) Elevation of the provitamin A content of transgenic tomato plants. *Nat. Biotechnol.*, **18**, 666–669.

Römer, S., Lubeck, J., Kauder, F., Steiger, S., Adomat, C. & Sandmann, G. (2002) Genetic engineering of a zeaxanthin-rich potato by antisense inactivation and co-suppression of carotenoid epoxidation. *Metab. Eng.*, **4**, 263–272.

Ronen, G., Cohen, M., Zamir, D. & Hirschberg, J. (1999) Regulation of carotenoid biosynthesis during tomato fruit development: expression of the gene for lycopene epsilon-cyclase is down-regulated during ripening and is elevated in the mutant *Delta. Plant J.*, **17**, 341–351.

Ronen, G., Carmel-Goren, L., Zamir, D. & Hirschberg, J. (2000) An alternative pathway to beta-carotene formation in plant chromoplasts discovered by map-based cloning of *Beta* and *old-gold* color mutations in tomato. *Proc. Natl. Acad. Sci. USA*, **97**, 11102–11107.

Rosati, C., Aquilani, R., Dharmapuri, S., Pallara, P., Marusic, C., Tavazza, R., Bouvier, F., Camara, B. & Giuliano, G. (2000) Metabolic engineering of beta-carotene and lycopene content in tomato fruit. *Plant J.*, **24**, 413–419.

Ruban, A. V., Lee, P. J., Wentworth, M., Young, A. J. & Horton, P. (1999) Determination of the stoichiometry and strength of binding of xanthophylls to the photosystem II light harvesting complexes. *J. Biol. Chem.*, **274**, 10458–10465.

Ruban, A. V., Young, A. J., Pascal, A. A. & Horton, P. (1994) The effects of illumination on the xanthophyll composition of the photosystem II light-harvesting complexes of spinach thylakoid membranes. *Plant Physiol.*, **104**, 227–234.

Sandmann, G. (2001) Genetic manipulation of carotenoid biosynthesis: strategies, problems and achievements. *Trends Plant Sci.*, **6**, 14–17.

Satoh, K. (1993) Isolation and properties of the photosystem II reaction center, in *The Photosynthetic Reaction Center* (eds J. Deisenhofer & J. Norris), Academic Press, San Diego, USA, pp. 289–318.

Schledz, M., Al-Babili, S., von Lintig, J., Haubruck, H., Rabbani, S., Kleinig, H. & Beyer, P. (1996) Phytoene synthase from *Narcissus pseudonarcissus*: functional expression, galactolipid requirement, topological distribution in chromoplasts and induction during flowering. *Plant J.*, **10**, 781–792.

Schmidt-Dannert, C., Umeno, D. & Arnold, F. H. (2000) Molecular breeding of carotenoid biosynthetic pathways. *Nat. Biotechnol.*, **18**, 750–753.

Schnurr, G., Misawa, N. & Sandmann, G. (1996) Expression, purification and properties of lycopene cyclase from *Erwinia uredovora. Biochem. J.*, **315**, 869–874.

Schwartz, S. H., Qin, X. Q. & Zeevaart, J. A. D. (2001) Characterization of a novel carotenoid cleavage dioxygenase from plants. *J. Biol. Chem.*, **276**, 25208–25211.

Schwartz, S. H., Tan, B. C., Gage, D. A., Zeevaart, J. A. D. & McCarty, D. R. (1997) Specific oxidative cleavage of carotenoids by VP14 of maize. *Science*, **276**, 1872–1874.

Schwartz, S. H., Tan, B. C., McCarty, D. R., Welch, W. & Zeevaart, J. A. D. (2003) Substrate specificity and kinetics for VP14, a carotenoid cleavage dioxygenase in the ABA biosynthetic pathway. *Biochim. Biophys. Acta-Gen. Subj.*, **1619**, 9–14.

Seddon, J. M., Ajani, U. A., Sperduto, R. D., Hiller, R., Blair, N., Burton, T. C., Farber, M. D., Gragoudas, E. S., Haller, J., Miller, D. T., Yannuzzi, L. A. & Willet, W. (1994) Dietary carotenoids, vitamins A, C, and E, and advanced age-related macular degeneration. *J. Am. Med. Assoc.*, **272**, 1413–1420.

Seibert, M. (1993) Biochemical, biophysical, and structural characterization of the photosystem II reaction center complex, in *The Photosynthetic Reaction Center*, Vol. 1 (eds J. Deisenhofer & J. R. Norris), Academic Press, San Diego, USA, pp. 319–356.

Seo, M. & Koshiba, T. (2002) Complex regulation of ABA biosynthesis in plants. *Trends Plant Sci.*, **7**, 41–48.

Shewmaker, C. K., Sheehy, J. A., Daley, M., Colburn, S. & Ke, D. Y. (1999) Seed-specific overexpression of phytoene synthase: increase in carotenoids and other metabolic effects. *Plant J.*, **20**, 401–412.

Siefermann, D. & Yamamoto, H. Y. (1975) Light-induced de-epoxidation of violaxanthin in lettuce chloroplasts. *Biochim. Biophys. Acta*, **387**, 149–158.

Siefermann-Harms, D., Joyard, J. & Douce, R. (1978) Light-induced changes of the carotenoid levels in the chloroplast envelopes. *Plant Physiol.*, **61**, 530–533.

Simkin, A. J., Zhu, C. F., Kuntz, M. & Sandmann, G. (2003) Light-dark regulation of carotenoid biosynthesis in pepper (*Capsicum annuum*) leaves. *J. Plant Physiol.*, **160**, 439–443.

Sommer, A. & Davidson, F. R. (2002) Assessment and control of vitamin A deficiency: the Annecy Accords. *J. Nutr.*, **132**, 2845S–2850S.

Sorefan, K., Booker, J., Haurogne, K., Goussot, M., Bainbridge, K., Foo, E., Chatfield, S., Ward, S., Beveridge, C., Rameau, C. & Leyser, O. (2003) *MAX4* and *RMS1* are orthologous dioxygenase-like genes that regulate shoot branching in Arabidopsis and pea. *Gene. Dev.*, **17**, 1469–1474.

Steinbrenner, J. & Linden, H. (2003) Light induction of carotenoid biosynthesis genes in the green alga *Haematococcus pluvialis*: regulation by photosynthetic redox control. *Plant Mol. Biol.*, **52**, 343–356.

Stickforth, P., Steiger, S., Hess, W. R. & Sandmann, G. (2003) A novel type of lycopene epsilon-cyclase in the marine cyanobacterium *Prochlorococcus marinus* MED4. *Arch. Microbiol.*, **179**, 409–415.

Strand, A., Asami, T., Alonso, J., Ecker, J. R. & Chory, J. (2003) Chloroplast to nucleus communication triggered by accumulation of Mg-protoporphyrinIX. *Nature*, **421**, 79–83.

Sun, W. H., Verhoeven, A. S., Bugos, R. C. & Yamamoto, H. Y. (2001) Suppression of zeaxanthin formation does not reduce photosynthesis and growth of transgenic tobacco under field conditions. *Photosyn. Res.*, **67**, 41–50.

Sun, Z. R., Gantt, E. & Cunningham, F. X. (1996) Cloning and functional analysis of the beta-carotene hydroxylase of *Arabidopsis thaliana*. *J. Biol. Chem.*, **271**, 24349–24352.

Susek, R. E., Ausubel, F. M. & Chory, J. (1993) Signal transduction mutants of Arabidopsis uncouple nuclear CAB and RBCS gene expression from chloroplast development. *Cell*, **74**, 787–799.

Tan, B. C. (1997) Genetic control of abscisic acid biosynthesis in maize. *Proc. Natl. Acad. Sci. USA*, **94**, 12235–12240.

Tan, B. C., Joseph, L. M., Deng, W. T., Liu, L., Li, Q. B., Cline, K. & McCarty, D. R. (2003) Molecular characterisation of the Arabidopsis 9-*cis*- epoxycarotenoid dioxygenase gene family. *Plant J.*, **35**, 44–56.

Tardy, F. & Havaux, M. (1997) Thylakoid membrane fluidity and thermostability during the operation of the xanthophyll cycle in higher-plant chloroplasts. *Biochim. Biophys. Acta*, **1330**, 179–193.

Taylor, H. F. & Smith, T. A. (1967) Production of plant growth inhibitors from xanthophylls: a possible source of dormin. *Nature*, **215**, 1513–1514.

Telfer, A. (2002) What is beta-carotene doing in the photosystem II reaction centre? *Philos. Trans. R. Soc. Lond. Ser. B-Biol. Sci.*, **357**, 1431–1439.

Tian, L. & DellaPenna, D. (2001) Characterization of a second carotenoid beta-hydroxylase gene from Arabidopsis and its relationship to the *LUT1* locus. *Plant Mol. Biol.*, **47**, 379–388.

Tian, L., Magallanes-Lundback, M., Musetti, V. & DellaPenna, D. (2003) Functional analysis of beta- and epsilon-ring carotenoid hydroxylases in Arabidopsis. *Plant Cell*, **15**, 1320–1332.

Tian, L., Musetti, V., Kim, J., Magallanes-Lundback, M. & DellaPenna, D. (2004) The Arabidopsis *LUT1* locus encodes a member of the cytochrome P450 family that is required for carotenoid ε-ring hydroxylation activity *Proc. Natl. Acad. Sci. USA*, **101**, 402–407.

Tomes, M. L., Quackenbush, F. L., Nelsom, O. E. & North, B. (1953) The inheritance of carotenoid pigment systems in tomato. *Genetics*, **38**, 117–127.

Verhoeven, A. S., Adams, W. W., Demmig-Adams, B., Croce, R. & Bassi, R. (1999) Xanthophyll cycle pigment localization and dynamics during exposure to low temperatures and light stress in *Vinca major*. *Plant Physiol.*, **120**, 727–737.

Verhoeven, A. S., Bugos, R. C. & Yamamoto, H. Y. (2001) Transgenic tobacco with suppressed zeaxanthin formation is susceptible to stress-induced photoinhibition. *Photosynth. Res.*, **67**, 27–39.

von Lintig, J., Welsch, R., Bonk, M., Giuliano, G., Batschauer, A. & Kleinig, H. (1997) Light-dependent regulation of carotenoid biosynthesis occurs at the level of phytoene synthase expression and is mediated by phytochrome in *Sinapis alba* and *Arabidopsis thaliana* seedlings. *Plant J.*, **12**, 625–634.

von Lintig, J. & Vogt, K. (2000) Molecular identification of an enzyme cleaving β-carotene to retinal. *J. Biol. Chem.*, **275**, 11915–11920.

Vrettos, J. S., Stewart, D. H., de Paula, J. C. & Brudvig, G. W. (1999) Low-temperature optical and resonance Raman spectra of a carotenoid cation radical in photosystem II. *J. Phys. Chem. B.*, **103**, 6403–6406.

Wahlberg, I. & Eklund, A.-M. (1998) Degraded carotenoids, in *Carotenoids: Biosynthesis and Metabolism*, Vol. III (eds. G. Britton, S. Liaaen Jensen & H. Pfander), Birkhäuser Verlag, Basel, Switzerland, pp. 195–216.

Wald, G. (1968) The molecular basis of visual excitation. *Nature*, **219**, 800–807.

Wang, C. W. & Liao, J. C. (2001) Alteration of product specificity of *Rhodobacter sphaeroides* phytoene desaturase by directed evolution. *J. Biol. Chem.*, **276**, 41161–41164.

Welsch, R., Beyer, P., Hugueney, P., Kleinig, H. & von Lintig, J. (2000) Regulation and activation of phytoene synthase, a key enzyme in carotenoid biosynthesis, during photomorphogenesis. *Planta*, **211**, 846–854.

Welsch, R., Medina, J., Giuliano, G., Beyer, P. & von Lintig, J. (2003) Structural and functional characterization of the phytoene synthase promoter from *Arabidopsis thaliana*. *Planta*, **216**, 523–534.

Wetzel, C. M. & Rodermel, S. R. (1998) Regulation of phytoene desaturase expression is independent of leaf pigment content in *Arabidopsis thaliana*. *Plant Mol. Biol.*, **37**, 1045–1053.

Woitsch, S. & Römer, S. (2003) Expression of xanthophyll biosynthetic genes during light-dependant chloroplast differentiation. *Plant Physiol.*, **132**, 1508–1517.

Wu, D. Y., Wright, D. A., Wetzel, C., Voytas, D. F. & Rodermel, S. (1999) The immutans variegation locus of Arabidopsis defines a mitochondrial alternative oxidase homolog that functions during early chloroplast biogenesis. *Plant Cell*, **11**, 43–55.

Wyss, A., Wirtz, G., Woggon, W., Brugger, R., Wyss, M., Friedlein, A., Bachmann, H. & Hunziker, W. (2000) Cloning and expression of β,β-carotene 15,15′-dioxygenase. *Biochem. Biophys. Res. Comm.*, **271**, 334–336.

Yamamoto, H. Y. (1962) Studies on the light and dark interconversions of leaf xanthophylls. *Biochim. Biophys. Acta*, **97**, 168–173.

Yamamoto, H. Y. (1979) Biochemistry of the xanthophyll cycle in higher plants. *Pure Appl. Chem.*, **51**, 639–648.

Ye, X. D., Al-Babili, S., Kloti, A., Zhang, J., Lucca, P., Beyer, P. & Potrykus, I. (2000) Engineering the provitamin A (beta-carotene) biosynthetic pathway into (carotenoid-free) rice endosperm. *Science*, **287**, 303–305.

Young, A. J. (1993) Factors that affect the carotenoid composition of higher plants and algae, in *Carotenoids in Photosynthesis* (eds A. J. Young & G. Britton), Chapman & Hall, London, UK, pp. 161–205.

Zeidler, J., Schwender, J., Muller, C., Wiesner, J., Weidemeyer, C., Beck, E., Jomaa, H. & Lichtenthaler, H. K. (1998) Inhibition of the non-mevalonate 1-deoxy-D-xylulose-5-phosphate pathway of plant isoprenoid biosynthesis by fosmidomycin. *Z. Naturforsch. (C)*, **53**, 980–986.

4 Flavonoids

Kathy E. Schwinn and Kevin M. Davies

4.1 Introduction

Flavonoids serve functions in plants as diverse as signalling to micro-organisms, protecting against pathogens, ameliorating biotic and abiotic stress, influencing auxin transport and enabling plant fertility. But their most obvious and best-characterised role is to provide floral visual cues for insect and animal pollinators. They are the most important floral pigments, occurring throughout the angiosperms and providing most colours in the visible spectrum.

Flower colour and, by extension, flavonoids are among the oldest subjects in formal plant science. In 1664, it was found that the purple pigment from *Viola* was a natural pH indicator (Boyle, 1664) and in 1835 the term 'anthocyan' was coined for the most important flavonoid pigment (Swain, 1976), a derivative of which is used today. Since then, floral pigmentation has been used as a model not only to help elucidate fundamental genetic principles, done by groundbreaking researchers such as Mendel, Darwin and McClintock, but also to contribute to our understanding of plant molecular biology and gene regulation. Further details on the history of flavonoid research can be found in Chapter 1 as well as in Swain (1976) and Bohm (1998). The long history of research means that present knowledge of the chemistry and biosynthesis of flavonoids is perhaps more advanced than for any other group of plant secondary metabolites.

Given the scientific interest in flavonoids that has spanned the last 100 years or so, there is an overwhelming amount of literature on these pigments. Thus, of course, this chapter is not meant to be exhaustive in its coverage of the subject. Our focus is on aspects of the chemistry and biosynthesis of flavonoids that are important to floral pigmentation, and approaches to modifying their production in transgenic plants. Human and animal health aspects of flavonoids feature in Chapters 5 and 8, and techniques for flavonoid structural analysis are covered in Chapter 10. Furthermore, we have confined ourselves, where possible, to referencing only key research papers and reviews, and we apologise to those researchers whose work we do not reference directly. There are more extensive reviews available on all aspects of flavonoids, including their history, chemistry, biochemistry and biology, and we recommend in particular the books of Harborne (1994) and Bohm (1998).

4.2 Flavonoids as pigments

A full understanding of flavonoid-based flower colour and its extraordinary diversity begins with an understanding of the chemistry of these pigments. Flavonoid pigmentation is a complex process involving pigments that are far from inert molecules. Indeed, while in most flowers colour begins with the production and accumulation of flavonoid chromophores, other factors, both intrinsic and extrinsic, come into play that determine the actual colour that is manifested by the flower. Flavonoid pigmentation is predominantly based on three components: the primary structures of the flavonoids, secondary structures of these molecules due to pH and other effects, and tertiary structures arising from self-association and inter- and intramolecular interactions. In some species, an additional component, metal complexation, is required for blue floral pigmentation.

4.2.1 Primary structures

Flavonoids have a 15-carbon (C_{15}) base structure comprised of two phenyl rings (called the A- and B-rings) connected by a three-carbon bridge that usually forms a third ring (called the C-ring) (Fig. 4.1). The various classes of flavonoids are determined by the degree of oxidation of the C-ring (Fig. 4.2). Despite the similarities in structure, only some flavonoids have the ability to absorb light in the visible region of the spectrum (between 400 and 800 nm) and are thus pigments. Pigment molecules are characterised by having unbound or loosely bound electrons. For such molecules, the energy required for excitation of the

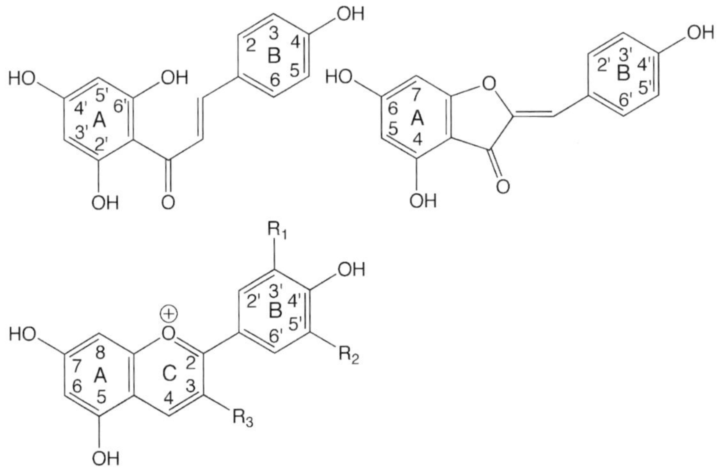

Fig. 4.1 Base structures of a chalcone (top left), an aurone (top right) and the main anthocyanidins (bottom). The lettering of the carbon rings is shown, as well as the numbering of the key carbons. Note that the numbering is different for each type of flavonoid. For the majority of flavonoid types the numbering is as for the anthocyanidins, and for the cinnamic acids it is as for the chalcones. R_1, R_2 and R_3 substitutions determine the various common anthocyanidins. The common 3-hydroxyanthocyanidins ($R_3 = OH$) are pelargonidin ($R_1 = H$ and $R_2 = H$), cyanidin ($R_1 = OH$ and $R_2 = H$), delphinidin ($R_1 = OH$ and $R_2 = OH$), peonidin ($R_1 = OCH_3$ and $R_2 = H$), petunidin ($R_1 = OCH_3$ and $R_2 = OH$) and malvidin ($R_1 = OCH_3$ and $R_2 = OCH_3$). The rare 3-deoxyanthocyanidins ($R_3 = H$) are apigeninidin ($R_1 = H$ and $R_2 = H$), luteolinidin ($R_1 = OH$ and $R_2 = H$) and tricetinidin ($R_1 = OH$ and $R_2 = OH$).

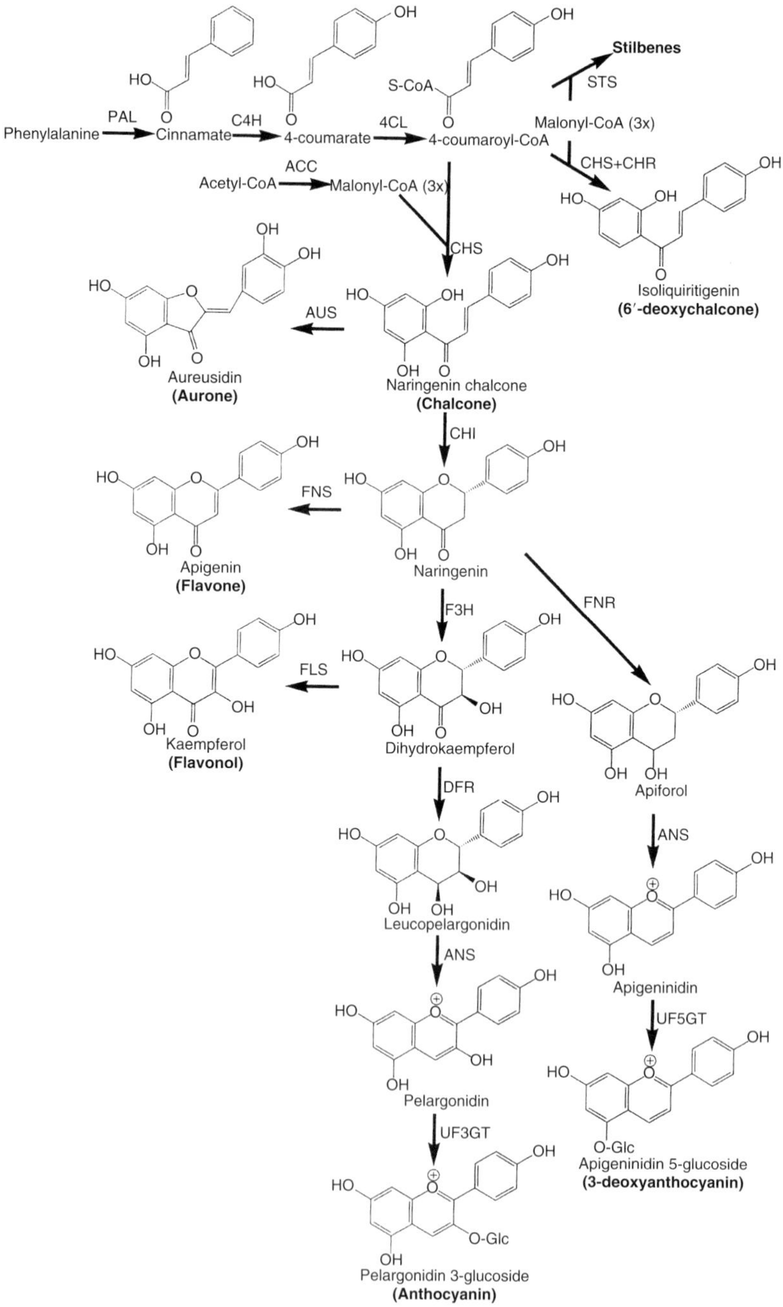

Fig. 4.2 Schematic of a section of the flavonoid biosynthetic pathway. Enzyme abbreviations are as given in the text. Only the routes to the production of flavonoids with 4′-hydroxylation are shown. The product of the ANS is shown in the traditional cation form (rather than the pseudobase).

electrons to a higher energy level is lowered, allowing the molecule to be energised by light within the visible range. The colour of a pigment is determined by the particular wavelengths of visible light that are absorbed by the molecule and those that are reflected or scattered. Key structural features of a flavonoid pigment are the degrees of double bond conjugation and oxygenation (in the form of hydroxylation). Increasing either causes light at longer wavelengths to be absorbed. The predominant flavonoid pigments are the anthocyanins, while chalcones, aurones and flavonols serve a more limited role.

4.2.1.1   *Anthocyanins*

Anthocyanins are the most abundant and widespread of the flavonoid pigments. They absorb light at the longest wavelengths, and are the basis for most orange, pink, red, magenta, purple, blue and blue-black floral colours. Key to providing such colour diversity is the degree of oxygenation of the anthocyanidins (the central chromophores of the anthocyanins) and the nature and number of substituents (e.g. sugar moieties) added to these chromophores.

At a primary level, the degree of oxygenation of the B-ring has the greatest impact on the colour of anthocyanin pigments. Most anthocyanins are derived from just three basic anthocyanidin types: pelargonidin, cyanidin and delphinidin. The difference between them is in the number of hydroxyl groups on the B-ring (Fig. 4.1). An increased number of hydroxyl groups on this ring has a blueing effect on the colour manifested by the anthocyanin. In general, there is a strong correlation between the flower colour and the predominant type of anthocyanin that accumulates. Orange and pink colours tend to be based on pelargonidin derivatives, magenta colours on cyanidin derivatives and purple and blue colours on delphinidin derivatives (Harborne, 1976). Flowers can also accumulate mixtures of anthocyanin types, providing further variation in colour.

A hydroxyl group or rather the lack thereof at the C-3 position in the C-ring also dramatically influences the colour of the pigment. The common anthocyanidins have 3-hydroxylation. 3-Deoxyanthocyanidins lack this hydroxyl group (Fig. 4.1), and show a marked difference in wavelength absorbance, with the derived pigments giving yellow, orange and bright red flower colours. 3-Deoxyanthocyanins are relatively rare. They are characteristic of New World species of the Gesneriaceae (e.g. the ornamental *Sinningia cardinalis*), but have also been found sporadically across the plant kingdom. There are reports of these pigments in some species of the Poaceae (*Zea mays* (maize) and *Sorghum bicolor*), *Camellia*, *Chiranthodendron*, as well as 'lower' plants (ferns and mosses) (Bohm, 1998).

Willstätter and Everest identified the first anthocyanin in 1913, from the blue cornflower *Centaurea cyanus* (cited in Bohm, 1998). Since then approximately 630 different anthocyanins have been structurally defined (see also Chapter 10). Secondary modifications to the core anthocyanidins are the basis for these diverse structures. In some cases, very complex anthocyanins may be formed, with multiple glycosyl and acyl groups, and an example of one of these complex structures is shown in Fig. 4.3.

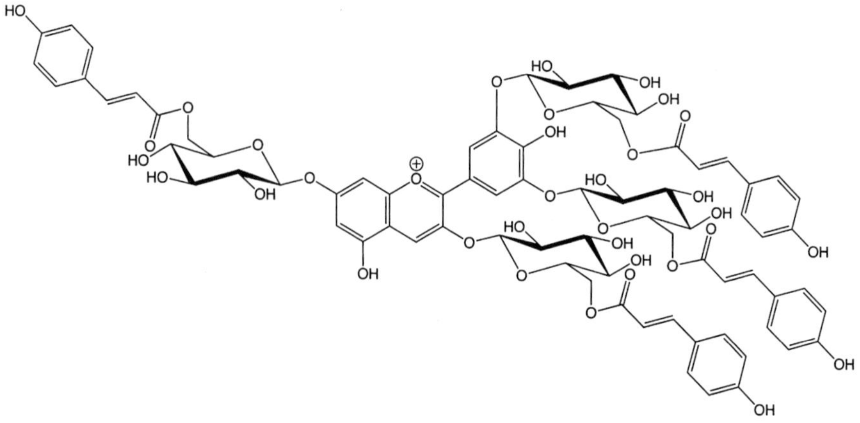

Fig. 4.3 Example of a complex anthocyanin structure; delphinidin 3,7,3′,5′-tetra-(6-*O*-*p*-coumaroyl-β-gluco-side) from *Dianella* berries (Bloor, 2001).

Anthocyanin modification typically involves *O*-glycosylation, *O*-acylation and *O*-methylation. Substitutions are targeted to one or more of the hydroxyl groups of the chromophore or to substituents attached to these groups. The hydroxylation pattern for most anthocyanins includes the C-3 position of the C-ring, the C-5 and C-7 positions of the A-ring and the C-4′ position of the B-ring (Fig. 4.1). This pattern is reflective of the precursors that are condensed to form the first product of the flavonoid pathway. Anthocyanins commonly have an additional hydroxyl group at the C-3′ position or at both the C-3′ and C-5′ positions. We will now discuss the impact of specific modifications on anthocyanin and flower colour.

O-glycosylation is the first modification to occur. For most anthocyanidins the initial glycosylation occurs at the C-3 position, while for the 3-deoxyanthocyani-dins it is the C-5 hydroxyl group that is targeted. Other glycosylations may then follow, either at other positions of the anthocyanin nucleus or to sugar moieties that are already attached. Glucose is the most common sugar that is attached, but other sugars such as galactose, glucuronic acid and rhamnose are also found.

The nature of a sugar residue itself does not affect the colour of the anthocyanin, although it may affect subsequent modifications. Rather it is the positions of glycosylation that are important. Most anthocyanins are 3-glycosides or 3,5-diglycosides, and in some ornamentals (e.g. *Pelargonium*, *Punica* and *Matthiola*) the latter produce a more intense colour than the corresponding 3-glycosides (Harborne, 1976; Forkmann, 1991). Other glycosylation patterns, although less common, can also have a large impact on colour. Anthocyanin 3′-*O*-glucosylation to give 3,5,3′-triglycosides in the *Bromeliaceae* causes a shift to brighter red colours (Saito & Harborne, 1983). 4′-*O*-glucosylation of cyanidin in *Allium cepa* (onion) shifts the spectral properties of the molecule towards those of pelargonidin 3-glycosides (Fossen *et al.*, 2003). In poppy (*Papaver* species), a change from 3- to 3,7-diglycosylation of pelargonidin shifts the flower colour from scarlet to orange/yellow (Harborne, 1976). Furthermore, in an unrelated study, 7-O-glyco-sylation (in a 3,7-di-O-glucoside) appeared to cause a profound reduction in the molar extinction coefficient of the pigment (Figueiredo *et al.*, 1999). The effect was nullified when glycosylation at the 7-hydroxyl was more complex.

Glycosylation position effects in many cases are tied to the effect of subsequent O-acylation. O-Acylation, a prevalent modification, is the addition of acid (acyl) groups to the sugar residues of the anthocyanin, and anthocyanins may have single or multiple acyl groups. It is particularly an important modification because it plays a significant role in the formation of pigment-stabilising tertiary structures, which are discussed further in section 4.2.3.

An acyl substituent is an aromatic acid (hydroxycinnamic, hydroxybenzoic) or aliphatic acid (most commonly malonic but also succinic or acetic), linked via an ester bond to a glycoside hydroxyl. Acylation involving hydroxycinnamic acids has been reported to be an advanced character because it occurs mainly in highly evolved families (Harborne, 1986). Floral anthocyanins having both aromatic and aliphatic acyl groups are known (e.g. in *Senecio cruentus* and *Commelina communis*) (Harborne, 1986).

O-methylation is a commonly encountered modification and it most frequently occurs on the B-ring hydroxyl groups. Methylation has a small reddening effect on the colour of the molecule, but this may be masked by the influence of other factors contributing to flower colour. However, O-methylation alone can alter flower colours in some cases. For example, in *Primula sinensis* mutants lacking co-pigments, lines containing only delphinidin anthocyanins had blue flowers while those with mainly methylated derivatives of delphinidin were pink or maroon (Harborne & Sherratt, 1961).

Although it may actually be a precursor rather than an anthocyanin modification, it is worth mentioning anthocyanins with increased hydroxylation of the A-ring through the presence of a hydroxyl group at the C-6 position (Fig. 4.1). 6-hydroxylation has a reddening effect on the pigment colour (Harborne, 1976). These pigments are rare in nature, but have been reported for flowers of *Impatiens aurantiaca* and *Alstromeria* species (Forkmann, 1991; Tatsuzawa *et al.*, 2003).

The final anthocyanin 'modification' that we will discuss is the attachment of other flavonoid glycosides to anthocyanins. These structures are formed through the bonding via ester linkages of an anthocyanin and a glycoside of a different flavonoid type (e.g. flavone) glycoside to the carboxyl functions of an organic di-acid (e.g. malonyl, succinic). Such structures have been found in flowers of orchids (Strack *et al.*, 1989), *Allium schoenoprasum* (chives) (Fossen *et al.*, 2000), *Eichhornia crassipes* (Toki *et al.*, 1994), *Agapanthus praecox* species *orientalis* (African lily) (Bloor & Falshaw, 2000) and *Lupinus* (garden lupin) (Takeda *et al.*, 1993).

Based on the variations in hydroxylation and O-methylation described above, at least 18 naturally occurring anthocyanidins have been identified (listed in Bohm, 1998). However, it should be noted that not all of these anthocyanidins are formed *in vivo*, as some of the modification activities may only use the anthocyanins as substrates. The historical and chemical definition is that anthocyanidin refers to the aglycone of the anthocyanin, i.e. the base structure of the anthocyanin without the glycosyl groups. A basic step in identifying the structure of an anthocyanin is the removal of the glycosyl groups and any acyl moieties attached to them through acid or enzyme hydrolysis, and the identification of the anthocyanidin.

4.2.1.2 Yellow flavonoid pigments

Carotenoids play the predominant role in yellow floral pigmentation; however, in some instances, flavonoids are responsible for the colour. More frequently, the yellow flavonoids co-occur with carotenoids. In these cases, the primary role of the yellow flavonoids is as UV-absorbing nectar or honey guides for pollinators. We touch briefly on nectar guides again in Section 4.2.5, but for the interested reader, we suggest Harborne and Grayer (1993) and Bohm (1998) for reviews on flavonoids in pollination biology.

The yellow flavonoid pigments are the chalcones, aurones and some flavonols. Their distribution is relatively limited. Aurones represent one of the smallest flavonoid classes in terms of the number of different structures that have been defined. They produce the strongest yellow colours owing to their absorbance at longer wavelengths compared to the other types. They have been reported to occur primarily in flowers of the Scophulariaceae (*Linaria* and *Antirrhinum*), Asteraceae (e.g. *Cosmos*, *Coreopsis* and *Zinnia*) and Plumbaginaceae (*Limonium*) (Bohm, 1994, 1998). *Antirrhinum majus* (snapdragon) is a particularly good source of aurones, which produce an impressive bright yellow flower colour, and in recent years it has become the model species for the study of aurone biosynthesis. It should be noted that aurones can be formed by the oxidation of chalcones during extraction and purification procedures, and older references in the literature reporting the co-occurrence of these compounds with chalcones are thus suspect.

Glycosides of 2′4′6′-trihydroxychalcones generate yellow flower colours in some cultivars of *Dianthus*, *Helichrysum* (everlasting flower), *Paeonia* and *Callistephus*; and aglycones and glycosides of 6′-deoxychalcones (chalcones lacking the 6′-hydroxyl group) have been reported as pigments in some yellow-flowered varieties of *Cosmos*, *Dahlia*, *Coreopsis* and *Bidens* (all members of the Asteraceae) (Harborne, 1966; Hosoki *et al.*, 1991; Bohm, 1993, 1998).

Regarding chalcones, of particular interest are modifications that stabilise the structures. Chalcones with a free or unsubstituted hydroxyl group at the 6′ position tend to spontaneously cyclise to the corresponding flavanone, a colourless flavonoid type. This conversion occurs at significant rates at physiological pH values (Miles & Main, 1985). Substitutions of the 6′-hydroxyl group with a glycosyl or methyl residue stabilise the chalcone structure and prevent C-ring formation because the 2′-hydroxyl (if present) is bound to the carbonyl group via an intramolecular hydrogen bond and so is unavailable for cyclisation. 6′-deoxychalcones for the same reason do not cyclise spontaneously to the flavanone isomers.

Most flavonols absorb at a shorter wavelength than chalcones, and therefore at best are only very weakly coloured. However, some modifications made to the chromophore can shift the wavelength of maximum absorbance so that the compound becomes a yellow pigment. The main yellow flavonols have an additional hydroxyl group on the A-ring, most commonly at the C-8 position, but occasionally at the C-6 position (Fig. 4.1). 8-hydroxyflavonols are the main floral pigments in *Gossypium* (cotton) and *Primula vulgaris*, and contribute to flower colour in different genera and species including *Rhododendron*, *Ranunculus*, *Mimulus luteus* and *Meconopsis paniculata* (Harborne, 1976). 6-hydroxyflavonols are primarily in the Asteraceae (Harborne, 1976). Concerning other modifications, it has been suggested that flavonols with *O*-glycosylation at the 7,4'-positions or *O*-methylation at the 3'- or 3',5'-positions may contribute to yellow colour in a few species (Harborne, 1976).

Finally, it is worth mentioning some exceptional cases in which yellow flower colour is derived from 'colourless' flavonoids. In *Camellia chrysantha* and *Lathyrus chrysanthus*, for example, the only pigments in the flowers are common flavonol glycosides (Hwang *et al.*, 1992; Markham *et al.*, 1992). Also, glycosides of a rare 5,7,2',4',5'-pentahydroxyflavone play a significant role in the flower colour of some species in the Cichorieae (Harborne, 1978). In both cases, how they provide yellow pigmentation is not satisfactorily known, although the answer may lie in the formation of secondary structures or interactions with other compounds in the cell.

4.2.2 Secondary structures

An important characteristic of anthocyanins is that they undergo pH-dependent changes in colour intensity and hue or even loss of colour. This is because an anthocyanin molecule undergoes complex rearrangements in aqueous solutions, existing in different coloured or colourless forms depending on the pH of its environment (Brouillard, 1988; Brouillard & Dangles, 1993). It is only at acidic pH values that an anthocyanin is in a stable coloured form. This is the flavylium cation form, which is red or orange. At higher pH values other coloured forms, often unstable, exist. As the pH increases towards neutral, neutral quinonoidal bases, usually purple, are formed, and are followed by the formation of blue anionic quinonoidal bases at alkaline pH values. Flavylium cation and quinonoidal base forms are both important to flower pigmentation (Brouillard, 1988).

Another facet of anthocyanin chemistry is that in mildly acidic conditions (pH3–6), anthocyanins essentially exist in colourless forms. This is because anthocyanin flavylium cations are vulnerable to hydration under these conditions, and a reaction readily converts them to colourless hemiacetal and (later) chalcone forms. Given that anthocyanins accumulate in the cell vacuole and that in many species petal cell vacuolar environments are mildly acidic, different mechanisms prevent hydration and stabilise the coloured anthocyanin forms. Most of these are based on the formation of tertiary structures (see Section 4.2.3). However, it has been observed that malonylated anthocyanins are more stable than other anthocyanins (Saito *et al.*, 1985), and Figueiredo *et al.* (1999) have proposed a mechanism in which malonylation can prevent anthocyanins from de-colourising due to the acidity caused by deprotonation of the malonyl groups. Given that malonylation occurs in many families (Harborne, 1986), this stabilising mechanism may have wide relevance.

Research is scant on the existence *in vivo* of secondary structures for the yellow flavonoid pigments and their relevance to flower colour. Although, it has long been observed that the colour of flavonols and flavones *in vitro* is shifted strongly towards yellow when they are ionised, for example, through exposure to acid or base. More recently Markham *et al.* (2001) have produced evidence that weakly coloured common flavonols (e.g. quercetin glycosides) can exist in yellower tautomer forms *in vivo* through flavonol–protein interactions (discussed further in Section 4.2.3). Concerning the case of the rare yellow flavone pigment in the Cichorieae (see Section 4.2.1), it is possible that the presence of the two *p*-hydroxyl groups in the B-ring allows it to exist in a yellow quinoid form (Harborne, 1978).

4.2.3   Tertiary structures

Research over the last several decades has revealed that flavonoid tertiary structures are of primary importance in flower colour (Brouillard, 1988; Brouillard & Dangles, 1993). They are crucial for stabilising anthocyanins. Equally important, they are also a mechanism for generating variety in flavonoid colour, particularly that of anthocyanins, due to the changes they cause in both the amount of light absorbed by the pigments and the wavelength of the light that is absorbed. There are several types of tertiary structures that have been identified. Furthermore, more than one type can be involved in pigmentation within a particular flower, and they can involve both primary and secondary structures of pigments (Brouillard & Dangles, 1993).

Robinson and Robinson (1931) and Lawrence (1932) independently discovered the involvement of flavonoid tertiary structures in flower colour (cited in Harborne, 1976). The phenomenon they described was intermolecular co-pigmentation. In addition to stabilising the anthocyanin, this type of interaction has a blueing effect and increases the intensity of the colour of the anthocyanin (Goto & Kondo, 1991). It is thought to be the most common mechanism for the formation of blue flower colours (Harborne & Williams, 2000).

In intermolecular co-pigmentation, a tertiary structure is formed by a vertical stacking of molecules, with an anthocyanin molecule sandwiched between two co-pigment molecules (Goto & Kondo, 1991). Co-pigments are commonly other flavonoids such as flavonols or flavones. Anthocyanin structures can affect this interaction. For example, anthocyanins having an aromatic acyl group form a more stable intermolecular co-pigmentation complex than those that do not (Goto & Kondo, 1991).

The structure of a co-pigment and its relative concentration also affect the degree of co-pigmentation and, consequently, the pigment colour and intensity. A nice example demonstrating these effects can be found in Bloor (1997). Furthermore, in relation to primarily blue flower colours, specific co-pigment structural features have been identified that enhance co-pigmentation ability. Flavone *C*-glycosides having a sugar residue attached directly to the carbon at the C-6 position seem to be particularly effective co-pigments in some circumstances. In *in vitro* experiments, the substitution of the usual flavonol co-pigments found in *Eustoma grandiflorum* (lisianthus) with flavone *C*-glycosides changed the colour of the delphinidin-based pigments present from purple to blue (Asen *et al.*, 1986). Flavone *C*-glycosides act as co-pigments in blue flowers of several species, including *Centura* and *C. communis* (Goto & Kondo, 1991), *Felicia amelloides* (blue marguerite daisy) (Bloor, 1998), *Iris* (Asen *et al.*, 1970) and *Limonium* (Asen *et al.*, 1973). The co-pigments in some of these species also are *O*-glycosylated on the B-ring, and it has been suggested that glycosylation at both 'ends' of the flavonoid may make the molecule a superior co-pigment (Bloor, 1998). Although they mainly serve to stabilise and modify anthocyanin colour, it should be noted that flavones and flavonols are the basis for cream-coloured flowers (e.g. lisianthus) (Davies *et al.*, 1993). They give some 'depth' or 'body' to acyanic petals, preventing them from appearing translucent.

Co-pigmentation of flavonoids other than anthocyanins is also possible, but it is either a rare or an understudied phenomenon. There has only been one report of its occurrence. In yellow *A. majus* flowers, the co-pigmentation of aurones with flavones results in the aurones absorbing light at a significantly longer wavelength (Asen *et al.*, 1972). However, this has a negligible effect on the flower colour perceived by the human eye.

Intramolecular co-pigmentation (sometimes referred to as intramolecular stacking) is a mechanism for stabilising more complex anthocyanins such as those that are polyacylated with aromatic acids (Goto & Kondo, 1991). The tertiary structure is formed by either the sandwiching of the anthocyanin nucleus between two of its acyl groups or by the anthocyanin nucleus being covered by one acyl group. For example, in red-purple flowers of the Orchidaceae, sugar-linked aromatic acyl groups at the 7 and 3' positions in anthocyanin 3,7,3'-triglycosides provide exceptional stability to the pigments (Figueiredo *et al.*, 1999). Also, the roles of each acyl group in colour development can differ. In *Gentiana triflora* (gentian) it is the acylated 3'-glycosyl group in the delphinidin 3,5,3'-triglycoside that contributes the most to the formation of the blue pigmentation (Yoshida *et al.*, 2000). Furthermore, Suzuki *et al.* (2002) showed that aliphatic acylation at the 6''-hydroxyl increased anthocyanin stability in buffer solutions. Intramolecular co-pigmentation can also occur in covalent anthocyanin–flavonol/flavone complexes (Figueiredo *et al.* 1995; Bloor & Falshaw, 2000; Fossen *et al.*, 2000).

Another stabilising mechanism is self-association, which requires high anthocyanin concentrations. It involves the vertical stacking of anthocyanin molecules, and is deduced to play at least some role in flower colour (Goto & Kondo, 1991). Certainly, anthocyanins accumulate to high levels in some species. For example, anthocyanins can account for up to 14% of dry flower weight (Harborne, 1976).

It has long been known that flavonoids can interact with proteins *in vitro*. On the basis of recent studies it appears that the concept of tertiary structures can be expanded to include complexes formed with or interactions between flavonoid pigments and proteins. In some species, stable or unstable complexes containing anthocyanins and protein are formed in the vacuole. These are termed anthocyanic vacuolar inclusions (AVIs) (Markham *et al.*, 2000a). Their formation can alter flower colour, but their biological function is not yet known. Section 4.2.5 presents an extensive discussion on these complexes.

It is possible that other flavonoid types interact with proteins *in vivo*. For some yellow-flowered species such as *L. chrysanthus* and *Dianthus caryophyllus* (carnation), a proportion of the petal flavonols are located in the cytoplasm (Markham *et al.*, 2001). The significance of this is not known, but it was proposed that aggregation of flavonoids such as the common flavonol glycosides on a cytoplasmic protein would allow the essentially colourless molecules to assume yellow-coloured secondary forms. This mechanism could provide the basis for yellow flower colour in such species as *L. chrysanthus*, which does not produce the archetypal yellow flavonoid pigments (Markham *et al.*, 1992).

4.2.4   Metal complexation

In a few species, anthocyanin–co-pigment complexes interact with metal ions such as iron, aluminium and magnesium to form stable blue pigment structures. In flowers of *C. communis*, for example, the pigment supra molecule (commelinin) is comprised of six molecules of a delphinidin-based anthocyanin with an aromatic acid-acylated glucosyl group at C-3 and an aliphatic acid-acylated glucosyl group at C-5, six molecules of a flavone co-pigment having *C*-glycosylation at C-6 and *O*-glycosylation at C-4′, and two Mg^{2+} ions (Kondo *et al.*, 1992 and references therein). Replacement studies changing the anthocyanin pigment or replacing Mg^{2+} with other divalent metal ions indicates that the metal chelation occurs with the *o*-dihydroxyl groups on the anthocyanin *B*-ring (Kondo *et al.*, 1992). Metal chelation, like co-pigmentation, causes a shift in absorbance by the pigment to longer wavelengths. This helps explain why the supra molecule protocyanin, which is comprised of a cyanidin-based anthocyanin, a malonylated flavone with *O*-glycosylation at the C-7 and C-4′ positions, Mg^{2+} and Fe^{2+} (in the ratio of 6:6:1:1), gives blue pigmentation in *C. cyanus* (Goto & Kondo, 1991). A complex similar to commelinin is the basis for blue pigmentation in *Salvia patens* (Takeda *et al.*, 1994), and metal complexation involving Al^{3+} ions is suggested to play a role in blue pigmentation in *Hydrangea macrophylla* (Takeda *et al.*, 1990; Goto & Kondo, 1991). Regarding other flavonoids, flavonols (and flavones) with an *o*-dihydroxyl group on the *B*-ring readily form yellow complexes with metal ions *in vitro*. However, there have been no reports of such metal complexation being involved in yellow flavonoid pigmentation *in vivo*.

4.2.5   Localisation

It is immediately obvious that petals generally are the most highly coloured of the floral organs. However, in some genera (e.g. *Hydrangea*, *Limonium*), it is the sepals that are the 'showy' organs. There are also genera in which it is not the floral organs but rather modified leaves or scales subtending the true flowers, such as the bracts of *Poinsettia* and spathes of *Anthurium* and *Zantedeschia*, which are brightly coloured.

Within petal tissue, flavonoid pigments generally are localised to the epidermal cells (Kay *et al.*, 1981), although anthocyanins can be localised sub-epidermally, as in many orchid and Boraginaceae species (Matsui *et al.*, 1984; Harborne, 1988; Matsui & Nakamura, 1988). Every plant cell is competent to produce flavonoids, as no specialised structures are required for their production. However, their synthesis is under strict spatial and temporal control involving a range of developmental and environmental factors, and this can give rise to a wide range of patterning of anthocyanin distribution in petals. These patterns include an apparent uniform distribution, irregular streaks or blushes, stripes, spots and splotches. Some patterns are due to biotic and abiotic stress (e.g. virus infection, light irradiation) or transposon activity. However, many patterns are developmentally controlled and may assist in pollinator attraction (see Chapter 1 and Bohm, 1998 for more details).

The patterning that can occur of flavonoids such as flavonols is less obvious. The existence of floral nectar guides, comprised of differential pigment markings at the bases of petals, has been known for a long time. But it was Thompson *et al.* (1972) who, by studying the composite flower *Rudbeckia hirta*, first determined a chemical basis for nectar guides visible only in UV light; the accumulation of strongly UV-absorbing flavonols in the basal regions of the petals. Flavonols, often in co-occurrence with carotenoids, are also responsible for nectar guides in flowers of other species, such as *Brassica rapa* (Sasaki & Takahashi, 2002).

Subcellularly, anthocyanins are deposited in the vacuole. In some species, typically in cells accumulating high levels of anthocyanins, highly pigmented bodies may form in the vacuole, which have been termed anthocyanoplasts (Pecket & Small, 1980) or AVIs (Markham *et al.*, 2000a). These may take the form of either circular, intensely pigmented globules (e.g. in *Ipomoea batatas* cell cultures) that are unstable upon extraction or more granular forms (e.g. in *E. grandiflorum*) that can be extracted as stable bodies *in vitro* (Nozue *et al.*, 1995; Markham *et al.*, 2000a). A high level of anthocyanin structural specificity in the AVI associations may occur. *E. grandiflorum* AVIs contain only cyanidin and delphinidin acylated 3,5-diglycosides and not the flavonols and other anthocyanins (acylated delphinidin triglycosides) present in the same cells (Markham *et al.*, 2000a). In *Vitis vinifera* (grape), preference occurs with regard to the acylation status of the anthocyanins incorporated into AVIs (Conn *et ali*, 2003).

In the case of both 'stable' and 'unstable' AVIs, specific proteins are associated with the formation of the bodies. Nozue *et al.* (1995, 1997) were able to isolate one of the associated proteins from the anthocyanin bodies of *I. batatas*, a 24 kDa protein termed VP24. Analysis of the corresponding cDNA revealed that VP24 is a fragment of a 96 kDa pre-protein, and that the mature VP24 domain has aminopeptidase activity when expressed in *E. coli* (Nozue *et al.*, 2003). Initial studies indicate the involvement of specific proteins in the formation of AVIs in *E. grandiflorum* and *V. vinifera* (Markham *et al.*, 2000a; Conn *et al.*, 2003).

The formation of AVIs can markedly alter flower colour. For example, the normally pink pelargonidin-based pigments of a *Dianthus* cultivar change to a blue-grey colouration as AVIs form during petal ageing (Markham *et al.*, 2000a); and AVI formation during petal ageing of the *Rosa hybrida* (rose) cultivar 'Rhapsody in Blue' changes flower colour from a vivid red-purple to a lighter and duller purple (Gonnet, 2003). However, the actual function, if any, of the protein–anthocyanin interaction has not been determined. Some of the proposed flavonoid vacuolar transport proteins, such as AN9 (Mueller *et al.*, 2000), can bind anthocyanins, and it is possible that AVI formation is related to a transport mechanism.

Other flavonoid pigments and co-pigments are also commonly deposited in the vacuole, although there have been no reports of subvacuolar 'bodies' associated with these. However, in contrast to anthocyanins, other flavonoids are occasionally found in other cellular locations. As mentioned previously, some of the petal flavonols of some lines of *E. grandiflorum*, *L. chrysanthus* and *D. caryophyllus* are located in the cytoplasm (Markham *et al.*, 2001). Furthermore, flavonol *O*-glycosides occur in the petal cell walls of *E. grandiflorum* (Markham *et al.*, 2000b). The function of these extravacuolar flavonoids is not known, although it has been postulated that they are involved in yellow floral pigmentation, at least in some species (Markham *et al.*, 2001; see also Section 4.2.3). Chalcones that accumulate in pollen grains may also acquire a strong yellow colour, e.g. in *Petunia* (van Tunen *et al.*, 1991; Davies *et al.*, 1998).

4.2.6 Extrinsic factors contributing to flower colour

In addition to features intrinsic to the flavonoid pigments, which we have already discussed, there are extrinsic factors contributing to flower colour. Structural features of pigmented tissue such as the shape of the individual cells are important to pigmentation due to their effect on light refraction and reflection. In *A. majus*, the shape of the pigmented epidermal cells is under control of the gene *Mixta*, encoding a MYB-type transcription factor (Noda *et al.*, 1994). In wild-type flowers the epidermal cells are conical in shape and the colour of the flower is a deep, rich magenta. In *Mixta* recessive lines, the cells are flattened resulting in increased light reflectance from the epidermal surface, as well as altered refraction of absorbed light (Noda *et al.*, 1994; Gorton & Vogelmann, 1996). This altered refraction focuses the light within the unpigmented mesophyll rather than in the pigment-containing vacuoles in the epidermis as occurs with the normal conical cells. The result of these effects is flower colour that is dulled. Functioning similarly to *Mixta* is the *Myb* gene *PhMyb1* in *Petunia* (Mur, 1995).

Many species produce flavonoids in the petal in combination with other classes of pigment. Most commonly, sub-epidermal carotenoids co-occur with the epidermally located anthocyanins. Such combinations can be important for generating particular colours, such as near-black petals.

4.3 Flavonoid biosynthesis

4.3.1 Overview

The flavonoid pathway is part of the larger phenylpropanoid pathway, which also produces a range of other secondary metabolites, such as lignins, lignans, stilbenes and hydroxycinnamic acids. The phenylpropanoid pathway starts from phenylalanine, itself derived from the shikimate/chorismate pathways. There are many branches to the flavonoid-specific pathway, and the products of many of these branches are generally colourless, such as those belonging to the large isoflavonoid class. In this chapter we focus only on those parts of the flavonoid pathway leading to pigmented compounds of flowers. Another group of pigmented flavonoids, the proanthocyanidins, which generally do not contribute to flower colour, is discussed in Chapter 5.

Flavonoids are synthesised in the cytosol and those involved in pigmentation are generally transported into the vacuole. The biochemistry of the core steps in both the general phenylpropanoid and flavonoid pathways has been elucidated. Furthermore, genes and/or cDNAs encoding the core biosynthetic enzymes have been cloned from many species, including a wide range of ornamentals. Mutations have been identified, often in multiple species, for all of the enzyme steps from chalcones to anthocyanins. Within the flavonoid pathway, many of the key biosynthetic enzymes carry out oxidative reactions and fall into two types: membrane-bound cytochrome P450-dependent mono-oxygenases (CytP450s) and soluble non-haem dioxygenases.

Recent advances in our understanding of flavonoid biosynthesis include characterisation of the key conversions that form anthocyanidin 3-*O*-glucoside pigment from colourless leucoanthocyanidin substrate, progress towards elucidating aurone and 3-deoxyanthocyanin formation, and the molecular characterisation of several enzymes modifying the pigment structures. Furthermore, the current wave of flavonoid biochemistry research is targeting the subcellular organisation of the biosynthetic enzymes within the cytosol, the fine detail of the structures of the biosynthetic enzymes and their active sites, the catalytic mechanisms and the bases for substrate specificity.

In this section, we describe briefly the biochemical steps involved in the formation of flavonoid precursors, followed by those involved in flavonoid pigment and co-pigment synthesis, focusing on recent advances and features important to flower colour. Included are a few flavonoid and related enzymes that are not involved in floral pigmentation, but have been used in biotechnological approaches to alter flavonoid-based flower colour. We also describe available data on the biochemical and molecular characterisation of modification sequences. We do not cover in detail cDNA and gene cloning studies for the major steps of flavonoid biosynthesis. Chapter 5 provides extensive details of the genes, molecular phylogeny and biochemistry of the core flavonoid enzymes that produce the anthocyanidins. Further information on the genes, mutants, molecular biology and enzymology of flavonoid biosynthesis can also be found in the reviews of Heller and Forkmann (1993), Martin and Gerats (1993), Davies and Schwinn (1997), Bohm (1998) and Springob *et al.* (2003).

4.3.2 Formation of flavonoid precursors

Phenylalanine is converted to the activated hydroxycinnamic acid *p*-coumaroyl-CoA, a key substrate that feeds into the flavonoid pathway, through a stepwise series of three enzymatic conversions catalysed by phenylalanine ammonia-lyase (PAL), cinnamate 4-hydroxylase (C4H) and 4-coumarate:CoA ligase (4CL). Malonyl-CoA is also required for flavonoid biosynthesis, both as an 'extender' molecule in the reaction that forms the first flavonoid of the pathway and as an acid moiety donor for acylation of flavonoid glycosides. Malonyl-CoA is formed in primary metabolism and is derived from citrate, via an acetyl-CoA intermediate, through sequential enzymatic reactions catalysed by ATP citrate lyase and acetyl-CoA carboxylase (ACC), respectively (Fig. 4.2) (Fatland *et al.*, 2002). ACCs synthesise malonyl-CoA in both plastids and the cytosol. Within the plastids, it is used in fatty acid biosynthesis, while it is the cytoplasmically located substrate that is used in secondary metabolism (Nikolau *et al.*, 2003).

4.3.3 Formation of chalcones

The entry point into the flavonoid pathway is the formation of chalcone, which establishes the flavonoid C_{15} backbone. The initial chalcone formed in most species is naringenin chalcone. It is synthesised from one molecule of *p*-coumaroyl-CoA and three acetate units derived from three molecules of mal-onyl-CoA through the action of the chalcone synthase (CHS) (Fig. 4.2), one of the best characterised enzymes in secondary metabolism. A series of sequential decarboxylation and condensation reactions produce a polyketide intermediate, which then undergoes cyclisation and aromatisation reactions that form the A-ring and the resultant chalcone structure (Fig. 4.2). All the reactions are carried out at a single active site (Tropf *et al.*, 1995; Preisig-Mueller *et al.*, 1997; Ferrer *et al.*, 1999). Research on enzyme structure, active site and reaction mechanism has been recently reviewed by Austin and Noel (2003).

CHS is a type III polyketide synthase (PKS). PKSs, of which there are three types, occur in bacteria, fungi and plants. They are characterised by their ability to catalyse the formation of polyketide chains from the sequential condensation of acetate units from malonate thioesters. PKSs produce a wide range of natural products with varied *in planta* and pharmacological properties (Springob *et al.*, 2003). Of particular note is resveratrol, formed in plants through the action of the stilbene synthase (STS), a type III PKS that is considered to have evolved from CHS more than once (Tropf *et al.*, 1994). STS amino acid sequences share high identity with those for CHS. STS also shares the same substrates as CHS, producing the same polyketide intermediate, but differs in its cyclisation mechanism and specificity due to amino acid substitutions in the backbone of the protein that subtly alter the active site (Austin & Noel, 2003).

A characteristic of some plants, particularly legumes, is the production of 5-deoxyflavonoids. They are derived from 6'-deoxychalcones (Fig. 4.2). The synthesis of these chalcones requires, in conjunction with CHS activity, the activity of a NADPH-dependent chalcone reductase (CHR) (also referred to as polyketide reductase or chalcone ketide reductase). It is believed that CoA-linked polyketide intermediates diffuse in and out of the CHS active site, and it is the unbound intermediates that are reduced to alcohols by the reductase (Austin & Noel, 2003). The resultant hydroxyl groups are removed from the CHR products in the final cyclisation and aromatisation steps catalysed by CHS. CHR can function with CHS from species that synthesise only the common 5-hydroxy-flavonoids (Welle & Grisebach, 1989; Davies et al., 1998).

Chalcones are directly or indirectly converted to a range of other flavonoids in a pathway of intersecting branches, with intermediate compounds being involved in the formation of more than one type of end product.

4.3.4   Formation of anthocyanidin 3-O-glycosides

The pathway to the formation of the simplest common anthocyanins, anthocyanidin 3-O-glycosides, is well conserved in plants, and is comprised of a minimum of five enzymatic steps subsequent to the formation of chalcones by CHS.

4.3.4.1   Formation of flavanones

In a reaction that establishes the flavonoid heterocyclic C-ring, naringenin chalcone is isomerised stereospecifically to the flavanone (2S)-naringenin through the action of chalcone isomerase (CHI) via an acid base catalysis mechanism (Jez et al., 2000; Jez & Noel, 2002). CHI appears unique to plants both in its structure and reaction (Jez et al., 2000).

6'-deoxychalcones are stable under physiological conditions due to the intramolecular hydrogen bond between the 2'-hydroxyl and the carbonyl group (Miles & Main, 1985). However, with 6'-hydroxychalcones such as naringenin chalcone, the isomerisation reaction can readily occur non-enzymically to form racemic (2R,2S) flavanone (Mol et al., 1985), and it has been reported to occur in vivo to the extent that moderate levels of anthocyanin are formed (e.g. in CHI mutants) (Forkmann & Dangelmayr, 1980). However, the enzyme accelerates ring closure to a 10^7-fold acceleration over the spontaneous reaction rate and ensures formation of the biosynthetically required (2S)-flavanones (Bednar & Hadcock, 1988; Jez et al., 2000; Jez & Noel, 2002).

There are at least two types of CHI: those that can catalyse isomerisation of both 6'-hydroxy- and 6'-deoxychalcones (e.g. in Cosmos sulphureus), and those that use only 6'-hydroxychalcone substrates (e.g. in Tulipa) (Chmiel et al., 1983). The occurrence of the latter type is more widespread. The exact structural basis for the differences in substrate specificity has not yet been determined, although amino acid residues that are potentially involved have been identified (Jez *et al.*, 2000).

4.3.4.2   Formation of dihydroflavonols

(2*S*)-naringenin and other flavanones are converted stereospecifically to the respective (2*R*,3*R*)-dihydroflavonols (DHFs) through a hydroxylation that is catalysed by flavanone 3β-hydroxylase (F3H), a 2-oxoglutarate-dependent (2OG) dioxygenase. Some progress is being made towards elucidating the tertiary structure of the enzyme, the active site and the binding of oxoglutarate, but the mechanism of action requires further investigation (Britsch *et al.*, 1993; Lukacin & Britsch, 1997; Lukacin *et al.*, 2000). Soluble non-haem dioxygenases, dependant on ferrous iron, O_2 and 2-oxoglutarate, are well represented in flavonoid biosynthesis, catalysing other key reactions, as will be discussed.

4.3.4.3   Formation of leucoanthocyanidins

DHFs are stereospecifically converted to (2*R*,3*S*,4*S*)-flavan-2,3-*trans*-3,4-*cis*-diols (leucoanthocyanidins) through a NADPH-dependent reduction reaction catalysed by dihydroflavonol 4-reductase (DFR), belonging to what has been termed the short chain dehydrogenase/reductase (SDR) superfamily (Johnson *et al.*, 2001). The biosynthetic step catalysed by DFR has special significance as it is one of the determinants of the type of anthocyanins formed, and hence, of flower colour. In some species DFR has a restricted substrate range. A well-known example is the *Petunia* DFR, which has significant activity only with DHFs having di- or trihydroxylated B-rings, dihydroquercetin (DHQ) and dihydromyricetin (DHM), respectively (Forkmann & Ruhnau, 1987). Thus pelargonidin-based anthocyanins derived from dihydrokaempferol (DHK) (mono B-ring hydroxylation) rarely accumulate in this ornamental. DFR in other commercially important crops, such as *Cymbidium hybrida* (cymbidium orchids), *Lycopersicon esculentum* (tomato) and *Vaccinium macrocarpon* (cranberry), also cannot efficiently reduce DHK (Johnson *et al.*, 1999; Polashock *et al.*, 2002). Furthermore, DFR enzymes of many ornamental species, such as *Chrysanthemum*, *Dahlia*, *Dianthus*, *Matthiola* and *Nicotiana* (tobacco), can use DHM even though delphinidin derivatives do not naturally occur in these ornamentals (Heller & Forkmann, 1993; Davies & Schwinn, 1997).

A study of chimeric DFRs of *Petunia* and *Gerbera hybrida* revealed a region of the DFR protein that determines substrate specificity, and it has been demonstrated that as little as one amino acid change in this region can alter the specificity of the enzyme (Johnson *et al.*, 2001). This region is postulated to be a binding pocket for the B-ring of DHF substrate. Johnson *et al.* (2001) also identified a region that potentially is part of the active site of the enzyme.

4.3.4.4   Formation of anthocyanins

In the final steps of anthocyanidin 3-*O*-glycoside biosynthesis, the pigments are formed from the colourless and unstable leucoanthocyanidins through the activity of the anthocyanidin synthase (ANS), also referred to as the leucoanthocyanidin dioxygenase or LDOX, and an anthocyanidin 3-*O*-glycosyltransferase (3GT). Furthermore, a key part of the process is their deposition in the vacuole. In contrast to studies of the other enzymes in the pathway, difficulties with biochemical approaches meant that the characterisation of the ANS proceeded first by the cloning and characterisation of the encoding gene, which was first isolated from *Z. mays* (Menssen *et al.*, 1990) and *A. majus* (Martin *et al.*, 1991). In a study of a gene unrelated to flavonoid biosynthesis, the sequence of the predicted protein from *Z. mays* was shown to have regions of extensive homology to F3H and other 2OG-dioxygenases (Matsuda *et al.*, 1991), and this preliminary identification was subsequently confirmed by *in vitro* assay of recombinant *Perilla frutescens* ANS (Saito *et al.*, 1999).

The product of the ANS catalysed reaction is the colourless anthocyanidin pseudobase, 3-flaven-2,3-diol (Nakajima *et al.*, 2001). The reaction most likely proceeds via a stereospecific hydroxylation at the C-3 position of the leucoanthocyanidin, the product of which undergoes spontaneous dehydration and isomerisation to form 2-flaven-3,4-diol, which itself spontaneously isomerises to the thermodynamically more stable final form (Nakajima *et al.*, 2001; Welford *et al.*, 2001; Wilmouth *et al.*, 2002). Information on the structure and mechanism of ANS can be found in Wilmouth *et al.* (2002). It should be noted that while chemists have identified at least 18 anthocyanidins, not all (in their pseudobase form) are formed by ANS. For example, in *Petunia*, peonidin, petunidin and malvidin are not natural anthocyanidins because the methyltransferases that modify the B-ring use cyanidin and delphinidin acylglycosides as substrates rather than intermediates of anthocyanin synthesis (Jonsson *et al.*, 1982; Brugliera *et al.*, 1994).

The pseudobase products of the ANS (or their derived anthocyanidins) are unstable and generally do not accumulate *in planta* in the free state. Rather, they are stabilised by an initial *O*-glycosylation, typically at the C-3 position, and transferred to the vacuole where conditions favour the generation of the coloured anthocyanin form. A 3GT such as uridine diphosphate (UDP)-glucose:flavonoid 3-*O*-glucosyltransferase (UF3GT) mediates the transfer of a glycosyl residue from the activated nucleotide sugar to the pseudobase (Nakajima *et al.*, 2001). The formation of the 3-flaven-2,3-diol pseudobase and action of the 3GT on this compound was previously hypothesised by Heller and Forkmann (1993). No other core enzymes are believed to be involved in the formation of the anthocyanidin 3-*O*-glycosides. Some earlier hypothesised reaction sequences included the involvement of a specific dehydratase (Heller & Forkmann, 1988). However, *in vitro* assays mimicking *in vivo* conditions by Nakajima *et al.* (2001) have demonstrated that an external dehydratase is not required. They showed that mild pH conditions, as would be expected in vacuoles, were sufficient for the colouring of the anthocyanin.

Curiously, there have been no reports in the literature of mutations specific for UF3GT giving a phenotype in petal tissue, which may reflect a certain amount of redundancy in GT activity. However, there is the well-known *bronze1* mutation in *Z. mays* kernels, which illustrates the importance of the initial glycosylation and its obligatory nature (Bohm, 1998). In this *UF3GT* mutant, the glycosylation does not occur and a brown pigment is formed by the condensation of anthocyanidins in the cytosol, giving a 'bronze' pigmentation phenotype. This mutation also revealed that this glycosylation step is required for proper sequestration of the anthocyanins in the vacuole, which will be discussed further in Section 4.3.11. Given the importance of the initial glycosylation, UF3GTs should not be viewed as belonging to the general class of modifying enzymes, but rather to the group catalysing the core reactions (Heller & Forkmann, 1993).

UF3GT enzymes belong to the UDP-glycosyltransferase superfamily (Vogt & Jones, 2000; Jones & Vogt, 2001; Ross et al., 2001). GTs in plants and mammals serve central roles in detoxifying or regulating the bioactivities of a range of endogenous and exogenous low–molecular weight compounds. The catalytic mechanism of GTs is not yet known, and no crystal structures of a plant enzyme have been obtained. However, they are thought to contain a substrate binding amino terminus and a UDP-sugar-binding carboxyl terminus.

UDP-glucose is the most common donor in the glycosylation reaction and UF3GT is the best-characterised flavonoid GT. It was designated as a flavonoid rather than an anthocyanidin glucosyltransferase because early studies assaying crude enzyme preparations from several species suggested that the enzyme could conjugate both anthocyanidin and flavonol substrates (Heller & Forkmann, 1988), which has been subsequently confirmed by expression of cDNAs in vitro (Yamazaki et al., 2002; Rosati et al., 2003) and in vivo (Schwinn et al., 1997). However, other studies suggest the existence in some species of UF3GT enzymes that specifically or primarily act on anthocyanidin substrate (Do et al., 1995; Tanaka et al., 1996; Ford et al., 1998).

4.3.5   Formation of 3-deoxyanthocyanins

There have been few studies on the changes to the common anthocyanin biosynthetic pathway that result in the formation of 3-deoxyanthocyanins. More information is available on production of the related tannin-like 3-deoxyflavonoid phlobaphenes in Z. mays; and the inducible production of 3-deoxyflavonoids as antifungal phytoalexins in S. bicolor (Hipskind et al., 1996; Chopra et al., 1999, 2002 and earlier references cited therein).

In Z. mays, 3-deoxyanthocyanins occur in various tissues, most obviously in the flower silks; however, it is phlobaphene biosynthesis that is best characterised. A MYB transcription factor, P, activates the biosynthetic genes CHS, CHI and DFR but not F3H (Grotewold et al., 1994), so that flavanones are formed but not converted to DHFs. Rather they are converted to the phlobaphene precursors flavan 4-ols by DFR, which has this flavanone 4-reductase (FNR) activity in addition to its usual activity on DHFs (Styles & Ceska, 1977; Halbwirth et al., 2003). The phlobaphenes are thought to form by polymerisation of the flavan 4-ols.

Enzymology studies suggest that a similar biosynthetic system as that for phlobaphenes occurs for other 3-deoxyflavonoids, i.e. the reduction of F3H activity and acquisition of an FNR activity by DFR. FNR activity has been shown from flowers of the 3-deoxyanthocyanin accumulating Gesneriads S. car-

dinalis and *Columnea hybrida* (Stich & Forkman, 1988a, 1988b). Furthermore, recombinant DFR proteins from *Malus domestica* and *Pyrus communis*, species that can produce 3-deoxyflavonoids under some circumstances, show both DFR and FNR activity (Fischer *et al.*, 2003). DFR/FNR proteins analysed to date still prefer DHF substrate to flavanone, supporting the need for a mechanism promoting flavanones as the available substrate (Fischer *et al.*, 2003; Halbwirth *et al.*, 2003). There is further evidence for this from 3-deoxyanthocyanin producing flower silks of *Z. mays*, as they have only low levels of F3H activity but show DFR/FNR activity (Halbwirth *et al.*, 2003). These enzymology studies on 3-deoxyanthocyanin biosynthesis are supported by gene expression data for *S. cardinalis*. Transcript levels for F3H in the petal are very low, but transcript for a cDNA encoding an FNR enzyme is found in great abundance (Davies *et al.*, 2002). The 3-deoxyanthocyanins are presumably formed from the flavan 4-ols by action of ANS and a 5-*O*-glycosyltransferase.

4.3.6   Hydroxylation

It is obvious from the preceding sections that hydroxylation has a key impact on anthocyanin colour. The most important variation in anthocyanin hydroxylation is with regard to the B-ring, with increased hydroxylation causing a shift in colour away from the red end of the spectrum towards the blue end. In a few species, the B-ring hydroxylation pattern is fully (*Silene*) (Kamsteeg *et al.*, 1981) or partially (*Verbena*) (Stotz *et al.*, 1984) incorporated through the hydroxycinnamic acid CoA ester used by CHS. However, increased B-ring hydroxylation is due more commonly to specific hydroxylase activity on flavonoid substrates. These hydroxylases are the flavonoid 3′-hydroxylase (F3′H) and the flavonoid 3′,5′-hydroxylase (F3′,5H), both of which are CytP450s that catalyse hydroxylation at the stated positions (Fig. 4.1). The activity of these enzymes is the major determinant of anthocyanin type and, consequently, flower colour.

The substrates of the hydroxylases are not the anthocyanins but rather their precursors. Flavanone and DHF are known substrates (Heller & Forkmann, 1993), and leucoanthocyanidins may also be used (Schwinn & Davies, unpublished data). As they can potentially act at multiple levels in the anthocyanin biosynthetic pathway, the hydroxylases may produce a strong biosynthetic drive to cyanidin and delphinidin derivatives. Given this important role the hydroxylases have in the type of anthocyanin produced, it is not surprising to find scientific and biotechnological interest in these enzymes and the encoding genes. Genes/cDNAs for both, sometimes informally referred to as the blue (*F3′,5′H*) and red (*F3′H*) genes, have been cloned first from *Petunia* (Holton *et al.*, 1993a; Brugliera *et al.*, 1999) and subsequently from several other species.

The hydroxylases require an electron donor to function. One such donor is NADPH-CytP450 reductase. A potential alternative electron donor is Cyt b_5. *In vitro* this type of protein can accept electrons from NADPH-CytP450 reductase and donate electrons to CytP450s (Vergères & Waskell, 1995). Based on transposon mutant analysis, a specific Cyt b_5 was shown to be required for full activity of F3′,5′H in *Petunia* flowers (de Vetten *et al.*, 1999). Interestingly, the protein apparently is not needed for F3′H activity, although it is possible that a Cyt b_5 protein encoded by a different locus may be involved in 3′-hydroxylation.

Little is known about the enzymes carrying out the rare 6- and 8-hydroxylation of the A-ring of anthocyanins and flavonols. A 2OG-dioxygenase activity catalysing 6-hydroxylation of a methylated flavonol has been identified in *Chrysosplenium americanum* (Anzellotti & Ibrahim, 2000). Also, a CytP450 cDNA isolated from elicitor-treated soyabean cells has been shown to be involved in the formation of 6-hydroxylated isoflavonoids (Latunde-Dada *et al.*, 2001). The *in vitro* expressed protein could accept flavanone and DHF substrates, suggesting that hydroxylation occurs prior to the shift of the B-ring in isoflavanoid formation. How this relates to the biosynthesis of 6-hydroxyanthocyanins and 6-hydroxyflavonols is not known.

4.3.7   Secondary modifications of flavonoids

As mentioned previously, the great diversity in flavonoid pigments is due to the number and type of modifications that are made to the different classes of flavonoids, which include glycosylation, methylation and acylation. The body of knowledge on the enzymes involved in producing the more complex anthocyanin structures is increasing. This is driven by efforts not only to clone genes that will aid genetic modification strategies to produce blue colours in commercially important ornamental species such as rose, but also to fully characterise anthocyanin production in selected species such as *Petunia* and *P. frutescens*. *Petunia* is one of the few models for anthocyanin synthesis in which modification reactions have been studied. Extensive genetic, biochemical and enzymatic studies of the production of methylated and acylated anthocyanidin 3,5-*O*-glycosides in *Petunia* has shown that the modifications occur in a sequential fashion and the enzymes involved show strict substrate specificity (Heller & Forkmann, 1993).

Genome and proteome research programmes will likely provide significant information on families of different enzymes, some of which will include flavonoid-modifying activities. For example, the genome analysis of *A. thaliana* has identified 99 GT-like sequences (Li *et al.*, 2001), and research is currently underway to characterise them by transgenic approaches and expression *in vitro*. However, such approaches are limited to identification of activities present in the major model species. Furthermore, activities of *in vitro* expressed proteins may not match *in vivo* roles of the enzyme, as has been found for some flavonoid enzymes.

In the following sections we principally discuss some of the activities studied for the coloured flavonoids. Often many other related activities have been studied that modify colourless flavonoids such as flavonols or isoflavonoids.

4.3.7.1   Glycosylation

The initial 3-*O*-glycosylation of 3-hydroxyanthocyanidins is an essential step in formation of anthocyanins. However, 3-glycosylation is often only the first of multiple sugar additions that may occur onto the anthocyanin. These include further addition at the 3-position and also glycosylation at hydroxyls of the A- and B-rings. Many of these activities have been studied biochemically, and genes have now been isolated for several glycosylation activities. The second anthocyanin GT enzyme for which a gene was isolated was the UDP-rhamnose: anthocyanin 3-glucoside rhamnosyltransferase from *P. hybrida* (3RT) (Brugliera *et al.*, 1994; Kroon *et al.*, 1994). Since then clones have also been obtained for UDP-Glu: anthocyanin 5-*O*-glucosyltransferases (UF5GT) (Yamazaki *et al.*, 1999, 2002) and a UDP-Glu: anthocyanin 3′-*O*-glucosyltransferase (Fukuchi-Mizutani *et al.*, 2003). The UF5GT protein from *P. frutescens* and *Verbena hybrida* expressed *in vitro* showed broad anthocyanin substrate acceptance, although prior 3-*O*-glucosylation was an essential requirement. In contrast, the *P. hybrida* UF5GT protein showed high specificity for the anthocyanidin 3-acylrutinoside.

Flavonols commonly accompany anthocyanins in flowers, and frequently show a similar pattern of glycosylation. Indeed some of the UF3GTs responsible for anthocyanidin 3-*O*-glucosylation have been shown to also accept flavonols as substrates (see Section 4.3.4.4). However, some of the anthocyanidin 3GTs show strong preference for the anthocyanidin substrates (Ford *et al.*, 1998). Furthermore, a range of GT cDNAs have been cloned that encode proteins active on non-pigmented flavonoids, although many of the characterised activities are for non-floral tissues. They include enzymes carrying out glucosylation, galactosylation or rhamnosylation at the C-3, C-7 or C-4′ hydroxyls of flavones, flavanones, flavonols and isoflavonoids (Miller *et al.*, 1999; Hirotani *et al.*, 2000; Jones *et al.*, 2003; Kramer *et al.*, 2003). They vary greatly in substrate specificity, as measured by expression *in vitro*, with some proteins accepting a wide range of flavonoids (Kramer *et al.*, 2003). The wide substrate acceptance of some GTs is well illustrated by two GTs from *Dorotheanthus bellidiformis*. The recombinant proteins not only add a glucose to either the 5- or 6-hydroxyl of the betalain pigment betanidin, but also glucosylate the 4′-hydroxyl of flavonols and flavones (Vogt *et al.*, 1999) and the 3-hydroxyl of flavonols and anthocyanidins, respectively (Vogt, 2002). Interestingly, the two enzymes show closer sequence similarity to other flavonoid glycosyltransferases than they do to each other, suggesting independent evolutionary origins of the activities on the 5- and 6-hydroxyl positions.

All the previously described activities are *O*-glycosyltransferases. However, *C*-glycosylation also occurs on flavonoids, and may be of significance to flower colour. *C*-glucosylflavones accumulate in the flowers of some species, including blue-flowered ones, and they are suggested to be particularly good co-pigments for generating blue flower colours (Asen *et al.*, 1986). To date, reports on genes for the enzyme(s) involved in formation of *C*-glucosylflavones have not been published, although a flavanone 2-hydroxylase has been suggested to be involved in flavone *C*-glycoside production (see Section 4.3.9).

4.3.7.2   Methylation

There are several hundred known *O*-methylated flavonoids in plants, and *O*-methylation is a common modification of anthocyanins, on which it has a reddening effect (see Section 4.2.1.1). O-methyltransferases acting on a range of flavonoids, in particular chalcones, flavonols, flavones and flavanones, have been studied extensively for both A- and B-ring modifications. Several have been characterised at the molecular level, and all are part of a large family of S-adenosyl-L-methionine (SAM)-dependent methyltransferases (Ibrahim et al., 1998; Ibrahim & Muzac, 2000). For anthocyanins the studied activities are for the 3'- and/or 5'-hydroxyl, although anthocyanins based on rare anthocyanidins are known that have methylation at the 5- or 7-hydroxyl. The isolation of cDNAs for anthocyanin O-methyltransferases was mentioned in Quattrocchio et al. (1993), but has been reported in full only in the patent literature (Brugliera et al., 2003). The patent relates to sequences from *P. hybrida*, *Fuchsia*, *Plumbago* or *Torenia* for O-methyltransferases that act on the 3'- or 3'- and 5'-hydroxyls, with the anthocyanin type accepted as a substrate dependent on the source species of the cDNA.

4.3.7.3 Acylation

Acyltransferases in flavonoid metabolism characterised to date catalyse transfer of acyl groups from a CoA-donor molecule to sugar hydroxyl residues, and they are part of a large family of enzymes involved in many primary and secondary metabolic pathways (St-Pierre & De Luca, 2000). Clones have been isolated encoding aromatic or aliphatic anthocyanin acyltransferase enzymes, and examples of the reactions are presented in Fig. 4.4. Fujiwara et al. (1998) isolated a cDNA from *G. triflora* that encoded an enzyme which, when expressed in *E. coli* or yeast, used either caffeoyl-CoA or *p*-coumaroyl-CoA to introduce a hydroxycinnamic acid group to the glucose at the C-5 position of anthocyanin 3,5-diglucoside (i.e. a hydroxycinnamoyl-CoA:anthocyanin 5-O-glucoside-6'''-O-acyltransferase). The

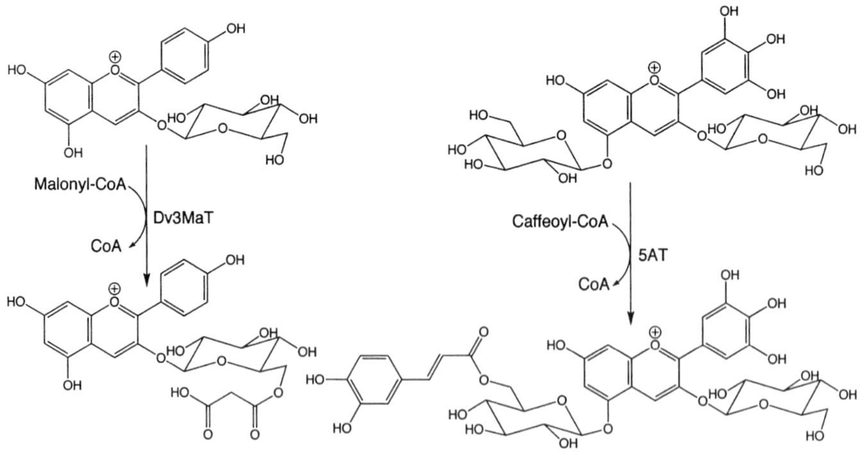

Fig. 4.4 Examples of aliphatic and aromatic acylation of anthocyanins, as carried out by the malonyl-CoA:anthocyanidin 3-O-glucoside-6''-O-malonyltransferase (Dv3MaT from *D. variabilis*) and hydroxycinnamoyl-CoA:anthocyanin 5-O-glucoside-6'''-O-acyltransferase (5AT from *G. triflora*), respectively.

enzyme would accept pelargonidin-, cyanidin- or delphinidin-derivatives, but showed specificity with regard to the anthocyanin glycosylation and acylation pattern. Yonekura-Sakakibara *et al.* (2000) isolated and characterised a cDNA from *P. frutescens* encoding a hydroxycinnamoyl-CoA:anthocyanin 3-*O*-glucoside-6″-*O*-acyltransferase, as based on analysis of the encoded product in *E. coli* and yeast. The recombinant enzyme could utilise cyanidin 3-glucoside and cyanidin 3,5-diglucoside, the putative substrates *in vivo*, as well as other anthocyanins. Details of the isolation and analysis of aromatic anthocyanin acyltransferase cDNAs from additional species are available in the patent literature (Ashikari *et al.*, 1996).

With regard to aliphatic acylation, cDNA clones have been isolated that encode a malonyl-CoA:anthocyanidin 3-*O*-glucoside-6″-*O*-malonyltransferase (Dv3MaT) (Suzuki *et al.*, 2002) and malonyl-CoA:anthocyanin 5-*O*-glucoside-6‴-*O*-malonyltransferase (Ss5MaT) (Suzuki *et al.*, 2001), from *Dahlia variabilis* and *Salvia splendens* (scarlet sage), respectively. The encoded enzyme activities were characterised by expression in *E. coli* and yeast. Although Dv3MaT accepted pelargonidin-, cyanidin- or delphinidin-3-*O*-glucosides, it would not use anthocyanin diglycosides. Ss5MaT showed high specificity, accepting only the endogenous 'bisdemalonylsalvianin' anthocyanin. Suzuki *et al.* (2003) has characterised the reaction mechanism of Ss5MaT in detail, including identification of crucial amino acid residues.

Serine carboxypeptidase-like (SCPL) enzymes, which use 1-*O*-acylglucosides as acyl group donors, also catalyse acylation reactions in plant secondary metabolism. Such an enzyme has been suggested to be involved in the acylation of cyanidin glycosides in wild carrot *Daucus carota* (discussed in Lehfeldt *et al.*, 2000). Some SCPL enzymes are located in the vacuole (e.g. sinapoylglucose:malate sinapoyltransferase) (Hause *et al.*, 2002); however, in *D. carota*, acylation appears to be required for uptake of the anthocyanins into the vacuole (Hopp & Seitz, 1987). It has also been proposed that anthocyanin malonylation may occur in the vacuole, and be related to anthocyanin sequestration and retention (Marrs *et al.*, 1995). However, anthocyanin acyltransferase sequences characterised to date do not feature vacuolar targeting sequences, and their localisation to the cytosol is thought much more likely (Suzuki *et al.*, 2001).

4.3.8   Formation of aurones

It is known from biochemical and flavonoid mutant studies of *A. majus* that aurones are produced from chalcones. Mutations in the gene encoding CHS prevent aurone production, while mutations of genes later in the flavonoid biosynthetic pathway do not. Furthermore, chalcones can be readily converted to aurones by chemical treatments. A cDNA has been isolated from *A. majus* corresponding to a polyphenol oxidase (PPO) variant thought to be the aureusidin synthase (AUS), which converts chalcones to aurones (Nakayama *et al.*, 2000, 2001; Davies *et al.*, 2001). The enzyme is a copper-containing glycoprotein that can use either 2′,4′,6′,4-tetrahydroxychalcone (naringenin chalcone) or 2′,4′,6′,3,4-pentahydroxychalcone to make aureusidin or bracteatin, respectively (Nakayama *et al.*, 2000). As aureusidin and bracteatin are 3′,4′- and 3′,4′,5′-B-ring-hydroxylated compounds, respectively, and the substrates are 4- and 3,4-hydroxylated (note that chalcone carbon numbering differs from aurones, and aurones are usually numbered differently to other flavonoids; see Fig. 4.1), the enzyme carries out both hydroxylation and cyclisation reactions. Indeed, 3-hydroxylation of the chalcone substrate may be a requirement for the formation of the aurone. Other chalcone structures can also be accepted as substrates, with a strict requirement for 2′- and 4-hydroxylation (Nakayama *et al.*, 2001). However, aurones have been reported with no B-ring hydroxylation (Bohm, 1998), and thus may arise through action of a variant enzyme or via a different biosynthetic route. The early enzymatic stages of B-ring deoxyaurone formation have been studied in cell cultures of *Cephalocereus senilis* (old man cactus), but not the specific aurone-forming step (Liu *et al.*, 1995). 4-deoxyaurones occur in a few species, and as they often co-occur with the 6′-deoxychalcones, they are likely to be synthesised from those precursors.

It is known that PPOs with sequences distantly related to that of the AUS can form aurones from chalcones. This has been demonstrated for a tyrosinase from *Neurospora crassa* expressed *in vitro* (Nakayama *et al.*, 2000). Specific variants of PPOs may also be involved in the biosynthesis of other plant pigments, in particular the initial reactions of betalain formation (see Chapter 6 for details).

There is a question as to the subcellular localisation of the AUS. All PPOs studied to date are localised in plastids. However, the deduced AUS peptide sequence lacks a typical PPO N-terminal plastid localisation signal, and it has been suggested that the AUS may be located to the vacuole (Nakayama *et al.*, 2000). How the action of the AUS is integrated with the action of other flavonoid enzymes is not clear, as activities such as hydroxylation and glycosylation of flavonoids are thought to occur prior to the localisation of flavonoids to the vacuole. The AUS can use chalcone 4′-glucosides as substrates *in vitro*, suggesting that aurone formation could occur on the glucosylated substrates *in vivo* (Nakayama *et al.*, 2000). However, confirmation of the role of AUS and further details of its mode of action await transgenic plant or genetic mutant studies. In *A. majus*, transcript for AUS increases in abundance during flower development, is not present in leaves or stem RNA samples, and is expressed specifically in the epidermis of the petal (Nakayama *et al.*, 2000; Davies *et al.*, 2001).

What other enzymatic steps are involved in aurone production is not certain, although glycosylation at the 4- and/or 6-hydroxyl is typical. In a patent application, Sakakibara *et al.* (2000) reported that a flavonoid 7-*O*-glucosyltransferase from *Scutellaria baicalensis* (Hirotani *et al.*, 2000) could also glucosylate aureusidin (at the eqivalent C-6). They also reported the cloning of similar 7-glucosyltransferase cDNAs from *A. majus* and *P. hybrida*. Whether these correspond to the previously characterised 6-glucosyltransferase activity from *Coreopsis grandiflora* (Halbwirth *et al.*, 1997) that acts on both 4-deoxyaurones and 6′-deoxychalcones is not known.

There are at least two genetic loci that affect aurone production in *A. majus*, *Sulfurea* and *Violacea*. Mutant alleles are known for both that, when present in the homozygous recessive state, alter the pattern of aurone production in the petal. These loci may encode AUS or other enzymatic steps in aurone biosynthesis, or more likely, regulatory factors of the aurone pathway. No genomic or cDNA clones corresponding to either loci have been published. As yet, besides CHS mutants, no *A. majus* mutants have been fully characterised that abolish aurone production, although a commercial white-flowered line was shown to lack *AUS* transcript in Nakayama *et al.* (2000), and a putative aurone-specific mutant has been isolated by Schwinn *et al.* (2002).

4.3.9   *Formation of flavones and flavonols*

A desaturation reaction forming a double bond between C-2 and C-3 of the *C*-ring is involved in the formation of both flavones and flavonols (Fig. 4.2). Flavones are formed from (2*S*)-flavanone substrate and two different types of enzymes have been identified that catalyse the reaction. In most plants studied to date the enzyme is a membrane-bound, NADPH-dependent CytP450 (a group-A P450), termed flavone synthase II (FNSII). Having a rare distribution in plants is the flavone synthase I (FNSI), another flavonoid 2OG-dioxygenase. FNSI was first identified in *Petroselinum crispum* (parsley) and it appears to occur only in the Apiaceae (Britsch, 1990; Martens *et al.*, 2001). Martens and Forkmann (1999) have postulated that FNSII and FNSI have analogous reaction mechanisms based on direct desaturation, and a recent study supports this for FNSI (Martens *et al.*, 2003a). In contrast, Akashi *et al.* (1999) have suggested that the FNSII reaction is likely to involve a hydroxylation step followed by dehydration of the hydroxylated intermediate. cDNAs for both types of the enzyme have been cloned (e.g. Akashi *et al.*, 1999; Martens & Forkmann, 1999; Martens *et al.*, 2001).

Another CytP450, (2*S*)-flavanone 2-hydroxylase (F2H), which shows high amino acid sequence similarity to FNSII (Akashi *et al.*, 1999), had been postulated to be involved in common flavone formation, in conjunction with a dehydratase (Akashi *et al.*, 1998). However, it has been suggested that F2H is actually involved in the formation of a subset of flavone compounds, the *C*-glycosylated types (Martens & Forkmann, 1999). In the reaction sequence for these compounds, the *C*-glycosylation step targets 2-hydroxyflavanone substrate, and thus occurs before formation of the flavone structure (Kerscher & Franz, 1987).

Flavonols are formed from (2*R*,3*R*)-dihydroflavonol substrate. The only enzyme identified to date responsible for this conversion is another 2OG-dioxygenase, flavonol synthase (FLS). However, Akashi *et al.* (1999) found that *Torenia* FNSII showed low but significant flavonol-forming activity with DHF substrate, causing them to question whether, analogous to flavone formation, two types of FLS exist. The reaction mechanism of the confirmed FLS involves *cis*-hydroxylation at C-3 followed by dehydration (Martens *et al.*, 2003a), which was proposed based on the similarity of FLS to ANS (Wilmouth *et al.*, 2002). FLS was first cloned by Holton *et al.* (1993b) from *P. hybrida*.

Recent studies of members of the flavonoid 2OG-dioxygenase family show overlapping substrate and product selectivities *in vitro*, as well as a division of the family into two subgroups, FLS/ANS and F3H/FNSI, with the former having wider substrate selectivity than the latter (see Turnbull *et al.*, 2004). For example, the ANS, like FLS, can convert DHF to flavonol (Turnbull *et al.*, 2000). ANS also has some overlapping activity with F3H, being able to catalyse the conversion of flavanone to DHF (Welford *et al.*, 2001). However, these *in vitro* activities may be less prevalent *in vivo*, as demonstrated by the lack of complementation of an *F3H* mutant by the still active *ANS* gene, e.g. in *A. majus*. The reasons for the observed difference *in vivo* are not clear, but may reflect the arrangement of the biosynthetic enzymes within complexes and the channeling of substrate.

4.3.10   Subcellular organisation of the enzymes

The formation of the flavonoid biosynthetic enzymes into a complex within the cell might facilitate the movement of substrate between enzymes, assisting reaction rates and protecting unstable, reactive intermediates from the general cellular environment. Furthermore, the specific organisation of a given 'globular' complex of multiple touching enzymes might allow the directed channeling of substrate to encourage the formation of one type of flavonoid over another alternative type. Such associations might vary from species to species, or between different cell types of a species or during development.

Early direct evidence for the formation of membrane-associated complexes of phenylpropanoid enzymes came from the studies of Hrazdina, Stafford and others in the 1970s and 1980s (reviewed in Hrazdina & Wagner, 1985). More recent evidence supporting the direct association of enzymes such as CHS, CHI, F3H and DFR has come from co-immunoprecipitation, affinity chromatography, yeast two-hybrid and other experimental approaches using *A. thaliana* (Burbulis & Winkel-Shirley, 1999). CHS and CHI in *A. thaliana* have been shown to be associated with the endoplasmic reticulum, as originally suggested by Hrazdina, and it is possible that the complexes are weakly membrane-associated through 'anchoring' by the cytP450 enzymes, such as C4H and F3'H (Saslowsky & Winkel-Shirley, 2001). Mutations affecting the structure of the F3'H protein in *A. thaliana* alter the localisation of CHS and CHI, again suggesting a direct association (Saslowsky & Winkel-Shirley, 2001). Also, there is increasing evidence supporting the metabolic channelling of intermediates in phenylpropanoid biosynthesis (Rasmussen & Dixon, 1999).

4.3.11   Import into the vacuole

The anthocyanin biosynthetic pathway should not be considered to end at the final modification of the anthocyanin molecule, but rather at the deposition of the anthocyanin in the vacuole where the environment allows full expression of the pigment colour. An important part of the pathway is, therefore, the mechanism that allows the transport of anthocyanins from the cytosol where they are produced into the vacuole. Data is lacking at present on the molecular basis of flavonoid transport to locations other than the vacuole, such as the cell wall, cytoplasm or external to the cell.

Anthocyanin transport into the vacuole shares many features with xenobiotic detoxification mechanisms that involve the addition of glycosyl, malonyl or glutathione residues to form stable water-soluble conjugates, and the sequestration of these conjugates by ATP binding cassette (ABC) transmembrane transporters (Marrs, 1996; Winefield, 2002). The importance of glycosylation for anthocyanin stability and vacuolar sequestration has been mentioned previously, and acylation has also been shown to be important for transport of some flavonoids (Matern *et al.*, 1986; Hopp & Seitz, 1987). Furthermore, in Z. *mays* a glutathione-S-transferase (GST) encoded by the *Bronze2* (*Bz2*) locus is required for anthocyanin transport into the vacuole, as mutations at *Bz2* result in oxidation and condensation of the anthocyanins in the cytosol, and bronze kernel pigmentation (Marrs *et al.*, 1995). In *Petunia*, *An9* encodes a functional homologue of BZ2, although it is a different class of GST, and mutations in it cause loss of flower colour (Alfenito *et al.*, 1998). Both BZ2 and AN9 can complement the pale pink phenotype of flowers of the *flavonoid3* mutant of *D. caryophyllus* (Larsen *et al.*, 2003). Further supporting the linkage to detoxification mechanisms, alternatively spliced *Bz2* mRNAs accumulate in a response specific to cadmium stress (Marrs & Walbot, 1997). However, many details of the transport process are still unclear. Although the proteins from the *Bz2* and *AN9* cDNAs can glutathionate cyanidin *in vitro*, no anthocyanin-glutathione conjugates have been found in plants. Furthermore, *in vivo* BZ2 and AN9 may act as anthocyanin binders and carriers, without glutathione addition occurring (Mueller *et al.*, 2000).

Support for a role by ABC transporters in anthocyanin sequestration comes from studies of the *A. thaliana* ABC transporter proteins AtMRP1 and AtMRP2. These proteins, which are homologues of the human multidrug resistance-associated protein (HmMRP1) (Cole *et al.*, 1992), are capable of transporting glutathionated anthocyanin *in vitro* as well as other glutathione *S*-conjugates and chlorophyll catabolites (Lu *et al.*, 1997, 1998). However, if formation of anthocyanin-glutathione conjugates does not occur, there may be variations to the known conjugate-dependent pumping mechanism, perhaps involving direct interaction with the GST-bound anthocyanin and the membrane transporter. Furthermore, alternative transport mechanisms may be used for different classes of flavonoids, or even different anthocyanin types. For example, a glutathione-independent mechanism involving a vacuolar H^+-ATPase has been implicated in transport of non-anthocyanin flavonoids in barley (Klein *et al.*, 1996), and the *TT12* gene of *A. thaliana* encodes a transporter of the multidrug and toxic compound extrusion (MATE)-type involved in sequestration of proanthocyanidins in seeds (Debeaujon *et al.*, 2001). Interestingly, mutation of the gene for ANS in

A. thaliana not only prevents proanthocyanidin accumulation but also prevents normal vacuole development (Abrahams *et al.*, 2003). Proanthocyanidin biosynthesis and sequestration is discussed in detail in Chapter 5.

4.4 Regulation of flavonoid biosynthesis in flowers

The production of flavonoids, both anthocyanins and co-pigments, during petal growth involves the coordinated induction of the genes for the biosynthetic enzymes in response to floral developmental signals. Furthermore, development may be only one component of a complex of signals the biosynthetic genes are responding to, as they are commonly expressed specifically in the petal epidermal cells, and may also respond to environmental factors such as the light quality and quantity. Biosynthesis of flavonoids in petals has now been studied in detail in a number of species, and the rate of gene transcription seems to be the key point for regulation of the biosynthetic genes amongst the various steps leading to protein production.

The rate of transcription is controlled by transcription factors (TFs). These are proteins that regulate gene activity by binding to motifs (*cis*-elements) within genes (usually in gene promoters) in a sequence-specific manner, increasing (as activators) or decreasing (as repressors) the rate of transcriptional initiation by the basal transcription machinery that includes RNA polymerase II (HannaRose & Hansen, 1996; Ranish & Hahn, 1996; Schwechheimer & Bevan, 1998). The activity of such TFs may be dependent on direct interaction with other proteins, which themselves do not bind directly to DNA (co-activators and co-repressors), competition with other TFs, or reversible post-translational modifications. In this section we review the role of TFs in regulating flavonoid biosynthesis in flowers, and the use of regulatory genes to modify flavonoid biosynthesis in transgenic plants. Further information on plant TF structure, their role in regulating production of flavonoids in other parts of the plant (especially for *A. thaliana* and *Z. mays*), and on TFs active in other biosynthetic pathways can be found in the reviews of Martin and Paz-Ares (1997), Schwechheimer and Bevan (1998), Eulgem *et al.* (2000), Memelink *et al.* (2000), Martin *et al.* (2001), Petroni *et al.* (2002), Vom Endt *et al.* (2002), Davies and Schwinn (2003), Heim *et al.* (2003) and Springob *et al.* (2003).

4.4.1 *Transcriptional regulation of flavonoid biosynthesis in flowers*

From studies of anthocyanin pigmentation in several dicotyledonous species a pattern has emerged in which the early biosynthetic genes (EBGs) and late biosynthetic genes (LBGs) in the pathway are regulated separately. The point of division is commonly at *F3H* or *DFR*, depending on whether there is predominant co-production with anthocyanins of flavones or flavonols, respectively (Martin *et al.*, 1991; Davies *et al.*, 1993; Martin & Gerats, 1993; Quattrocchio *et al.*, 1993;

Rosati *et al.*, 2003). There are exceptions; for example, *ANS* may be a key regulatory target in *Viola cornuta* (Farzad *et al.*, 2003), and in *M. domestica*, the flowers of which pigment very early in development, concomitant induction of both EBGs and LBGs is found (Dong *et al.*, 1998). There are few studies on flowers of monocotyledonous species, but in the coloured floral parts of *Anthurium andraeanum* the regulatory groupings seem distinct to those described for many dicotyledonous plants (Collette *et al.*, 2004). The genes involved in transport to the vacuole, such as GSTs, are likely to be regulated with other LBGs, based on studies in *Z. mays* (Marrs *et al.*, 1995), *L. esculentum* (Mathews *et al.*, 2003) and *P. hybrida* (Alfenito *et al.*, 1998).

Much is understood about both the specific *cis*-elements in the biosynthetic genes that respond to the regulatory signals, and the TFs involved. Some of the functional interactions between the different types of TF are now being elucidated, and upstream regulators of the genes encoding the TFs are being identified. A major common feature that has emerged is the key role of interacting MYB and basic helix-loop-helix (bHLH)-type TFs as activators, and the involvement of assisting WD40 proteins. The TFs in petals have been studied in detail for *A. majus*, *G. hybrida* and *Petunia*. Similar regulatory systems have also been described for non-petal systems, in particular the vegetative tissues and seeds for *A. thaliana*, *P. frutescens*, *Z. mays*, all of which also involve MYB and bHLH proteins as positive regulatory factors (reviewed in Martin *et al.*, 2001; Saito & Yamazaki, 2002; and Davies & Schwinn, 2003).

For *A. majus* and *Petunia*, multiple MYB and bHLH proteins have been identified that regulate anthocyanin biosynthesis in petals, with the LBGs being the prime regulatory targets. Recent studies on *A. majus* also cast light on the mechanisms involved in formation of the complex pigmentation phenotypes and patterns that often occur in flowers. In *A. majus*, *Rosea1*, *Rosea2* and *Venosa* form an anthocyanin *Myb* gene family (Martin *et al.*, 2001), and *Delila* (Goodrich *et al.*, 1992) and *Mutabilis* (Martin *et al.*, 2001) encode bHLH factors. The combined action of these genes provides for both control of the temporal production of anthocyanins in the petals and the complex patterns of spatial distribution. *Delila* is expressed in both the petal lobes and tube, while *Mutabilis* is expressed only in the lobes. *Rosea1* and *Rosea2* result in differing pigmentation patterns in the petal, and *Venosa* expression generates pigmentation only in epidermal cells overlying veins, producing a striking venation pattern (Martin *et al.*, 2001).

In *Petunia*, *Jaf13* and *Anthocyanin1* (*An1*) encode bHLH factors, and *An2*, and probably *An4*, encode MYBs (Quattrocchio *et al.*, 1998, 1999; Spelt *et al.*, 2000, 2002). All control the expression of the LBGs. The AN1 amino acid sequence is most similar to INTENSIFIER1 of *Z. mays* and that of JAF13 is more similar to DELILA of *A. majus* and R of *Z. mays*. AN1 and JAF13 may also differ in function, with the suggestion that hierarchical control may be operating within the regulatory cascade controlling anthocyanin biosynthesis in *Petunia* (Spelt *et al.*, 2000). This is supported by the observation that *Jaf13* expression does not compensate for loss of *An1* activity in *Petunia* petals, and that *An1* gene expression, but not *Jaf13* gene expression, is dependent on the activity of *An2*. Studies on transgenic *Petunia* expressing AN1-glucocorticoid receptor constructs have demonstrated AN1 is a direct activator of the LBGs and also of an *Myb* gene of unknown function (*Pmyb27*).

Mutant *Petunia* lines lacking activity of a WD-repeat (WD40) protein encoded by the gene *An11* (de Velten *et al.*, 1997) lose the ability to produce anthocyanins. Transgenic introduction of TFs active in the anthocyanin pathway can partially complement the phenotype of similar WD40 mutations in *A. thaliana*, suggesting the WD40 is required for their activity. However, the *An11* product does not function through transcriptional control of the *Myb* and *bHLH* genes, as mRNA levels for *An2* are not reduced in the *an11* mutant (de Vetten *et al.*, 1997). Rather, WD40 proteins may promote the activity of MYB and/or bHLH proteins by physical interaction, perhaps by stabilising the MYB:bHLH transcription complexes. This is supported by yeast two-hybrid assays showing interactions between anthocyanin-related WD40 and bHLH proteins (Payne *et al.*, 2000; Sompornpailin *et al.*, 2002). As yet, there are few candidates for direct regulators of the *Myb* and *bHLH* genes.

Less information is available on regulation of anthocyanin biosynthesis in flowers of species other than *A. majus* and *Petunia*. However, a *bHLH* gene has been isolated from gerbera, *Gmyc1*, which regulates *DFR* gene activity in the corolla and carpel (Elomaa *et al.*, 1998). Also, a white-flowered mutant of *Matthiola incana* named '*G*' has been shown to be the result of a mutation in a *bHLH* gene, and the flower phenotype can be complemented by expression of *bHLH* cDNAs from *A. thaliana* (Ramsay *et al.*, 2003). A second *M. incana* mutant, '*E*', corresponds to a WD40 protein required for anthocyanin production (Ramsay *et al.*, 2003).

The genes encoding *FLS*, *FNS* and some hydroxylation and secondary modification enzymes appear to be regulated separately from the LBGs in some species, although limited data are available as yet. For example, the expression of the *Ht1* gene for F3'H in *Petunia* is not affected by the *an1*, *an2* and *an11* mutations (de Vetten *et al.*, 1999). A pair of *Myb* genes has been identified in *Z. mays* that controls production of flavone *C*-glycosides in the silks (Zhang *et al.*, 2003), but their equivalents have not been reported for regulation of flavonols or flavones in petals.

Similarly, there is little data on the regulation of the EBGs in petals, although their expression in response to light and pathogen stimuli in other plant tissues has been studied extensively (reviewed in Weiss, 2000 and Davies & Schwinn, 2003). However, there have been some *Myb* genes isolated that are expressed in flowers and that may affect the activity of genes in the early phenylpropanoid biosynthetic pathway. Jackson *et al.* (1991) isolated six *Myb* cDNAs from *A. majus* petals (*AmMYBs*), and the subsequent functional studies of gene expression patterns, promoter-binding activities and phenotypes resulting from their expression in transgenic plants suggest that some of these may have a role in either regulating EBGs and/or flavone or flavonol biosynthesis. The AmMYB305 protein and the equivalent protein from *Nicotiana* have been shown to bind and activate a 'P-box' MYB recognition element that is linked to petal-enhanced expression of phenylpropanoid genes (Sablowski *et al.*, 1994). Furthermore, AmMYB305 activates the promoters of the *CHI* and *F3H* genes of *A. majus*, suggesting it regulates flavonol biosynthesis (Moyano *et al.*, 1996). AmMYB340 may serve a related function to AmMYB305, as they are structurally related and AmMYB340 binds to the P-box and regulates *CHI* (Moyano *et al.*, 1996). In yeast and plant protoplast studies with target promoters, AmMYB305 had stronger promoter-binding affinity but weaker transactivation strength than AmMYB340. The difference in binding affinity was associated with phosphorylation of AmMYB340 but not AmMYB305. AmMYB340 could potentially be out-competed by the weaker activator for promoter sites, providing for fine gearing of gene expression levels.

There is evidence that some of the anthocyanin TFs regulate multiple pathways. The *Rosea1* (Schwinn, unpublished data) and *An1*, *An2* and *An11* genes (Spelt *et al.*, 2002) regulate vacuolar pH in addition to anthocyanin production, and *An1* also is involved in seed coat epidermal cell development. *A. thaliana* provides an excellent model for studying such overlapping pathways. The anthocyanin and proanthocyanidin regulatory pathways share components with those controlling developmental processes such as trichome formation, root epidermal cell development and seed mucilage production (Vom Endt *et al.*, 2002; Davies & Schwinn, 2003). The regulation of proanthocyanidin biosynthesis is reviewed in Chapter 5, and the regulation of the biosynthesis of the proanthocyanidin-like phlobaphenes is discussed in Section 4.3.5.

4.4.2   Biotechnology applications of transcription factors regulating flavonoid biosynthesis

The identification of defined TF genes provides tools for modulating both the amount of flavonoids in plants and their temporal and spatial occurrence. The validity of this approach has been well established with studies involving transgenic plants of several species. These experiments have also illustrated the ability to apply TF gene technology in heterologous species. Another advantage of targeting regulation of pathways, for example to increase flavonoid levels, is that the use of TF genes may overcome the need to identify a rate-limiting biosynthetic step.

In the experiments to date, *Myb* or *bHLH* cDNAs, in particular *C1* and *Lc* from *Z. mays*, have been placed under the control of the *CaMV35S* promoter and introduced into a range of species. Anthocyanin production was successfully increased in transgenic plants of several species, but the transgenes also showed no effect in some species (Table 4.1). Levels of flavonols in *L. esculentum* (Bovy *et al.*, 2002; Le Gall *et al.*, 2003), proanthocyanidins in *Lotus corniculatus* (Robbins *et al.*, 2003) and various flavonoids and other metabolites in *Z. mays* cell lines (Grotewold *et al.*, 1998) have also been increased using anthocyanin-related TF transgenes. Furthermore, other TF transgenes have been used successfully to downregulate branches of the phenylpropanoid pathway (Tamagnone *et al.*, 1998; Damiani *et al.*, 1999; Jin *et al.*, 2000; Aharoni *et al.*, 2001).

Table 4.1 Genetic modification of plant colour by introduction of genes encoding transcription factors that regulate the anthocyanin biosynthetic pathway

Species	Transgene (all sense orientation)	TF type	Effect on anthocyanin levels[1]	Reference
Arabidopsis thaliana				
	C1	Myb	No visible change	Lloyd *et al.*, 1992
	Lc	bHLH	Increased	Lloyd *et al.*, 1992
	Lc and *C1*	bHLH and Myb	Increased	Lloyd *et al.*, 1992
	Delila	bHLH	Possible slight increase	Mooney *et al.*, 1995
Eustoma grandiflorum				
	Lc	bHLH	No visible change	Bradley *et al.*, 1999
	Rosea1	Myb	Increased	Schwinn *et al.*, 2001
Gerbera hybrida				
	Delila	bHLH	Increased	Elomaa, 1996
Lotus corniculatus				
	Sn	bHLH	Increased	Robbins *et al.*, 2003
Lycopersicon esculentum				
	ant1	Myb	Increased	Mathews *et al.*, 2003
	Delila	bHLH	Increased	Mooney *et al.*, 1995
Medicago sativa				
	Lc	bHLH	Increased	Ray *et al.*, 2003
	B-Peru	bHLH	No visible change	Ray *et al.*, 2003
	C1	Myb	No visible change	Ray *et al.*, 2003
Nicotiana				
	C1	Myb	No visible change	Lloyd *et al.*, 1992
	Lc	bHLH	Increased	Lloyd *et al.*, 1992
	ant1	Myb	Increased	Mathews *et al.*, 2003
	Delila	bHLH	Increased	Mooney *et al.*, 1995
	GMYB10	Myb	Increased	Elomaa *et al.*, 2003
Pelargonium				
	Lc	bHLH	No visible change	Bradley *et al.*, 1999
Petunia				
	Lc	bHLH	Increased	Bradley *et al.*, 1998
Trifolium repens				
	B-Peru	bHLH	Increased	de Majnik *et al.*, 2000
	C1	Myb	Increased	de Majnik *et al.*, 2000
	Myb.Ph2	Myb	Increased	de Majnik *et al.*, 2000
Zea mays (cell cultures)				
	C1 and *R*	bHLH and Myb	Increased	Grotewold *et al.*, 1998

[1] Only a general indication of phenotype is given, as results were variable and included small increases in anthocyanin in tissues already producing anthocyanins, production of anthocyanin earlier in flower development than normal, novel anthocyanin production in particular cells or tissues, novel anthocyanin production only under stress conditions.

Combining transgenes for both MYB and bHLH factors can result in high levels of anthocyanin production, or induction of anthocyanins in tissues that lacked them in single transgene plants. For example, *A. thaliana* plants with both *CaMV35S::C1* and *CaMV35S::Lc* transgenes had increased anthocyanin levels and novel tissue distribution patterns not seen in plants with either transgene on its own, with anthocyanins appearing in the petals, roots and stamens (Lloyd *et al.*, 1992).

Obviously, flavonoid TF genes may be used to modulate both the amount and distribution of anthocyanins in plants. However, the results to date also highlight problems with the available technology, or the lack of understanding of the exact roles of the TFs encoded by the transgenes in the target plant species. Experiments with the related bHLH TFs LC, SN, DELILA, GL3 and MYC-146 have shown that similar TFs produce quite different transgenic phenotypes, and furthermore that the same gene constructs may have markedly different effects in transgenic plants of different species (Mooney *et al.*, 1995; Bradley *et al.*, 1998, 1999; Damiani *et al.*, 1999; de Majnik *et al.*, 2000; Payne *et al.*, 2000; Ramsay *et al.*, 2003; Ray *et al.*, 2003; Robbins *et al.*, 2003). Phenotypic variation in transgenics of different species containing transgenes for the same TF may reflect differences in the presence or absence of interacting endogenous TFs and/or differences in the promoter sequences of the endogenous biosynthetic genes. Furthermore, TFs with very similar amino acid sequences may differ in their activities *in planta*, so that transgenes for such TFs may generate different phenotypes in transgenic plants of the same species. The influence of endogenous factors on the transgenic phenotypes is illustrated by the markedly different phenotypes of some anthocyanin regulatory gene transgenics under varying environmental conditions. *Medicago sativa* *CaMV35S::Lc* transgenics only showed a strong vegetative pigmentation pheno-type when placed in stress conditions or exposed to high light levels (Ray *et al.*, 2003). Similarly, *CaMV35S::Lc Petunia* plants grown in greenhouses had pigmen-tation levels barely above those of non-transgenic plants, but the same transgenics lines acquired a strongly pigmented phenotype within days of transplanting to the field (Schwinn and co-workers, unpublished data – a photograph of the transgenics is in Davies & Schwinn, 2003). As with *M. sativa*, the induction of pigmentation in the *Petunia* transgenics has been linked to high light levels.

4.5   Genetic modification of flower colour

4.5.1   Overview

Not surprisingly, flower colour is a key consumer trait for ornamentals. As some leading ornamental crops have only a narrow colour spectrum, or lack specific colours such as blue or yellow, several research groups have identified genetic modification of pigment biosynthesis as a means to introduce colours outside the existing range. The extensive information available on the genetics and biochem-istry of flavonoid biosynthesis provides a strong foundation to such research, increasing the probability of success. Combined with the importance of some flavonoids to plant defence and human health, this has led to the flavonoid pathway being the target of more biotechnology research than any of the other pigment pathways of plants, and indeed, probably more than any other plant secondary metabolite pathway. Thus, there are numerous examples of modification of plant or flower colour by changing flavonoid biosynthesis in transgenic plants (Tables 4.2 and 4.3). There have been several field trials of potential commercial products arising from this research, and transgenic *Dianthus* cultivars with novel flower colours are being grown in South America and Australia for sale in Australia, Japan and the USA (Lu *et al.*, 2002). In some examples two or three different transgenes altering flavonoid biosynthesis have been used in combination to achieve the desired result. In this section we present an overview of the major approaches that have been taken to modify plant colour through alteration of flavonoid production through expression of biosynthetic transgenes. The use of transcription factor genes to modify flavonoid biosynthesis is discussed in Section 4.4.2.

Production of stably transformed plants is expensive and time-consuming. Two approaches are useful for obtaining prior information on the potential success of major transgenic experiments. Firstly, simple chemical and biochemical studies can provide information on the biosynthetic pathway and the substrate specificity of relevant enzymes of the target species. They also have the potential to give an indication of new flower colours that may be formed. An example of this is the use of enzyme inhibitors or flavonoid substrate feeding to generate pelargonidin- or delphinidin-derived anthocyanins, respectively, in *Chrysanthemum* florets (Schwinn *et al.*, 1994; Davies & Schwinn, 1997) or flowers of other species (Martens *et al.*, 2003b). Secondly, transient gene overexpression or RNAi-based gene inhibition in petals, typically using the gene gun, is often an effective way to get an indication of the transgene effect. Such approaches may be useful in a wide range of plant species for designing new strategies to modify flavonoid biosynthesis and flower colours.

Of course any genetic modification approach for producing a new ornamental crop requires a transformation system. For many years this was a major hurdle to ornamental biotechnology, and still remains so for some species. However, the number of ornamental species that can be transformed has increased rapidly over the last few years. Deroles *et al.* (2002) lists over 30 species of ornamental plants for which transgenic plants have been produced, and reports on others have been published since that review was prepared. The transformable species include major cut flower crops (e.g. *Chrysanthemum*, *Dianthus*, *G. hybrida* and *R. hybrida*), bedding plants (e.g. *Pelargonium*), bulbs, pot plants (e.g. *C. persicum*) and groundcover plants.

4.5.2   *Preventing anthocyanin production*

Perhaps the simplest approach for modifying flower colour is to prevent pigment formation by inhibiting production of a key flavonoid biosynthetic enzyme, such as CHS. Indeed, the first published account of antisense RNA (van der Krol *et al.*, 1988) or sense RNA (Napoli *et al.*, 1990; van der Krol *et al.*, 1990) inhibition of plant gene expression was for *CHS* in *Petunia*. Subsequently, similar experiments

Table 4.2 Genetic modification of flower colour in ornamentals using inhibition of flavonoid biosynthetic gene activity by sense suppression or antisense constructs

Species	Transgene	Colour change	Reference
Chrysanthemum			
	Sense and antisense *CHS*	Pink to white	Courtney-Gutterson *et al.*, 1994
Dianthus caryophyllus			
	Sense *CHS*	Pink to pale pink	Gutterson, 1995
	antisense *F3H*	Orange to white	Zuker *et al.*, 2002
Eustoma grandiflorum			
	Antisense *CHS*	Purple to white and patterns	Deroles *et al.*, 1998
	Antisense *FLS*	Change in purple shade	Nielsen *et al.*, 2002
Gerbera hybrida			
	Antisense *CHS*	Red to pink/cream	Elomaa *et al.*, 1993
	Antisense *DFR*	Red to pink	Elomaa *et al.*, 1993
Petunia			
	Sense *DFR*	Purple to white and patterns	Jorgensen *et al.*, 2002
	Antisense *CHS*	Red to white and patterns	van der Krol *et al.*, 1988
	Antisense *CHS*	Purple to pale purple/white	Tanaka *et al.*, 1998
	Sense *CHS*	Purple to white and patterns	Napoli *et al.*, 1990; van der Krol *et al.*, 1990
	Antisense *FLS*	Purple to red	Holton *et al.*, 1993b
	Antisense *FLS*	White to pale pink-purple	Davies *et al.*, 2003a
	Antisense *FLS* and sense *DFR*	White to pink	Davies *et al.*, 2003a
	Antisense *3RT*	Purple to pink and patterns	Brugliera *et al.*, 1994
	Sense *F3′,5′H*	Dark blue to pale blue or pink	Shimada *et al.*, 2001; Jorgensen *et al.*, 2002
Rosa hybrida			
	Sense *CHS*	Red to pale red	Firoozabady *et al.*, 1994
Torenia			
	Sense *CHS* or *DFR*	Blue to white, patterns and yellow	Suzuki *et al.*, 2000
	Sense and antisense *CHS* or *DFR*	Violet to pale violet and patterns	Aida *et al.*, 2000a, 2000b
	Sense *F3′,5′H*	Blue to pink	Suzuki *et al.*, 2000
	FNSII	Paler flower colour	Ueyama *et al.*, 2002

have been carried out in a range of species, including *Chrysanthemum, D. caryophyllus, E. grandiflorum, G. hybrida, Torenia fournieri, Petunia* and *R. hybrida*, with varied results (reviewed in van Blokland *et al.*, 1993; Elomaa & Holton, 1994; Davies & Schwinn, 1997; Tanaka *et al.*, 1998; Ben-Meir *et al.*, 2002). Inhibition of flavonoid biosynthesis may result in sterility in some (but not all) species, examples including male (Taylor & Jorgensen, 1992) or female (Jorgensen *et al.*, 2002) sterility in *Petunia*, and male sterility in *Nicotiana* (Fischer *et al.*, 1997).

Often the predicted white-flower phenotype is produced, but in some species transgenic lines with complex pigment patterning are also generated. Both ordered and erratic pigmentation patterns have been obtained for sense and/or antisense experiments with *CHS* and *DFR* in *Petunia* (van der Krol *et al.*, 1988, 1990; Napoli *et al.*, 1990), *CHS* in *E. grandiflorum* (Deroles *et al.*, 1998) and *CHS* and *DFR* in *T. fournieri* (Aida *et al.*, 2000a). The different types of pattern show variations in their stability throughout a particular plant and through inheritance, and this may relate to varied responses to morphological and environmental signals, and also to differences in transgene structure (Jorgensen, 1995). Striking patterns in *E. grandiflorum*, such as colour only at the petal tips or stripes and swirls of colour, were found to be too unstable to be of easy commercial application (Bradley *et al.*, 2000). Interestingly, no floral pigmentation patterning was found with antisense experiments with *Chrysanthemum, G. hybrida* and *R. hybrida*, species that traditionally lack patterned varieties.

An alternative approach for controlled reduction of flavonoid enzyme activity, for example in a tissue-specific manner, is the expression of single-chain antibody fragments developed against the target enzyme. This has been attempted for enzymes such as DFR (De Jaeger *et al.*, 1999), but without clear phenotype effects.

4.5.3 Redirecting substrate in the flavonoid pathway

The colourless flavonols are abundant in many flower petals and also occur in foliage. Flavonols are produced from DHFs, which are also indirect precursors of the coloured anthocyanins. If present at the same stage of development, the FLS and DFR compete for the DHF substrate. Altering the balance of these activities and the conversion of DHF to either flavonols or leucoanthocyanins may enable an increase in anthocyanin levels in petals or foliage or even the introduction of coloured flowers into white-flowered cultivars or species. Introduction of antisense *FLS* (Holton *et al.*, 1993b) or *DFR* sense (Polashock *et al.*, 2002) constructs into *Nicotiana tabacum* increased anthocyanin content of some flower parts and changed petal colour from light pink to red. Increased anthocyanin production also resulted from the introduction of an antisense *FLS* construct into *E. grandiflorum* (Davies *et al.*, 1997; Nielsen *et al.*, 2002). In a white-flowered line of *Petunia* (cultivar Mitchell), either reducing *FLS* transcript abundance using antisense RNA or increasing *DFR* transcript abundance restored anthocyanin biosynthesis to the flower limb (Davies *et al.*, 2003a). In *Forsythia* anthocyanins accumulate in some tissues, such as the stems, petioles, leaf veins and sepals, but only flavonol glycosides are found in petals. Introduction of an *A. majus DFR* cDNA, under the control of the *CaMV35S* promoter into *Forsythia* X*intermedia*, resulted in increased levels of anthocyanins in tissues that were already producing them, demonstrating that DFR is rate-limiting for anthocyanin biosynthesis in these tissues (Rosati *et al.*, 1997). The addition of a second transgene, for *M. incana ANS*, extended anthocyanin pigmentation to the flowers of the double transgenics (Rosati *et al.*, 2003).

In some species, strong substrate preference is shown by DFR with regard to the degree of B-ring hydroxylation of the DHF substrate. Notable from a biotechnology perspective is the absent or low selectivity for DHK by the DFR of *C. hybrida*, *Nicotiana*, *Petunia*, *L. esculentum* and *V. macrocarpon* (see Section 4.3.4.3), which markedly limits the production of pelargonidin-derived anthocyanins in these species. In the first published case of genetic modification of flower colour, Meyer *et al.* (1987) overcame this biosynthetic block in *P. hybrida* by introducing a *Z. mays CaMV35S::DFR* transgene, leading to production of pelargonidin-derived anthocyanins in the petals. Subsequent crosses of the transgenics to the 'Red titan' cultivar lead to a range of F_2 lines that had some flower colours novel to *Petunia*, including orange. Similar results in *P. hybrida* have been obtained with *DFR* cDNAs from *G. hybrida* (Helariutta *et al.*, 1993) and *R. hybrida* (Tanaka *et al.*, 1995).

Introduction of enzymes that compete with the anthocyanin biosynthetic enzymes and make weakly coloured or colourless compounds can be used to reduce anthocyanin biosynthesis and change flower colour. STS is directly responsible for the production of resveratrol-type phytoalexins in species such as *V. vinifera* and *Arachis hypogaea* (peanut). The enzyme from *V. vinifera* uses the same substrates as CHS, malonyl-CoA and 4-coumaroyl CoA. Thus, if both CHS and STS are active in the same tissue they will compete for substrate. Fischer *et al.* (1997) introduced a cDNA encoding the *V. vinifera* STS under the control of the *CaMV35S* promoter into *Nicotiana*. Flowers of the transgenics were near white rather than the usual dark pink, due to reduced anthocyanin levels. Similarly, Davies *et al.* (1998) and Joung *et al.* (2003) used *CaMV35S::CHR* transgenes to divert substrate into 6′-deoxychalcone production in *Petunia* and *Nicotiana*, respectively, significantly reducing floral anthocyanin biosynthesis and resulting in pale flower colours. Introducing competing activities at other branchpoints in the flavonoid pathway can have a similar effect. Overexpression of the anthocyanidin reductase (*ANR*) cDNA from *A. thaliana* in *N. tabacum* changed the flower colour of some transgenics from pink to white through competition with UF3GT (Xie *et al.*, 2003). Thus, the introduction of *STS*, *CHR* or *ANR* transgenes offers a route for controlled reduction of anthocyanin biosynthesis.

4.5.4   *Introducing novel flavonoid compounds*

4.5.4.1   *Chalcones, aurones and flavonols*
Good yellow-flowered varieties are lacking for many commercial ornamentals such as *Cyclamen*, *E. grandiflorum*, *Impatiens*, *Pelargonium* and *Petunia*. Thus, a

Table 4.3 Genetic modification of flavonoid biosynthesis in flowers by introduction of novel flavonoid biosynthetic activities or increasing endogenous activities (all 'sense' transgenes)

Species	Transgene	Flower colour change	Reference
Arabidopsis thaliana			
	AUS	None (seed colour changed)[1]	Davies *et al*., 2003b
Dianthus caryophyllus			
	F3',5'H	Pink to blue/purple	Lu *et al*., 2003
	DFR and *F3',5'H*	Pink/white to blue/purple	Holton, 1996; Lu *et al*., 2003
Eustoma grandiflorum			
	UF3GT	None[2]	Schwinn *et al*., 1997
Forsythia X*intermedia*			
	DFR	Vegetative pigmentation increased	Rosati *et al*., 1997
	DFR and *ANS*	Vegetative and flower pigmentation increased	Rosati *et al*., 2003
Nicotiana			
	CHR	Pink to pale pink	Joung *et al*., 2003
	STS	Pink to pale pink	Fischer *et al*., 1997
	DFR	Pink to dark pink	Polashock *et al*., 2002
	F3',5'H	Change in pink shade	Shimada *et al*., 1999; Okinaka *et al*., 2003
	ANR	Pink to white	Xie *et al*., 2003
Petunia			
	CHR	White to pale yellow	Davies *et al*., 1998
	DFR	White to pink	Davies *et al*., 2003
	DFR	Pale pink to orange and red	Meyer *et al*., 1987; Helariutta *et al*., 1993; Tanaka *et al*., 1995; Johnson *et al*., 2001
	F3'H	Lilac to pink	Brugliera *et al*., 1999
	F3',5'H	Pale pink to magenta and patterns	Holton *et al*., 1993a; Shimada *et al*., 2001
	UF3'GT and *UF5GT*	None[3]	Fukuchi-Mizutani *et al*., 2003
	Dv3MaT	None[4]	Suzuki *et al*., 2002
Torenia			
	F3'H	Increased cyanidin-derivatives	Ueyama *et al*., 2002

[1] Seed colour was restored in the *tt5* mutant.

[2] A change in flavonoid glycosylation and acylation occurred.

[3] One new anthocyanin type, delphinidin 3,5,3'-triglucoside, was found.

[4] Anthocyanins with novel malonylation were formed.

gene technology approach for generating yellow flower colours in target species would be of much commercial interest. Three groups of flavonoids are known to provide yellow colours: chalcones, aurones and flavonols (see Section 4.2.1.2 for further details).

Chalcones are pale yellow intermediates in the biosynthesis of all flavonoids but in petals their accumulation is usually ephemeral. Surprisingly, attempts so far to reduce expression of *CHI* genes in *Petunia* by antisense RNA or sense suppression, which would encourage chalcone accumulation, have been unsuccessful (van Blokland *et al.*, 1993). However, Davies *et al.* (1998) developed a biotechnology route for generating chalcone accumulation in transgenics using CHR. As described in Section 4.3.3, when CHR is active in conjunction with CHS, it generates 6′-deoxychalcones, which are more stable than 6′-hydroxychalcones and may not be substrates for the endogenous CHI of the target species. By introducing a *CHR* cDNA from *M. sativa* under the control of the *CaMV35S* promoter into a white-flowered line of *Petunia*, transgenic lines were generated that accumulated up to 50% of their petal flavonoids as 6′-deoxychalcones, changing the flower colour from white to pale yellow. Why chalcones resulted in pale colours in these transgenic lines but can produce strong colours in, for example, some *Dianthus* cultivars and *Petunia* pollen is not clear. However, Markham *et al.* (2001) have shown how differences in cellular localisation, and thus chemical environment, can impact strongly on the colour of the yellow flavonoid pigments (see Section 4.2.5).

Aurones offer very good prospects for biotechnology approaches to generating yellow colours. They are bright yellow, their substrates (chalcones) are common in many flowers, and their biosynthesis is thought to require the activity of only the AUS to form the base chromophore from the chalcone (see Section 4.3.8). However, no experiments showing the use of the *AUS* cDNA to control aurone production in transgenic plants have been published.

4.5.4.2   3-deoxyanthocyanins
Although the regulation of 3-deoxyanthocyanin production has been elucidated for *Z. mays*, and cDNAs encoding enzymes with FNR activities have been isolated from a number of species (see Section 4.3.5), no success has been reported for the introduction of 3-deoxyanthocyanin biosynthesis into species that normally lack it. It would be predicted that in addition to introducing a FNR activity, a loss of F3H activity would be required in the target cultivar. Furthermore, a specific ANS may be required to form the 3-deoxyanthocyanidin, or a specific glycosyltransferase (e.g. a UF5GT) to form the anthocyanin. Despite this, the bright orange-red colours known to result from 3-deoxyanthocyanins means they are still a good prospect for biotechnology approaches for generating new flower colours.

4.5.4.3   Altering B-ring hydroxylation
If one looks at the occurrence of the major flavonoid types in flowers of important ornamental crops from a biotechnology perspective, some key issues stand out. Firstly, some of the major species lack anthocyanins based on one or more of the

major anthocyanidin types. In particular, significant numbers lack delphinidin-derived anthocyanins and thus blue colours. Also, a smaller number of species lack pelargonidin-derived anthocyanins and consequently anthocyanin-based orange colours. Secondly, some species lack blue flower colours even though they produce delphinidin-derived anthocyanins. An obvious approach for biotechnology is to introduce or inhibit the activity of the F3'H and F3',5'H, cDNAs for which are available from a range of species. *F3',5'H* transgenes have been introduced into several species that lack the activity, including *Chrysanthemum*, *R. hybrida* and *Dianthus*, and indeed, transgenic *Dianthus* with novel colours based on delphinidin-derived anthocyanins are now commercially available (Elomaa & Holton, 1994; Lu *et al.*, 2002).

However, as discussed earlier, the presence of delphinidin-derived anthocyanins does not necessarily result in a blue flower colour, as seen with the transgenic *Dianthus* mentioned above, which have violet rather than blue flower colour. Similarly, *C. persicum*, *E. grandiflorum*, *Impatiens*, *Pelargonium* and *Tulipa* are all important ornamental crop species that produce delphinidin derivatives yet lack (true) blue flower colours. Further, there are a few examples in which blue flower colours derive from pigments other than delphinidin: e.g. the main pigment in *C. cyanus* is cyanidin-based (see Section 4.2.4) and in blue-flowered *Ipomoea* it is peonidin with six molecules of glucose and three molecules of caffeic acid attached (Goto & Kondo, 1991). Chemical studies on several blue-flowering species have suggested that other factors besides the presence of delphinidin-derived anthocyanins are usually involved in producing blue. The major factors are vacuolar pH, intramolecular interactions and co-pigmentation (see earlier sections). The efficiency of intramolecular interaction or co-pigmentation in generating blue colour from the anthocyanin is dependent on the type of co-pigments present and the secondary modifications of both the anthocyanins and co-pigments. Furthermore, a suitable vacuolar pH is required, and indeed, it is possible to obtain good blue colours in the absence of some of these other factors if the pH is sufficiently high. The transgenic *Dianthus* accumulate complex delphinidin-derived anthocyanins in good amounts, have flavone co-pigments and a relatively high petal vacuolar pH of 5.5, yet are still not true blue in colour (Fukui *et al.*, 2003). Thus, successful generation of novel blue colours by introduction of the F3',5'H will depend on either the selection of a recipient cultivar with a suitable chemical background, or subsequent further genetic modification of the secondary factors. To date, there are no reports on changing co-pigment biosynthesis to assist in generating blue flower colours. Section 4.5.4.4 discusses progress in developing gene technology to control some of the required secondary modification activities.

Accumulation of high amounts of delphinidin-derived anthocyanins in the absence of pelargonidin- or cyanidin-derived anthocyanins is preferable for true blue colours. However, the efficiency of delphinidin production in some of the early *F3',5'H* transgenics was low, due to substrate competition from the F3'H and the substrate preference of the endogenous DFR for DHK or DHQ. One approach taken to overcome this problem has been the combination of a *F3′,5′H* transgene with a transgene for a *DFR* that 'prefers' DHM as substrate, such as that from *Petunia* (Holton, 1996). In particular, the ideal target for this approach is a plant background that lacks DFR or F3′H activity, and so accumulates DHK. As the *Petunia* DFR greatly prefers DHM to DHK as a substrate, delphinidin production is promoted. In plant lines that have an active F3′H, cyanidin pigments may also be produced, but delphinidin production will still be favoured. A second prospective approach has been suggested by Okinaka *et al.* (2003), who reported the isolation of a *F3′,5′H* cDNA from *Campanula medium* that was particularly efficient at generating delphinidin precursors. When expressed in *Nicotiana*, over 99% of the anthocyanins accumulated were delphinidin-based.

A small number of leading ornamental crops, such as *Chrysanthemum*, *Cymbidium* and *C. persicum*, lack pelargonidin-derived anthocyanins. Except in a few cases in which DFR specificity also plays a role, this is most probably due solely to the dominant nature of the F3′H and F3′,5′H activities. A high activity of these enzymes on naringenin and/or DHK substrates could completely convert the available substrate to a 3′4′-hydroxylated form before it becomes committed to the formation of pelargonidin. Inhibition of the activities of the F3′H and F3′,5′H could enable production of pelargonidin-derived anthocyanins, although results of transgenic experiments demonstrating this approach have not been published. Use of the *F3′H* cDNA to increase the amount of cyanidin-derived anthocyanins has been reported for *T. hybrida* (Ueyama *et al.*, 2002).

4.5.4.4 Secondary modifications
Many cDNAs are now available for enzymes carrying out a range of secondary modifications of anthocyanins, and some of these may have biotechnology applications.

Methylation has a slight reddening effect on the anthocyanin, so alteration of the methylation status of the anthocyanin would be expected to result in a small change in flower colour. To date, no transgenic results targeting anthocyanin methylation have been published. However, methylation has been altered through an indirect route. Brugliera *et al.* (1994) inhibited the activity of the *3RT* gene in *Petunia* by introducing an antisense RNA construct. Flowers of the transgenics had reduced levels of malvidin and increased levels of delphinidin and petunidin pigments, presumably due to the methyltransferase being unable to use the non-rhamnosylated substrate. Flower colour shade was changed in some of the transgenic lines, as well as pigmentation patterns in some cases. Altering glycosylation has also been shown to alter other secondary modification patterns. Introduction of a *CaMV35S::UF3GT* transgene into *E. grandiflorum* resulted in a change from flavonoid 3-galactosylation to 3-glucosylation, and an associated reduction in acylation of the flavonoids (Schwinn *et al.*, 1997). No change in flower colour was observed, however.

As discussed in Section 4.2, glycosylation of the A- and B-rings, and associated acylation, are often key to generating true blue colours from anthocyanins, and the genes for such activities have been the research targets of companies wanting to introduce blue flower colours to specific ornamental species. As a result, there has been much recent progress on isolation of cDNAs for secondary modification enzymes, and in theory, the necessary biotechnology tools are available to introduce production of complex delphinidin structures and appropriate co-pigments for formation of blue colours in ornamentals such as *Dianthus*, *G. hybrida*, *R. hybrida* and *Chrysanthemum*. However, published transgenic experiments to date have concentrated on model species rather than the major commercial ornamentals.

Fukuchi-Mizutani *et al.* (2003) cloned a cDNA for a UDP-Glu:anthocyanin 3'-*O*-glucosyltransferase from *G. triflora* that is involved in the formation of the blue anthocyanin gentiodelphin (Yoshida *et al.*, 2000). The cDNA was expressed in *P. hybrida* in conjunction with a *UF5GT* cDNA. One new anthocyanin type, delphinidin 3,5,3'-triglucoside, was found, but only at low levels (2–6% of total anthocyanins), and no change in flower colour occurred. The lack of other new anthocyanin types, such as delphinidin 3,3'-diglucoside, was probably related to the specificity of the enzyme for a 3,5-glucosylated anthocyanin substrate. The identification and availability of aromatic and aliphatic anthocyanin acyltransferase cDNAs (see Section 4.3.7.3) should prove useful for modification of flower colour. However, transgenic plant results are available only for the Dv3MaT, which was overexpressed in *Petunia*. The transgenics produced new malonylated anthocyanins in the petals but flower colour was unaffected.

4.5.5   Modifying the cellular environment of the flavonoid

Altering the immediate environment of a flavonoid pigment can have dramatic effects on the colour the pigment manifests for reasons discussed earlier. The environment may be altered by changing the pH of the cellular compartment, altering the subcellular site of flavonoid accumulation or altering the other cellular components with which the flavonoid interacts. In theory, gene technology is available for targeting all of these aspects, but it has yet to be proven in transgenic plants.

For example, a gene encoding a vacuolar Na^+/H^+ exchanger has been isolated that results in a change in petal vacuolar pH in *Ipomoea nil* of about 0.7 units (Fukada-Tanaka *et al.*, 2000; Yamaguchi *et al.*, 2001). Other *pH* genes are known from genetic studies in *Petunia*, and some of the anthocyanin regulatory genes of *Petunia* and *A. majus* also regulate vacuolar pH (see Section 4.4.1). Also, one gene has been isolated that is associated with the formation of AVIs (see Section 4.2.5). However, no transgenic plant applications with pH or AVI genes have been published.

Alteration of the subcellular localisation of flavonoids has been reported for mutations affecting the activity of GTs or GSTs that act on flavonoids. These mutations result in changes in the colour of the pigments due to their accumulation in the cytoplasm rather than the vacuole. Similarly, it is thought that some yellow flower colours are caused by accumulation of flavonols in the cytoplasm (see

Section 4.2.5). Thus, it should be possible to achieve colour changes in transgenics by applying approaches to cause accumulation of flavonoids in the cytoplasm.

4.5.6 Prospects for new flavonoid gene technology

As discussed previously, there are less common types of flavonoid that differ in patterns of glycosylation or the oxygenation of the A- or C-rings, which can result in novel flower colours, including strong yellow, orange, red and blue. The presence of 6-hydroxyanthocyanins produces striking colours in some species, and 6- and 8-hydroxyflavonols may contribute to yellow flower colours, most notably in *Gossypium*. However, the enzyme(s) involved in the 6- and 8-hydroxylation of these flavonoids have yet to be characterised. However, given the level of genomic research into *Gossypium*, one might expect progress in this area in the near future. Also, it may be possible to apply the 6-hydroxylase gene of isoflavonoid biosythesis (Latunde-Dada *et al.*, 2001) to the biosynthesis of other 6-hydroxylated flavonoids.

Anthocyanin *O*-methyltransferase cDNAs could potentially be used to introduce novel activities into target species, which should result in a change in colour shade. The highly specific substrate usage of the enzymes from species such as *P. hybrida* may be a limiting factor (Brugliera *et al.*, 1994), although cDNAs from a range of species have been isolated (Brugliera *et al.*, 2003). Variant patterns of glycosylation for which no molecular studies on the biosynthesis have been published yet include cyanidin 3′- or 4′-glycosylation, anthocyanins with 3′-glycosylation in the absence of the common 3-glucosylation, and the rare flavonol and flavone glycosides, such as quercetin 7- or 4′-glucosides, 3′-methoxy-quercetin glycosides and 3′,5′-dimethoxy-myricetin glycosides (see earlier sections and Harborne, 1976; Fossen & Andersen, 1999; Fossen *et al.*, 2003).

4.6   Concluding comments

Knowledge of flavonoid chemistry and biosynthesis is very well advanced, with most of the biosynthetic enzymes now characterised and associated genes isolated. This has allowed a wide range of genetic modification approaches of flavonoid biosynthesis in transgenic plants. Furthermore, anthocyanin biosynthesis has been at the forefront of research on the regulation of gene transcription in plants. Genes for TFs regulating flavonoid biosynthesis have become important tools for increasing the levels of anthocyanins and other flavonoids in ornamental and food crops. However, there are still areas of flavonoid biosynthesis that are not well understood. In particular, the localisation of flavonoids to their varied subcellular locations, the potential metabolic channelling of substrates within flavonoid enzyme complexes, flavonoid turnover, and biosynthesis of some of the rarer flavonoid types. With regard to genetic modification of flavonoid biosynthesis for novel flower colours, there have been many published successes and transgenic flowers are in the marketplace. However, the generation in transgenic plants of commercially viable yellow or blue flower colours has not been reported.

Acknowledgements

Our thanks to Dr Ken Markham for his comments on the manuscript.

References

Abrahams, S., Lee, E., Walker, A. R., Tanner, G. J., Larkin, P. J. & Ashton, A. R. (2003) The Arabidopsis *TDS4* gene encodes leucoanthocyanidin dioxygenase (LDOX) and is essential for proanthocyanidin synthesis and vacuole development. *Plant Cell*, **35**, 624–636.

Aharoni, A., De Vos, C. H., Wein, M., Sun, Z., Greco, R., Kroon, A., Mol, J. N. & O'Connell, A. P. (2001) The strawberry FaMYB1 transcription factor suppresses anthocyanin and flavonol accumulation in transgenic tobacco. *Plant J.*, **28**, 319–332.

Aida, R., Kishimoto, S., Tanaka, Y. & Shibata, M. (2000a) Modification of flower colour in torenia (*Torenia fournieri* Lind.) by genetic transformation. *Plant Sci.*, **153**, 33–42.

Aida, R., Yoshida, K., Kondo, T., Kishimoto, S. & Shibata, M. (2000b) Copigmentation gives bluer flowers on transgenic torenia plants with the antisense dihydroflavonol-4-reductase gene. *Plant Sci.*, **160**, 49–56.

Akashi, T., Aoki, T. & Ayabe, S. (1998) Identification of a cytochrome P450 cDNA encoding (2S)-flavanone 2-hydroxylase of licorice (*Glycyrrhiza echinata* L. Fabaceae) which represents licodione synthase and flavone synthase II. *FEBS Lett.*, **431**, 287–290.

Akashi, T., Fukuchi-Mizutani, M., Aoki, T., Ueyama, Y., Yonekura-Sakakibara, K., Tanaka, Y., Kusumi, T. & Ayabe, S.-I. (1999) Molecular cloning and biochemical characterization of a novel cytochrome P450, flavone synthase II, that catalyzes direct conversion of flavanones to flavones. *Plant Cell Physiol.*, **40**, 1182–1186.

Alfenito, M. R., Souer, E., Goodman, C. D., Buell, R., Mol, J., Koes, R. & Walbot, V. (1998) Functional complementation of anthocyanin sequestration in the vacuole by widely divergent glutathione S-transferases. *Plant Cell*, **10**, 1135–1149.

Anzellotti, D. & Ibrahim, R. K. (2000) Novel flavonol 2-oxyglutarate dependent dioxygenase: affinity purification, characterization, and kinetic properties. *Arch. Biochem. Biophys.*, **382**, 161–172.

Asen, S., Stewart, R. N, Norris, K. H. & Massie, D. R. (1970) A stable blue non-metallic copigment complex of delphanin and C-glycosylflavones in Prof. Blaauw iris. *Phytochemistry*, **9**, 619–627.

Asen, S., Norris, K. H. & Stewart, R. N. (1972) Copigmentation of aurone and flavone from petals of *Antirrhinum majus*. *Phytochemistry*, **11**, 2739–2741.

Asen, S., Norris, K. H., Stewart, R. N. & Semeniuk, P. (1973) Effect of pH, anthocyanin, and flavonoid co-pigments on the color of statice flowers. *J. Am. Soc. Hort. Sci.*, **98**, 174–176.

Asen, S., Griesbach, R. J., Norris, K. H. & Leonhard, B. A. (1986) Flavonoids from *Eustoma grandiflorum* flower petals. *Phytochemistry*, **25**, 2509–2513.

Ashikari, T., Tanaka, Y., Fujiwara, H., Nakao, M., Fukui, Y., Sakaibara, K., Mizutani, M. & Kusumi, T. (1996) Genes coding for proteins having acyl transfer activity. International Patent Application, WO96/ 25500.

Austin, M. B. & Noel, J. P. (2003) The chalcone synthase superfamily of type III polyketide synthases. *Nat. Prod. Rep.*, **20**, 79–110.

Bednar, R. A. & Hadcock, J. R. (1988) Purification and characterization of chalcone isomerase from soybeans. *J. Biol. Chem.*, **263**, 9582–9588.

Ben-Meir, H., Zucker, A., Weiss, D. & Vainstein, A. (2002) Molecular control of floral pigmentation: anthocyanins, in *Breeding for Ornamentals: Classical and Molecular Approaches* (ed. A. Vainstein), Kluwer Academic Publishers, Dordrecht, The Netherlands, pp. 253–272.

Bloor, S. J. (1997) Blue flower colour derived from flavonol–anthocyanin co-pigmentation in *Ceanothus papillosus. Phytochemistry*, **45**, 1399–1405.

Bloor, S. J. (1998) Novel pigments and copigmentation in the blue marguerite daisy. *Phytochemistry*, **50**, 1395–1399.

Bloor, S. J. (2001) Deep blue anthocyanins from blue *Dianella* berries. *Phytochemistry*, **58**, 923–927.

Bloor, S. J. & Falshaw, R. (2000) Covalently linked anthocyanin-flavonol pigments from blue *Agapanthus* flowers. *Phytochemistry*, **53**, 575–579.

Bohm, B. A. (1994) The minor flavonoids, in *The Flavonoids: Advances in Research Since 1986* (ed. J.B. Harborne), Chapman & Hall, London, UK, pp. 387–440.

Bohm, B. A. (1998) *Introduction to Flavonoids.* Harwood Academic Publishers, Amsterdam, The Netherlands.

Bovy, A., de Vos, R., Kemper, M., Schijlen, E., Pertejo, M. A., Muir, S., Collins, G., Robinson, S., Verhoeyen, M., Hughes, S., Santos-Buelga, C. & van Tunen, A. (2002) High-flavonol tomatoes resulting from the heterologous expression of the maize transcription factor genes *LC* and *C1. Plant Cell*, **14**, 2509–2526.

Boyle, R. (1664) *Experiments and Considerations Touching Colours: First Occasionally Written Among Some Other Essays to a Friend and now Suffer'd to Come Abroad as the Beginning of an Experimental History of Colours.* Henry Herringman, London, UK. Available on microfilm at *Early English Books Online*, http://www.lib.umich.edu/eebo/

Bradley J. M., Davies K. M., Deroles S. C., Bloor S. J. & Lewis D. H. (1998) The maize *Lc* regulatory gene up-regulates the flavonoid biosynthetic pathway of *Petunia. Plant J.*, **13**, 381–392.

Bradley J. M., Deroles S. C., Boase M. R., Bloor S., Swinny E. & Davies K. M. (1999) Variation in the ability of the maize *Lc* regulatory gene to upregulate flavonoid biosynthesis in heterologous systems. *Plant Sci.*, **140**, 31–39.

Bradley, J. M., Rains, S. R., Manson, J. L. & Davies, K. M. (2000) Flower pattern stability in genetically modified lisianthus (*Eustoma grandiflorum*) under commercial growing conditions. *N. Z. J. Crop Hort. Sci.*, **28**, 175–184.

Britsch, L. (1990) Purification and characterization of flavone synthase I, a 2-oxoglutarate-dependent desaturase. *Arch. Biochem. Biophys.*, **276**, 348–354.

Britsch, L., Dedio, J., Saedler, H. & Forkmann, G. (1993) Molecular characterization of flavanone 3β-hydroxylase: consensus sequence, comparison with related enzymes and the role of conserved histidines. *Eur. J. Biochem.*, **217**, 745–754.

Brouillard, R. (1988) Flavonoids and flower colour, in *The Flavonoids: Advances in Research Since 1980* (ed. J. B. Harborne), Chapman & Hall, London, UK, pp. 525–538.

Brouillard, R. & Dangles, O. (1993) Flavonoids and flower colour, in *The Flavonoids: Advances in Research Since 1986* (ed. J. B. Harborne), Chapman & Hall, London, UK, pp. 565–587.

Brugliera, F., Holton, T. A., Stevenson, T. W., Farcy, E., Lu, C. Y. & Cornish, E. C (1994) Isolation and characterization of a cDNA clone corresponding to the *Rt* locus of *Petunia hybrida. Plant J.*, **5**, 81–92.

Brugliera, F., Barri-Rewell, G., Holton, T. A. & Mason, J. G. (1999) Isolation and characterization of a flavonoid 3′-hydroxylase cDNA clone corresponding to the *Ht1* locus of *Petunia hybrida. Plant J.*, **19**, 441–451.

Brugliera, F., Demelis, L., Tanaka, Y. & Koes, R. (2003) Genetic sequences having methyltransferase activity and uses therefor. International Patent Application, WO03/062428.

Burbulis, I. E. & Winkel-Shirley, B. (1999) Interactions among enzymes of the *Arabidopsis* flavonoid biosynthetic pathway. *Proc. Natl. Acad. Sci. USA*, **96**, 12929–12934.

Chmiel, E., Sütfeld, R. & Wiermann, R. (1983) Conversion of phloroglucinol-type chalcones by purified chalcone isomerase from tulip anthers and from *Cosmos* petals. *Biochem. Physiol. Pflanzen*, **178**, 139–146.

Chopra, S., Brendel, V., Zhang, J., Axtell, J. D. & Peterson, T. (1999) Molecular characterization of a mutable pigmentation phenotype and isolation of the first active transposable element from *Sorghum bicolor. Proc. Natl. Acad. Sci. USA*, **96**, 5330–5335.

Chopra, S., Gevens, A., Svabek, C., Wood, K. V., Peterson, T. & Nicholson R. L. (2002) Excision of the *Candystripe1* transposon from a hyper-mutable *Y1-cs* allele shows that the sorghum *Y1* gene controls the biosynthesis of both 3-deoxyanthocyanidin phytoalexins and phlobaphene pigments. *Physiol. Mol. Plant Path.*, **60**, 321–330.

Cole, S. P. C., Bhardwaj, G., Gerlach, J. H., Mackie, J. E., Grant, C. E., Almquist, K. C., Stewart, A. J., Kurz, E. U., Duncan, A. M. V. & Deeley, R. G. (1992) Overexpression of a transporter gene in a multidrug-resistant human lung cancer cell line. *Science*, **258**, 1650–1654.

Collette, V. E., Jameson, P. E., Schwinn, K. E., Umaharan, P. & Davies, K. M. (2004) Temporal and spatial expression of flavonoid biosynthetic genes in flowers of *Anthurium andraeanum. Physiol. Plant.*, in press.

Conn, S., Zhang, W. & Franco, C. (2003) Anthocyanic vacuolar inclusions (AVIs) selectively bind acylated anthocyanins in *Vitis vinifera* L. (grapevine) suspension culture. *Biotech. Lett.*, **25**, 835–839.

Courtney-Gutterson, N., Napoli, C., Lemieuz, C., Morgan, A., Firoozabady, E. & Robinson, K. E. P. (1994) Modification of flower color in florists chrysanthemum – production of a white-flowering variety through molecular-genetics. *Biotechnology*, **12**, 268–271.

Damiani, F., Paolocci, F., Cluster, P. D., Arcioni, S., Tanner G. J., Joseph R. J., Li, Y. G., de Majnik J. & Larkin P. J. (1999) The maize transcription factor *Sn* alters proanthocyanidin synthesis in transgenic *Lotus corniculatus* plants. *Aust. J. Plant Phys.*, **26**, 159–169.

Davies, K. M. & Schwinn, K. E. (1997) Flower colour, in *Biotechnology of Ornamental Plants* (eds R. L. Geneve, J. E. Preece & S. A. Merkle), CAB International, Wallingford, UK, pp. 259–294.

Davies, K. M. & Schwinn, K. E. (2003) Transcriptional regulation of secondary metabolism. *Funct. Plant Biol.*, **30**, 913–925.

Davies, K. M., Bradley, J. M., Schwinn, K. E., Markham, K. R. & Podivinsky, E. (1993) Flavonoid biosynthesis in flower petals of five lines of lisianthus (*Eustoma grandiflorum* Grise.). *Plant Sci.*, **95**, 67–77.

Davies, K., Winefield, C., Lewis, D., Nielsen, K., Bradley, J. M., Schwinn, K., Deroles, S., Manson, D. & Jordan, B. (1997) Research into flower colour and flowering time with *Eustoma grandiflorum* (lisianthus). *Flowering News.*, **23**, 24–32.

Davies, K. M., Bloor, S. J., Spiller, G. B. & Deroles, S. C. (1998) Production of yellow colour in flowers: redirection of flavonoid biosynthesis in *Petunia. Plant J.*, **13**, 259–266.

Davies, K. M., Spiller, G. B., Bradley, J. M., Winefield, C. S., Schwinn, K. E., Martin, C. R. & Bloor, S. J., (2001) Genetic engineering of yellow flower colours. *Acta Hort.*, **560**, 39–44.

Davies, K. M., Winefield, C. S., Swinny, E. E., Lewis, D. H., Marshall, G. B., Schwinn, K. E. & Boase, M. R. (2002) Biosynthetic mechanism of 3-deoxyanthocyanins in *Sinningia cardinalis*, in *Polyphenols Communications 2002*, Vol. 2 (ed. I. El Hadrami), Imprimerie Papeterie el Awatanya, Marrakech, Morocco, pp. 35–36.

Davies, K. M., Schwinn, K. E., Deroles, S. C., Manson, D. G., Lewis, D. H., Bloor, S. J. & Bradley, J. M. (2003a) Enhancing anthocyanin production by altering competition for substrate between flavonol synthase and dihydroflavonol 4-reductase. *Euphytica*, **131**, 259–268.

Davies, K. M., Marshall, G. B., Lewis, D. H., Winefield, C. S., Deroles, S. C., Boase, M. R., Zhang, H., Nielsen, K. M., Schwinn, K. E., Bloor, S. J., Swinny, E. & Martin, C. R. (2003b). Generation of new ornamental varieties through genetic modification of pigment biosynthesis. *Acta Hort.*, **624**, 435–447

Debeaujon, I., Peeters, A. J., Leon-Kloosterziel, K. M. & Koornneef, M. (2001) The TRANSPARENT TESTA12 gene of Arabidopsis encodes a multidrug secondary transporter-like protein required for flavonoid sequestration in vacuoles of the seed coat endothelium. *Plant Cell*, **13**, 853–871.

De Jaeger, G., Buys, E., Eeckhout, D., De Wilde, C., Jacobs, A., Kapila, J., Angenon, G., Van Montagu, M., Gerats, T. & Depicker, A. (1999) High level accumulation of single-chain variable fragments in the cytosol of transgenic *Petunia hybrida. Eur. J. Biochem.*, **259**, 426–434.

de Majnik, J., Weinman, J. J., Djordjevic, M. A., Rolfe, B. G., Tanner, G. J., Joseph, R. G. & Larkin, P. J. (2000) Anthocyanin regulatory gene expression in transgenic white clover can result in an altered pattern of pigmentation. *Aus. J. Plant Physiol.*, **27**, 659–667.

Deroles, S. C., Bradley, J. M., Schwinn, K. E., Markham, K. R., Bloor, S. J., Manson D. G. & Davies, K. M. (1998) An antisense chalcone synthase gene leads to novel flower patterns in lisianthus (*Eustoma grandiflorum*). *Mol Breed.*, **4**, 59–66.

Deroles, S. C., Boase, M. R., Lee, C. E. & Peters, T. A. (2002) Gene transfer to plants, in *Breeding for Ornamentals: Classical and Molecular Approaches* (ed. A. Vainstein), Kluwer Academic Publishers, Dordrecht, The Netherlands, pp. 155–196.

de Vetten, N., Quattrocchio, F., Mol, J. & Koes, R. (1997) The *an11* locus controlling flower pigmentation in petunia encodes a novel WD-repeat protein conserved in yeast, plants, and animals. *Genes Dev.*, **11**, 1422–1434.

de Vetten, N., ter Horst, J., van Schaik, H.-P., de Boer, A., Mol, J. & Koes, R. (1999) A cytochrome b_5 is required for full activity of flavonoid 3′,5′-hydroxylase, a cytochrome P450 involved in the formation of blue flower colours. *Proc. Natl. Acad. Sci. USA*, **96**, 778–783.

Do, C. B., Cormier, F. & Nicholas, Y. (1995) Isolation and characterization of a UDP-glucose:cyanidin 3-*O*-glucosyltransferase from grape cell suspension cultures (*Vitis vinifera* L.). *Plant Sci.*, **112**, 43–51.

Dong, Y-.H., Beuning, L., Davies, K., Mitra, D., Morris, B. & Kootstra, A. (1998) Expression of pigmentation genes and photoregulation of anthocyanin biosynthesis in developing Royal Gala apple flowers. *Aus J. Plant Physiol.*, **25**, 245–252.

Elomaa, P. (1996) Genetic modification of flavonoid pathway in ornamental plants. PhD Thesis, University of Helsinki, Finland.

Elomaa, P. & Holton, T. (1994) Modification of flower colour using genetic engineering. *Biotech. Genet. Eng. Rev.*, **12**, 63–88.

Elomaa, P., Honkanen, J., Puska, R., Seppanen, P., Helariutta, Y., Mehto, M., Kotilainen, M., Nevalainen, L. & Teeri, T. H. (1993) *Agrobacterium*-mediated transfer of antisense chalcone synthase cDNA to *Gerbera hybrida* inhibits flower pigmentation. *Biotechnology*, **11**, 508–511.

Elomaa, P., Mehto, M., Kotilainen, M., Helariutta, Y., Nevalainen, L. & Teeri, T. H. (1998) A bHLH transcription factor mediates organ, region and flower type specific signals on dihydroflavonol-4-reductase (DFR) gene expression in the inflorescence of *Gerbera hybrida* (Asteraceae). *Plant J.*, **16**, 93–99.

Elomaa, P., Uimari, A., Mehto, M., Albert, V. A., Laitinen, R. A. & Teeri, T. H. (2003) Activation of anthocyanin biosynthesis in *Gerbera hybrida* (Asteraceae) suggests conserved protein-protein and protein-promoter interactions between the anciently diverged monocots and eudicots. *Plant Physiol.*, **133**, 1831–1842.

Eulgem, T., Rushton, P. J., Robatzek, S. & Somssich, I. E. (2000) The WRKY superfamily of plant transcription factors. *Trends Plant Sci.*, **5**, 199–206.

Farzad, M., Griesbach, R., Hammond, J., Weiss, M. R. & Elmendorf, H. G. (2003) Differential expression of three key anthocyanin biosynthetic genes in a color-changing flower, *Viola cornuta* cv. Yesterday, Today and Tomorrow. *Plant Sci.*, **165**, 1333–1342.

Fatland, B. L., Ke, J., Anderson, M. D., Mentzen, W. I., Cui, L. W., Allred, C. C., Johnston, J. L., Nikolau, B. J. & Wurtele, E. S. (2002) Molecular characterization of a heteromeric ATP-citrate lyase that generates cytosolic acetyl-coenzyme A in Arabidopsis. *Plant Physiol.*, **130**, 740–756.

Ferrer, J.-L., Jez., J. M., Bowman, M. E., Dixon, R. A. & Noel, J. P. (1999) Structure of chalcone synthase and the molecular basis of plant polyketide biosynthesis. *Nature Struc. Biol.*, **6**, 775–784.

Figueiredo, P., Elhabiri, M., Toki, K., Saito, N., Dangles, O. & Brouillard, R. (1996) New aspects of anthocyanin complexation. Intramolecular copigmentation as a means for colour loss? *Phytochem*istry, **41**, 301–308.

Figueiredo, P., George, F., Tatsuzawa, F., Toki, K., Saito, N. & Brouillard, R. (1999) New features of intramolecular copigmentation by acylated anthocyanins. *Phytochemistry*, **51**, 125–132.

Firoozabady, E., Moy, Y., Courtney-Gutterson, N. & Robinson, K. (1994) Regeneration of transgenic rose (*Rosa hybrida*) plants from embryogenic tissue. *Biotechnology*, **12**, 609–613.

Fischer, R., Budde, I. & Hain, R. (1997) Stilbene synthase gene expression causes changes in flower colour and male sterility in tobacco. *Plant J.*, **11**, 489–498.

Fischer, T. C., Halbwirth, H., Meisel, B., Stich, K. & Forkmann, G. (2003) Molecular cloning, substrate specificity of the functionally expressed dihydroflavonol 4-reductases from *Malus domestica* and *Pyrus communis* cultivars and the consequences for flavonoid metabolism *Arch. Biochem. Biophys.*, **412**, 223–230.

Ford, C. M., Boss, P. K. & Høj, P. B. (1998) Cloning and characterization of *Vitis vinifera* UDP-glucose:-flavonoid 3-*O*-glucosyltransferase, a homologue of the enzyme encoded by the maize *Bronze-1* locus that may primarily serve to glucosylate anthocyanidins *in vivo. J. Biol. Chem.*, **273**, 9224–9233.

Forkmann, G. (1991) Flavonoids as flower pigments: the formation of the natural spectrum and its extension by genetic engineering. *Plant Breed.*, **106**, 1–26.

Forkmann, G. & Dangelmayr, B. (1980) Genetic control of chalcone isomerase activity in flowers of *Dianthus caryophyllus. Biochem. Genet.*, **18**, 519–527.

Forkmann G. & Ruhnau B. (1987) Distinct substrate specificity of dihydroflavonol 4-reductase from flowers of *Petunia hybrida. Z. Naturforschung*, **42**, 1146–1148.

Fossen, T. & Anderson, Ø. M. (1999) Delphinidin 3′-galloylgalactosides from blue flowers of *Nymphaéa caerulea. Phytochemistry*, **50**, 1185–1188.

Fossen, T., Slimestad, R., Øvstedal D. O. & Anderson, Ø. M. (2000) Covalent anthocyanin–flavonol complexes from flowers of chive, *Allium schoenoprasum. Phytochemistry*, **54**, 317–323.

Fossen, T., Slimestad, R. & Anderson, Ø. M. (2003) Anthocyanins with 4′-glucosidation from red onion, *Allium cepa. Phytochemistry*, **64**, 1367–1374.

Fujiwara, H., Tanaka, Y., Yonekura-Sakakibara, K., Fukuchi-Mizutani, M., Nakao, M., Fukui, Y., Yamaguchi, M., Ashikari, T. & Kusumi, T. (1998) cDNA cloning, gene expression and subcellular localization of anthocyanin 5-aromatic acyltransferase from *Gentiana triflora. Plant J.*, **16**, 421–431.

Fukada-Tanaka, S., Inagaki, Y., Yamaguchi, T., Saito, N. & Iida, S. (2000) Colour-enhancing protein in petals. *Nature*, **407**, 581.

Fukuchi-Mizutani, M., Okuhara, H., Fukui, Y., Nakao, M., Katsumoto, Y., Yonekura-Sakakibara, K., Kusumi, T., Hase, T. & Tanaka, Y. (2003) Biochemical and molecular characterization of a novel UDP-glucose:anthocyanin 3′-*O*-glucosyltransferase, a key enzyme for blue anthocyanin biosynthesis, from gentian. *Plant Physiol.*, **132**, 1652–1663.

Fukui, Y., Tanaka, Y., Kusumi, T., Iwashita, T. & Nomoto, K. (2003) A rationale for the shift in colour towards blue in transgenic carnation flowers expressing the flavonoid 3′,5′-hydroxylase gene. *Phytochemistry*, **63**, 15–23.

Gonnet, J. F. (2003) Origin of the color of cv. rhapsody in blue rose and some other so-called 'Blue' roses. *J. Agric. Food Chem.*, **51**, 4990–4994.

Goodrich, J., Carpenter, R. & Coen, E. S. (1992) A common gene regulates pigmentation pattern in diverse plant species. *Cell*, **68**, 955–964.

Gorton, H. L. & Vogelmann, T. C. (1996) Effects of epidermal cell shape and pigmentation on optical properties of *Antirrhinum* petals at visible and ultraviolet wavelengths. *Plant Physiol.*, **112**, 879–888.

Goto, T. & Kondo, T. (1991) Structure and molecular stacking of anthocyanins – flower color variation. *Ang. Chemie – Int. Ed. Eng.*, **30**, 17–33.

Grotewold, E., Drummond, B. J., Bowen, B. & Peterson, T. (1994) The *myb*-homologous *P* gene controls phlobaphene pigmentation in maize floral organs by directly activating a flavonoid biosynthetic gene subset. *Cell*, **76**, 543–553.

Grotewold, E., Chamberlin, M., Snook, M., Siame, B., Butler, L., Swenson, J., Maddock, S., St.Clair, G. & Bowen, B. (1998) Engineering secondary metabolism in maize cells by ectopic expression of transcription factors. *Plant Cell*, **10**, 721–740.

Gutterson, N. (1995) Anthocyanin biosynthetic genes and their application to flower colour modification through sense suppression. *Hort Sci.*, **30**, 964–966.

Halbwirth, H., Wimmer, G., Wurst, F., Forkmann, G. & Stich, K. (1997) Enzymatic glucosylation of 4-deoxyaurones and 6′-deoxychalcones with enzyme extracts of *Coreopsis grandiflora*, Nutt. I.. *Plant Sci.*, **122**, 125–131.

Halbwirth, H., Martens, S., Wienand, U., Forkmann, G. & Stich, K. (2003) Biochemical formation of anthocyanins in silk tissue of *Zea mays. Plant Sci.*, **164**, 489–495.

HannaRose, W. & Hansen, U. (1996) Active repression mechanisms of eukaryotic repressors. *Trends Genet.*, **12**, 229–234.

Harborne, J. B. (1966) Comparative biochemistry of flavonoids I. Distribution of chalcone and aurone pigments in plants. *Phytochemistry*, **5**, 111–115.

Harborne, J. B. (1976) Functions of flavonoids in plants, in *Chemistry and Biochemistry of Plant Pigments*, Vol. 1 (ed. T. W. Goodwin), Academic Press, New York, USA, pp. 736–778.

Harborne, J. B. (1978) The rare flavone isoetin as a yellow flower pigment in *Heywoodiella oligocephala* and in other Cichorieae. *Phytochemistry*, **17**, 915–917.

Harborne, J. B. (1986) The natural distribution in angiosperms of anthocyanins acylated with aliphatic dicarboxylic acids. *Phytochemistry*, **25**, 1887–1894.

Harborne J. B. (1988) The flavonoids: recent advances, in *Plant Pigments* (ed. T. W. Goodwin), Academic Press, London, UK, pp. 299–343.

Harborne, J. B. (ed.) (1994) *The Flavonoids: Advances in Research Since 1986*, Chapman & Hall, London, U.K.

Harborne, J. B. & Grayer, R. J. (1994) Flavonoids and insects, in *The Flavonoids: Advances in Research Since 1986* (ed. J. B. Harborne), Chapman & Hall, London, UK, pp. 589–618.

Harborne, J. B. & Sherratt, H. S. A. (1961) Plant polyphenols: 3. Flavonoids in genotypes of *Primula sinensis*. *Biochem. J.*, **78**, 298–306.

Harborne, J. B. & Williams, C. A. (2000) Advances in flavonoid research since 1992. *Phytochemistry.*, **55**, 481–504.

Harborne, J. B., Greenham, J. & Eagles, J. (1990) Malonylated chalcone glycosides in *Dahlia*. *Phytochemistry*, **29**, 2899–2900.

Hause, B., Meyer, K., Viitanen, P. V., Chapple, C. & Strack, D. (2002) Immunolocalization of 1-*O*-sinapoylglucose:malate sinapoyltransferase in *Arabidopsis thaliana*. *Planta*, **215**, 26–32.

Heim, M. A., Jakoby, M., Werber, M., Martin, C., Weisshaar, B. & Bailey, P. C. (2003) The basic helix-loop-helix transcription factor family in plants: a genome-wide study of protein structure and functional diversity. *Mol. Biol. Evol.*, **20**, 735–747.

Helariutta, Y., Elomaa, P., Kotilainen, M., Seppanen, P. & Teeri, T. H. (1993) Cloning of a cDNA coding for dihydroflavonol-4-reductase (DFR) and characterization of *dfr* expression in corrollas of *Gerbera hybrida* var. Regina (Compositae). *Plant Mol. Biol.*, **22**, 183–193.

Heller, W. & Forkmann, G. (1988) Biosynthesis of flavonoids, in *The Flavonoids: Advances in Research Since 1980* (ed. J. B. Harborne), Chapman & Hall, London, UK, pp. 399–425.

Heller, W. & Forkmann, G. (1993) Biosynthesis of flavonoids, in *The Flavonoids: Advances in Research Since 1986* (ed. J. B. Harborne), Chapman & Hall, London, UK, pp. 499–536.

Hipskind, J. D., Goldsbrough, P. B., Urmeev, F. & Nicholson, R. L. (1996) Synthesis of 3-deoxyanthocyanidin phytoalexins in sorghum does not occur via the same pathway as 3-hydroxylated anthocyanidins and phlobaphenes. *Maydica*, **41**, 155–166.

Hirotani, M., Kuroda, R., Suzuki, H. & Yoshikawa, T. (2000) Cloning and expression of UDP-glucose: flavonoid 7-*O*-glucosyltransferase from hairy root cultures of *Scutelliaria baicalensis*. *Planta*, **210**, 1006–1013.

Holton, T. A. (1996) Transgenic plants exhibiting altered flower color and methods for producing same. International Patent Application, WO96/36716.

Holton, T. A., Brugliera, F., Lester, D. R., Tanaka, Y., Hyland, C. D., Menting, J. G., Lu, C. Y., Farcy, E., Stevenson, T. W. & Cornish, E. C (1993a) Cloning and expression of cytochrome P450 genes controlling flower colour. *Nature*, **366**, 276–279.

Holton, T. A., Cornish, E. C., & Tanaka, Y. (1993b) Cloning and expression of flavonol synthase from *Petunia hybrida*. *Plant J.*, **4**, 1003–1010.

Hopp, W. & Seitz, H. U. (1987) The uptake of acylated anthocyanin into isolated vacuoles from a suspension culture of *Daucus carota*. *Planta*, **170**, 74–85.

Hosoki, T., Hamada, M., Kando, T., Moriwaki, R. & Inaba, K. (1991) Comparative study of anthocyanins in tree peony flowers. *J. Jap. Soc. Hort. Sci.*, **60**, 395–403.

Hrazdina, G. & Wagner, G. J. (1985) Metabolic pathways as enzyme complexes: evidence for the synthesis of phenylpropanoids and flavonoids on membrane associated enzyme complexes. *Arch. Biochem. Biophys.*, **237**, 88–100.

Hwang, Y. J., Yoshikawa, K., Miyajima, I. & Okubo, H. (1992) Flower colours and pigments in hybrids with *Camellia chrysantha*. *Sci. Hort.*, **51**, 251–259.

Ibrahim, R. K. & Muzac, I. (2000) The methyltransferase gene superfamily: a tree with multiple branches, in *Evolution of Metabolic Pathways* (eds J. T. Romeo, R. Ibrahim, L. Varin & V. De Luca), Pergamon, Amsterdam, The Netherlands, pp. 349–384.

Ibrahim, R. K., Bruneau, A. & Bantignies, B. (1998) Plant *O*-methyltransferases: molecular analysis, common signature and classification. *Plant Mol. Biol.*, **36**, 1–10.

Jackson, D., Culianez-Macia, F., Prescott, A. G., Roberts, K. & Martin, C. (1991) Expression patterns of *myb* genes from *Antirrhinum* flowers. *Plant Cell*, **3**, 115–125.

Jez, J. M. & Noel, J. P. (2002) Reaction mechanism of chalcone isomerase. *J. Biol. Chem.*, **277**, 1361–1369.

Jez, J. M., Bowman, M. E., Dixon, R. A. & Noel, J. P. (2000) Structure and mechanism of the evolutionarily unique plant enzyme chalcone isomerase. *Nature Struc. Biol.*, **7**, 786–791.

Jin, H., Cominelli, E., Bailey, P., Parr, A., Mehrtens, F., Jones, J., Tonelli, C., Weisshaar, B. & Martin, C. (2000) Transcriptional repression by AtMYB4 controls production of UV-protecting sunscreens in *Arabidopsis*. *EMBO J.*, **19**, 6150–6161.

Johnson, E. T., Yi, H., Shin, B., Oh, B.-J., Cheong, H. & Choi, G. (1999) *Cymbidium hybrida* dihydroflavonol 4-reductase does not efficiently reduce dihydrokaempferol to produce orange pelargonidin-type anthocyanins. *Plant J.*, **19**, 81–85.

Johnson, E. T., Ryu, S., Yi H. Shin, B., Cheong, H. & Choi, G. (2001) Alteration of a single amino acid changes the substrate specificity of dihydroflavonol 4-reductase. *Plant J.*, **25**, 325–333.

Jones, P. & Vogt, T. (2001) Glucosyltransferases in secondary metabolism: tranquilizers and stimulant controllers. *Planta*, **213**, 164–174.

Jones, P., Messner, B., Nakajima, J.-I., Schäffner, A. R. & Saito, K. (2003) UGT73C6 and UGT78D1, glycosyltransferases involved in flavonol glycoside biosynthesis in *Arabidopsis thaliana*. *J. Biol. Chem.*, **278**, 43910–43918.

Jonsson, L. M. V., Aarsman, M. E. G., Schram, A. W. & Bennink, G. J. H. (1982) Methylation of anthocyanins by cell-free extracts of flower buds of *Petunia hybrida*. *Phytochemistry*, **21**, 2457–2459.

Jorgensen, R. (1995) Cosuppression, flower color patterns, and metastable gene expression states. *Science*, **268**, 686–691.

Jorgensen, R. A., Que, Q. D. & Napoli, C. A. (2002) Maternally-controlled ovule abortion results from cosuppression of dihydroflavonol-4-reductase or flavonoid-3′,5′-hydroxylase genes in *Petunia hybrida*. *Funct. Plant Biol.*, **29**, 1501–1506.

Joung, J. Y., Kasthuri, G. M., Park J. Y., Kang, W. J. Kim, H. S., Yoon, B. S., Joung, H. & Jeon, J. H. (2003) An overexpression of chalcone reductase of *Pueraria montana* var. lobata alters biosynthesis of anthocyanin and 5′-deoxyflavonoids in transgenic tobacco. *Biochem. Biophys. Res. Comm.*, **303**, 326–331.

Kamsteeg, J., van Brederode, J., Verschuren, P. M. & van Nigtevecht, G. (1981) Identification, properties and genetic control of *p*-coumaroyl coenzyme A 3-hydroxylase isolated from petals of *Silene dioica*. *Z. Pflanzenphysiol.*, **102**, 435.

Kay, Q. O. N., Daoud, H. S. & Stirton, C. H. (1981) Pigment distribution, light reflection and cell structure in petals. *Bot. J. Linnean Soc.*, **83**, 57–84.

Kerscher, F. & Franz, G. (1987) Biosynthesis of vitexin and isovitexin: enzymatic synthesis of the C-glucosylflavones vitexin and isovitexin with an enzyme preparation from *Fagopyrum esculentum* M. seedlings. *Z. Naturforsch.*, **42c**, 519–524.

Klein, M., Weissenbock, G., Dufaud, A., Gaillard, C., Kreuz, K. & Martinoia, E. (1996) Different energization mechanisms drive the vacuolar uptake of a flavonoid glucoside and a herbicide glucoside. *J. Biol. Chem.*, **271**, 29666–29671.

Kondo, T., Yoshida, K., Nakagawa, A., Kawai, T., Tamura, H. & Goto, T. (1992) Structural basis of blue-colour development in flower petals from *Commelina communis*. *Nature*, **358**, 515–518.

Kramer, C. M., Prata, R. T. N., Willits, M. G., De Luca, V., Steffens, J. C. & Graser, G. (2003) Cloning and regiospecificity studies of two flavonoid glucosyltransferases from *Allium cepa*. *Phytochemistry*, **64**, 1069–1076.

Kroon, J., Souer, E., de Graaff, A., Xue, Y., Mol, J. & Koes, R. (1994) Cloning and structural analysis of the anthocyanin pigmentation locus *Rt* of *Petunia hybrida*: characterization of insertion sequences in two mutant alleles. *Plant J.*, **5**, 69–80.

Larsen, E. S., Alfenito, M. R., Briggs, W. R. & Walbot, V. (2003) A carnation anthocyanin mutant is complemented by the glutathione S-transferases encoded by maize *Bz2* and petunia *An9*. *Plant Cell Rep.*, **21**, 900–904.

Latunde-Dada, A. O., Cabello-Hurtado, F., Czittrich, N., Didierjean, L., Schopfer, C., Hertkorn, N., Werck-Reichhart, D. & Ebel J. (2001) Flavonoid 6-hydroxylase from soybean (*Glycine max* L.), a novel plant P450 monooxygenase. *J. Biol. Chem.*, **276**, 1688–1695.

Le Gall, G., DuPont, M. S., Mellon, F. A., Davis, A. L., Collins, G. J., Verhoeyen, M. E. & Colquhoun, I. J. (2003) Characterization and content of flavonoid glycosides in genetically modified tomato (*Lycopersicon esculentum*) fruits. *J. Agric. Food Chem.*, **51**, 2438–2446.

Lehfeldt, C., Shirley, A. M., Meyer, K., Ruegger, M. X., Cusumano, J. C., Viitanen, P. V., Strack, D. & Chapple, C. (2000) Cloning of the *SNG1* gene of arabidopsis reveals a role for a serine carboxypeptidase-like protein as an acyltransferase in secondary metabolism. *Plant Cell*, **12**, 1295–1306.

Li, Y., Baldauf, S., Lim, E.-K. & Bowles, D. J. (2001) Phylogenetic analysis of the UDP-glycosyltransferase multigene family of *Arabidopsis thaliana*. *J. Biol. Chem.*, **276**, 4338–4343.

Liu, Q., Bonness, M. S., Liu, M., Seradge, E., Dixon, R. A. & Mabry, T. J. (1995) Enzymes of B-ring-deoxy flavonoid biosynthesis in elicited cell cultures of 'old man' cactus (*Cephalocereus senilis*). *Arch. Biochem. Biophys.*, **321**, 397–404.

Lloyd, A. M., Walbot, V. & Davis, R. W (1992) *Arabidopsis* and *Nicotiana* anthocyanin production activated by maize regulators *R* and *C1*. *Science*, **258**, 1773–1775.

Lu, C., Chandler, S. F., Mason, J. G. & Brugliera, F. (2003) Florigene flowers: From laboratory to market, in *Plant Biotechnology 2002 and Beyond* (ed. I. K. Vasil), Kluwer Academic Publishers, Dordrecht, The Netherlands, pp. 333–336.

Lu, Y.-P., Li, Z.-S. & Rea, P. A. (1997) *AtMRP1* gene of *Arabidopsis* encodes a glutathione *S*-conjugate pump: isolation and functional definition of a plant ATP-binding cassette transporter gene. *Proc. Natl. Acad. Sci. USA*, **94**, 8243–8248.

Lu Y.,-P., Li Z.-S., Drozdowicz, Y. M. Hörtensteiner, S., Martinoia, E. & Rea, P. A. (1998) AtMRP2, an Arabidopsis ATP binding cassette transporter able to transport glutathione S-conjugates and chlorophyll catabolites: functional comparisons with AtMRP1. *Plant Cell*, **10**, 267–282.

Lukacin, R. & Britsch, L., (1997) Identification of strictly conserved histidine and arginine residues as part of the active site in *Petunia hybrida* flavanone 3β-hydroxylase. *Eur. J. Biochem.*, **249**, 748–757.

Lukacin, R., Gröning, I., Pieper, U. & Matern, U. (2000) Site-directed mutagenesis of the active site serine290 in flavanone 3β-hydroxylase from *Petunia hybrida*. *Eur. J. Biochem.*, **267**, 853–860.

Markham, K. R., Gould, K. S. & Ryan, K. G. (2001) Cytoplasmic accumulation of flavonoids in flower petals and its relevance to yellow flower colouration. *Phytochemistry*, **58**, 403–413.

Markham, K. R., Gould, K. S., Winefield, C. S., Mitchell, K. A., Bloor, S. J. & Boase, M. R. (2000a) Anthocyanic vacuolar inclusions – their nature and significance in flower colouration. *Phytochemistry*, **55**, 327–336.

Markham, K. R., Hammett, K. R. W. & Ofman, D. J. (1992) Floral pigmentation in two yellow-flowered *Lathyrus* species and their hybrid. *Phytochemistry*, **31**, 549–554.

Markham, K. R., Ryan, K. G., Gould, K. S. & Rickards, G. K. (2000b) Cell wall sited flavonoids in lisianthus flower petals. *Phytochemistry*, **54**, 681–687.

Marrs, K. A. (1996) The functions and regulation of glutathione S-transferases in plants. *Ann. Rev. Plant Physiol. Plant Mol. Biol.*, **47**, 127–158.

Marrs K. A. & Walbot V. (1997) Expression and RNA splicing of the maize glutathione S-transferase *Bronze2* gene is regulated by cadmium and other stresses. *Plant Physiol.*, **113**, 93–102.

Marrs, K. A., Alfenito, M. R., Lloyd, A. M. & Walbot, V. (1995) A glutathione *S*-transferase involved in vacuolar transfer encoded by the maize gene *Bronze-2*. *Nature*, **375**, 397–400.

Martens, S. & Forkmann, G. (1999) Cloning and expression of flavone synthase II from *Gerbera* hybrids. *Plant J.*, **20**, 611–618.

Martens, S., Forkmann, G., Britsch, L., Wellmann, F., Matern, U. & Lukacin, R. (2003a) Divergent evolution of flavonoid 2-oxoglutarate-dependent dioxygenases in parsley. *FEBS Lett.*, **544**, 93–98.

Martens, S., Forkmann, G., Matern, U. & Lukacin, R. (2001) Cloning of parsley flavone synthase I. *Phytochemistry*, **58**, 43–46.

Martens, S., Knott, J., Seitz, C.A., Janvari, L., Yu S.N. & Forkmann, G. (2003b) Impact of biochemical pre-studies on specific metabolic engineering strategies of flavonoid biosynthesis in plant tissues. *Biochem. Eng. J.*, **14**, 227–235.

Martin, C. & Gerats, T. (1993) Control of pigment biosynthesis genes during petal development. *Plant Cell*, **5**, 1253–1264.

Martin, C. & Paz-Ares, J. (1997) MYB transcription factors in plants. *Trends Genet.*, **13**, 67–73.

Martin, C., Prescott, A., Mackay, S., Bartlett, J. & Vrijlandt, E. (1991) Control of anthocyanin biosynthesis in flowers of *Antirrhinum majus*. *Plant J.*, **1**, 37–49.

Martin, C., Jin, H. & Schwinn, K. (2001) Mechanisms and applications of transcriptional control of phenylpropanoid metabolism, in *Regulation of Phytochemicals by Molecular Techniques* (eds J. T. Romeo, J. A, Saunders & B. F. Matthews), Elsevier Science, Oxford, UK, pp. 155–170.

Matern, U., Reichenbach, C. & Heller, W. (1986) Efficient uptake of flavonoids into parsley (*Petroselinum hortense*) vacuoles requires acylated glycosides. *Planta*, **167**, 183–189.

Mathews, H., Clendennen, S. K., Caldwell, C. G., Liu, X. L., Connors, K., Matheis, N., Schuster, D. K., Menasco, D. J., Wagoner, W., Lightner, J. & Wagner, D. R. (2003) Activation tagging in tomato identifies a transcriptional regulator of anthocyanin biosynthesis, modification, and transport. *Plant Cell*, **15**, 1689–1703.

Matsuda, J., Souichi, O., Hashimoto, T. & Yamada, Y. (1991) Molecular cloning of hyoscyamine 6β-hydroxylase, a 2-oxoglutarate-dependent dioxygenase, from cultured roots of *Hyoscyamus niger*. *J. Biol. Chem.*, **266**, 9460–9464.

Matsui, S. & Nakamura, M. (1988) Distribution of flower pigments in perianth of *Cattleya* and allied genera. I. Species. *J. Jap. Soc. Hort. Sci.*, **57**, 222–232.

Matsui, S., Suzuki, T. & Nakamura, M. (1984) Distribution of flower pigments in perianth of *Vandeae* orchids. *Res. Bull. Fac. Ag. Gifu Uni.*, **49**, 361–369.

Memelink, J., Menke, F. L. H., van der Fits, L. & Kijne, J. W. (2000) Transcriptional regulators to modify secondary metabolism, in *Metabolic Engineering of Plant Secondary Metabolism* (eds R. Verpoorte & W. Alfermann), Kluwer Academic Publishers, Dordrecht, The Netherlands, pp. 111–125.

Menssen, A., Hohmann, S., Martin, W., Schnable, P. S., Peterson, P. A., Saedler, H. & Gierl, A. (1990) The *En/Spm* transposable element of *Zea mays* contains splice sites at the termini generating a novel intron from a *dSpm* element in the *A2* gene. *EMBO J.*, **9**, 3051–3057.

Meyer, P., Heidmann, I., Forkmann, G. & Saedler, H. (1987) A new petunia flower colour generated by transformation of a mutant with a maize gene. *Nature*, **330**, 677–678.

Miles, C. O. & Main, L. (1985) Kinetics and mechanism of the cyclisation of 2′,6′-dihydroxy-4,4′-dimethoxychalcone; influence of the 6′-hydroxyl group on the rate of cyclisation under neutral conditions. *J. Chem. Soc. Perkin Trans. II*, 1639–1642.

Miller, K. D., Guyon, V., Evans, J. N. S., Shuttleworth, W. A. & Taylor, L. P. (1999) Purification, cloning, and heterologous expression of a catalytically efficient flavonol-3-*O*-galactosyltransferase expressed in the male gametophyte of *Petunia hybrida*. *J. Biol. Chem.*, **274**, 34011–34019.

Mol, J. N. M., Robbins, M. P., Dixon, R. A. & Veltkamp, E. (1985) Spontaneous and enzymatic rearrangement of naringenin chalcone to flavanone. *Phytochemistry*, **24**, 2267–2269.

Mooney, M., Desnos, T., Harrison, K., Jones, J., Carpenter, R. & Coen, E. (1995) Altered regulation of tomato and tobacco pigmentation genes caused by the *delila* gene of *Antirrhinum*. *Plant J.*, **7**, 333–339.

Moyano E, Martinez-Garcia J. F. & Martin C. (1996) Apparent redundancy in *myb* gene function provides gearing for the control of flavonoid biosynthesis in *Antirrhinum* flowers. *Plant Cell*, **8**, 1519–1532.

Mueller, L. A., Goodman, C. D., Silady, R. A. & Walbot, V. (2000) AN9, a petunia glutathione S-transferase required for anthocyanin sequestration, is a flavonoid-binding protein. *Plant Physiol.*, **123**, 1561–1570.

Mur, L. (1995) Characterization of members of the *Myb* gene family of transcription factors from *Petunia hybrida*. PhD Thesis, Vrije Universiteit, Amsterdam.

Nakajima, J.-I., Tanaka, Y., Yamazaki, M. & Saito K. (2001) Reaction mechanism from leucoanthocyanidin to anthocyanidin 3-glucoside, a key reaction for coloring in anthocyanin biosynthesis. *J. Bio. Chem.*, **276**, 25797–25803.

Nakayama, T., Yonekura-Sakakibara, K., Sato, T., Kikuchi, S., Fukui, Y., Fukuchi-Mizutani, M., Ueda, T., Nakao, M., Tanaka, Y., Kusumi, T. & Nishino, T. (2000) Aureusidin synthase: a polyphenol oxidase homolog responsible for flower coloration. *Science*, **290**, 1163–1166.

Nakayama, T., Sato, T., Fukui, Y., Yonekura-Sakakibara, K., Hayashi, H., Tanaka, Y., Kusumi, T. & Nishino, T. (2001) Specificity analysis and mechanism of aurone synthesis catalyzed by aureusidin synthase, a polyphenol oxidase homolog responsible for flower coloration. *FEBS Lett.*, **499**, 107–111.

Napoli, C., Lemieux, C. & Jorgensen, R. (1990) Introduction of a chimeric chalcone synthase gene into *Petunia* results in reversible co-suppression of homologous genes *in trans*. *Plant Cell*, **2**, 279–289.

Nielsen, K. M., Deroles, S. C., Markham, K. R., Bradley, J. M., Podivinsky, E. & Manson, D. (2002) Antisense flavonol synthase alters copigmentation and flower color in lisianthus. *Mol. Breed.*, **9**, 217–229.

Nikolau, B. J., Ohlrogge, J. B. & Wurtele, E. S. (2003) Plant biotin-containing carboxylases. *Arch. Biochem. Biophys*, **414**, 211–222.

Noda, K.-I., Glover, B. J., Linstead, P. & Martin C. (1994) Flower colour intensity depends on specialized cell shape controlled by a Myb-related transcription factor. *Nature*, **369**, 661–664.

Nozue, M., Kubo, H., Nishimura, M. & Yasuda, H. (1995) Detection and characterization of a vacuolar protein (VP24) in anthocyanin-producing cells of sweet potato in suspension culture. *Plant Cell Physiol.*, **36**, 883–889.

Nozue, M., Yamada, K., Nakamura, T., Kubo, H., Kondo, M. & Nishimura, M. (1997) Expression of a vacuolar protein (VP24) in anthocyanin-producing cells of sweet potato in suspension culture. *Plant Physiol.*, **115**, 1065–1072.

Nozue, M., Baba, S., Kitamura, Y., Xu, W. X., Kubo, H., Nogawa, M., Shioiri, H. & Kojima, M. (2003) VP24 found in anthocyanic vacuolar inclusions (AVIs) of sweet potato cells is a member of a metalloprotease family. *Biochem. Eng. J.*, **14**, 199–205.

Okinaka, Y., Shimada, Y., Nakano-Shimada, R., Ohbayashi, M., Kiyokawa, S. & Kikuchi, Y. (2003) Selective accumulation of delphinidin derivatives in tobacco using a putative flavonoid 3′,5′-hydroxylase cDNA from *Campanula medium*. *Biosci. Biotech. Biochem.*, **67**, 161–165.

Payne, C. T, Zhang, F. & Lloyd, A. M. (2000) *GL3* encodes a bHLH protein that regulates trichome development in *Arabidopsis* through interaction with GL1 and TTG1. *Genetics*, **156**, 1349–1362.

Pecket, R. C. & Small, C. J. (1980) Occurrence, location and development of anthocyanoplasts. *Phytochemistry*, **9**, 2571–2576.

Petroni, K., Tonelli, C. & Paz-Ares, J. (2002) The MYB transcription factor family: from maize to Arabidopsis. *Maydica*, **47**, 213–232.

Polashock, J. J., Griesbach, R. J., Sullivan, R. F. & Vorsa, N. (2002) Cloning of a cDNA encoding the cranberry dihydroflavonol-4-reductase (DFR) and expression in transgenic tobacco. *Plant Sci.*, **163**, 241–251.

Preisig-Mueller, R., Gehlert, R., Melchior, F., Stietz, U. & Kindl, H. (1997) Plant polyketide synthases leading to stilbenoids have a domain catalyzing malonyl-CoA:CO$_2$ exhange, malonyl-CoA decarboxylation, and covalent enzyme modification and a site for chain lengthening. *Biochemistry*, **36**, 8349–8358.

Quattrocchio, F., Wing, J. F., Leppen, H. T. C., Mol, J. N. M. & Koes, R. E. (1993) Regulatory genes controlling anthocyanin pigmentation are functionally conserved among plant species and have distinct sets of target genes. *Plant Cell*, **5**, 1497–1512.

Quattrocchio, F., Wing, J. F., van der Woude, K., Mol, J. N. M. & Koes, R. (1998) Analysis of bHLH and MYB–domain proteins: Species–specific regulatory differences are caused by divergent evolution of target anthocyanin genes. *Plant J.*, **13**, 475–488.

Quattrocchio, F., Wing, J., van der Woude, K., Souer, E, de Vetten, N., Mol, J. & Koes, R. (1999) Molecular analysis of the *anthocyanin2* gene of petunia and its role in the evolution of flower color. *Plant Cell*, **11**, 1433–1444.

Ramsay, N. A., Walker, A. R., Mooney, M. & Gray, J. C. (2003) Two basic-helix-loop-helix genes (*MYC-146* and *GL3*) from *Arabidopsis* can activate anthocyanin biosynthesis in a white-flowered *Matthiola incana* mutant. *Plant Mol. Biol.*, **52**, 679–688.

Ranish, J. A. & Hahn, S. (1996) Transcription: basal factors and activation. *Cur. Opin. Genes Dev.*, **6**, 151–158.

Rasmussen, S. & Dixon, R. A. (1999) Transgene-mediated and elicitor-induced perturbation of metabolic channeling at the entry point into the phenylpropanoid pathway. *Plant Cell*, **11**, 1537–1551.

Ray, H., Yu, M., Auser, P., Blahut-Beatty, L., McKersie, B., Bowley, S., Westcott, N., Coulman, B., Lloyd, A. & Gruber, M. Y. (2003) Expression of anthocyanins and proanthocyanidins after transformation of alfalfa with maize *Lc*. *Plant Physiol.*, **132**, 1448–1463.

Robbins, M. P., Paolocci, F., Hughes, J. W., Turchetti, V., Allison, G., Arcioni, S., Morris, P. & Damiani, F. (2003) *Sn*, a maize bHLH gene, modulates anthocyanin and condensed tannin pathways in *Lotus corniculatus*. *J. Exp. Bot.*, **54**, 239–248

Rosati, C., Cadic, A., Duron, M., Renou, J. P. & Simoneau, P. (1997) Molecular cloning and expression analysis of dihydroflavonol 4-reductase gene in flower-organs of *Forsythia x intermedia*. *Plant Mol. Biol.*, **35**, 303–311.

Rosati, C., Simoneau, P., Treutter, D., Poupard, P., Cadot, Y., Cadic, A. & Duron, M. (2003) Engineering of flower colour in forsythia by expression of two independently-transformed dihydroflavonol 4-reductase and anthocyanidin synthase genes of flavonoid pathway. *Mol. Breed.*, **12**, 197–208.

Ross, J., Li, Y., Lim, E.-K. & Bowles, D. J. (2001) Higher plant glycosyltransferases. *Genome Biol.*, **2**, 3004.1–3004.6.

Sablowski, R. W. M., Moyano, E., Culianez-Macia, F. A., Schuch, W., Martin, C. & Bevan, M. (1994) A flower-specific Myb protein activates transcription of phenylpropanoid biosynthetic genes. *EMBO J.*, **13**, 128–137.

Saito, K. & Yamazaki, M. (2002) Biochemistry and molecular biology of the late-stage of biosynthesis of anthocyanin: lessons from *Perilla frutescens* as a model plant. *New Phytol.*, **155**, 9–23.

Saito, K., Kobayashi, M., Gong Z.-Z., Tanaka, Y. & Yamazaki, M. (1999) Direct evidence for anthocyanidin synthase as a 2-oxoglutarate-dependent oxygenase: molecular cloning and functional expression of cDNA from a red forma of *Perilla frutescens*. *Plant J.*, **17**, 181–189.

Saito, N. & Harborne, J. B. (1983) A cyanidin glycoside giving scarlet coloration in plants of the *Bromeliaceae*. *Phytochemistry*, **22**, 1735–1740.

Saito, N., Abe, K., Honda, T., Timberlake, C. F. & Bridle, P. (1985) Acylated delphinidin glucosides and flavonols from *Clitoria ternatea*. *Phytochemistry*, **24**, 1583.

Sakakibara, K., Yuko, F., Tanaka, Y., Kusumi, T. & Yoshikawa, T. (2000) Genes encoding proteins having activity of transferring sugar onto aurone. International Patent Application, WO00/49155.

Sasaki, K. & Takahashi, T. (2002) A flavonoid from *Brassica rapa* flower as the UV-absorbing nectar guide. *Phytochemistry*, **61**, 339–343.

Saslowsky, D. E. & Winkel-Shirley, B. (2001) Localisation of flavonoid enzymes in Arabidopsis roots. *Plant J.*, **27**, 37–48.

Schwechheimer, C. & Bevan, M. (1998) The regulation of transcription factor activity in plants. *Trends Plant Sci.*, **3**, 378–383.

Schwinn, K. E., Markham, K. R. & Given, N. K. (1994) Floral flavonoids and the potential for pelargonidin biosynthesis in commercial chrysanthemum cultivars. *Phytochemistry*, **35**, 145–150.

Schwinn, K. E., Davies, K. M., Deroles, S. C., Markham, K., Miller, R. M., Bradley, M., Manson, D. G. & Given, N. K. (1997) Expression of an *Antirrhinum majus* UDP-glucose:flavonoid-3-O-glucosyltransferase transgene alters flavonoid glycosylation and acylation in lisianthus (*Eustoma grandiflorum* Grise.). *Plant Sci.*, **125**, 53–61.

Schwinn, K., Alm, V., Mackay, S., Davies, K. & Martin C. (2001) Regulation of anthocyanin biosynthesis in *Antirrhinum*. *Acta Hort.*, **560**, 201–206.

Schwinn, K., Davies, K., Marshall, G., Bradley, J. M., Winefield, C., Deroles, S., Martin, C. & Bloor, S. (2002) Elucidating aurone biosynthesis for the production of yellow flower colours, in *Polyphenols Communications 2002*, Vol. 2 (ed. I. El Hadrami), Imprimerie Papeterie el Awatanya, Marrakech, pp. 65–66.

Shimada, Y., Nakano-Shimada, R., Ohbayashi, M., Okinaka, Y., Kiyokawa, S. & Kikuchi, Y. (1999) Expression of chimeric P450 genes encoding flavonoid-3′,5′-hydroxylase in transgenic tobacco and petunia plants. *FEBS Lett.*, **461**, 241–245.

Shimada, Y., Ohbayashi, M., Nakano-Shimada, R., Okinaka, Y., Kiyokawa, S. & Kikuchi, Y. (2001) Genetic engineering of the anthocyanin biosynthetic pathway with flavonoid-3′,5′-hydroxylase: specific switching of the pathway in petunia. *Plant Cell Rep.*, **20**, 456–462.

Sompornpailin, K., Makita, Y., Yamazaki, M. & Saito, K. (2002) A WD-repeat-containing putative regulatory protein in anthocyanin biosynthesis in *Perilla frutescens*. *Plant Mol. Biol.*, **50**, 485–495.

Spelt, C., Quattrocchio, F., Mol, J. N. M. & Koes, R. (2000) *Anthocyanin1* of petunia encodes a basic helix-loop-helix protein that directly activates transcription of structural anthocyanin genes. *Plant Cell*, **12**, 1619–1631.

Spelt, C., Quattrocchio, F., Mol, J. & Koes, R. (2002) ANTHOCYANIN1 of petunia controls pigment synthesis, vacuolar pH, and seed coat development by genetically distinct mechanisms. *Plant Cell*, **14**, 2121–2135.

Springob, K., Nakajima, J., Yamazaki, M. & Saito, K. (2003) Recent advances in the biosynthesis and accumulation of anthocyanins. *Nat. Prod. Rep.*, **20**, 288–303.

St-Pierre, B. & De Luca, V. (2000) Evolution of acyltransferase genes: origin and diversification of the BAHD superfamily of acyltransferases involved in secondary metabolism, in *Evolution of Metabolic Pathways* (eds J. T. Romeo, R. Ibrahim, L. Varin & V. De Luca), Pergamon, Amsterdam, The Netherlands, pp. 285–316.

Stich, K. & Forkmann, G. (1988a) Biosynthesis of 3-deoxyanthocyanins with flower extracts from *Sinningia cardinalis*. *Phytochemistry*, **27**, 785–789.

Stich, K. & Forkmann, G. (1988b) Studies on Columnidin biosynthesis with flower extracts from *Columnea hybrida*. *Z. Naturforschung*, **43c**, 311–314.

Stotz, G., Spribille, R. & Forkmann, G. (1984) Flavonoid biosynthesis in flowers of *Verbena hybrida*. *J. Plant. Physiol.*, **116**, 173.

Strack, D., Busch, E. & Klein, E. (1989) Anthocyanin patterns in European orchids and their taxonomic and phylogenetic relevance. *Phytochemistry*, **28**, 2127–2139.

Styles E. D. & Ceska, O. (1977) The genetic control of flavonoid synthesis in maize. *Can. J. Genet. Cytol.*, **19**, 289–302.

Suzuki, H., Nakayama, T., Yonekura-Sakakibara, K., Fukui, Y., Nakamura, N., Nakao, N., Tanaka, Y., Yamaguchi, M., Kusumi, T. & Nishino, T. (2001) Malonyl-CoA: anthocyanin 5-*O*-glucoside-6‴-*O*-malonyltransferase from scarlet sage (*Salvia splendens*) flowers. *J. Biol. Chem.*, **276**, 49013–49019.

Suzuki, H., Nakayama, T., Yonekura-Sakakibara, K., Fukui, Y., Nakamura, N., Yamaguchi, M., Tanaka, Y., Kusumi, T. & Nishino, T. (2002) cDNA cloning, heterologous expressions, and functional characterization of malonyl-coenzyme A: anthocyanidin 3-*O*-glucoside-6″-*O*-malonyltransferase from dahlia flowers. *Plant Physiol.*, **130**, 2142–2151.

Suzuki, H., Nakayama, T. & Nishino, T. (2003) Proposed mechanism and functional amino acid residues of malonyl-CoA: anthocyanin 5-*O*-glucoside-6‴-*O*-malonyltransferase from flowers of *Salvia splendens*, a member of the versatile plant acyltransferase family. *Biochemistry*, **42**, 1764–1771.

Suzuki, K., Xue, H., Tanaka, Y., Fukui, Y., Fukuchi-Mizutani, M., Katsumoto, Y., Tsuda, S. & Kusumi, T. (2000) Flower color modifications of *Torenia hybrida* by cosuppression of anthocyanin biosynthesis genes. *Mol. Breed.*, **6**, 239–246.

Swain, T. (1976) Nature and properties of flavonoids, in *Chemistry and Biochemistry of Plant Pigments*, Vol. 1 (ed. T.W. Goodwin), Academic Press, New York, pp. 425–463.

Takeda, K., Yamashita, T., Takahashi, A. & Timberlake, C. F. (1990) Stable blue complexes of anthocyanin-aluminium-3-*p*-coumaroyl- or 3-caffeoyl-quinic acid involved in the blueing of *Hydrangea* flower. *Phytochemistry*, **29**, 1089–1091.

Takeda, K., Harborne, J. B. & Waterman, P.G. (1993) Malonylated flavonoids and blue flower colour in lupin. *Phytochemistry*, **34**, 421–423.

Takeda, K., Yanagisawa, M., Kifune, T., Kinoshita, T. & Timberlake, C. F. (1994) A blue pigment complex in flowers of *Salvia patens*. *Phytochemistry*, **35**, 1167–1169.

Tamagnone, L., Merida, A., Parr, A., Mackay, S., Culianez-Macia, F. A., Roberts, K. & Martin, C. (1998) The AmMYB308 and AmMYB330 transcription factors from *Antirrhinum* regulate phenylpropanoid and lignin biosynthesis in transgenic tobacco. *Plant Cell*, **10**, 135–154.

Tanaka, Y., Fukui, Y., Fukuchi-Mizutani, M., Holton, T. A., Higgins, E. & Kusumi, T. (1995) Molecular cloning and characterization of *Rosa hybrida* dihydroflavonol 4–reductase gene. *Plant Cell Physiol.*, **36**, 1023–1031.

Tanaka, Y., Tsuda, S. & Kusumi, T. (1998) Metabolic engineering to modify flower colour. *Plant Cell Physiol.*, **39**, 1119–1126.

Tanaka, Y., Yonekura, K., Fukuchi-Mizutani, M., Fukui, Y., Fujiwara, H., Ashikari, T. & Kusumi, T. (1996) Molecular and biochemical characterization of three anthocyanin synthetic enzymes from *Gentiana triflora*. *Plant Cell Physiol.*, **37**, 711–716.

Tatsuzawa, F., Saito, N., Murata, N., Shinoda, K., Shigihara, A. & Honda, T. (2003) 6-Hydroxypelargonidin glycosides in the orange-red flowers of *Alstroemeria*. *Phytochemistry*, **62**, 1239–1242.

Taylor, L. P. & Jorgensen, R. (1992) Conditional male fertility in chalcone synthase-deficient petunia. *J. Hered.*, **83**, 11–17.

Thompson, W. R., Meinwald, J., Aneshansley, D. & Eisner, T. (1972) Flavonols: pigments responsible for ultraviolet absorption in nectar guide of flower. *Science*, **177**, 528–530.

Toki, K., Saito, N., Iimura, K., Suzuki, T. & Honda, T. (1994) (Delphinidin 3-gentiobiosyl) (apigenin 7-glucosyl) malonate from the flowers of *Eichhornia crassipes*. *Phytochemistry*, **36**, 1181–1183.

Tropf, S., Kaercher, B., Schroeder, J. & Schroeder, G. (1995) Reaction mechanisms of homodimeric plant polyketide synthases (stilbene and chalcone synthase): a single active site for the condensing reaction is sufficient for synthesis of stilbenes, chalcones, and 6'-deoxychalcones. *J. Biol. Chem.*, **270**, 7922–7928.

Tropf, S., Lanz, S., Rensing, S. A., Schröeder, J. & Schröeder, G. (1994) Evidence that stilbene synthases have developed from chalcone synthases several times in the course of evolution. *J. Mol. Evol.*, **38**, 610–618.

Turnbull, J. J., Nakajima, J.-I., Welford, R. W. D., Yamazaki, M., Saito, K. & Scholfield, C. J. (2004) Mechanistic studies on three 2-oxoglutarate dependent oxygenases of flavonoid biosynthesis: anthocyanidin synthase, flavonol synthase and flavanone 3β-hydroxylase. *J. Biol. Chem.*, **279**, 1206–1216.

Turnbull, J. J., Sobey, W. J., Aplin, R. T., Hassan, A., Firmin, J. L., Scholfield, C. J. & Prescott, A. G. (2000) Are anthocyanidins the immediate products of anthocyanidin synthase? *Chem. Comm.*, 2473–2474.

Ueyama, Y., Suzuki, K., Fukuchi-Mizutani, M., Fukui, Y., Miyazaki, K., Ohkawa, H., Kusumi, T. & Tanaka, Y. (2002) Molecular and biochemical characterization of torenia flavonoid 3'-hydroxylase and flavone synthase II and modification of flower color by modulating the expression of these genes. *Plant Sci.*, **163**, 253–263.

van Blokland, R., de Lange, P., Mol, J. N. M. & Kooter, J. M. (1993) Modulation of gene expression in plants by antisense genes, in *Antisense Research and Applications* (ed. S. T. Crooke & B. Lebleu), CRC Press, London, UK, pp. 125–148.

van der Krol, A. R., Lenting, P. E., Veenstra, J., Van der Meer, I. M., Koes, R. E., Gerats, A. G. M., Mol, J. N. M. & Stuitje, A. R. (1988) An anti-sense chalcone synthase gene in transgenic plants inhibits flower pigmentation. *Nature*, **333**, 866–870.

van der Krol, A. R., Mur, L. A., Beld, M., Mol, J. N. M. & Stuitje, A. R (1990) Flavonoid genes in petunia: addition of a limited number of gene copies may lead to suppression of gene expression. *Plant Cell.*, **2**, 291–299.

van Tunen, A. J., Mur, L. A., Recourt, K., Gerats, A. G. & Mol, J. N. (1991) Regulation and manipulation of flavonoid gene expression in anthers of petunia: the molecular basis of the *Po* mutation. *Plant Cell*, **3**, 39–48.

Vergères, G. & Waskell, L. (1995) Cytochrome *b*5, its functions, structure and membrane topology. *Biochimie*, **77**, 604–620.

Vogt, T. (2002) Substrate specificity and sequence analysis define a polyphyletic origin of betanidin 5- and 6-*O*-glucosyltransferase from *Dorotheanthus bellidiformis. Planta*, **214**, 492–495.

Vogt, T. & Jones, P. (2000) Glycosyltransferases in plant natural product synthesis: characterization of a supergene family. *Trends Plant Sci.*, **5**, 380–386.

Vogt, T., Grimm, R. & Strack, D. (1999) Cloning and expression of a cDNA encoding betanidin 5-*O*-glucosyltransferase, a betanidin- and flavonoid-specific enzyme with high homology to inducible glucosyltransferases from the Solanaceae. *Plant J.*, **19**, 509–519.

Vom Endt, D., Kijne, J. W. & Memelink, J. (2002) Transcription factors controlling plant secondary metabolism: what regulates the regulators? *Phytochemistry*, **61**, 107–114.

Weiss, D. (2000) Regulation of flower pigmentation and growth: multiple signaling pathways control anthocyanin synthesis in expanding petals. *Physiol. Plant.*, **110**, 152–157.

Welford, R. W. D., Turnbull, J. J., Claridge, T. D. W., Prescott, A. G. & Schofield, C. J. (2001) Evidence for oxidation at C-3 of the flavonoid C-ring during anthocyanin biosynthesis. *Chem. Comm.*, 1828–1829.

Welle, R. & Grisebach, H. (1989) Phytoalexin synthesis in soybean cells: elicitor induction of reductase involved in biosynthesis of 6′-deoxychalcone. *Arch. Biochem. Biophys.*, **272**, 97–102.

Wilmouth, R. C., Turnbull, J. J., Welford, R. W. D., Clifton, I. J., Prescott, A. G. & Schofield, C. J. (2002) Structure and mechanism of anthocyanidin synthase from *Arabidopsis thaliana. Structure*, **10**, 93–103.

Winefield, C. (2002) The final steps in anthocyanin formation: a story of modification and sequestration. *Adv. Bot. Res.*, **37**, 55–74.

Xie, D.-Y., Sharma, S. B., Pavia, N. L., Ferreira, D. & Dixon, R. A. (2003) Role of anthocyanidin reductase, encoded by *BANYULS* in plant flavonoid biosynthesis. *Science*, **299**, 396–399.

Yamaguchi, T., Fukada-Tanaka, S., Inagaki, Y., Saito, N., Yonekura-Sakakibara, K., Tanaka, Y., Kusumi, T. & Iida, S. (2001) Genes encoding the vacuolar Na^+/H^+ exchanger and flower coloration. *Plant Cell Physiol.*, **42**, 451–461.

Yamazaki, M., Gong, Z., Fukuchi-Mizutani, M., Fukui, Y., Tanaka, Y., Kusumi, T. & Saito, K. (1999) Molecular cloning and biochemical characterization of a novel anthocyanin 5-*O*-glucosyltransferase by mRNA differential display for plant forms regarding anthocyanin. *J. Biol. Chem.*, **274**, 7405–7411.

Yamazaki, M., Yamagishi, E., Gong, Z., Fukuchi-Mizutani, M., Fukui, Y., Tanaka, Y., Kusumi, T., Yamaguchi, T. & Saito, K. (2002) Two flavonoid glucosyltransferases from *Petunia hybrida*: molecular cloning, biochemical properties and developmentally regulated expression. *Plant Mol. Biol.*, **48**, 401–411.

Yonekura-Sakakibara, K., Tanaka, Y., Fukuchi-Mizutani, M., Fujiwara, H., Fukui, Y., Ashikari, T., Murakami, Y., Yamaguchi, M. & Kusumi, T. (2000) Molecular and biochemical characterization of a novel hydroxycinnamoyl-CoA: anthocyanin 3-*O*-glucoside-6″-*O*-acyltransferase from *Perilla frutescens. Plant Cell Physiol.*, **41**, 495–502.

Yoshida, K., Toyama, Y., Kameda, K. & Kondo T. (2000) Contribution of each caffeoyl residue of the pigment molecules of gentiodelphin to blue color development. *Phytochemistry*, **54**, 85–92.

Zhang, P. F., Wang, Y. B., Zhang, J. B., Maddock, S., Snook, M. & Peterson, T. (2003) A maize QTL for silk maysin levels contains duplicated *Myb*-homologous genes which jointly regulate flavone biosynthesis. *Plant Mol. Biol.*, **52**, 1–15.

Zuker, A., Tzfira, T., Ben-Meir, H., Ovadis, M., Schklarman, E., Itzhaki, H., Forkmann, G., Martens, S., Neta-Sharir, I., Weiss, D. & Vainstein, A. (2002) Modification of flower colour and fragrance by antisense suppression of the flavanone 3-hydroxylase gene. *Mol. Breed.*, **9**, 33–41.

5   Condensed tannins

Greg Tanner

5.1   Introduction

Flavonoids are nonessential but abundant and easily visualized secondary metabolites that accumulate in many plant tissues. There are three main metabolic sinks where the metabolic flux of flavonoid biosynthesis accumulates:

(1) flavones,
(2) anthocyanins, and
(3) proanthocyanidins (condensed tannins).

Flavones are yellow pigments that accumulate in leaf tissues, petals and seed coats; anthocyanins are pigments responsible for the attractive and distinctive color of many plants; and proanthocyanidins are colorless polymers, *in vivo*, that oxidize upon desiccation to produce the pigments responsible for the dark color of most seed coats. The proanthocyanidins are polymers of flavan-3-ols and should not be confused with hydrolysable tannins, which are biosynthetically unrelated esters of gallic acid and which share some of the protein-precipitating properties of proanthocyanidins. Red polymers of flavan-4-ols, the deoxy-proanthocyanidins or phlabophenes are found in maize kernels but will not be discussed (Grotewold *et al.*, 1994).

The sequence of biochemical intermediates which leads to the synthesis of anthocyanins is largely the same as that which leads to the proanthocyanidins. Thus progress in these two fields has gone hand in hand. The enzymology of biosynthesis and the manipulation of (pro)anthocyanidins will be reconsidered and reviewed here, while the role of transcriptional regulation of anthocyanin biosynthesis will be separately reviewed elsewhere in this volume.

Proanthocyanidins are mixtures of flavan-3-ol oligomers (Fig. 5.1) with a degree of polymerization (Dp) ranging from 2 to about 10, although much larger polymers have been reported in fruit (Labarbe *et al.*, 1999). Each oligomer is composed of linear (C4–C8) linkages between 2,3-*trans*-flavonoids (e.g. catechin-like; 2*R*,3*S*) or 2,3-*cis*-flavonoids (e.g. epicatechin-like; 2*R*,3*R*). The interflavan bond at C4 is always *trans* with respect to the hydroxyl group at C3. This stereochemistry is maintained following acid hydrolysis in the presence of phlor-oglucinol, allowing the determination of the subunit composition (Kennedy & Jones, 2001). Flavonoid subunits with the alternate 2*S* configuration, designated by the *enantio* (*ent*) prefix are rarely encountered. The stereochemistry of each interflavan linkage, specified as for carbohydrates, is usually either (4α-8) or

(4β-8) for linkages from 2,3-*trans*-flavonoids or 2,3-*cis*-flavonoids, respectively (Porter, 1993). Linkages involving (C4–C6) or other C atoms are uncommon and nothing is known of their biosynthesis. There are several possible flavan-3-ol isomers with either two (catechin, epicatechin) or three (gallocatechin, epigallo-catechin) B-ring hydroxyls. The heterogeneity of proanthocyanidins is clearly shown by mass spectroscopy, which visualizes heteropolymeric series (Fig. 5.1). Propelargonidins with subunits carrying a single B-ring hydroxyl are rare; how-ever, inhibition of the *Arabidopsis thaliana TT7* gene encoding flavonoid 3'-hydroxylase (F3'H) (Fig. 5.2) forces their accumulation by redirecting meta-bolic flux through dihydrokaempferol (Schoenbohm *et al.*, 2000).

The flavan-3-ol that initiates polymerization is here called the 'initiating sub-unit', although it has often been confusingly referred to as the 'terminal subunit', which more correctly refers to the last biosynthetic unit added to the growing polymer chain. Each additional monomer added to the growing polymer becomes an 'extension subunit' (Fig. 5.1).

All of the structural genes required for anthocyanidin synthesis have been cloned, often from several tissues (Winkel-Shirley, 2002); however, few ortholo-gous genes have been cloned from tissues that accumulate proanthocyanidins.

Analysis of proanthocyanidin-free mutants has increased our understanding of proanthocyanidin biosynthesis. There are two model plant systems, *Hordeum vulgare* and *A. thaliana*, in which only a single locus encodes each biosynthetic enzyme and which accumulate relatively simple polymers. These simple polymers are the result of expression of the pathway, which leads to the accumulation of either the 2,3-*cis*-flavanols, e.g. epicatechin (Xie *et al.*, 2003) or the 2,3-*trans*-flavanols, e.g. catechin (Tanner *et al.*, 2003). In plants where both pathways operate, or where enzymes are encoded by multiple loci, mutation of one step of either branch may not lead to a recognizable tannin-deficient phenotype. Fortu-nately, these metabolic simplifications, together with the nonessential character of proanthocyanidins, has allowed the identification of seed coat mutants where the accumulation of proanthocyanidin is blocked by a mutation in a structural or a regulatory gene required in the biochemical pathway.

H. vulgare accumulates oligomers of the 2,3-*trans*-flavan-3-ols, catechin and gallocatechin (Fig. 5.1), in the seed testa (Aastrup *et al.*, 1984). An extensive mutation and mass screening program for plants lacking leaf anthocyanins, or with decreased testa proanthocyanidins, identified over 700 proanthocyanidin-free mutants. These were localized to 28 different complementation groups or *Ant* genes, and several *H. vulgare* genes involved in the (pro)anthocyanidin pathway have been cloned (Jende-Strid, 1993).

Developing *A. thaliana* seeds accumulate polymers of the alternate 2,3-*cis*-flavan-3-ol, epicatechin, in the endothelial layer of the seed coat (Abrahams *et al.*, 2003; Fig. 5.1). Mutants with defective flavonoid accumulation leading to color-less seed coats – *transparent testa* (*tt*) (Koornneef, 1990; Shirley *et al.*, 1995) or reduced testa proanthocyanidins – *tannin-deficient seed* (*tds*) (Abrahams *et al.*, 2002) have been identified. Several of these genes appear to be involved in the

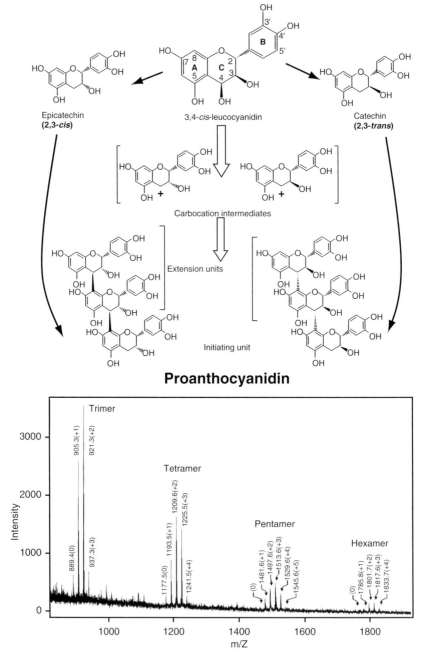

Fig. 5.1 Polymerization of proanthocyanidins. There are two pathways that lead to the synthesis of either 2,3-*trans*-flavan-3-ols or 2,3-*cis*-flavan-3-ols. The location of the A, B, and C rings and the general numbering scheme of the flavonoid carbon atoms is shown for 3,4-*cis*-leucocyanidin. Flavan-3-ols initiate

(continued)

final stages of proanthocyanidin synthesis, and for the first time provide an experimental path to define the last steps of proanthocyanidin biosynthesis.

5.1.1 Economic importance

Economically, proanthocyanidins are arguably the third most important plant polymer after cellulose and starch and exert their main biological effect through an ability to bind to proteins and either precipitate or alter their biological effects. Proanthocyanidins exert positive effects in at least five systems.

(1) *Reduced pasture bloat, increased protein absorption and reduced intestinal parasite burden in grazing ruminants.* Moderate levels of proanthocyanidins (0.5–5% dry weight) reduce voluntary intake and absorption of protein and minerals in monogastric animals (Santos-Buelga & Scalbert, 2000). In contrast, similar concentrations of proanthocyanidins are beneficial in the diet of grazing ruminants, through reduced pasture bloat, increased protein uptake and reduced intestinal parasite burdens (Aerts *et al.*, 1999a; Min *et al.*, 2003).

Bloat is caused by protein foam, formed in the rumen when animals graze protein-rich legume pastures such as *Trifolium repens* (white clover) or *Medicago sativa* (lucerne; alfalfa). The causes of this foam are not well understood and the onset of bloat symptoms is unpredictable and rapid. Rumen foams trigger the involuntary closure of a sphincter that normally opens periodically to allow the release of fermentation gases. Cattle can produce over 2 l of gas per minute and if this gas is not released, the rumen expands quickly, crushing the heart and lungs (Clarke & Reid, 1974). In 1989, in terms of animal deaths alone, bloat was estimated to cost US$100m. in the USA and US$25m. in Canada each year (Goplen, 1989). In Australia, bloat concerns the beef and dairy industries in the higher rainfall areas of Victoria and New South Wales. On average 2–5% of the herd is lost each year to pasture bloat (Cameron & Malmo, 1993). On this basis, the current cost of pasture bloat, in terms of lost animals alone, in Australia in 1994/95 was A$60m.–148m./year.

Not all animals are equally prone to bloat and attempts to breed bloat-safe cattle are ongoing (Morris *et al.*, 1997). There has also been considerable interest in producing bloat-safe pastures. Not all pasture legumes cause bloat –

Fig. 5.1 (continued)
proanthocyanidin polymerization by attacking a carbocation intermediate (or the quinone methide equivalent). The 2,3-*trans* extension units may be derived from leucocyanidin; however, the source of the 2,3-*cis* extension units is unknown. In species where both pathways operate, the polymer contains both 3C epimers. For simplicity, only dihydroxylated B-ring intermediates are shown, but in practice both di- and trihydroxylated B-rings are found. The inset shows the heterogeneity of purified *H. vulgare* proanthocyanidin following MALDI-TOFF mass spectroscopy as in Hedqvist *et al.* (2000). Groups of heterogenic polymers, up to hexamers, are resolved as the Na$^+$ ions (dimers are not shown). Within each group of oligomers, masses corresponding to heteropolymers containing 0, 1 (+1), 2 (+2) or more trihydroxylated B-rings can be seen. Mass spectroscopy does not distinguish between 2,3-*trans*- and 2,3-*cis*-flavan-3-ols; however, these can be differentiated following acid hydrolysis in the presence of nucleophiles such as phloroglucinol (Kennedy & Jones, 2001).

some are bloat-safe as a result of tough leaf tissue that reduces the initial rate of digestion, and hence the tendency to foam. Selection for lower initial rates of digestion has led to the release of alfalfa (cultivar AC Grazeland Br) with lower bloating tendency (Berg *et al.*, 2000). However, most bloat-safe pastures contain foliar proanthocyanidins that destabilize rumen foams, thus preventing bloat (Jones & Lyttleton, 1971; Li *et al.*, 1996).

Proanthocyanidins also increase protein uptake. Up to 30% of dietary protein may be lost following degradation by rumen microflora (Barry & Reid, 1985). Proanthocyanidins precipitate protein at neutral rumen pH, increasing the proportion of dietary protein that escapes rumen degradation (so called *by-pass protein*) and enters the intestine. At the lower intestinal pH, protein is released from the complex and digested by the animal. This by-pass protein increases growth rates, wool and milk production (Min *et al.*, 2003).

Dietary proanthocyanidins also reduce the parasite burden of deer, goats and sheep through reduced numbers and fertility of nematodes (Molan *et al.*, 2002).

(2) *Chilling haze in beer.* Proanthocyanidins are extracted from barley during mashing and precipitate soluble proteins, leading to the formation of a 'chilling haze' on storage. This haze must be removed by filtration or adsorption to maintain clarity of the final product. Proanthocyanidin-free mutants of barley have been successfully used to reduce chilling haze without changing the taste of brewed beer (von Wettstein *et al.*, 1985).

(3) *Inhibition of fungal, bacterial and bird and insect attack on plants.* Proanthocyanidins do not have a consistent effect on herbivorous insects, with similar numbers of studies finding negative, positive or neutral effects depending on the physiology of the insect gut and the proanthocyanidin tested (Ayres *et al.*, 1997).

High concentrations of proanthocyanidins found in sorghum seeds may act as a feeding deterrent to birds (Asquith *et al.*, 1983). Proanthocyanidins also have well-documented antimicrobial and antifungal properties (Scalbert, 1991; Schultz *et al.*, 1992), e.g. barley seed coat proanthocyanidins are important in *Fusarium* resistance (Skadhauge *et al.*, 1997b). Proanthocyanidins have also been reported to accumulate in zones surrounding sites of bacterial infection as a result of a dynamic plant defense reaction (Feucht *et al.*, 1992).

(4) *Probiotics for human consumption.* Even a rudimentary search of the literature shows numerous publications linking dietary proanthocyanidins with a range of beneficial effects in humans including anti-inflammatory responses, antioxidant activity (Keli *et al.*, 1996), and the so-called 'French paradox', i.e. the reduced atherosclerosis, associated with red wine consumption, of French populations when compared to other populations with similar fat intake (Frankel *et al.*, 1993). The majority of these effects appear due to flavonoid monomers rather than polymers.

Proanthocyanidin polymers do not cross the intestinal wall in humans, sheep or goats but are degraded to low–molecular weight phenolic acids in the gastrointestinal tract and excreted (Scalbert *et al.*, 2002).

Smaller oligomers up to trimers cross membranes (Deprez *et al.*, 2001) and have the potential to act as free radical scavengers or moderate biochemical processes inside the cell. Procyanidin dimers, trimers and tetramers are among the most potent plant-derived inhibitors of protein kinases with IC_{50}s in the micromolar range (Polya & Foo, 1994). Proanthocyanidin oligomers may inhibit protein kinases involved in enzyme cascades that initiate and control various processes such as inflammatory responses.

(5) *Soil ecology.* Proanthocyanidins affect decomposition rates and nutrient cycling in soil ecosystems indirectly by binding and increasing the bioavailability of phosphate and organic nitrogen (Northup *et al.*, 1995). Proanthocyanidin oligomers and polymers may also affect soil ecology directly as they leach from root exudates by acting on soil micro- and macroflora (Erikson *et al.*, 2000; Bais *et al.*, 2002; Kraus *et al.*, 2003).

5.2   Biosynthesis from coumaroyl-CoA to leucoanthocyanidin

The conversion of radiolabeled intermediates to (pro)anthocyanins has established that a condensation of three malonyl-CoA molecules with one 4-coumaroyl-CoA molecule by chalcone synthase leads to both anthocyanidins and proanthocyanidins (Fig. 5.2). Anthocyanidin biosynthesis has been regularly reviewed over the past decade (Dooner *et al.*, 1991; Lancaster, 1992; Forkmann, 1993; Heller & Forkmann, 1993; Martin & Gerats, 1993; Koes *et al.*, 1994; Holton & Cornish, 1995; Mol *et al.*, 1998; Rausher *et al.*, 1999; Winkel-Shirley, 2001, 2002) as has proanthocyanidin biosynthesis (Stafford, 1990; Jende-Strid, 1993; Porter, 1993; Koes *et al.*, 1994; Tanner *et al.*, 2000; Marles *et al.*, 2003). These reviews have concentrated on the molecular biological aspects of proanthocyanidin biosynthesis. The proteins involved in (pro)anthocyanidin biosynthesis beyond chalcone synthase are reconsidered and reviewed here. The key enzymes are listed in Table 5.1.

Details of these enzymes may be viewed at http://www.chem.qmul.ac.uk/iubmb/enzyme/. The BRENDA website – a relational database of information on enzymes – is open to nonprofit users upon application at http://www.brenda.uni-koeln.de/ and kinetic constants have been obtained from this site unless otherwise acknowledged.

Enzymes purified from tissues that accumulate anthocyanins have been well studied but there are relatively few studies of enzymes from tissues that accumulate proanthocyanidins. Given that the same enzyme purified from different anthocyanin-positive species can have quite different substrate specificities, it is likely that conclusions drawn from a study of enzymes involved in anthocyanidin biosynthesis may not entirely predict the properties of proanthocyanidin-specific enzymes.

Table 5.1 Enzymes of proanthocyanidin biosynthesis

Protein	Also	Acronym	EC number	Protein family
Chalcone synthase		CHS	2.3.1.74	Polyketide synthase
Chalcone isomerase		CHI	5.5.1.6	No other related sequences
Flavanone 3-hydroxylase		FHT (also F3H)	1.14.11.9	Oxoglutarate dioxygenase
Flavonoid 3′-monooxygenase	Flavonoid 3′-hydroxylase	F3′H	1.14.13.21	P450-dependent monooxygenase
Flavonoid 3′,5′-hydroxylase		F3′5′H		P450-dependent monooxygenase
Dihydroflavonol 4-reductase		DFR	1.1.1.219	RED
Leucoanthocyanidin 4-reductase		LAR (also LCR)	1.17.1.3	RED
Anthocyanidin synthase	Leucoanthocyanidin dioxygenase	ANS (also LDOX)	1.14.11.19	Oxoglutarate dioxygenase
Anthocyanidin reductase	BANYULS	ANR		RED
Intravacuolar transport	TT19			GST
Intravacuolar transport	TT12			MATE
Enzymes that may affect flux				
Flavonol synthase		FLS (also FS)		Oxoglutarate dioxygenase
Flavone synthase I		FNS I		Oxoglutarate dioxygenase
Flavone synthase II		FNS II		P450-dependent monooxygenase

Source: Nomenclature according to Heller & Forkmann (1993).

Considerable advances have been made following heterologous expression and purification of enzymes of anthocyanidin biosynthesis. These studies indicate that some enzymes have surprisingly broad substrate specificities, but for simplicity only the main reactions are indicated (see Figs. 5.2 & 5.3).

5.2.1 CHS

Chalcone synthase (CHS; EC 2.3.1.74) is a remarkable enzyme which catalyses the sequential condensation of one molecule of coumaroyl CoA with three molecules of malonyl CoA in a series of decarboxylation, condensation, cyclization and aromatization reactions within a single active site to produce naringenin chalcone (Fig. 5.2). The enzyme has no requirement for other cofactors. The enzyme mechanism resembles that of the fatty acid and polyketide synthases but instead of using acyl carrier proteins to anchor intermediates and substrates in the active site,

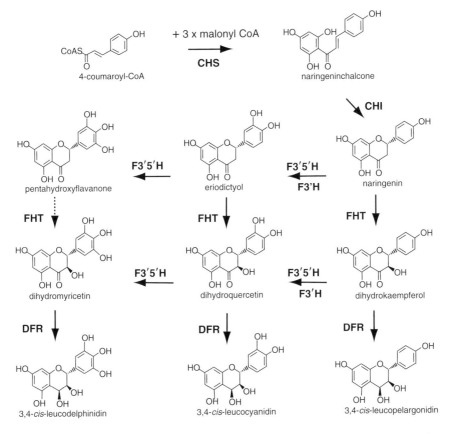

Fig. 5.2 Biosynthesis of leucoanthocyanidins. The intermediates involved in the formation of the leucoanthocyanidins are shown for the enzymes beyond CHS, described in Table 5.1. Pentahydroxy-flavanone was not reduced by purified *Matthiola incana* FHT (Britsch & Grisebach, 1986), but this may not be true for all FHT enzymes and so this step is shown dotted. A 'metabolic grid' of intermediates may be involved in the final production of the leucoanthocyanidins. The exact pathway taken by the metabolic flux depends on the substrate specificity of the enzymes involved.

CHS uses CoA thioesters (Preisig-Mueller *et al.*, 1997). CHS is a member of the polyketide synthase family, which also includes stilbene synthase, bibenzyl synthase, 2-pyrone synthase and acridone synthase (Eckermann *et al.*, 1998).

The native CHS enzyme is a homodimer with subunits of about 43 kDa and has been purified from many tissues and is one of the most widely studied enzymes of secondary metabolism (Heller & Forkmann, 1988). The preferred substrates are 4-coumaroyl CoA (Km 0.6–1.0 μM) and malonyl-CoA (Km 2 μM); however, reactions occur with other substrates including caffeoyl-CoA (Km 1.4–7.0 μM) to produce eriodictyol chalcone, and feruloyl-CoA (Heller & Forkmann, 1993). The formation of 6′-deoxychalcone involved in legume phytoalexin synthesis proceeds through the concerted action of CHS with a polyketide reductase, purified and cloned from *Glycine max* (Welle *et al.*, 1991).

The first flavonoid biosynthetic gene to be cloned and sequenced encoded *Petroselinum hortense CHS* (Kreuzaler *et al.*, 1983; Reimold *et al.*, 1983). *CHS* genes have subsequently been cloned from several hundred species. Complementation with the maize *CHS* gene has shown that the *A. thaliana TT4* locus encodes the single *CHS* gene (Dong *et al.*, 2001).

The structure of the recombinant alfalfa CHS2 protein has been determined and the residues of the active site defined (Ferrer *et al.*, 1999; Shimada *et al.*, 2003). Acridone synthase and CHS have an almost identical spatial structure. Site-directed mutagenesis was used to replace three amino acids conserved in acridone synthase, with the corresponding residues conserved in CHS, effectively transforming an acridone synthase to a functional CHS enzyme (Lukacin *et al.*, 2001).

5.2.2   CHI

Chalcone isomerase (CHI; EC 5.5.1.6) catalyses the cyclization of naringenin chalcone (6′-hydroxychalcone) to (2S)-naringenin (Fig. 5.2). This reaction takes place spontaneously (Mol *et al.*, 1985) to produce an enantomeric mixture. The enzymatic rate exceeds the spontaneous rate by 10^7-fold, approaching the diffusion-controlled limit, and stereospecifically produces the (2S)-isomer (Bednar & Hadcock, 1988). CHI sequences are found only in higher plants and are highly related, displaying no homology to other proteins. This raises the question of how this enzyme function evolved (Jez *et al.*, 2000).

CHI has been purified from many species and is a monomer of 27–29 kDa (Dixon *et al.*, 1988; Heller & Forkmann, 1988; Shimada *et al.*, 2003). CHI proteins have been divided into two classes based on sequence homology and substrate specificity: CHI-1, found in nonlegumes, has activity only towards naringenin chalcone (Km 4–10 μM), while CHI-2, found in legumes, converts both naringenin chalcone (Km 4–10 μM) to (2S)-naringenin and isoliquiriteginen (6′-deoxychalcone) (Km 9–12 μM) to (2S)-5-deoxyflavanone.

Genes for CHI have been cloned from numerous species and in *A. thaliana* there is only one functional *CHI* gene, *TT5* (Dong *et al.*, 2001). The flavonoid-free phenotype of the *tt5* mutant confirms that in *A. thaliana* the CHI reaction is not spontaneous. This may not be true of all tissues since *Zea mays* cultures synthesize anthocyanin in the absence of detectable CHI activity (Grotewold *et al.*, 1998).

The structure of the recombinant alfalfa CHI protein has been determined (Jez *et al.*, 2000).

5.2.3   FHT (F3H)

Flavanone 3-hydroxylase (FHT; 1.14.11.9) is a soluble nonheme iron enzyme, dependent on Fe^{2+}, O_2, 2-oxoglutarate and ascorbate, and adds a single hydroxyl to the 3C of flavanones (Fig. 5.2). FHT is a member of the 2-oxoglutarate-dependent dioxygenase protein family, which also includes ANS, FLS, FNSI and isopenicillin N-synthase (see Fig. 5.5). Curiously, FLS appears to have a dual FHT/FLS activity (see below).

FHT catalyses the reduction of (2S)-naringenin and (2S)-eriodictyol to (2R,3R)-dihydrokaempferol or -dihydroquercetin, respectively. This activity was first demonstrated in crude extracts of *Matthiola incana* (Forkmann *et al.*, 1980). The partially purified *Petunia* enzyme converted both naringenin (Km 5.6 μM) and eriodictyol (Km 12 μM) in the presence of 2-oxoglutarate (Km 20 μM). The nonphysiological (2R)-enantiomer of naringenin and 5,7,3',4',5'-pentahydroxy-flavanone were not substrates (Britsch & Grisebach, 1986).

The cDNA encoding FHT was first isolated from *Antirrhinum majus* and *Petunia* (Britsch *et al.*, 1992). There appears to be a single gene for FHT in *A. majus*, *Petunia* and *A. thaliana* (Pelletier & Shirley, 1996). In *A. thaliana* the *TT6* locus encodes FHT (Pelletier & Shirley, 1996; Wiseman *et al.*, 1998). The gene for FHT has now been cloned from 11 other species including *H. vulgare* (Meldgaard, 1992); *Malus* (Davies, 1993); *Vitis* (Sparvoli *et al.*, 1994); *Medicago* (Charrier *et al.*, 1995) and *Perilla frutescens* (Gong *et al.*, 1997) with the size of the predicted proteins averaged at 40.9 kDa.

Sequence correlation of cloned *FHT* genes (Britsch *et al.*, 1993) and mutation analysis of the *Petunia FHT* gene (Lukacin *et al.*, 2000a) confirmed that Arg288 and Ser290 are essential for 2-oxoglutarate-binding, and His220, His278 and Asp222 are involved in iron binding in the active site.

Expression of the *Petunia* gene in yeast showed that the native 41 655 Da enzyme was subject to rapid proteolysis and loss of a 3800 Da C-terminal peptide during previous purifications (Britsch & Grisebach, 1986; Lukacin *et al.*, 2000b). The native monomeric enzyme may associate with itself or other enzymes to form macromolecular structures *in vivo* (Lukacin *et al.*, 2000). FHT was rapidly purified following expression of the *Petunia* cDNA in yeast with 5% yield and a final specific activity of 32 mkat/kg, some 100-fold greater than previously attained (Lukacin *et al.*, 2000b). The kinetic properties of this new full-length enzyme have not yet been reported.

5.2.4   *F3'H and F3',5'H*

Flavonoid-3'-hydroxylase (F3'H; EC 1.14.13.21) and the closely related enzyme, flavonoid-3',5'-hydroxylase (F3',5'H) introduce 3'- or 3'- and 5'-hydroxyl groups respectively into the B-ring of a number of flavonoids (Fig. 5.2). They are both microsomal, heme-containing, cytochrome P450-dependent monooxygenases that require NADPH and O_2 for activity, and are characteristically inhibited by carbon monoxide (Forkmann, 1993). These enzymes are part of the P450-dependent mono-oxygenase family which includes flavone synthase II, isoflavone synthase, isoflavone 2'-hydroxylase, isoflavone 3'-hydroxylase, *trans*-cinnamate 4-hydroxylase, (2S)-flavanone 2-hydroxylase and ferulate 5-hydroxylase (see Fig. 5.6; Kitada *et al.*, 2001).

F3'H, F3',5'H, FHT, DFR and FLS produce a metabolic grid of possible products with varying hydroxylation states (Stafford, 1990); however, only the intermediates leading to proanthocyanidin biosynthesis are considered here (Fig. 5.2).

F3'H enzyme activity was demonstrated in microsomal preparations from *Haplopappus* (Fritsch & Grisebach, 1975), *Petroselinum* (Hagmann *et al.*, 1983), *Z. mays* (Larson & Bussard, 1986) and *Citrus sinensis* (Doostdar *et al.*, 1995). F3'H has a broad substrate specificity and introduces a 3'-hydroxyl to convert the flavanone (2S)-naringenin to eriodictyol; the dihydroflavonol dihydrokaempferol to dihydroquercetin; the flavonol kaempferol to quercetin; or the flavone apigenin to luteolin. However, it does not act on 4-coumarate or 4-coumaroyl-CoA, or glycosylated flavonoids (Hagmann *et al.*, 1983). The *Z. mays* F3'H specifically requires NADPH (Km 5.8 μM) and has little activity with NADH (Larson & Bussard, 1986).

Recombinant *P. frutescens* F3'H activity expressed in yeast microsomes had a similar Km (about 20 μM) for naringenin, apigenin and dihydrokaempferol. The ratio Vmax/Km for all three substrates was also similar, indicating that there was no significant preference for any of these substrates (Kitada *et al.*, 2001). *C. sinensis* F3'H does not significantly convert kaempferol or apigenin (Doostdar *et al.*, 1995); whereas the *Z. mays* enzyme converted naringenin and kaempferol (Km 7 μM), but not dihydrokaempferol (Larson & Bussard, 1986).

In *Petunia*, F3'H enzyme activity is controlled by two loci, *Ht1* and *Ht2*. The cDNA corresponding to the *Ht1* locus was cloned from *Petunia* and later by homology from *A. thaliana*, *P. frutescens* and *G. max* (Brugliera *et al.*, 1999; Schoenbohm *et al.*, 2000; Kitada *et al.*, 2001; Toda *et al.*, 2002). The *A. thaliana* *TT7* locus encodes F3'H activity (Schoenbohm *et al.*, 2000). The *A. thaliana* and *Petunia* F3'H proteins are predicted to be 56.8 and 56.9 kDa, respectively.

F3',5'H enzyme activity was first demonstrated in crude extracts of *Verbena hybrida* (Stotz & Forkmann, 1982). F3',5'H also has a broad substrate specificity and hydroxylates both 3'C and 5'C of naringenin and the 5'C of eriodictyol to produce pentahydroxyflavanone. Similarly, both apigenin and luteolin are converted to tricetin; kaempferol and quercetin to myricetin; and dihydrokaempferol and dihydroquercetin to dihydromyricetin (Stotz *et al.*, 1984; Holton *et al.*, 1993; Menting *et al.*, 1994; Kaltenbach *et al.*, 1999). *Petunia* F3',5'H is sensitive to detergents, well below the concentration required to solubilize the enzyme, thus preventing isolation of F3',5'H activity from microsomal membranes (Menting *et al.*, 1994). Activity has been successfully demonstrated following heterologous expression in yeast membranes. *Petunia* and *Catharanthus roseus* F3',5'H activity was expressed in yeast microsomes and had the greatest activity with naringenin (Km 7 μM) and apigenin. Kaempferol and dihydroquercetin were less effective substrates showing 50% and 30%, respectively, of the activity with naringenin. These enzymes preferentially used the 4'-hydroxylated compounds naringenin, apigenin, kaempferol and dihydrokaempferol before the corresponding 3',4'-hydroxylated compounds eriodictyol, luteolin, quercetin and dihydroquercetin (Kaltenbach *et al.*, 1999; Shimada *et al.*, 1999).

Two cDNAs corresponding to the genetic loci *Hf1* and *Hf2*, which control F3',5'H enzyme activity in *Petunia*, were cloned using degenerative PCR approaches and confirmed by mutant complementation and expression in yeast (Holton *et al.*, 1993). F3',5'H cDNAs were subsequently isolated from several species including *Gentiana triflora*, *Eustoma grandiflorum* (Lisianthus or Prairie gentian) and *C. roseus* (Tanaka *et al.*, 1996; Nielsen & Podivinsky, 1997; Kaltenbach *et al.*, 1999; Shimada *et al.*, 1999). In *Petunia* an additional gene *difF* that encodes a Cyt b_5 appears to be specifically required for expression of F3',5'H enzyme activity (de Vetten *et al.*, 1999).

5.2.5   DFR

Dihydroflavonol 4-reductase (DFR; EC 1.1.1.219) stereospecifically reduces the 4-carbonyl of (+)-(2*R*,3*R*)-dihydroflavonol in the presence of NADPH to give the respective (2*R*,3*S*,4*S*)-flavan-2,3-*trans*-3,4-*cis*-leucoanthocyanidin (Fig. 5.2). DFR is a member of the single-domain reductase/epimerase/dehydrogenase (RED) protein family, which includes LAR, ANR, isoflavone reductase, pinoresinol/lariciresinol reductase, phenylcoumaran benzylic ether reductase, GDP-4-keto-6-deoxy-D-mannose epimerase/reductase, UDP-galactose-4-epimerase, cholesterol dehydrogenase, and short-chain dehydrogenases such as 3β-hydroxysteroid dehydrogenase (see Fig. 5.7; Baker & Blasco, 1992; Labesse *et al.*, 1994; Rizzi *et al.*, 1998; Gang *et al.*, 1999).

DFR activity was first demonstrated in extracts of the proanthocyanidin-accumulating *Pseudotsuga menziesii* (Stafford & Lester, 1982) and *H. vulgare* (Kristiansen, 1986). *Dahlia variabilis* DFR was purified to homogeneity and is a monomer of 41 kDa (Fischer *et al.*, 1988), which stereospecifically transfers the pro-S hydrogen of NADPH (Km 42 μM) to (+)-dihydrokaempferol (Km 10 μM) and (+)-dihydroquercetin (Km 15 μM). A DFR with similar kinetic properties was also partially purified from *Cryptomeria japonica* but had a molecular weight of 133 kDa (Ishikura *et al.*, 1988).

The specificity of DFR for dihydroflavonols varies markedly between species. The DFRs of *Z. mays*, *D. variabilis*, *M. incana*, *Dianthus caryophyllus* and *Gerbera hybrida* accept dihydrokaempferol, dihydroquercetin and dihydromyricetin. Conversely, *Petunia*, *Nicotiana*, *Lycopersicon esculentum* and *Cymbidium* DFRs do not efficiently reduce dihydrokaempferol and subsequently do not produce orange pelargonidin anthocyanins (Heller & Forkmann, 1993). Substrate specificity appears to be determined by a region of 26 amino acids, and mutagenesis of a single amino acid in this region either abolished or altered the specificity of *G. hybrida* DFR (Johnson *et al.*, 2001).

In addition to accepting dihydroflavonols, purified *D. variabilis* DFR also converted the (2*S*)-flavanones naringenin (Km 2.3 μM) and eriodictyol (Km 2 μM) to the corresponding flavan-4-ols apiferol and luteoferol, respectively, which are involved in deoxy-(pro)anthocyanidin synthesis. The V/Km values indicate that dihydroflavonols are the preferred substrates (Fischer *et al.*, 1988).

Similar substrate specificity was observed following heterologous expression of the *Malus domestica, Pyrus communis* and *Z. mays* DFRs in yeast (Fischer *et al.*, 2003; Halbwirth *et al.*, 2003). It appears that many DFR enzymes have a flavanone-4-reductase (FNR)-like activity arising from their ability to accept flavanones.

The locus-encoding DFR was first cloned from *Z. mays* and *A. majus* (Martin *et al.*, 1985; O'Reilly *et al.*, 1985) and the identity confirmed by *in vitro* translation of the *Z. mays* cDNA (Reddy *et al.*, 1987). The gene has now been cloned from many species with the size of the predicted protein averaging about 38 kDa. Complementation with the *Z. mays* A1 gene has shown that the *A. thaliana TT3* locus encodes the single *DFR* gene in *A. thaliana* (Dong *et al.*, 2001).

5.2.6   *FLS*

Flavonol synthase (FLS) is a member of the 2-oxogluatrate-dependent dioxygenases (Fig. 5.5) and converts (+)(2R,3R)-dihyroquercetin and -dihydrokaempferol to the corresponding flavonols, quercetin and kaempferol. A 37.9 kDa recombinant *Citrus unshiu* FLS purified to homogeneity has Km values of 45 μM, 272 μM, 11 μM and 36 μM for dihydrokaempferol, dihydroquercetin, Fe^{2+}, and 2-oxoglutarate, respectively (Wellmann *et al.*, 2002). *C. unshiu* FLS also appears to have a dual FHT/FLS activity. The purified, recombinant *C. unshiu* enzyme converts (2S)-naringenin first to (+)(2R,3R)-dihydrokaempferol and then to kaempferol. The nonphysiological (2R)-naringenin was converted to (−)(2S,3S)-dihydrokaempferol (Lukacin *et al.*, 2003).

5.3   Synthesis of catechin (2,3-*trans*-flavan-3-ol)

Flavan-3-ols initiate the formation of proanthocyanidins (Fig. 5.3). Beyond the 3,4-*cis*-leucoanthocyanidins there are two complementary metabolic pathways which lead to the accumulation of either the 2,3-*trans*-flavan-3-ols (e.g. catechin) via the enzyme LAR (Tanner *et al.*, 2003) or the 2,3-*cis*-flavan-3-ols (e.g. epicatechin) via the concerted action of ANS and ANR (Xie *et al.*, 2003).

5.3.1   *LAR (LCR)*

Leucoanthocyanidin reductase (LAR, E.C. 1.17.1.3) removes the 4-hydroxyl from (2R,3S,4S)-2,3-*trans*-3,4-*cis*-leucoanthocyanidins in a single step to produce the corresponding flavan-3-ols. Specifically, (+)-2,3-*trans*-3,4-*cis*-[4-^{3}H]-leucocyanidin (Tanner & Kristiansen, 1993) is reduced to (+)-[4-^{3}H]-catechin in the presence of NADPH (Fig. 5.3).

LAR activity was first demonstrated in extracts of *P. menziesii* and *Ginkgo biloba* and subsequently in *H. vulgare* and other legumes (Stafford & Lester, 1984, 1985; Kristiansen, 1986; Tanner & Kristiansen, 1993; Singh *et al.*, 1997; Skadhauge *et al.*, 1997a). The enzyme was purified to homogeneity from *Desmo-*

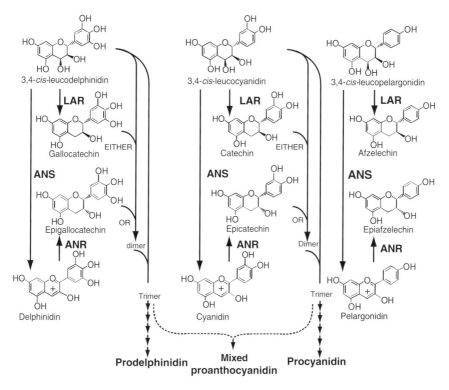

Fig. 5.3 Biosynthesis of flavan-3-ols and proanthocyanidins. The intermediates involved in the formation of both 2,3-*trans*- and 2,3-*cis*-flavan-3-ols are shown with the enzymes described in Table 5.1. Flavan-3-ols initiate dimer formation by attacking an extension unit derived from the leucoanthocyanidins. Subsequent addition of extension units leads to higher order oligomers.

dium uncinatum (Jacq.) DC with a final specific activity of 10 kat/kg, partially sequenced and the corresponding cDNA cloned and expressed in *Escherichia coli*, *T. repens* and *Nicotiana tabacum* (Tanner *et al.*, 2003). LAR is a monomer of 42.7 kDa, most closely related to the isoflavone reductase group of plant enzymes that are part of the RED protein family (Fig. 5.7; Gang *et al.*, 1999).

The preferred substrate of purified *D. uncinatum* LAR is 3,4-*cis*-leucocyanidin (Km 6 μM). The alternate substrates 3,4-*cis*-leucodelphinidin (Km 5 μM) and 3,4-*cis*-leucopelargonidin (Km 26 μM) were less effective, showing 20% and 4% respectively of the Vmax with leucocyanidin. LAR has an unusually low Km for NADPH (0.4 μM) and only partial activity with NADH (Km 60 μM). *D. uncinatum* LAR was inhibited strongly by the 2,3-*trans*-flavan-3-ol products afzelechin, catechin and gallocatechin (I_{50} of 14, 12 and 280 μM, respectively) but not by the epimeric 2,3-*cis*-flavan-3-ol products, epicatechin and epigallocatechin (I_{50} 1 and 1.4 mM, respectively), indicating that LAR is probably not involved in epicatechin synthesis.

There is a single copy of the LAR gene in *D. uncinatum* and there is no LAR homologue in *A. thaliana* (Tanner *et al.*, 2003) consistent with the accumulation of epicatechin and not catechin in seed coats of this plant (Abrahams *et al.*, 2002).

5.4 Synthesis of epicatechin (2,3-*cis*-flavan-3-ol)

5.4.1 ANS (LDOX)

Anthocyanidin synthase (ANS, also leucoanthocyanidin dioxygenase or LDOX, E.C. 1.14.11.19) has always been considered an anthocyanidin biosynthetic enzyme; however, recent biochemical and genetic evidence has redefined the role of ANS as relevant to both anthocyanidin and proanthocyanidin pathways (Abrahams *et al.*, 2003; Xie *et al.*, 2003). ANS catalyses the reduction of (2R,3S,4S)-2,3-*trans*-3,4-*cis*-leucoanthocyanidins to a pseudobase (3-flavene-3,4-diol) which rearranges in mild acid to anthocyanidin, without the need for a separate dehydratase enzyme (Fig. 5.3; Nakajima *et al.*, 2001). Mutation of the *ANS* gene blocks proanthocyanidin accumulation in the endothelial layer of *A. thaliana* seed coat (Abrahams *et al.*, 2003) revealing for the first time its essential role in both (pro)anthocyanidin pathways. ANS is a soluble nonheme iron enzyme, dependent on Fe^{2+}, oxygen, 2-oxoglutarate and ascorbate, and is a member of the 2-oxoglutarate-dependent dioxygenase protein family (Fig. 5.5) along with FHT and FLS. The *P. frutescens* ANS protein is predicted to be 40.5 kDa, similar to that predicted for other ANS genes (Saito *et al.*, 1999; Nakajima *et al.*, 2001).

Z. *mays A2* and *A. majus Candica* genes, cloned by transposon tagging, were thought to encode ANS activity (Menssen *et al.*, 1990; Martin *et al.*, 1991). Homologues were subsequently isolated from a number of other species. However, ANS activity was not demonstrated until 1999, following expression of the *P. frutescens* gene in *E. coli* and purification of a recombinant ANS-fusion protein (Saito *et al.*, 1999). Equimolar stoichiometry was observed between the formation of anthocyanidin (assayed by HPLC at A_{520}), and the leucocyanidin-dependent CO_2 liberation from 2-oxoglutarate in the presence of Fe^{2+}, and ascorbate. Curiously, the nonphysiological (2R,3S,4R)-3,4-*trans*-leucocyanidin (Km 39 μM) and (2R,3S,4R)-3,4-*trans*-leucodelphinidin (Km 110 μM) were used as substrates in the presence of 2-oxoglutarate (Km 60 mM). Saito *et al.* (1999) claim that the nonphysiological 3,4-*trans*-leucocyanidins readily isomerize to the physiological 3,4-*cis*-isomers during the reaction; however, isomerization of 3,4-*trans*-leucocyanidins was not observed with radiolabeled substrates (Tanner *et al.*, 2003). The highest specific activity of cyanidin formation for recombinant *P. frutescens* ANS was 8.6 μkat/kg, three orders of magnitude lower than that seen for *Petunia* FHT, another protein from the same protein family (Lukacin *et al.*, 2000b). This may be due to inherent instability of ANS, limited proteolysis as observed for FHT, or the use of inappropriate substrates. The requirement for oxoglutarate was also unusually high compared to other members of the same protein family; e.g. *Petunia* FHT and *C. unshiu* FLS had Km values of 20 μM and 36 μM for 2-oxoglutarate, respectively (Lukacin *et al.*, 2000b; Wellmann *et al.*, 2002).

Similar results were obtained for four recombinant ANS proteins derived from *A. majus*, *Petunia*, *Torenia* and *Z. mays* (Nakajima *et al.*, 2001). Specific activities of cyanidin formation of 0.97 and 0.22 μkat/kg were observed for the purified recombinant *Petunia* and *Z. mays* enzymes, respectively – some four orders of magnitude smaller than that observed for *Petunia* FHT. The *in vitro* formation of cyanidin-3-*O*-glucoside from the nonphysiological (2*R*,3*S*,4*R*)-3,4-*trans*-leuco-cyanidin in the presence of recombinant *Petunia* ANS and UDP-glucose:flavonoid 3-*O*-glucosyltransferase (3-GT) was also shown, confirming that ANS and 3-GT were necessary and sufficient for this reaction.

Product formation *in vitro* with *A. thaliana* ANS was more complicated when followed by HPLC at A_{287} instead of A_{520} as above, with cyanidin only a minor product (Turnbull *et al.*, 2000). The physiological (+)(2*R*,3*R*)-2,3-*trans*-dihydro-quercetin (Fig. 5.2) was reduced to quercetin – the same reaction catalysed by FLS (Turnbull *et al.*, 2003). The physiological ANS substrate (2*R*,3*S*,4*S*)-3,4-*cis*-leucocyanidin (Fig. 5.3) was converted predominantly to quercetin (85%) and lesser amounts of (2*R*,3*S*)-2,3-*cis*-dihydroquercetin, i.e. (−)epi-dihydroquercetin (10%), (+)(2*R*,3*R*)-2,3-*trans*-dihydroquercetin (3%) and cyanidin (2%). Conversely the nonphysiological substrate (2*R*,3*S*,4*R*)-3,4-*trans*-leucocyanidin was converted predominantly to (2*R*,3*S*)-2,3-*cis*-dihydroquercetin (55%), quercetin (30%) and lesser amounts of (2*R*,3*R*)-2,3-*trans*-dihydroquercetin (11%) and cya-nidin (4%) (Turnbull *et al.*, 2003). These other reactions may explain the low activity observed for cyanidin formation by (Nakajima *et al.*, 2001). There is also the possibility that (2*R*,3*S*)-epi-dihydroquercetin is synthesized *in vivo* by ANS. DFR has a broad substrate specificity and may be able to reduce epi-dihydroquer-cetin to the hypothetical epi-leucocyanidin (2*R*,3*R*,4*S*)-2,3-*cis*-3,4-*trans*-leucoanthocyanidin, which may be a possible source of the (2,3)-*cis*-extension units seen in *A. thaliana* proanthocyanidin.

A. thaliana ANS was crystallized and the structure determined (Turnbull *et al.*, 2001). An enzyme mechanism involving stereoselective hydroxylation at C3 has been suggested (Wilmouth *et al.*, 2002). *A. thaliana* TDS4 and TT18 both encode ANS (Winkel-Shirley, 2001; Abrahams *et al.*, 2003; Xie *et al.*, 2003; Kitamura *et al.*, 2004).

5.4.2 ANR (BANYULS)

Anthocyanidin reductase (ANR) converts anthocyanidins to the corresponding 2,3-*cis*-flavan-3-ols such as epicatechin in the presence of NADPH or NADH (Fig. 5.3; Xie *et al.*, 2003). ANR is a member of the RED protein family, which includes DFR and LAR (Fig. 5.7).

The *A. thaliana* BANYULS gene is named after a French red wine because mutations in this gene caused precocious accumulation of anthocyanins in the seed coat endothelium (Albert *et al.*, 1997). There was a single copy of the gene in *A. thaliana*, which was cloned and shown to encode a cytosolic DFR-like protein of 38 kDa expressed only in the endothelium of developing seed coats (Devic *et al.*, 1999). Mutation of *BANYULS* also prevents accumulation of proanthocyanidins in the seed coat endothelium, suggesting that metabolic flux is diverted away from proanthocyanidins and towards anthocyanidins. On this basis it was suggested that the gene may encode LAR.

Recombinant BANYULS from *A. thaliana* and *M. truncatula* was expressed in *E. coli* and did not produce LAR activity (Xie *et al.*, 2003). However, extracts containing BANYULS protein reduced the anthocyanidins, pelargonidin, cyanidin and delphinidin to the corresponding (2*R*,3*R*)-2,3-*cis*-flavan-3-ols, epiafzelechin, epicatechin and epigallocatechin, thus solving the riddle of the synthesis of these epimers. Minor production of the nonphysiological (2*S*)-isomers, ent-epiafzelechin, ent-epicatechin and ent-epigallocatechin, was also identified. The specific ANR activity was not reported. Ectopic expression of BANYULS in *Nicotiana* petals led to reduced anthocyanidin accumulation and increased accumulation of 4-dimethylamino-cinnamaldehyde (DMACA)-positive compounds (Xie *et al.*, 2003). Four additional DMACA-positive compounds were identified when TLCs of flavonoid extracts from *N. tabacum* petals expressing BANYULS were compared to control extracts. These compounds are unstable and have not been identified. There was no evidence of proanthocyanidin polymer formation in these BANYULS tobacco petals (Tanner & Abrahams, unpublished data), suggesting that further enzymatic steps may exist beyond ANR.

5.5 Intravacuolar transport and polymerization

5.5.1 Intravacuolar transport

The initiating flavan-3-ols are synthesized in the cytoplasm, while proanthocyanidin accumulates inside the vacuole. The initiating units must therefore be transported into the vacuolar compartment, where polymerization occurs (Fig. 5.4).

Intravacuolar transport of anthocyanins involves glutathione *S*-transferase and ABC-like transporter proteins (Mueller *et al.*, 2001). The *Z. mays* BZ2 and *P. hybrida* AN9 proteins conjugate cyanidin-3-*O*-glucoside with glutathione (Alfenito *et al.*, 1998; Mueller *et al.*, 2000). The conjugated anthocyanin is believed to be subsequently transported across the tonoplast via an Mg-ATP-requiring glutathione pump (Lu *et al.*, 1997). Similar ATP-dependent systems transport flavone glucuronides across the tonoplast in rye vacuoles (Klein *et al.*, 2000) and medicarpin across mung bean vacuolar membrane vesicles (Li *et al.*, 1997).

In contrast, intravacuolar transport of proanthocyanidins appears to require at least two specific transport proteins. The *A. thaliana TT12* locus (Fig. 5.4) encodes a member of the multidrug and toxic compound extrusion (MATE) family of transport proteins. Mutation of this locus prevents accumulation of proanthocyanidins

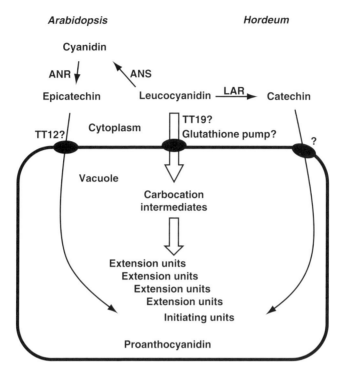

Fig. 5.4 Intravacuolar transport and polymerization of proanthocyanidins. The intracellular distribution of intermediates involved in proanthocyanidin polymerization are shown for the model plants *Arabidopsis thaliana* (left) or *Hordeum vulgare* (right) for the enzymes described in Table 5.1. For simplicity, only dihydroxylated B-ring flavonoids are shown. Synthesis of the intermediates prior to leucocyanidin occurs in the cytoplasm of cells. Proanthocyanidins ultimately accumulate inside the vacuole. The transport of initiating and extension units into the vacuole is dependent upon specific transport proteins (e.g. TT12 in *A. thaliana*). It is not clear if initiating and extension units are transported by separate proteins. The metabolic requirement for extension units is greater than the requirement for initiating units. Mutation of ten *A. thaliana* and four *H. vulgare* loci beyond intravacuolar transport blocks accumulation of proanthocyanidins, and these genes may code for condensing enzymes or other regulatory proteins.

(Debeaujon *et al.*, 2001). The *A. thaliana TT19* locus encodes a Phi glutathione *S*-transferase that is required for normal vacuolar accumulation of both proanthocya-nidins and anthocyanins. TT19 has 50% homology with *P. hybrida* AN9; and anthocyanin, but not proanthocyanidin, accumulation is restored in the *tt19* muta-tion by expression of the *AN9* cDNA (Kitamura *et al.*, 2004). Most glutathione *S*-transferases are cytosolic (Frova, 2003), therefore the action of TT19 is presum-ably followed by the subsequent action of a membrane-bound transporter (Fig. 5.4) such as a glutathione pump (Mueller *et al.*, 2001). It is not clear if TT12 and TT19 are involved in the transport of initiating or extension units or both.

However, *in vivo*, the process may not simply involve transport proteins on the tonoplast. Vacuole formation involves small pro-vacuolar vesicles that bud from plasma membrane, endoplasmic reticulum or Golgi bodies and fuse with the tonoplast (Bethke & Jones, 2000; Grotewold, 2001). Some small pro-vacuoles already contain osmium-staining material such as proanthocyanidins and mono-mers (Parham & Kaustinen, 1977), which may be carried into the vacuole as a result of fusion with the main vacuole. There are many types of pro-vacuole, identified by different tonoplast-intrinsic proteins decorating the vesicle surface. These intrinsic proteins correlate with the ultimate function of the vesicles such as protein or pigment storage, lysis or autophagosis (Juah et al., 1999; Hinz & Herman, 2003). Thus, the proanthocyanidin transporters may be on the surface of specific vesicles involved in the formation of proanthocyanidin storage vacuoles. There is a further level of complication. Mutation of A. thaliana TDS4, the gene for ANS, not only blocks proanthocyanidin accumulation but also prevents normal vacuole develop-ment. The failure of vacuoles to develop may be due to failure to produce epica-techin, since other mutations, which also prevent proanthocyanidin accumulation but do not block epicatechin accumulation, do not interfere with vacuole develop-ment. This implies a coupling may exist at this step between vacuole development and epicatechin biosynthesis (Abrahams et al., 2003). However, the vacuole is conceptually represented here by the one surface (Fig. 5.4).

5.5.2   Polymerization (condensation)

Proanthocyanidin polymerization occurs sequentially when the C8 of the initiating flavan-3-ol captures the protonated carbocation (or the quinone methide equiva-lent) on C4 of an intermediate, presumably formed from a flavan-3,4-diol such as leucocyanidin (Jacques et al., 1977). The dimer in turn captures another inter-mediate to form a trimer, and so on (Fig. 5.4; Haslam, 1977). The protonated benzylic carbocation formed by loss of the 4-hydroxyl is unusually stable due to charge delocalization over nearby A and B benzene rings (Fig. 5.1; Ferreira et al., 1992) that may allow the intermediate sufficient half-life to react with the initiat-ing nucleophilic flavanol.

   The carbocation intermediates could arise either as a result of the activity of a specific polymerizing enzyme, or as a result of premature release of the intermedi-ate from the active site of an enzyme like LAR, ANS or ANR. However, these enzymes are believed to be cytosolic, and transport of unstable charged intermedi-ates across the tonoplast would be unlikely. The enzymes of proanthocyanidin biosynthesis may form a membrane-bound, multienzyme complex with a transport protein on the outer face of the tonoplast to facilitate directional transport across the tonoplast. This transport may be coupled to polymerization and release from the inner face of the tonoplast (Stafford, 1990). This is supported by the observa-tion that proanthocyanidins tend to form first as an electron-dense layer on the inner tonoplast surface in banana roots and Onobrychis viciifolia leaf (Mueller & Beckman, 1974; Lees et al., 1995; Tanner, unpublished).

   Purified CHS, which does not naturally make 6'-deoxychalcone, readily does so in the presence of purified polyketide reductase (Welle et al., 1991) indicating that the active sites of the two enzymes interact intimately allowing synthesis of a novel product. Similar physical complexes between other flavonoid enzymes may allow efficient 'channeling' of substrates between enzymes and manage competition between alternate enzyme pathways (Stafford, 1974). Channeling experiments, gel filtration and immunocytochemical co-labeling have all shown that at least a small proportion of the total cellular activity of enzymes such as phenylalanine ammonia lyase, cinnamate-4-hydroxylase, CHS and 3-GT is associated with the endoplasmic reticulum (Hrazdina & Jensen, 1992). The cytosolic protein concentration is very high, encouraging protein–protein interactions; however, the process of cell disruption involves enormous dilution and mitigates against the biochemical isolation and purification of such multienzyme complexes in high yields.

More convincingly, yeast double hybrid studies and immunoprecipitation have shown that recombinant CHS, CHI and DFR enzymes interact (Burbulis & Winkel-Shirley, 1999; Winkel-Shirley, 1999). A high–molecular weight subcellular fraction isolated from *O. viciifolia* leaves catalysed the two-step NADPH-dependent reduction of (+)-dihydromyricetin to (+)-gallocatechin, presumably through the concerted action of a complex involving DFR and LAR. [^{14}C](+)-dihydromyricetin was preferentially utilized compared to [^{3}H]-3,4-*cis*-leucodelphinidin, suggesting preferential substrate channeling of the [^{14}C]-dihydromyricetin through DFR to leucodelphinidin and then without release to the cytoplasm, directly via LAR to gallocatechin (Singh *et al.*, 1997).

In developing *H. vulgare* testa, free catechin peaks at about 16 days after flowering, at 100 nmol/seed and then decreases to 40 nmol/seed (Kristiansen, 1984). This decrease in free catechin is accompanied by a concomitant increase in catechin dimers (proanthocyanidin B3) and trimers (proanthocyanidin C3), suggesting that the free catechin is being used to provide the initiating unit, by reacting with 3,4-*cis*-leucocyanidin, which provides the extension units (Fig. 5.4).

Hydrolysis of purified *A. thaliana* proanthocyanidin in the presence of phloroglucinol produces free epicatechin (originating from the initiating unit) and epicatechin-phloroglucinol (originating from the extension units) indicating that *A. thaliana* proanthocyanidin consists exclusively of epi-catechin units (Abrahams *et al.*, 2003). Free epicatechin is present in developing *A. thaliana* siliques (Abrahams *et al.*, 2002) and is the source of the initiating units; however, the source of the extension units in *A. thaliana* is not clear. They might arise from epi-leucocyanidin (2*R*,3*R*,4*S*)-2,3-*cis*-3,4-*trans*-leucocyanidin. This compound has not been identified or synthesized but it could arise from epimerization of the physiological (2*R*,3*S*,4*S*)-3,4-*cis*-leucocyanidin by an unknown enzyme. ANS produces epi-dihydroquercetin in fivefold excess compared to cyanidin (Turnbull *et al.*, 2003). DFR has broad substrate specificity, and could reduce epi-dihydroquercetin to epi-leucocyanidin.

The metabolic requirement for initiating and extension units is not equal (Fig. 5.4). Only a relatively small amount of flavan-3-ol is needed to prime the formation of oligomers; however, a larger number of extension units must be provided for the chain to grow. During development of *H. vulgare* (Kristiansen, 1984), a pool of flavan-3-ol continues to exist as the amount of proanthocyanidin polymer increases. This suggests that carbocation intermediates are preferentially attacked by already existing oligomers rather than the flavan-3-ol pool. This is supported by differential labeling of initiating and extension units. Extension units incorporated a tenfold higher specific radioactivity than initiating units following feeding of $[U^{14}C]$-phenylalanine to *P. menziesii* cell cultures over 6 h. This indicates that the majority of the metabolic flux was directed towards production of extension units rather than initiating units (Stafford *et al.*, 1982). Similar results were also reported for whole plants (Jacques *et al.*, 1977).

It is not clear if polymerization occurs enzymatically *in vivo*, as suggested by Kristiansen (1984). Flavan-3-ols and leucoanthocyanidins polymerize spontaneously, *in vitro*, at pH 5 to yield predominantly (4–8) linked proanthocyanidin polymers, studied in series of biomimetic syntheses (Haslam, 1977; Delcour *et al.*, 1983; Saunders *et al.*, 1996). However, the possibility of a nonenzymatic reaction does not prove the absence of a condensing enzyme, as there are sufficient examples where reactions that occur spontaneously *in vitro* are facilitated by an enzyme *in vivo*, e.g. CHI. In tissues that accumulate both 2,3-*cis*-flavanols and 2,3-*trans*-flavanols, the 2,3-*cis*-flavanols usually account for the majority of the extension subunits (Foo & Porter, 1980). The hydroxylation state of initiating units and extension units also differs. Many tissues accumulate both di- and trihydroxylated B-ring flavan-3-ols, such as *O. viciifolia* (Koupai-Abyazani *et al.*, 1993) and *D. uncinatum* leaves (Tanner & Downey unpublished), and grape skins (Downey *et al.*, 2003). In these tissues, catechin and epicatechin (carrying two B-ring hydroxyls) account for the bulk of the initiating units while the extension units are predominantly gallocatechin or epigallocatechin (three B-ring hydroxyls). It is not clear if these differences in proanthocyanidin composition are a result of an enzymatic mechanism of polymerization, or due to increased reactivity or availability of these initiating units.

A hypothetical polymerizing enzyme would also have to cope with an increasingly hostile environment as the vacuole fills with proanthocyanidin. Protein–proanthocyanidin interactions are strongly affected by protein composition and pH (Jones & Mangan, 1977; Hagermann & Butler, 1981), so that a polymerizing enzyme may have evolved which is insensitive to proanthocyanidin precipitation at the vacuolar pH of around 5. A membrane-bound complex may also protect a polymerizing enzyme from the phytotoxic action of newly synthesized proanthocyanidins.

Perhaps the strongest indication that condensing enzymes do exist comes from the number of mutations that block proanthocyanidin accumulation and appear to occur after monomer formation. In *A. thaliana*, double mutant analysis has confirmed that *TDS4*, encoding ANS, occurs before the *BANYULS* gene which encodes ANR (Fig. 5.4); similarly *TDS3* acts after *BANYULS* and before a group containing *TDS1*, *TDS2*, *TDS5* and *TDS6* whose mutant phenotypes could not be distinguished (Abrahams *et al.*, 2003). Similarly the *TT12* transporter acts before

TT9, TT10, TT11, TT13 and *TT14* (Debeaujon *et al.*, 2001). Presumably some of the *TDS* genes are allelic to the *TT* genes, reducing the number of *A. thaliana* genes required for proanthocyanidin synthesis beyond *TT12* to about five. In *H. vulgare*, there are at least four mutations, *ant25*, *ant26*, *ant27* and *ant28*, beyond LAR that prevent proanthocyanidin accumulation (Jende-Strid, 1993). Some of these genes may encode polymerizing enzymes, or proteins involved in vacuole formation. They may also encode dirigent proteins, which do not have any metabolic activity but are involved in directing the stereochemistry of products, e.g. in lignin synthesis (Davin & Lewis, 2000). Similar proteins may be involved in specifying the stereospecificity of proanthocyanidin biosynthesis and polymerization.

5.6   Evolution of biochemical pathways

The distribution of flavonoids throughout the plant kingdom suggests that there has been a sequential acquisition of the ability to accumulate these compounds during evolution. The flavones, flavonols and chalcones appeared about 500 million years ago (mya), the proanthocyanidins about 370 mya and the anthocyanins most recently about 120 mya (Koes *et al.*, 1994; Rausher *et al.*, 1999).

Enzymes of the proanthocyanidin pathway seemed to have evolved from a limited number of basic templates:

- Polyketide reductases, e.g. CHS (Schroder, 2000)
- Dioxygenases, e.g. FHT, LDOX, FLS (Fig. 5.5) (Prescott, 2000)
- Microsomal P450-dependent monooxygenases, e.g. F3'H, F3',5'H, FNSII (Fig. 5.6) (Kitada *et al.*, 2001).
- RED family, e.g. DFR, LAR, BAN (Fig. 5.7) (Gang *et al.*, 1999)

Flavonoids containing dihydroxy- and trihydroxy-B-rings absorb UV, and are potent free radical scavengers (Markham *et al.*, 1998). This may have provided sufficient selection pressure for the evolution of a functional chalcone synthase from an ancestral polyketide synthase involved in fatty acid biosynthesis. Enzymes such as FHT and F3'H have broad substrate specificity and could have evolved from ancestral enzymes involved in primary metabolism, further increasing the suite of available flavonoids. Continued selection pressures to produce a wider array of flavonoids could have arisen from their ability to act as specific deterrents to insects, animals and bacteria, or as signals to encourage symbiotic rhizobia, or arbuscular mycorrhizal partners (Stafford, 2000).

5.7   Regulation of biosynthesis

Proanthocyanidin biosynthesis may be regulated at substrate level by feedback inhibition of enzymes. For example, both DFR and LAR are inhibited by μM concentrations of their respective products (Kristiansen, 1986; Tanner *et al.*,

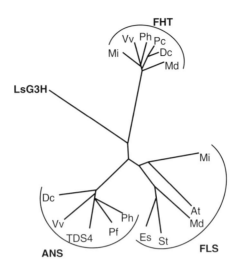

Fig. 5.5 Phylogenetic tree showing the 2-oxoglutarate-dependent dioxygenases. The clustering of protein sequences with species and accession are shown for representative flavonone 3-hydroxylases (FHT): Mi, *Matthiola incana*, CAA51192.1; Vv, *Vitis vinifera*, CAA53579.1; Ph, *Petunia hybrida*, Q07353; Pc, *Petroselinum crispum*, AAP57394.1; Dc, *Daucus carota*, AAD56577.1; Md, *Malus* species, CAA49353.1; flavonol synthases (FLS): Mi, *Matthiola incana*, O04395; At, *Arabidopsis thaliana*, BAB10451.1; Md, *Malus* species, Q9XHG2; St, *Solanum tuberosum*, Q41452; Es, *Eustoma grandiflorum*, Q9M547; anthocyanidin synthases (ANS): Ph, *Petunia hybrida*, S36233; Pf, *Perilla frutescens*, BAA20143.1; TDS4, *Arabidopsis thaliana*, CAD91994.1; Vv, *Vitis vinifera*, CAA53580.1; Dc, *Daucus carota*, AAD56580.1; and gibberelin 3-β-hydroxylase (LsG3H) from *Lactuca sativa*, BAA37129.1. The protein sequences of these dioxygenases were clearly related following analysis by CLUSTAL W (Thompson *et al.*, 1994), which showed conservation of 27 invariant amino acids distributed across the full length of the proteins.

2003). This would provide a means of matching the metabolic flux through the pathway to the accumulation of the end product. The normal development of proanthocyanidin storage vacuoles in *A. thaliana* also appears to be linked to proanthocyanidin biosynthesis at one biochemical step (Abrahams *et al.*, 2003). This indicates that the regulation of this pathway may be more complicated than we realize at present.

There now appears to be six classes of regulatory proteins that control the genes of the proanthocyanidin pathway in *A. thaliana*: myc- or bHLH-like, e.g. *TT8*; myb-like, e.g. *TT2*; MADS, e.g. *TT16*; WIP, e.g. *TT1*; WD40, e.g. *TTG1*; WRKY, e.g. *TTG2*. The interaction of these genes and the molecular genetics of control were recently reviewed by Marles *et al.* (2003) and Winkel-Shirley (2002).

5.8   Opportunities for manipulation of proanthocyanidins

Foliar proanthocyanidins commonly occur in legume tribes such as *Hedysareae* and *Loteae*, but are rare in *Trifolium* species and absent in *Medicago* species

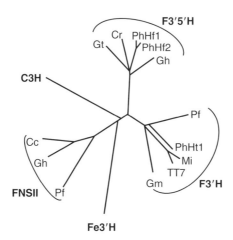

Fig. 5.6 Phylogenetic tree showing the P450 monooxygenases (redrawn with permission from Kitada *et al.* (2001)). The clustering of protein sequences with species and accession are shown for representative flavonoid 3′-hydroxylases (F3′H): Pf, *Perillia frutescens* BAB59005.1; PhHt1, *Petunia hybrida* Q9SBQ9; Mi, *Matthiola incana* AAG49301.1; TT7, *Arabidopsis thaliana* Q9SD85 (Schoenbohm *et al.*, 2000); Gm, *Glycine max* BAB83261.1; flavonoid 3′,5′-hydroxylases (F3′,5′H): Gt, *Gentiana triflora* BAA12735.1; Cr, *Catharanthus roseus* CAA09850.1; PhHf1 S38985 and PhHf2 S38984 from *Petunia hybrida*: Gh, *Gossypium hirsutum* AAP31058.1. Related P450 enzymes are shown including flavone synthase II (FNSII): Cc, *Callistephus chinensis* AAF04115.1; Gh, *Gerbera hybrida* AAD39549.1; Pf, *Perilla frutescens* BAB59004.1; ferulate 3′-hydroxylase (Fe3′H), from *Populus balsamifera* CAB65335.1; and p-coumarate 3-hydroxylase (C3H), from *Pinus taeda* AAL47685.1. The protein sequences of these P450-monooxygenases were clearly related following analysis by CLUSTAL W (Thompson *et al.*, 1994), which showed conservation of 57 invariant amino acids distributed across the full length of the proteins.

(Marshall *et al.*, 1979; Rumbaugh, 1979; Goplen *et al.*, 1980; Fay & Dale, 1993). For the last 30 years there has been sustained interest in producing leaf proanthocyanidins in *Trifolium* and *Medicago* forages but conventional breeding is unlikely to introduce the required trait. Asymmetric intraspecies protoplast fusion was of limited success as a means of transferring genes from the proanthocyanidin-positive species *O. viciifolia* to *M. sativa*. Many hybrid plants were identified, but they did not continue to produce proanthocyanidin in the glasshouse (Li *et al.*, 1993).

A transgenic approach using either structural or regulatory genes has been used to manipulate preexisting levels of proanthocyanidins in forages. Regulatory transgenes also have the potential to introduce proanthocyanidin biosynthesis into *Trifolium* and *Medicago* leaves.

5.8.1 Manipulation of structural genes

Expression of an antisense *DFR* construct, comprising the 5′ half of the *A. majus DFR cDNA*, in root cultures of *Lotus corniculatus* significantly lowered the content of proanthocyanidin and of trihydroxylated B-rings (Carron *et al.*,

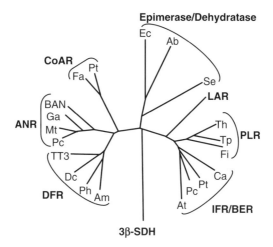

Fig. 5.7 Phylogenetic tree showing various RED proteins after CLUSTAL W analysis. The clustering of protein sequences with species and accession are shown for LAR CAD79341; with representative epimerases/dehydratases: Ec, *E. coli* UDP-galactose-4-epimerase 1UDC; Ab, *Azospirillum brasilense* UDP-glucose 4'-epimerase Q59083; Se, *Salmonella enterica* dTDP-D-Glucose 4,6-dehydratase 1G1AC; pinoresinol-lariciresinol reductases (PLR): Th, *Tsuga heterophylla* AAF64184; Tp, *Thuja plicata* AAF63507; Fi, *Forsythia x intermedia* AAC49608; isoflavone/benzylic ether reductases (IFR/BER): Ca, *Cicer arietinum* CAA43167; Pc, *Pyrus communis* AAC24001; At, *Arabidopsis thaliana* NP_173385; and Pt, *Pinus taeda* phenylcoumaran benzylic ether reductase AAF64173; dihydroflavonol 4-reductases (DFR): Am, *Antirrhinum majus* P14721; Ph, *Petunia hybrida* P14720; Dc, *Dianthus caryophyllus* P51104; TT3, *Arabidopsis thaliana* P51102; anthocyanidin reductases (ANR): Pc, *Phaseolus coccineus* CAD91909; Mt, *Medicago truncatula* AAN77735; Ga, *Gossypium arboretum* CAD91910; BAN, *Arabidopsis thaliana BANYULS* Q9SEV0. The location of sequences from cinnamoyl-CoA reductases (CoAR): Fa, *Fragaria x ananassa* AAP46143; Pt, *Pinus taeda* AAL47684; and a representative of the short-chain dehydrogenases: *Homo sapiens* 3-β hydroxysteroid dehydrogenase (3β-SDH) P14060 are also shown. The RED family is extremely diverse and many residues are conserved in proteins when compared to members of neighboring subgroups, indicating they are clearly related. The conservation is weaker when representatives of the full family were analyzed by CLUSTAL W (Thompson *et al.*, 1994). This showed partial conservation of six amino acids distributed across the full length of the proteins.

1994). Transgenic *L. corniculatus* plants carrying this antisense *A. majus DFR* construct were identified that had either increased or decreased proanthocyanidin levels compared to nontransgenic controls (Robbins *et al.*, 1998).

Expression of a full-length sense DFR construct from the same gene in hairy root cultures of *L. corniculatus* resulted in 30% of lines with decreased proanthocyanidin content but unchanged composition, presumably due to sense suppression of one or more endogenous DFRs. One of these lines had increased proanthocyanidin and a doubling of propelargonidin content compared to control lines. This is consistent with the increased substrate specificity for dihydrokaempferol of the *A. majus* DFR transgene (Bavage *et al.*, 1997).

Expression of a full-length antisense construct from a *Phaseolus vulgaris CHS* gene resulted in an inexplicable increase in proanthocyanidins of both root cultures and leaves of regenerated plants (Morris & Robbins, 1997).

Ectopic expression of ANR in tobacco diverted metabolic flux away from anthocyanin and towards flavan-3-ol production (Xie *et al.*, 2003).

5.8.2 *Manipulation of regulatory genes*

Transformation of *L. corniculatus* with the maize *Sn* gene affected proanthocyanidin accumulation in leaves of some lines. Proanthocyanidin increased in roots. Plants carrying a single copy of *Sn* had increased foliar proanthocyanidins, while plants with multiple copies of *Sn* had decreased foliar proanthocyanidin levels (Robbins *et al.*, 2003). These changes were mirrored by changed LAR activity (Damiani *et al.*, 1999). Clearly a single copy of the maize *Sn* transgene (a *bHLH*-like gene; Consonni *et al.* (1992)) was able to recruit an endogenous MYB-like partner (Ludwig *et al.*, 1989), which is usually required for the formation of an active complex. Conversely, multiple copies of *Sn* apparently led to co-suppression of an endogenous regulator.

Expression of a *bHLH*-like maize *B-Peru* gene, alone or in combination with a *Myb*-like maize *C1* gene, resulted in increased anthocyanin in *T. repens* (de Majnik *et al.*, 2000). No increase in anthocyanins was seen in transgenic *M. sativa* carrying both genes; there was also no detectable leaf proanthocyanidin in either case (Larkin, unpublished). In contrast, expression of maize *Lc* (a *bHLH*-like gene, similar to *B-Peru*) in *M. sativa* plants resulted in increased anthocyanidins and proanthocyanidins (Ray *et al.*, 2003). The increased proanthocyanidins were below the level required for bloat safety in *M. sativa* (Li *et al.*, 1996).

5.9   Conclusion

The identification of two complementary metabolic pathways that lead to the accumulation of either the 2,3-*trans*- or the 2,3-*cis*-flavan-3-ols has advanced our understanding of proanthocyanidin biosynthesis considerably. This, together with a number of mutations in the model plant *A. thaliana* that appear to block the proanthocyanidin pathway after the synthesis of flavan-3-ols, may provide for the first time an experimental path to the isolation and cloning of genes involved in the last steps of intravacuolar transport and polymerization of the proanthocyanidins.

Thirty years have now passed since the search began for flavonoid-free mutants of barley and proanthocyanidin-positive *Medicago* germplasm (Jende-Strid, 1975; Marshall *et al.*, 1979), which were the ultimate stimuli for much of the work reviewed here. The elucidation of proanthocyanidin biosynthesis now appears within reach. The identification and cloning of specific regulatory genes involved in proanthocyanidin biosynthesis appears to provide the most likely tool for the addition of this trait to the forage legumes.

Acknowledgments

I thank the many authors who kindly provided prepublication details of their work as well my colleagues Drs Abrahams, Ashton, Larkin and Watson for helpful suggestions on the manuscript.

References

Aastrup, S., Outtrup, H. & Erdal, K. (1984) Location of the proanthocyanidins in the barley grain. *Carlsberg Res. Comm.*, **49**, 105–109.

Abrahams, S., Lee, E., Walker, A. R., Tanner, G. J., Larkin, P. J. & Ashton, A. R. (2003) The *Arabidopsis TDS4* gene encodes leucoanthocyanidin dioxygenase (LDOX) and is essential for proanthocyanidin synthesis and vacuole development. *Plant Cell*, **35**, 624–636.

Abrahams, S., Tanner, G. J., Larkin, P. J. & Ashton, A. R. (2002) Identification and biochemical characterization of mutants in the proanthocyanidin pathway in Arabidopsis. *Plant Physiol.*, **130**, 561–576.

Aerts, R. J., Barry, T. N. & McNabb, W. C. (1999a) Polyphenols and agriculture: beneficial effects of proanthocyanidins in forages. *Agric. Ecosystems Environ.*, **75**, 1–12.

Albert, S., Delseny, M. & Devic, M. (1997) *Banyuls*, a novel negative regulator of flavonoid biosynthesis in the *Arabidopsis* seed coat. *Plant J.*, **11**, 289–299.

Alfenito, M. R., Souer, E., Goodman, C. D., Buell, R., Mol, J., Koes, R. & Walbot, V. (1998) Functional complementation of anthocyanin sequestration in the vacuole by widely divergent glutathione S-transferases. *Plant Cell*, **10**, 1135–1149.

Asquith, T. N., Izuno, C. C. & Butler, L. G. (1983) Characterization of the condensed tannin (proanthocyanidin) from a Group II sorghum. *J. Agric. Food Chem.*, **31**, 1299–1303.

Ayres, M. P., Clausen, T. P., MacLean, S. F., Redman, A. M. & Reichardt, P. B. (1997) Diversity of structure and antiherbivore activity in condensed tannins. *Ecology*, **78**, 1696–1712.

Bais, H. P., Walker, T. S., Stermitz, F. R., Hufbauer, R. A. & Vivanco, J. M. (2002) Enantiomeric-dependant phytotoxic and antimicrobial activity of (+)catechin. A rhizosecreted racemic mixture from spotted knapweed. *Plant Physiol.*, **128**, 1173–1179.

Baker, M. E. & Blasco, R. (1992) Expansion of the mammalian 3β-hydroyysteroid dehydrogenase/plant dihydroflavonol reductase superfamily to include a bacterial cholesterol dehydrogenase, a bacterial UDP-galactose-4-epimerase, and open reading frames in *vaccinia* virus and fish lymphocystis disease virus. *FEBS Lett.*, **301**, 89–93.

Barry, T. N. & Reid, C. S. W. (1985) Nutritional effects attributable to condensed tannins, cyanogenic glycosides and oestrogenic compounds in New Zealand forages, in *Forage Legumes for Energy-Efficient Animal Production* (eds R. F. Barnes, P. R. Ball, P. R. Brougham, G. G. Martin & D. J. Minson), USDA Agricultural Research Service, Washington, USA, pp. 251–259.

Bavage, A. D., Davies, I. G., Robbins, M. P. & Morris, P. (1997) Expression of an *Antirrhinum* dihydroflavonol reductase gene results in changes in condensed tannin structure and accumulation in root cultures of *Lotus corniculatus* (bird's foot trefoil). *Plant Mol. Biol.*, **35**, 443–458.

Bednar, R. A. & Hadcock, J. R. (1988) Purification and characterization of chalcone isomerase from soybeans. *J. Biol. Chem.*, **263**, 9582–9588.

Berg, B. P., Majak, W., McAllister, T. A., Hall, J. W., McCartney, D., Coulman, B. E., Goplen, B. P., Acharya, S. N., Tait, R. M. & Cheng, K. J. (2000) Bloat in cattle grazing alfalfa cultivars selected for a low initial rate of digestion: a review. *Can. J. Plant Sci.*, **80**, 493–502.

Bethke, P. C. & Jones, R. L. (2000) Vacuoles and provacuolar compartments. *Curr. Opin. Plant Biol.*, **3**, 469–475.

Britsch, L., Dedio, J., Saedler, H. & Forkmann, G. (1993) Molecular characterization of flavanone 3 beta-hydroxylases. Consensus sequence, comparison with related enzymes and the role of conserved histidine residues. *Eur. J. Biochem.*, **217**, 745–754.

Britsch, L. & Grisebach, H. (1986) Purification and characterization of (2S)-flavanone 3-hydroxylase from *Petunia hybrida*. *Eur. J. Biochem.*, **156**, 569–577.

Britsch, L., Ruhnau-Brich, B. & Forkmann, G. (1992) Molecular cloning, sequence analysis, and *in vitro* expression of flavanone 3 beta-hydroxylase from *Petunia hybrida*. *J. Biol. Chem.*, **267**, 5380–5387.

Brugliera, F., Barri-Rewell, G., Holton, T. A. & Mason, J. G. (1999) Isolation and characterization of a flavonoid 3'-hydroxylase cDNA clone corresponding to the *Ht1* locus of *Petunia hybrida*. *Plant J.*, **19**, 441–451.

Burbulis, I. E. & Winkel-Shirley, B. (1999) Interactions among enzymes of the *Arabidopsis* flavonoid biosynthetic pathway. *Proc. Natl. Acad. Sci. USA*, **96**, 12929–12934.

Cameron, A. R. & Malmo, J. (1993) A survey of the efficacy of sustained-release monensin capsules in the control of bloat in dairy cattle. *Aust. Vet. J.*, **70**, 1–4.

Carron, T. R., Robbins, M. P. & Morris, P. (1994) Genetic modification of condensed tannin biosynthesis in *Lotus corniculatus*. 1. Heterologous antisense dihydroflavonol reductase down-regulates tannin accumulation in 'hairy root' cultures. *Theor. Appl. Genet.*, **87**, 1006–1015.

Charrier, B., Coronado, C., Kondorosi, A. & Ratet, P. (1995) Molecular characterization and expression of alfalfa (*Medicago sativa L.*) flavanone-3-hydroxylase and dihydroflavonol-4-reductase encoding genes. *Plant Mol. Biol.*, **29**, 773–786.

Clarke, R. T. J. & Reid, C. S. W. (1974) Foamy Bloat of Cattle. A review. *J. Dairy Sci.*, **57** 753–85.

Consonni, G., Viotti, A., Dellaporta, S. L. & Tonelli, C. (1992) cDNA nucleotide sequence of *Sn*, a regulatory gene in maize. *Nuc. Acids Res.*, **20**, 373.

Damiani, F., Paolocci, F., Cluster, P. D., Arcioni, S., Tanner, G. J., Joseph, R. G., Li, Y. G., deMajnik, J., & Larkin, P. J. (1999) The maize transcription factor *Sn* alters proanthocyanidin synthesis in transgenic *Lotus corniculatus* plants. *Aust. J. Plant Physiol.*, **26**, 159–169.

Davies, K. M. (1993) A cDNA clone for flavanone 3-hydroxylase from *Malus. Plant Physiol.*, **103**, 291.

Davin, L. B. & Lewis, N. G. (2000) Dirigent proteins and dirigent sites explain the mystery of specificity of radical precursor coupling in lignan and lignin biosynthesis. *Plant Physiol.*, **123**, 453–461.

Debeaujon, I., Peeters, A. J. M., Leon-Kloosterziel, K. M. & Koorneef, M. (2001) The TRANSPARENT TESTA 12 gene of *Arabidopsis* encodes a multidrug secondary transporter-like protein required for flavonoid sequestration in vacuoles of the seed coat endothelium. *Plant Cell*, **13**, 853–871.

Delcour, J. A., Ferreira, D. & Roux, D. G. (1983) Synthesis of condensed tannins. Part 9. The condensation sequence of leucocyanidin with (+)-catechin and with the resultant procyanidin. *J. Chem. Soc. Perkin Trans. I*, 1711–1717.

de Majnik, J., Weinman, J. J., Djordjevic, M. A., Rolfe, B. G., Tanner, G. J., Joseph, R. G. & Larkin, P. J. (2000) Anthocyanin regulatory gene expression in transgenic white clover can result in an altered pattern of pigmentation. *Aust. J. Plant Physiol.*, **27**, 659–667.

Deprez, S., Mila, I., Huneau, J. F., Tome, D. & Scalbert, A. (2001) Transport of proanthocyanidin dimer, trimer, and polymer across monolayers of human intestinal epithelial Caco-2 cells. *Antioxid. Redox Signal.*, **3**, 957–967.

de Vetten, N., Horst, J. T., van Schaik, H. P., Boer, A. D., Mol, J. & Koes, R. (1999) A cytochrome b5 is required for full activity of flavonoid 3',5'-hydroxylase, a cytochrome P450 involved in the formation of blue flower colors. *Proc. Natl. Acad. Sci. USA*, **96**, 778–783.

Devic, M., Guilleminot, J., Debeaujon, I., Bechtold, N., Bensaude, E., Koornneef, M., Pelletier, G. & Deiseny, M. (1999) The *BANYULS* gene encodes a DFR-like protein and is a marker of early seed coat development. *Plant J.*, **19**, 387–398.

Dixon, R. A., Blyden, E. R., Robbins, M. P., van Tunen, A. J. & Mol, J. N. (1988) Comparative biochemistry of chalcone isomerases. *Phytochemistry*, **27**, 2801–2808.

Dong, X. Y., Braun, E. L. & Grotewold. E. (2001) Functional conservation of plant secondary metabolic enzymes revealed by complementation of *Arabidopsis* flavonoid mutants with maize genes. *Plant Physiol.*, **127**, 46–57.

Dooner, H. K., Robbins, T. P. & Jorgensen, R. A. (1991) Genetic and developmental control of anthocyanin biosynthesis. *Ann. Rev. Genet.*, **25**, 173–199.

Doostdar, H., Shapiro, J. P., Niedz, R., Burke, M. D., McCollum, T. G., McDonald, R. E. & Mayer, R. T. (1995) A cytochrome P450 mediated naringenin 3'-hydroxylase from sweet orange cell cultures. *Plant Cell Physiol.*, **36**, 69–77.

Downey, M. O., Harvey, J. S. & Robinson, S. P. (2003) Analysis of tannins in seeds and skins of Shiraz grapes throughout berry development. *Aust. J. Grape Wine Res.*, **9**, 15–27.

Eckermann, S., Schroeder, G., Schimdt, J., Strack, D., Edrada, R. A., Helariutta, Y., Elomaa, P., Kotilainen, M., Kippelainen, I., Proksch, P., Teeri, T. H. & Schroder, J. (1998) New pathways to polyketides in plants. *Nature*, **396**, 387–390.

Erikson, A. J., Ramsewak, R. S., Smucker, A. J. & Nair, M. G. (2000) Nitrification inhibitors from roots of *Leucaena leucocephala. J Agric. Food Chem.*, **48**, 6174–6177.

Fay, M. F. & Dale, P. J. (1993) Condensed tannins in *Trifolium* species and their significance for taxonomy and plant breeding. *Genet. Resour. Crop Evol.*, **40**, 7–13.

Ferreira, D., Steynberg, J. P., Burger, J. F. W. & Bezuidenhoudt, B. C. B. (1992) Synthesis and base-catalysed transformations of proanthocyanidins, in *Phenolic Metabolism in Plants* (eds H. A. Stafford & R. K. Ibrahim), Plenum, New York, USA, pp. 255–295.

Ferrer, J.-L., Jez, J. M., Bowman, M., Dixon, R. A. & Noel, J. P. (1999) Structure of chalcone synthase and the molecular basis of plant polyketide biosynthesis. *Nat. Struct. Biol.*, **6**, 775–784.

Feucht, W., Treutter, D. & Christ, E. (1992) The precise localisation of catechins and proanthocyanidins in protective layers around fungal infections. *J. Plant Dis. Protect.*, **99**, 404–413.

Fischer, D., Stich, K., Britsch, L. & Grisebach, H. (1988) Purification and characterization of (+)-dihydro-flavonol (3-hydroxyflavanone) 4-reductase from flowers of *Dahlia variabilis. Arch. Biochem. Biophys.*, **264**, 40–47.

Fischer, T., Halbwirth, H., Meisel, B., Stich, K. & Forkmann, G. (2003) Molecular cloning, substrate specificity of the functionally expressed dihydroflavonol 4-reductases from *Malus domestica*, and *Pyrus communis* cultivars and the consequences for flavonoid metabolism. *Arch. Biochem. Biophys.*, **412**, 223–230.

Foo, L. Y. & Porter, L. J. (1980) The phytochemistry of proanthocyanidin polymers. *Phytochemistry*, **19**, 1747–1754.

Forkmann, G., Heller, W. & Grisebach, H. (1980) Anthocyanin biosynthesis in flowers of *Matthiola incana* flavanone 3- and flavonoid 3'-hydroxylases. *Z. Naturforsch.*, **C35**, 691–695.

Forkmann, G. (1993) Genetics of flavonoids, in *The Flavonoids: Advances in Research Since 1986* (ed. J. B. Harborne), Chapman & Hall, London, UK, pp. 537–364.

Frankel, E. N., Kanner, J., German, J. B., Parks, E. & Kinsella, J. E. (1993) Inhibition of oxidation of human low-density lipoprotein by phenolic substances in red wine. *Lancet*, **341**, 454–457.

Fritsch, H. & Grisebach, H. (1975) Biosynthesis of cyanidin in cell cultures of *Haplopappus gracilis. Phytochemistry*, **14**, 2437–2442.

Frova, C. (2003) The plant glutathione transferase gene family: genomic structure, functions, expression and evolution. *Physiol. Plant.*, **119**, 469–479.

Gang, D. R., Kasahara, H., Xia, Z.-Q., Mijnsbrugge, K. V., Bauw, G., Boerjan, W., Montagu, M. V., Davin, L. B. & Lewis, N. G. (1999) Evolution of plant defense mechanisms. *J. Biol. Chem.*, **274**, 7516–7527.

Gong, Z. Z., Yamazaki, M., Sugiyama, M., Tanaka, Y. & Saito, K. (1997) Cloning and molecular analysis of structural genes involved in anthocyanin biosynthesis and expressed in a forma-specific manner in *Perilla frutescens. Plant Mol. Biol.*, **35**, 915–927.

Goplen, B. P. (1989) Breeding bloat-safe alfalfa. *Forage Focus*, **2**, 3.

Goplen, B. P., Howarth, R. E., Sarkar, S. K. & Lesins, K. (1980) A search for condensed tannins in annual and perennial species of *Mediago, Trigonella, & Onobrychis. Crop Sci.*, **20**, 801–804.

Grotewold, E. (2001) Subcellular trafficking of phytochemicals. *Recent Res. Dev. Plant Physiol.*, **2**, 31–48.

Grotewold, E., Chamberlin, M., Snook, M., Siame, B., Butler, L., Swenson, J., Maddock, S., St-Clair, G. & Bowen, B. (1998) Engineering secondary metabolism in maize cells by ectopic expression of transcription factors. *Plant Cell*, **10**, 721–740.

Grotewold, E., Drummond, B. J., Bowen, B. & Peterson, T. (1994) The *myb*-homologous P gene controls phlobaphene pigmentation in maize floral organs by directly activating a flavonoid biosynthetic gene subset. *Cell*, **76**, 543–553.

Hagermann, A. E. & Butler, L. G. (1981) The specificity of proanthocyanidin-protein interactions. *J. Biol. Chem.*, **256**, 4494–4497.

Hagmann, M. L., Heller, W. & Grisebach, H. (1983) Induction and characterisation of a microsomal flavonoid 3′-hydroxylase from parsley cell cultures. *Eur. J. Biochem.*, **134**, 547–554.

Halbwirth, H., Martens, S., Wienand, U., Forkmann, G. & Stich, K. (2003) Biochemical formation of anthocyanins in silk tissue of *Zea mays*. *Plant Sci.*, **164**, 489–495.

Haslam, E. (1977) Review. Symmetry and promiscuity in procyanidin biochemistry. *Phytochemistry*, **16**, 1625–1640.

Hedqvist, H., Mueller-Harvey, I., Reed, J. D., Krueger, C. G. & Murphy, M. (2000) Characterisation of tannins and *in vitro* protein digestibility of several *Lotus corniculatus* varieties. *Anim. Feed Sci. Tech.*, **87**, 41–56.

Heller, W. & Forkmann, G. (1988) Biosynthesis, in *The Flavonoids. Advances in Research Since 1980* (ed. J. B. Harborne), Chapman & Hall, London, UK, pp. 399–425.

Heller, W. & Forkmann, G. (1993) Biosynthesis of flavonoids, in *The Flavonoids: Advances in Research Since 1986* (ed. J. B. Harborne), Chapman & Hall, London, UK, pp. 499–535.

Hinz, G. & Herman, E. M. (2003) Sorting of storage proteins in the plant Golgi apparatus, in *The Golgi Apparatus and the Plant Secretory Pathway* (ed. D. G. Robinson), Blackwell, London, UK and CRC Press, Boca Raton, Florida, USA, pp. 141–164.

Holton, T. A., Brugliera, F., Lester, D. R., Tanaka, Y., Hyland, C. D., Menting, J. G. T., Lu, C. Y., Farcy, E., Stevenson, T. W. & Cornish, E. C. (1993) Cloning and expression of cytochrome P450 genes controlling flower colour. *Nature*, **366**, 276–279.

Holton, T. A. & Cornish, E. C. (1995) Genetics and biochemistry of anthocyanin biosynthesis. *Plant Cell*, **7**, 1071–1083.

Hrazdina, G. & Jensen, R. A. (1992) Spatial organisation of enzymes in plant metabolic pathways. *Annu. Rev. Plant Physiol. Plant Mol. Biol.*, **43**, 241–267.

Ishikura, N., Murakami, H. & Fujii, Y. (1988) Conversion of (+)-dihydroquercetin to 3,4-*cis*-leucocyanidin by a reductase extracted from cell suspension cultures of *Cryptomeria japonica*. *Plant Cell Physiol.*, **29**, 795–799.

Jacques, D., Opie, C. T., Porter, L. J. & Haslam, E. (1977) Plant proanthocyanidins. Part 4. Biosynthesis of procyanidins and observations on the metabolism of cyanidin in plants. *J. Chem. Soc. Perkin Trans. I*, 1637–1643.

Jende-Strid, B. (1975) Mutations affecting flavonoid synthesis in barley, in *Barley Genetics III. Proc. 3rd Int. Barley Genet. Sympos.* Verlag Karl Thiemig, Munich, Germany, p. 36.

Jende-Strid, B. (1993) Genetic control of flavonoid biosynthesis in barley. *Hereditas*, **119**, 187–204.

Jende-Strid, B. & Moller, B. L. (1981) Analysis of proanthocyanidins in wild-type and mutant barley (*Hordeum vulgare* L.). *Carlsberg Res. Comm.*, **46**, 53–64.

Jez, J. M , Bowman, M. E., Dixon, R. A. & Noel, J. P. (2000) Structure and mechanism of the evolutionarily unique plant enzyme chalcone isomerase. *Nat. Struct. Biol.*, **7**, 786–791.

Johnson, E. T., Ryu, S., Yi, H., Shin, B., Cheong, H. & Choi, G. (2001) Alteration of a single amino acid changes the substrate specificity of dihydroflavonol 4-reductase. *Plant J.*, **25**, 325–333.

Jones, W. T. & Lyttleton, J. W. (1971) Bloat in cattle. XXXIV. A survey of legume forages that do and do not produce bloat. *N. Z. J. Agric. Res.*, **14**, 101–107.

Jones, W. T. & Mangan, J. L. (1977) Complexes of the condensed tannins of sainfoin (*Onobrychis viciifolia* Scop.) with fraction 1 leaf protein and with submaxillary mucoprotein, and their reversal by polyethylene glycol and pH. *J. Sci. Food Agric.*, **28**, 126–136.

Juah, G. Y., Phillips, T. E. & Rogers, J. C. (1999) Tonoplast intrinsic protein isoforms as markers for vacuolar functions. *Plant Cell*, **11**, 1867–1882.

Kaltenbach, M., Schroder, G., Schmelzer, E., Lutz, V. & Schroder, J. (1999) Flavonoid hydroxylase from *Catharanthus roseus*: cDNA, heterologous expression, enzyme properties and cell-type specific expression in plants. *Plant J.*, **19** 183–93.

Keli, S. O., Hertog, M. G. L., Feskens, E. J. M. & Kromhout, D. (1996) Dietary flavonoids, antioxidant vitamins, and incidence of stroke. The Zutphen study. *Arch. Intern. Med.*, **156**, 637–642.

Kennedy, J. A. & Jones, G. P. (2001) Analysis of proanthocyanidin cleavage products following acid-catalysis in the presence of excess phloroglucinol. *J. Agric. Food Chem.*, **49**, 1740–1746.

Kitada, C., Gong, Z., Tanaka, Y., Yamazaki, M. & Saito, K. (2001) Differential expression of two cytochrome P450s involved in the biosynthesis of flavones and anthocyanins in chemo-varietal forms of *Perilla frutescens*. *Plant Cell Physiol.*, **42**, 1338–1344.

Kitamura, S., Shikazono, N. & Tanaka, A. (2004) *Transparent testa 19* is involved in the accumulation of both anthocyanins and proanthocyanidins in *Arabidopsis*. *Plant J.*, **37**, 104–114.

Klein, M., Martinoia, E., Hoffmann-Thoma, G. & Weissenbock, G. (2000) A membrane-potential dependant ABC-like transporter mediates the vacuolar uptake of rye flavone glucuronides: regulation of glucuronide uptake by glutathione and its conjugates. *Plant J.*, **21**, 289–304.

Koes, R. E., Quattrocchio, F. & Mol, J. N. M. (1994) The flavonoid biosynthetic pathway in plants: function and evolution. *BioEssays*, **16**, 123–132.

Koornneef, M. (1990) Mutations affecting the testa colour in Arabidopsis. *Arabid. Inf. Serv.*, **27**, 1–4.

Koupai-Abyazani, M. R., McCallum, J., Muir, A. D., Bohm, B. A., Towers, G. H. N. & Gruber, M. Y. (1993) Developmental changes in the composition of proanthocyanidins from leaves of sainfoin (*Onobrychis viciifolia Scop.*) as determined by HPLC analysis. *J. Agric. Food Chem.*, **41**, 1066–1070.

Kraus, T. E. C., Dahlgren, R. A. & Zasoski, R. J. (2003) Tannins in nutrient dynamics of forest ecosystems: a review. *Plant and Soil*, **256**, 41–66.

Kreuzaler, F., Ragg, H., Fautz, E., Kuhn, D. N. & Hahlbrock, K. (1983) UV-induction of chalcone synthase mRNA in cell suspension cultures of *Petroselinum hortense*. *Proc. Natl. Acad. Sci. USA*, **80**, 2591–2593.

Kristiansen, K. N. (1984) Biosynthesis of proanthocyanidins in barley: genetic control of the conversion of dihydroquercetin to catechin and procyanidins. *Carlsberg Res. Comm.*, **49**, 503–524.

Kristiansen, K. N. (1986) Conversion of (+)-dihydroquercetin to (+)-2,3-*trans*-3,4-*cis*-leucocyanidin and (+)-catechin with an enzyme extract from maturing grains of barley. *Carlsberg Res. Comm.*, **51**, 51–60.

Labarbe, B., Cheynier, V., Broussaud, F., Souquet, J.-M. & Moutounet, M. (1999) Quantitative fractionation of grape proanthocyanidins according to their degree of polymerisation. *J. Agric. Food Chem.*, **47**, 2719–2723.

Labesse, G., Vidal-Cros, A., Chomilier, J., Gaudry, M. & Mornon, J.-P. (1994) Structural comparisons lead to the definition of a new superfamily of NAD(P)(H)-accepting oxidoreductases: the single-domain reductases/ epimerases/ dehydrogenases (the 'RED' family). *Biochem. J.*, **304**, 95–99.

Lancaster, J. E. (1992) Regulation of skin color in apples. *Crit. Rev. Plant Sci.*, **10**, 487–502.

Larson, R. L. & Bussard, J. B. (1986) Microsomal flavonoid 3′-monooxygenase from maize seedlings. *Plant Physiol.*, **80**, 483–486.

Lees, G. L., Gruber, M. Y. & Suttill, N. H. (1995) Condensed tannins in sainfoin. II. Occurrence and changes during leaf development. *Can. J. Bot.*, **73**, 1540–1547.

Li, Y. G., Tanner, G. J., Delves, A. C. & Larkin, P. J. (1993) Asymmetric somatic hybrid plants between *Medicago sativa* L. (alfalfa, lucerne) and *Onobrychis viciifolia Scop.* (sainfoin). *Theor. Appl. Genet.*, **87**, 455–463.

Li, Y. G., Tanner, G., & Larkin, P. (1996) The DMACA-HCl protocol and the threshold proanthocyanidin content for bloat safety in forage legumes. *J. Sci. Food Agric.*, **70**, 89–101.

Li, Z.-S., Alfenito, M., Rea, P. A., Walbot, V., Dixon, R. A. & Li, Z. S. (1997) Vacuolar uptake of the phytoalexin medicarpin by the glutathione conjugate pump. *Phytochemistry*, **45**, 689–693.

Lu, Y. P., Li, Z. S. & Rea, P. A. (1997) *AtMRP1* gene of Arabidopsis encodes a glutathione S-conjugate pump: isolation and functional definition of a plant ATP-binding cassette transporter gene. *Proc Natl. Acad. Sci. USA*, **94**, 8243–8248.

Ludwig, S. R., Habera, L. F., Dellaporta, S. L. & Wessler, S. R. (1989) *Lc*, a member of the maize *R* gene family responsible for tissue specific anthocyanin production, encodes a protein similar to transcription activators and contains the *myc*-homology region. *Proc Natl. Acad. Sci. USA*, **86**, 7092–7096.

Lukacin, R., Groning, I., Pieper, U. & Matern, U. (2000a) Site-directed mutagenesis of the active site serine290 in flavanone 3β-hydroxylase from *Petunia hybrida*. *Eur. J. Biochem.*, **267**, 853–860.

Lukacin, R., Groning, I., Schiltz, E., Britsch, L. & Matern, U. (2000b) Purification of recombinant flavanone 3β-hydroxylase from *Petunia hybrida* and assignment of the primary site of proteolytic degradation. *Arch Biochem. Biophys.*, **375**, 364–370.

Lukacin, R., Schreiner, S. & Matern, U. (2001) Transformation of acridone synthase to chalcone synthase. *FEBS Lett.*, **508**, 413–417.

Lukacin, R., Urbanke, C., Groning, I. & Matern, U. (2000) The monomeric polypeptide comprises the functional flavanone 3β-hydroxylase from *Petunia hybrida*. *FEBS Lett.*, **467**, 353–358.

Lukacin, R., Wellmann, F., Britsch, L., Martens, S. & Matern, U. (2003) Flavonol synthase from *Citrus unshiu* is a bifunctional dioxygenase. *Phytochemistry*, **62**, 287–292.

Markham, K. R., Tanner, G. J., Caasi-Lit, M., Whitecross, M. I., Nayudu, M. & Mitchell, K. A.(1998) Possible protective role for 3′,4′-dihydroxyflavones induced by enhanced UV-B in a UV-tolerant rice cultivar. *Phytochemistry*, **49**, 1913–1919.

Marles, M. A., Ray, H. & Gruber, M. Y. (2003) New perspectives on proanthocyanidin biochemistry and molecular regulation. *Phytochemistry*, **64**, 367–383.

Marshall, D. R., Broue, P. & Munday, J. (1979) Tannins in pasture legumes. *Aust. J. Exp. Agric. Anim. Husb.*, **19**, 192–197.

Martin, C. & Gerats, T. (1993) The control of flower coloration, in *The Molecular Biology of Flowering* (ed. B. R. Jordan), CAB International, Wallingford, UK, pp. 219–255.

Martin, C., Carpenter, R., Sommer, H., Saedler, H. & Coen, E. S. (1985) Molecular analysis of instability in flower pigmentation of *Antirrhinum majus*, following isolation of the *pallida* locus by transposon tagging. *EMBO J.*, **4**, 1625–1630.

Martin, C., Prescott, A., Mackay, S., Bartlett, J. & Vrijlandt, E. (1991) Control of anthocyanin biosynthesis in flowers of *Antirrhinum majus*. *Plant J.*, **1**, 37–49.

Meldgaard, M. (1992) Expression of chalcone synthase, dihydroflavonol reductase, and flavanone-3-hydroxylase in mutants of barley deficient in anthocyanin and proanthocyanidin biosynthesis. *Theor. Appl. Genet.*, **83**, 695–706.

Menssen, A., Hohmann, S., Martin, W., Schnable, P. S., Peterson, P. A., Saedler, H. & Gierl, A. (1990) The *En/Spm* transposable element of *Zea mays* contains splice sites at the termini generating a novel intron from a *dSpm* element in the *A2* gene. *EMBO J.*, **9**, 3051–3057.

Menting, J. G. T., Scopes, R. K. & Stevenson, T. W. (1994) Characterization of flavonoid 3′,5′-hydroxylase in microsomal membrane fraction of *Petunia hybrida* flowers. *Plant Physiol.*, **106**, 633–642.

Min, B. R., Barry, T. N., Attwood, G. T. & Mcnabb, W. C. (2003) The effect of condensed tannins on the nutrition and health of ruminants fed fresh temperate forages: a review. *Anim. Feed Sci. and Technology*, **106**, 3–19.

Mol, J., Grotewold, E. & Koes, R. (1998) How genes paint flowers and seeds. *Trends Plant Sci.*, **3**, 212–217.

Mol, J. N. M., Robbins, M. P., Dixon, R. A. & Veltkamp, E. (1985) Spontaneous and enzymatic rearrangement of naringenin chalcone to flavanone. *Phytochemistry*, **24**, 2267–2269.

Molan, A. L., Waghorn, G. C. & McNabb, W. C. (2002) Effect of condensed tannins on egg hatching and larval development of *Trichostrongylus colubriformis in vitro*. *Vet. Rec.*, **150**, 65–69.

Morris, C. A., Cullen, N. G. & Geertsema, H. G. (1997) Genetic studies of bloat susceptibility in cattle. *N. Z. Soc. Anim. Prod.*, **57**, 19.

Morris, P. & Robbins, M. P. (1997) Manipulating condensed tannins in forage legumes, in *Biotechnology and the Improvement of Forage Legumes* (eds B. D. McKersie & D. C. W. Brown), CAB International, Wallingford, UK, pp. 147–173.

Mueller, L. A., Goodman, C. D., Silady, R. A. & Walbot, V. (2000) AN9, a petunia glutathione S-transferase required for anthocyanin sequestration, is a flavonoid-binding protein. *Plant Physiol.*, **123**, 1561–1570.

Mueller, L. A., Walbot, V., Romeo, J. T., Saunders, J. A. & Matthews, B. F. (2001) Models for vacuolar sequestration of anthocyanins. *Rec. Adv. Phytochem.*, **35**, 297–312.

Mueller, W. C. & Beckman, C. H. (1974) Ultrastructure of the phenol-storing cells in the roots of banana. *Physiol. Plant Path.*, **4**, 187–190.

Nakajima, J.-I., Tanaka, Y., Yamazaki, M. & Saito, K. (2001) Reaction mechanism from leucoanthocyanidin to anthocyanidin 3-glucoside, a key reaction for coloring in anthocyanidin biosynthesis. *J. Biol. Chem.*, **276**, 25797–25803.

Nielsen, K. M. & Podivinsky, E. (1997) cDNA cloning and endogenous expression of a flavonoid 3'5'-hydroxylase from petals of Lisianthus (*Eustoma grandiflorum*). *Plant Sci.*, **129**, 167–174.

Northup, R. R., Yu, Z., Dahlgren, R.A. & Vogt, K. A. (1995) Polyphenol control of nitrogen release from pine litter. *Nature*, **377**, 227–229.

O'Reilly, C., Sheperd, N., Pereira, A., Schwarz-Somer, Z., Bertram, I., Robertson, D.S., Peterson, P.A. & Saedler, H. (1985) Molecular cloning of the *A1* locus of *Zea mays* using the transposable elements *En* and *Mu1*. *EMBO J.*, **4**, 877–882.

Parham, R. A. & Kaustinen, H. M. (1977) On the site of tannin synthesis in plant cells. *Bot. Gaz.*, **138**, 465–467.

Pelletier, M. K. & Shirley, B. W. (1996) Analysis of flavanone 3-hydroxylase in Arabidopsis seedlings. Coordinate regulation with chalcone synthase and chalcone isomerase. *Plant Physiol.*, **111**, 339–345.

Polya, G. M. & Foo, L. Y. (1994) Inhibition of eukaryote signal-regulated protein kinases by plant-derived catechin-related compounds. *Phytochemistry*, **35**, 1399–1405.

Porter, L. J. (1993) Flavans and proanthocyanidins, in *The Flavanoids: Advances in Research Since 1986*. (ed. J. B. Harbourne), Chapman & Hall, London, UK, pp. 23–55.

Preisig-Mueller, R., Gehlert, R., Melchior, F., Stietz, U. & Kindl, H. (1997) Plant polyketide synthases leading to stilbenoids have a domain catalysing malonyl-CoA:CO_2 exchange, malonyl-CoA decarboxylation, and covalent enzyme modification and a site for chain lengthening. *Biochemistry*, **36**, 8349–8358.

Prescott, A. G. (2000) Two-oxoacid-dependant dioxygenases: inefficient enzymes or evolutionary driving force, in *Evolution of Metabolic Pathways* (eds J. T. Romeo, R. Ibrahim, L. Varin & V. deLuca), Pergamon, New York, USA, pp. 249–284.

Rausher, M. D., Miller, R. E. & Tiffin, P. (1999) Patterns of evolutionary rate variation among genes of the anthocyanin biosynthetic pathway. *Mol. Biol. Evol.*, **16**, 266–274.

Ray, H., Yu, M., Auser, P., Blahut-Beatty, L., McKersie, B., Bowley, S., Westcott, N., Coulman, B., Lloyd, A. & Gruber, M. (2003) Expression of anthocyanins and proanthocyanidins after transformation of alfalfa with maize *Lc*. *Plant Physiol.*, **132**, 1448–1463.

Reddy, A. R., Britsch, L., Salamini, F., Saedler, H. & Rohde, W. (1987) The A1 (*anthocyanin-1*) locus in *Zea mays* encodes dihydroquercetin reductase. *Plant Sci.*, **52**, 7–13.

Reimold, U., Kroger, M., Kreuzaler, F. & Hahlbrock, K. (1983) Coding and 3' non-coding nucleotide sequence of chalcone synthase mRNA and assignement of amino acid sequence of the enzyme. *EMBO J.*, **2**, 1801–1805.

Rizzi, M., Tonetti, M., Vigevani, P., Sturla, L., Bisso, A., de Flora, A., Bordo, D. & Bolognesi, M. (1998) GDP-4-keto-6-deoxy-D-mannose epimerase/reductase from *Escherichia coli*, a key enzyme in the biosynthesis of GDP-fucose, displays the structural characteristics of the RED protein homology superfamily. *Structure*, **6**, 1453–1465.

Robbins, M. P., Bavage, A. D., Strudwicke, C. & Morris, P. (1998) Genetic manipulation of condensed tannins in higher plants. II. Analysis of birdsfoot trefoil plants harboring antisense dihydroflavonol reductase constructs. *Plant Physiol.*, **116**, 1133–1144.

Robbins, M. P., Paolocci, F., Hughes, J.-W., Turchette, V., Allison, G., Arcioni, S., Morris, P. & Damiani, F. (2003) *Sn*, a maize bHLH gene, modulates anthocyanin and condensed tannin pathways in *Lotus corniculatus*. *J. Ex. Bot.*, **54**, 239–248.

Rumbaugh, M. D. (1979) The search for condensed tannins in the genus *Medicago*. *Agronomy abstracts*. Abstracts of 1979 annual meeting of American Society of Agronomy; Crop Science Society of America; and Soil Science Society of America; Fort Collins, Colorado, August 5–10, 1979, p. 75.

Saito, K., Kobayashi, M., Gong, Z. Z., Tanaka, Y. & Yamazaki, M. (1999) Direct evidence for anthocyanidin synthase as a 2-oxoglutarate-dependant oxygenase: molecular cloning and functional expression of cDNA from a red forma of *Perilla frutescens*. *Plant J.*, **17**, 181–189.

Santos-Buelga, C. & Scalbert, A. (2000) Proanthocyanidins and tannin-like compounds – nature, occurence, dietary intake and effects on nutrition and health. *J. Sci. Food Agric.*, **80**, 1094–1117.

Saunders, C. M., Bonnett, S. L., Steynberg, J. P. & Ferreira, D. (1996) Oligomeric flavanoids. Part 24. Controlled biomimetic synthesis of profisetinidin triflavanoid related phlobatannins. *Tetrahedron*, **52**, 6003–6010.

Scalbert, A. (1991) Antimicrobial properties of tannins. *Phytochemistry*, **30**, 3875–3883.

Scalbert, A., Morand, C., Manach, C. & Remesy, C. (2002) Absorption and metabolism of polyphenols in the gut and impact on health. *Biomedicine Pharmacotherapy*, **56**, 276–282.

Schoenbohm, C., Martens, S., Eder, C., Forkmann, G. & Weisshaar, B. (2000) Identification of the *Arabisopsis thaliania* flavonoid 3′-hydroxylase gene and functional expression of the encoded P450 enzyme. *Biol. Chem.*, **381**, 749–753.

Schroder, J. (2000) The family of chalcone synthase-related proteins: functional diversity and evolution, in *Evolution of Metabolic Pathways* (eds J. T. Romeo, R. Ibrahim, L. Varin & V.deLuca), Pergamon, New York, USA, pp. 55–89.

Schultz, J. C., Hunter, M. D. & Appel, H. M. (1992) Antimicrobial activity of polyphenols mediates plant-herbivore interactions, in *Plant Polyphenols: Synthesis, Properties, Significance* (eds R. W. Hemmingway, P. E. Laks & S. J. Branham), Plenum, New York, USA, pp. 621–637.

Shimada, N., Aoki, T., Sato, S., Nakamura, Y., Tabata, S. & Ayabe, S. (2003) A cluster of genes encodes the two types of chalcone isomerase involved in the biosynthesis of general flavonoids and legume-specific 5-deoxy(iso)flavonoids in *Lotus japonicus*. *Plant Physiol.*, **131**, 941–951.

Shimada, Y., Nakano-Shimada, R., Ohbayashi, M., Okinaka, Y., Kiyokawa, S. & Kikuchi, Y. (1999) Expression of chimeric P450 genes encoding flavonoid-3′5′-hydroxylase in transgenic tobacco and *Petunia* plants. *FEBS Lett.*, **461**, 241–245.

Shirley, B. W., Kubasek, W. L., Storz, G., Bruggemann, E., Koornneef, M., Ausubel, F. M. & Goodman, H. M. (1995) Analysis of *Arabidopsis* mutants deficient in flavonoid biosynthesis. *Plant J.*, **8**, 659–671.

Singh, S., McCallum, J., Gruber, M. Y., Towers, G. H. N., Muir, A. D., Bohm, B. A., Koupai-Abyazani, M. R., Glass, A. D. M. & Singh, S. (1997) Biosynthesis of flavan-3-ols by leaf extracts of *Onobrychis viciifolia*. *Phytochemistry*, **44**, 425–432.

Skadhauge, B., Gruber, M. Y., Thomsen, K. K., & von Wettstein, D. (1997a) Leucocyanidin reductase activity and accumulation of proanthocyanidins in developing legume tissues. *Am. J. Bot.*, **84**, 494–503.

Skadhauge, B., Thomsen, K. K., & von Wettstein, D. (1997b) The role of the barley testa layer and its flavonoid content in resistance to *Fusarium* infections. *Hereditas*, **126**, 147–160.

Sparvoli, F., Martin, C., Scienza, A., Gavazzi, G. & Tonelli, C. (1994) Cloning and molecular analysis of structural genes involved in flavonoid and stilbene biosynthesis in grape (*Vitis vinifera* L.). *Plant Mol. Biol.*, **24**, 743–755.

Stafford, H. A. (1974) Possible multi-enzyme complexes regulating the formation of C6-C3 phenolic compounds and lignans in higher plants. *Recent Adv. Phytochem.*, **8**, 53–79.

Stafford, H. A. (1990) *Flavonoid metabolism*, CRC Press, Boca Raton, Florida, USA.

Stafford, H. A. (2000) The evolution of phenolics in plants, in *Evolution of Metabolic Pathways* (eds J. T. Romeo, R. Ibrahim, L. Varin & V. deLuca), Pergamon, New York, USA, pp. 25–54.

Stafford, H. A. & Lester, H. A. (1984) Flavan-3-ol biosynthesis. The conversion of (+)dihydroquercetin and flavav-3,4-*cis*-diol (leucocyanidin) to (+)-catechin by reductases extracted from cell suspension cultures of Douglas Fir. *Plant Physiol.*, **76**, 184–186.

Stafford, H. A. & Lester, H. A. (1985) Flavan-3-ol biosynthesis: the conversion of (+)-dihydromyrecetin to its flavan-3,4-diol (leucodelphinidin) and to (+)-gallocatechin by reductases extracted from tissue cultures of *Ginkgo biloba* and *Pseudotsuga menziesii* . *Plant Physiol.*, **78**, 791–794.

Stafford, H. A. & Lester, H. H. (1982) Enzymic and non-enzymic reduction of (+)-dihydroquercetin to its 3,4-diol. *Plant Physiol.*, **70**, 695–698.

Stafford, H. A., Shimamoto, M. & Lester, H. H. (1982) Incorporation of [^{14}C]phenylalanine into flavan-3-ols and procyanidins in cell suspension cultures of Douglas Fir. *Plant Physiol.*, **69**, 1055–1059.

Stotz, G. & Forkmann, G. (1982) Hydroxylation of the B-ring of flavonoids in the 3′- and 5′-positions with enzyme extracts from flowers of *Verbena hybrida*. *Z. Naturforsch.*, **37c**, 19–23.

Stotz, G., Spribille, R. & Forkmann, G. (1984) Flavonoid biosynthesis in flowers of *Verbena hybrida*. *J. Plant Physiol.*, **116**, 173–183.

Tanaka, Y., Yonekura, K., Fukuchi-Mizutani, M., Fukui, Y., Fujiwara, H., Ashikari, T. & Kusumi, T. (1996) Molecular and biochemical characterization of three anthocyanin synthetic enzymes from *Gentiana triflora*. *Plant Cell Physiol.*, **37**, 711–716.

Tanner, G. J., Abrahams, S. A. & Larkin, P. J. (2000) Biosynthesis of condensed tannins in cereals and other plants, in *Tannins in Livestock and Human Nutrition* (ed. J. D. Brooker), ACIAR, Canberra, http://www.aciar.gov.au/downloads/publications/tannins/TAN08BTA.PDF.

Tanner, G. J., Francki, K. T., Abrahams, S., Watson, J. M., Larkin, P. J. & Ashton, A. R. (2003) Proantho-cyanidin biosynthesis in plants. Purification of legume leucoanthocyanidin reductase and molecular cloning of its cDNA. *J. Biol. Chem.*, **278**, 31647–31656.

Tanner, G. J. & Kristiansen, K. N. (1993) Synthesis of 3,4,-*cis*-[^{3}H]leucocyanidin and enzymatic reduction to catechin. *Anal. Biochem.*, **209**, 274–277.

Thompson, J. D., Higgins, D. G. & Gibson, T. J. (1994) CLUSTAL W: improving the sensitivity of progressive multiple sequence alignment through sequence weighting, position-specific gap penalties and weight matrix choice. *Nuc. Acids Res.*, **22**, 4673–4680.

Toda, K., Yang, D., Yamanaka, N., Watanabe, S., Harada, K. & Takahashi, R. (2002) A single-base deletion in soybean flavonoid 3′-hydroxylase gene is associated with gray pubescence color. *Plant Mol. Biol.*, **50**, 187–196.

Turnbull, J. J., Nagle, M. J., Seibel, J. F., Welford, R. W. D., Grant, G. & Schofield, C. J. (2003) The C-4 stereochemistry of leucocyanidin substrates for anthocyanidin synthase affects product selectivity. *Bioorg. Med. Chem. Lett.*, **13**, 3853–3857.

Turnbull, J. J., Sobey, W. J., Alpin, R. T., Hassan, A., Firmin, J. L., Schofield, C. J. & Prescott, A. G. (2000) Are anthocyanidins the immediate products of anthocyanidin synthase? *Chem. Commun.*, **24**, 2473–2474.

Turnbull, J. J., Prescott, A. G., Schofield, C. J. & Wilmouth, R. C. (2001) Purification, crystallization and preliminary X-ray diffraction of anthocyanidin synthase from *Arabidopsis thaliana*. *Acta Cryst. Sec. D, Biol. Cryst.*, **57**, 425–427.

von Wettstein, D., Nilan, R. A., Ahrenst-Larsen, B., Erdal, K., Ingversen, J., Jende-Strid, B., Kristiansen, K. N., Larsen, J., Outtrup, H. & Ullrich, S. E. (1985) Proanthocyanidin-free barley for brewing: progress in breeding for high yield and research tool in polyphenol chemistry. *MBAA Tech. Quart.*, **22**, 41–52.

Welle, R., Schroder, G., Schiltz, E., Grisebach, H. & Schroder, J. (1991) Induced plant responses to pathogen attack. Analysis and heterologous expression of the key enzyme in the biosynthesis of phytoalexins in soybean (*Glycine max* L. Merr. cv. Harosoy 63). *Eur. J. Biochem.*, **196**, 423–430.

Wellmann, F., Lukacin, R., Moriguchi, T., Britsch, L., Schiltz, E. & Matern, U. (2002) Functional expression and mutational analysis of flavonol synthase from *Citrus unshiu*. *Eur. J. Biochem.*, **269**, 4134–4142.

Wilmouth, R., Turnbull, J. J., Welford, R. W. D., Clifton, I. J., Prescot, A. G. & Schofield, C. J. (2002) Structure and mechanism of anthocyanidin synthase from *Arabidopsis thaliania*. *Structure*, **10**, 93–103.

Winkel-Shirley, B. (1999) Evidence for enzyme complexes in the phenylpropanoid and flavonoid pathways. *Physiol. Plant.*, **107**, 142–149.

Winkel-Shirley, B. (2001) Flavonoid biosynthesis. A colorful model for genetics, biochemistry, cell biology, and biotechnology. *Plant Physiol.*, **126**, 485–493.

Winkel-Shirley, B. (2002) Molecular genetics and control of anthocyanin expression, in *Advances in Botanical Research* (eds K. S. Gould, D. W. Lee, & J. A. Callow), Academic Press, London, UK, pp. 76–94.

Wiseman, E., Hartmann, U., Sagasser, M., Baumann, E., Palme, K., Hahlbrock, K., Saedler, H. & Weisshaar, B. (1998) Knockout mutants from an En-1 mutagenised *Arabidopsis thaliania* population generate phenylpropanoid biosynthesis phenotypes. *Proc. Natl. Acad. Sci. USA*, **95**, 12432–12437.

Xie, D. Y., Sharma, S. B., Paiva, N. L., Ferreira, D. & Dixon, R. A. (2003) Role of anthocyanidin reductase, encoded by *BANYULS* in plant flavonoid biosynthesis. *Science*, **299**, 396–399.

6 Betalains

Jean-Pierre Zrÿd and Laurent Christinet

6.1 Betalain pigments

Betalain pigments are water-soluble vacuolar yellow (betaxanthins) and violet (betacyanins) pigments that replace anthocyanins in most plant families of the order Caryophyllales. They are also found in some species of the fungal genera *Amanita* and *Hygrocybe*. Betalains are conjugates of the chromophore betalamic acid, which derives from 3-(3,4-dihydroxyphenyl)alanine (DOPA) by an oxidative 4,5-extradiol ring-opening mechanism. The term betalain was introduced to describe these pigments as derivatives from betalamic acid (Wohlpart & Mabry, 1966). Betalains were erroneously named 'nitrogenous anthocyanins' in the past and are today often referred to as 'chromo-alkaloids' due to the presence of a nitrogen atom in the chromophore. It is still common to find erroneous references, even in contemporary textbooks, to anthocyanins in lieu of betalains (Fig. 6.1).

6.2 Chemistry

The main pigment of red beet, betanin (Schudel, 1919), has been used as a model for the determination of the structure and biosynthetic pathway of betalains. Our knowledge of the chemistry of betalain started in the late 1950s with the pioneering work of Dreiding's laboratory at the University of Zürich (Wyler *et al.*, 1963). The short but fascinating *compte rendu* by Mabry (2001) of his early years as a post-doc

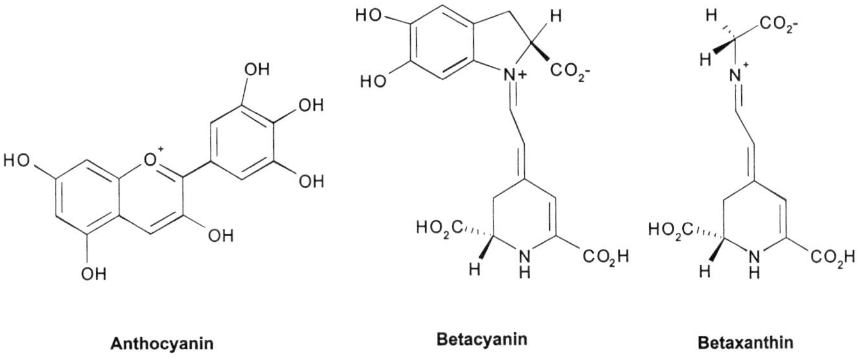

Anthocyanin   **Betacyanin**   **Betaxanthin**

Fig. 6.1 The structure of anthocyanins (delphinidin), betacyanins (betanidin) and betaxanthins. R = amino acid or amine.

gives an idea of the atmosphere surrounding this work and the progress made by the group. Since then, works by Piatelli in Italy, Mabry in the USA and Wyler in Switzerland have added to our knowledge of this unique family of pigments.

The common characteristic of all betalains is the presence of the betalamic acid chromophore, a dihydropyridine moiety attached via a vinyl group to another nitrogenous group (Miller *et al.*, 1968). Its lemon-yellow colour (λmax 424 nm) results from the resonance system induced by the presence of three conjugated double bonds (Fig. 6.2). Betaxanthins are formed by the condensation of an amino acid or an amine with the aldehyde group of betalamic acid, resulting in a Schiff-base (Fig. 6.2). This structure is responsible for the strong yellow or yellow-orange colours of betaxanthins and the maximum absorbance between 470 and 486 nm. Betacyanins are also formed of a betalamic acid unit linked to a molecule of *cyclo*-DOPA (Wyler & Dreiding, 1961) (Fig. 6.2). The latter highly aromatic structure is responsible for the deep violet colour of this pigment; this aromatic structure also induces a strong batochrome shift of 60–70 nm (λmax 534–554 nm). The basic betacyanin and betaxanthin structures can be modified in numerous ways; conjugation reactions like glycosylation or acylation are common. A detailed view on new structures and the methods used for their characterisation can be found in the recent review by Strack *et al.* (2003).

6.2.1 Betacyanins

All betacyanins need two molecules of tyrosine as precursors. The simplest natural betacyanins are the non-glycosylated betanidin or isobetanidin chromophores obtained by the condensation of *cyclo*-DOPA with betalamic acid (Wyler *et al.*, 1963). Both molecules differ only by the absolute configuration of their C_{15} chiral centre (Wilcox *et al.*, 1965) (Fig. 6.3). In beet hypocotyls the two compounds have a concentration ratio of about 4:1.

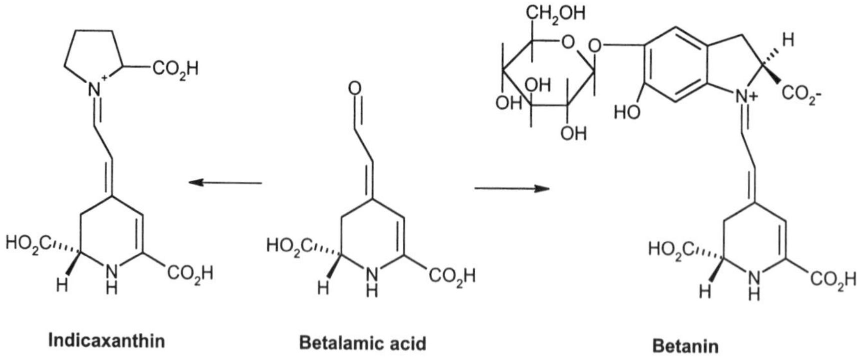

Fig. 6.2 Chemical structure of betalamic acid, the main chromophore of betalains. This structure is present in all betaxanthins associated to an amino acid or an imino compound (e.g. indicaxanthin from *Opuntia ficus-indica*) and in all betacyanins associated to *cyclo*-DOPA (e.g. betanin from *Beta vulgaris*).

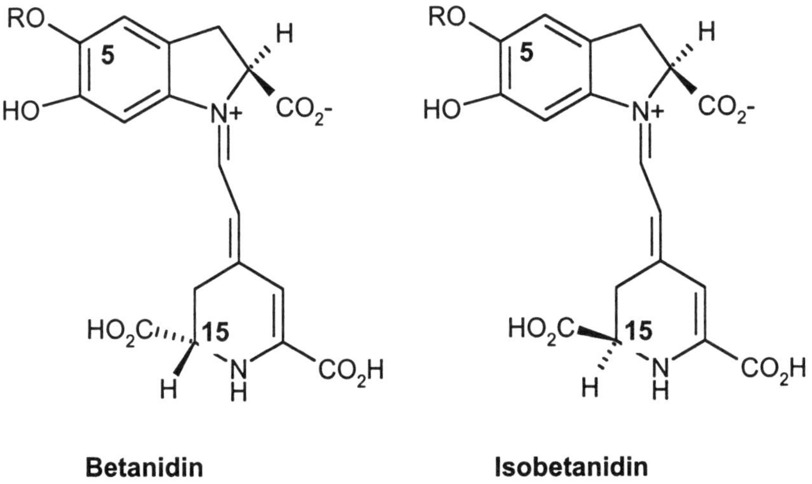

Betanidin   **Isobetanidin**

Fig. 6.3 Betacyanins non-glycosylated chromophores differ only by the C_{15} chiral centre (R=H).

Other betacyanins derive from these two isomers by *O*-glycosylation on one of the two free hydroxyl groups of *cyclo*-DOPA (Table 6.1). Glucosylation in position 5 is called betanin and is the major red beet pigment (Fig. 6.4). Some rare pigments like gomphrenins have been identified as 6-*O*-glucosides (Minale, 1967). Until recently, the only known non-glycosylated betacyanin was 2-descarboxy-betanidin (Fig. 6.4), a minor pigment from *Carpobrotus acinaciformis* flowers (Aizoaceae) containing a decarboxylated *cyclo*-DOPA moiety (Piattelli & Imperato, 1970). Lately, a small amount of this pigment was also identified in yellow beet (Schliemann *et al.*, 1999) accompanied by two new similar pigments, the 2-descarboxy-betanin and the 6′-*O*-malonyl-2-descarboxy-betanin (Kobayashi *et al.*, 2001).

The majority of betacyanins are acylated by ferulic acid, or less frequently by cinnamic acid, on their glycoside part via an ester linkage. Malonylation is also present in betacyanin structure; it is known for stabilising pigments in flowers, preventing anthocyanins from β-glycosidase attacks (Suzuki *et al.*, 2002). A new type of acylated betacyanin containing both an aliphatic and a hydroxycinnamoyl aromatic acyl residue has been detected in *Phytolacca americana* (Schliemann *et al.*, 1996). This kind of acylation is also observed in complex anthocyanins.

6.2.2 Betaxanthins

The yellow betaxanthins are immonium conjugates of betalamic acid with an amine or an amino acid. All protein amino acids and any of the 220 known non-protein amino acids found in plants can participate in the making of the

Table 6.1 Some betacyanins arranged according to their structures

Aglycones	
Betanidin	(Wyler *et al.*, 1963)
Isobetanidin	(Wilcox *et al.*, 1965)
2-descarboxy-betanidin group	
2-descarboxy-betanidin	(Piattelli & Imperato, 1970)
2-descarboxy-betanin	(Kobayashi *et al.*, 2000)
6′-*O*-malonyl-2-descarboxy-betanin	(Kobayashi *et al.*, 2001)
Glycosides	
5-O-glycosylation	
Betanin group	
Betanin: betanidin 5-*O*-glucoside	(Wyler & Dreiding, 1961)
Phyllocactin: 6′-*O*-malonyl-betanin	(Piattelli *et al.*, 1969)
2′-apiosyl-phyllocactin	(Kobayashi *et al.*, 2000)
2′-(5″-*O*-e-feruloylapiosyl)-betanin	(Schliemann *et al.*, 1996)
2′-(5″-*O*-e-feruloylapiosyl)-phyllocactin	(Schliemann *et al.*, 1996)
Hylocerenin:6′-*O*-(3″-hydroxy-3″-methyl)-betanin	(Wybraniec *et al.*, 2001)
Amaranthin group	
Betanidin 5-*O*-(glucuronide)-glucoside-amaranthin	(Sciuto *et al.*, 1974)
Iresinin I : hydroxymethylglutaryl-amaranthin	(Cai *et al.*, 2001)
Celosianin I, II : coumaroyl and feruloyl-amaranthin	(Cai *et al.*, 2001)
6-O-glycosylation	
Gomphrenin group	
Gomphrenin I: betanidin 6-*O*-glucoside	(Piattelli & Minale, 1964)
Gomphrenin II: coumaroyl derivative of gomphrenin I	(Heuer *et al.*, 1992)
Gomphrenin III: feruloyl derivative of gomphrenin I	(Heuer *et al.*, 1992)
Betanidin 6-*O*-(hydroxycinnamoyl)-β-sophoroside derivatives	(Heuer *et al.*, 1994)

betaxanthins molecules (Trezzini & Zrÿd, 1991b), therefore numerous betax-anthins can be found in plants, but only some have been fully characterised, most of them being present only in trace amounts. The classification of betax-anthins distinguishes the amino acid–derived compounds from the amine-derived conjugates (Table 6.2). The first isolated structure was indicaxanthin (Fig. 6.2) from the fruit of cactus pear (*Opuntia ficus-indica*) (Piattelli *et al.*, 1964). Numer-ous new structures have been identified during the last 15 years (Fig. 6.5). In *Portulaca grandiflora*, two pigments, one containing tyrosine (portulacaxanthin II) and the other glycine (portulacaxanthin III), were characterised (Trezzini & Zrÿd, 1991a). In Amaranthaceae, tryptophan-betaxanthin and the first methylated betaxanthin, 3-methoxytyramine-betaxanthin have been isolated from *Celosia argentea* (Schliemann *et al.*, 2001). The latter compound seems to be methylated already at the catecholic stage rather than at the betaxanthins stage.

Gomphrenin I   **2-descarboxy-betanidin**

Fig. 6.4 Different types of glycosylation of betacyanins: betanidin 6-*O*-glucoside (gomphrenin I) from *Gomphrena globosa* and the non-glycosylated 2-descarboxybetanidin from *Carpobrotus acinaciformis*.

Table 6.2 A selection of betaxanthin representative structures distributed in two groups according to the type of molecule conjugated with betalamic acid (amines or amino acids)

Amino acid–derived conjugates	Amino acids	Identification
Dopaxanthin	DOPA	(Impellizzeri *et al.*, 1973)
Indicaxanthin	Proline	(Piattelli *et al.*, 1964)
Miraxanthin I	Methionine sulfoxide	(Piattelli *et al.*, 1965a)
Miraxanthin II	Aspartic acid	(Piattelli *et al.*, 1965a)
Portulacaxanthin I	Hydroxyproline	(Piattelli *et al.*, 1965b)
Portulacaxanthin II	Tyrosine	(Trezzini & Zrÿd, 1991a)
Portulacaxanthin III	Glycine	(Trezzini & Zrÿd, 1991a)
Vulgaxanthin I	Glutamine	(Piattelli *et al.*, 1965c)
Vulgaxanthin II	Glutamic acid	(Piattelli *et al.*, 1965c)
Tryptophan-betaxanthin	Tryptophan	(Schliemann *et al.*, 2001)
Amine-derived conjugates	**Amines**	
Miraxanthin III	Tyramine	(Piattelli *et al.*, 1965a)
Miraxanthin V	Dopamine	(Piattelli *et al.*, 1965a)
3-methoxytyramine-betaxanthin	Methoxytyramine	(Schliemann *et al.*, 2001)
Humilixanthin	Hydroxynorvaline	(Strack *et al.*, 1987)
Miraxanthin I	Methionine sulfoxide	(Schliemann *et al.*, 2001)

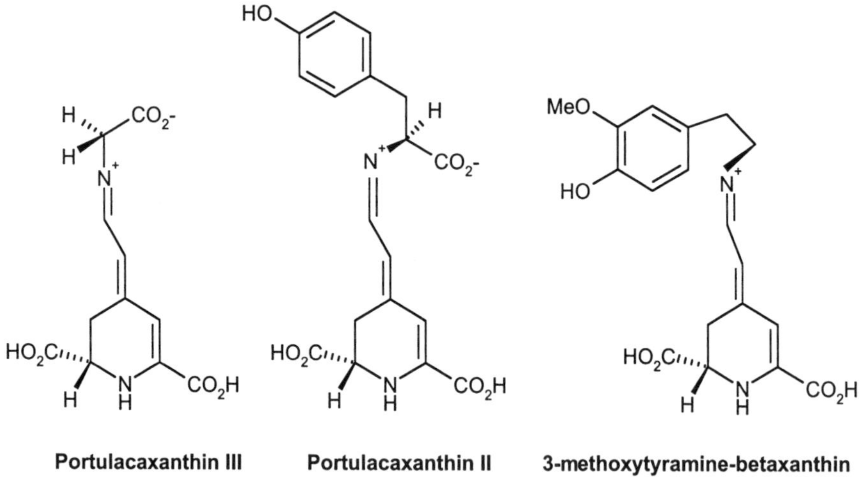

Portulacaxanthin III   **Portulacaxanthin II**   **3-methoxytyramine-betaxanthin**

Fig. 6.5 Chemical structure of some betaxanthins: portulacaxanthin III (glycin-betaxanthin), portulacax-anthin II ((S)-tyrosine-betaxanthin) from *Portulaca grandiflora* and 3-methoxytyramine-betaxanthin (derived from dopamine-betaxanthin) from *Celosia* species.

Recently, seven new pigments from cactus pear were characterised, six of which were new in plants (Stintzing *et al.*, 2002). That brings the question of the choice of the methodology to identify betaxanthins often present in trace amount. According to Stintzing (Stintzing *et al.*, 2002), enrichment of the plant extract in betaxanthins, after separation from sugars and pectins, gives a more detailed profile close to the *in vivo* chemical diversity.

Several betaxanthins have been identified in the fungus *Amanita muscaria*, including seven orange musca-aurins (λmax 480 nm) (Döpp & Musso, 1973a, 1973b; Trezzini & Zrÿd, 1991b). *A. muscaria* does not contain betacyanins but red and yellow betalamic acid–derived compounds, called respectively musca-purpurin (λmax 540 nm) and musca-flavin (λmax 420 nm) (Terradas & Wyler, 1991a; Mueller *et al.*, 1997b).

Betaxanthins can be easily synthesised non-enzymatically *in vitro* by mixing betalamic acid and the desired amine at neutral or slightly acidic pH (Trezzini, 1990; Trezzini & Zrÿd, 1991b); this occurs spontaneously in the acidic plant vacuole (Schliemann *et al.*, 1999) (Fig. 6.6).

6.3 Physiology of betalains

6.3.1 Light

Synthesis of betacyanins or betaxanthins does not require light in most species examined until now, although light is one major factor controlling the quantities of

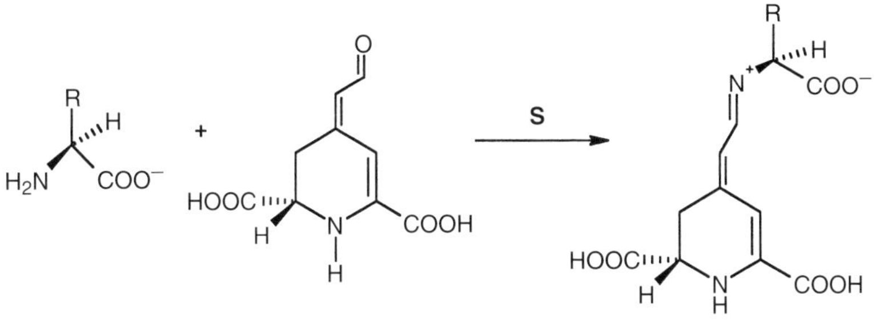

S–spontaneous aldimine formation

Fig. 6.6 Betaxanthins are formed by a spontaneous, non-enzymatic reaction at neutral and slightly acidic pH.

pigments produced (Wohlpart & Mabry, 1968). Phytochrome is involved in the photocontrol of amaranthin synthesis in *Amaranthus paniculatus*, one of the few species that have an absolute light requirement (Rast *et al.*, 1972); the red stimulation of pigment accumulation is made partially reversible by a far-red treatment. In *Amaranthus tricolor* seedlings, it has been shown that the photo-control occurs at the level of the formation of the dihydropyridine moiety of the molecule (Giudici De Nicola *et al.*, 1975). This photocontrol involves both blue light and red light and is thought to be mediated both by phytochrome and cryptochrome (Kochhar *et al.*, 1981). It was shown recently that UV-A alone could induce the production of betacyanins and flavonol glycosides in bladder cells of *Mesembryanthemum crystallinum* leaves (Vogt *et al.*, 1999b); this induc-tion is maximum at wavelengths between 305 and 320 nm (Ibdah *et al.*, 2002).

Girod and Zrÿd (1987) observed the appearance of pigmented cells on green calli from *Beta vulgaris* when they were transferred from dim to high light intensity; this trans-differentiation process is dependent on light intensity. Blue light induces pigmentation in *P. grandiflora* callus (Kishima *et al.*, 1995).

Light is known to affect the competition between the dopamine pathway leading to catecholamines (adrenaline) and the betalain pathway; in *Portulaca* callus, catecholamines are only synthesised from dopamine in the dark (Endress *et al.*, 1984).

6.3.2   Cytokinins and other hormones

In *Amaranthus caudatus*, betacyanins are synthesised in the dark in the presence of adenine-derived compounds (Bigot, 1968). A bio-assay for the determination of cytokinins (6-benzyl aminopurine) in plant extracts based on the cytokinin-induced formation of betacyanins in the dark-grown *A. caudatus* seedlings in the presence of tyrosine has been developed (Biddington & Thomas, 1973). This bio-assay is one of the most sensitive methods still available for the quantification of natural cytokinins in plants (Kubota *et al.*, 1999). Cytokinins can mimic light treatment in the control of tyrosine utilisation (Stobart & Kinsman, 1977). Studies on the molecular mode of action of cytokinin in the *Amaranthus* betacyanin assay with actinomycin D, alpha-amanitin and cycloheximide showed that rapid transcript induction and their translation were needed for amaranthin formation (Romanov *et al.*, 2000). The mean intervals between transcription, translation and pigment formation were estimated to be approximately 2 h.

In *P. grandiflora*, auxin (2,4-D) is needed for the appearance of a stable pigmentation; in the absence of auxin, pigmentation disappears almost entirely and becomes unstable (Trezzini, 1990). In cultured cells from *B. vulgaris*, the ratio of auxin (2,4-D) to cytokinin (6-BAP) is an important factor in the regulation of betalain biosynthesis (Girod & Zrÿd, 1991b).

6.3.3   Cell and hairy root cultures

Cell and hairy root cultures of betalain-producing plants have been established long ago, either for basic research on the biosynthetic pathways or with the purpose of producing high-quality violet and yellow pigments for the food industry. The low market price of beet extract nevertheless put a limit to the possible factory production of such pigments. Specific requests for pure yellow water-soluble pigments (betaxanthins) to replace potentially carcinogenic compounds have never materialised.

Beet cell cultures can be manipulated to produce a high amount of pure pigment (Leathers *et al.*, 1992); yields of total betacyanins can be as high as 28 mg/g dry weight, which is well above the content of cultivated red beet. Hairy root cultures have been shown to produce similar amount of pigments in air-lift reactors (Shin *et al.*, 2002). In an experiment with *B. vulgaris* cells cultivated on solid medium (callus), it was shown that the amount of total betacyanins in the violet cell lines is slightly higher (28 μmol/g DW) than the amount present in storage organs of the plant (21 μmol/g DW). In the orange or red cell lines the amount of total betaxanthin can reach more than 12 μmol/g DW (Girod & Zrÿd, 1991b); violet betacyanin-producing cell cultures contain a large amount of dopamine; it would be probably necessary to shut down the biosynthetic pathway leading to dopamine to further increase fluxes from DOPA toward betalain biosynthesis.

Regulatory mechanisms of betacyanin biosynthesis in suspension cultures of *P. americana* were investigated in relation to cell division activity. By inhibiting cell division, incorporation of radioactivity from labelled tyrosine into betacyanin is reduced (Sakuta *et al.*, 1994); the conclusion was that conversion of tyrosine to DOPA is associated to cell division. By using the DNA-methylation inhibitor 5-azacytidine, Girod & Zrÿd (1991b) demonstrated that methylation plays a key role in the repression of genes encoding enzymes involved in betacyanin biosynthesis. Yellow cells (accumulating only betaxanthins) could be switched to red cells (containing both betacyanins and betaxanthins) by reducing the concentration of the auxin 2,4-D in the culture medium. The chain of events occurring during this change is stimulated by the methylation inhibitor; this indicates gene activation of normally repressed genes. 3-methoxybenzamide, a poly (ADP-ribose) polymerase inhibitor, was found in some experiments to inhibit colour transition. That chromatin rearrangement and DNA-methylation are both involved in the regulation of betalain colour genes does not come as a surprise.

Betalain synthesis had been reported to be modified in diseased plants after insect, pathogen or virus attacks. Phytopathologists (Steddom *et al.*, 2003) used multi-spectral canopy reflectance to study the physiological differences between healthy sugar beets and beets infested with Beet Necrotic Yellow Vein Virus. It was shown that the ratio of betacyanins to chlorophyll, estimated from canopy spectra, was increased in symptomatic beets at 4:7 sampling dates. Differences in betacyanin levels appeared to be related to disease impact and to the development of rhizomania associated with this virus attack. Wounding and infiltration with *Pseudomonas syringae* or *Agrobacterium tumefaciens* also induced the synthesis of betacyanins in red beet leaves (Sepúlveda Jiménez *et al.*, 2003).

6.4   Genetics of betalains

We will consider here only two of the main studied species with respect to betalain biosynthesis: *P. grandiflora* and *B. vulgaris*.

6.4.1   *Portulaca grandiflora* – large-flowered purslane

Large-flowered purslane (*P. grandiflora*) was described for the first time by Hooker (1829). This plant originates from South America and can be found there in two forms: one with deep orange flowers and the other with deep red flowers. The description by Hooker is lively: ' ... *On the western side of the Rio Desaguadero plants were in great profusion, giving to the ground over which they were spread a rich purple hue, here and there marked with spots of orange color, from the orange-colored variety which grew intermixed with the other.*' He concludes: ' ... *I am happy [...] having the opportunity of giving an analysis of what I cannot but think a new species of the genus, and one, the beauty of whose flowers must render it a desirable inhabitant of the cool stove or greenhouse.*'

The beauty, large spectrum of colour and shape of the flower attracted Japanese geneticists from the beginning of the twentieth century (Yashi, 1920; Ikeno, 1921; Adachi, 1972). Most of those works were done at a time when the nature of the pigments was unknown. More recently, based on an increasing biochemical knowledge, a genetic model of betalain pigmentation based on the analysis of crosses of inbred lines of *P. grandiflora* was published (Trezzini & Zrÿd, 1990). The resultant segregation patterns indicated that a minimum of three loci, *C* (*colour*), *R* (*red*) and *I* (*inhibitory*) were involved in petal pigmentation (Table 6.3).

Table 6.3 A three-gene model of *Portulaca grandiflora* flower colour

	Violet	Orange-red	Deep yellow	Pale yellow	White
C	*C-*	*C-*	*C-*	*C-*	*cc*
R	*R-*	*R-*	—	—	—
I	*I-*	*ii*	*ii*	*I-*	—

The expression of the dominant *C* locus, responsible for the conversion of DOPA to betalamic acid, leads to plants with coloured shoots and petals. The expression of the dominant *R* locus is necessary for the production of violet betacyanins correlated with the availability of *cyclo*-DOPA or its glucosylated form. The dominant *I* locus inhibits (but does not prevent) the accumulation of betaxanthins. In the absence of the dominant *R* locus, *Portulaca* flowers are either pale yellow in presence of the dominant *I* or deep yellow in the homozygous recessive *ii* background. The presence of the dominant *I* locus is essential for the production of pure violet hue; otherwise, the flower will display an orange-red colour due to the high level of yellow betaxanthin over a violet betacyanin background. The *I* and *R* loci are strongly linked at ~5.3 cM (Trezzini, 1990).

The *I* locus prevents the conjugation of amino acid or imino residues (others than *cyclo*-DOPA) with betalamic acid, thus decreasing the synthesis of betaxanthins and leading to the accumulation of betacyanins in the violet phenotype (dominant *R* locus). It is hypothesised that the chromophore is formed in the cytoplasm, then transported to the vacuole where conjugation to betaxanthins will take place spontaneously (Trezzini, 1990). This hypothesis postulates the existence of a specific carrier for betalamic acid, which would be partially or totally shut down in presence of the product of locus *I*. A small amount of yellow pigment was always produced despite the expression of the inhibitor *I* in the pale yellow *CCrrII* phenotype. In those plants a high amount of free betalamic acid is present, as well as traces of dopaxanthin, but no miraxanthin V (dopamine) can be detected (Trezzini & Zrÿd, 1991a). The identity of the products of the *R* and *I* loci are still unknown and will need further investigations.

Colouration of *Portulaca* petals shows an instability, which is modulated by the activity of transposable elements (Rossi-Hassani & Zrÿd, 1994, 1995).

6.4.2 *Beta vulgaris* – table beet

Beet fleshy hypocotyl is the major source of betalain pigments for the food industry. Betalains are stored in the huge fleshy hypocotyl that is the storage organ of the plant (often misidentified as root); they can also be found in the shoot and leaf veins (Swiss chard varieties). Many pigmented cultivars have been selected during the long history of beet domestication. Since the cultivation of beet as an alternative source of sucrose at the beginning of the nineteenth century, huge effort has been made to improve sucrose content and disease resistance. The genetics of *B. vulgaris* pigmentation has been comparatively less studied. The presence of dominant alleles at two linked loci (*R* and *Y*) conditions the qualitative production of betalain pigment in the beet plant (Wolyn & Gabelman, 1989; Goldman & Austin, 2000). Red-pigmented hypocotyls are observed only in the presence of dominant alleles at both the *R* and *Y* loci, while white hypocotyls are conditioned by recessive alleles at the *Y* locus, and yellow hypocotyls by the genotype *rrY-*. Therefore the beet *Y* locus should correspond to the *C* locus and the beet *R* locus to the *R* locus of *P. grandiflora*. Alleles at the *R* locus determine the ratio of betacyanin and betaxanthin (Wolyn & Gabelman, 1989).

In *B. vulgaris* cell lines cultivated *in vitro*, careful clonal selection leads to five phenotypes that mimic purslane flower colours (Girod & Zrÿd, 1991b). The red and violet cell phenotypes were shown to contain two to three times more pigments than the orange and yellow cell phenotypes. This is due mainly to the accumulation of a high amount of betanin over a quite stable background content of betaxanthins. Girod and Zrÿd suggested that the synthesis of *cyclo*-DOPA (gene *R*) is coordinated with that of betalamic acid (gene *C*).

A regulatory gene named *blotchy* (*bl*) that conditions a blotchy or irregular pigment patterning in either red or yellow hypocotyls has been characterised (Goldman & Austin, 2000). There are linkage relationships between the *R* and *Y* loci and the *bl* gene. The two-point linkage estimate between the *R* and *Y* loci was estimated at ~7.4 cM. The *R-Y-bl* genomic region is therefore important in the genetic control of betalain biosynthesis in *B. vulgaris*. The linkage relationships demonstrated both in purslane (*R* and *I*) and beet plants (*R* and *Y*) could indicate a strong pressure for co-evolution of the betalain biosynthesis syndrome. The *bl* gene shows some transmission distortion suggesting among others an epigenetic control (Austin & Goldman, 2001).

Careful field selection of red beet populations for high total betalain pigment concentration (Goldman & Austin, 2000) lead to a total pigment increase of about 200%; betalain pigment concentration responds favorably to recurrent selection. Since betalain pigments are formed following glycosylation of *cyclo*-DOPA and betalamic acid, it could be possible to associate sugar biosynthesis with pigment biosynthesis. RAPD markers associated with genes controlling pigment in red beet have been identified (Eagen & Goldman, 1996).

In one habituated red beet cell line it has been shown that the block in betalain biosynthesis (leading to white cells) is at the level of tyrosine hydroxylation (Zrÿd *et al.*, 1982) – no DOPA is formed. This observation suggests a possible 'silencing' of a gene coding for a multifunctional polyphenol oxidase in the white cell lines.

6.5   Enzymology and biotechnology

Betalains derive from tyrosine; this distinguishes them clearly from the phenyl-alanine-derived anthocyanins (Miller *et al.*, 1968). The betalain biosynthetic pathway is rather simple; only three to four enzymes are needed for the synthesis of the most simple betaxanthin or betacyanin (see Fig. 6.6): they will catalyse (a) a ring-opening reaction leading to the chromophore betalamic acid, (b) the formation of *cyclo*-DOPA, (c) a further glycosylation step in betacyanin synthesis, and finally (d) the transport of betaxanthins in the vacuole. Betacyanins synthesis requires a minimum of two tyrosines as precursors, whereas one is sufficient for betaxanthins. Feeding experiments with C_{14}-radiolabelled tyrosine showed that the entire C_6–C_3 skeleton of this amino acid is incorporated into betalamic acid and *cyclo*-DOPA molecules (Liebisch & Bohm, 1981). Tyrosine hydroxylation by a tyrosinase produces DOPA – a reaction that is very common in the plant kingdom.

6.5.1   Polyphenol – tyrosinase reactions

Tyrosinases (polyphenoloxidases, PPOs) are widespread among plants and fungi. They are copper-containing enzymes that catalyse hydroxylation of phenols to *o*-diphenols (EC 4.14.18.1 – monophenol: mono-oxygenase) and their subsequent oxidation to *o*-quinones (EC 1.10.3.1. – *o*-diphenol: oxygen oxidoreductase). Another PPO called catechol oxidase catalyses only the oxidation of *o*-diphenols.

The first step in betalain biosynthesis is the formation of DOPA from tyrosine. DOPA is an important metabolic product of a large number of plant families and accumulates in large amount in some Leguminosae species (e.g. *Vicia faba*) (Guggenheim, 1913). In betalain plants DOPA is then oxidised through a further diphenolase reaction to dopaquinone (rearranged to *cyclo*-DOPA) and modified by a ring-opening extradiol dioxygenase to *seco*-DOPA (rearranged to betalamic acid) (Fig. 6.7).

A tyrosinase activity closely linked to the biosynthesis of betalamic acid and muscaflavin has been characterised in *A. muscaria* (Mueller *et al.*, 1996); this tyrosinase was not specific for tyrosine but has higher affinity and lower Km value (0.3 mM) for this compound compared to analogues; the protein is a heterodimer with a molecular weight of $\sim 50\,000$ kDa. It is strictly localised in the coloured part of the fungus where DOPA and betalains accumulate and is not found anywhere else. In plants a PPO transcript correlating with betalain biosynthesis was found in *P. americana* fruits (Joy *et al.*, 1995). Recently, a plant tyrosinase involved in betalain synthesis has also been isolated from callus cultures of *P. grandiflora* and *B. vulgaris* (Steiner *et al.*, 1996, 1999); this enzyme has a molecular weight of 53 000 kDa. It was shown that both the fungal and plant PPOs are truly bifunctional enzymes producing not only DOPA but also able to use DOPA as a substrate. A bifunctional PPO catalysing the hydroxylation and oxidation of chalcones to aurones has also been found in flowers of the anthocyanin-producing plant *Anthirrhinum majus* (Nakayama *et al.*, 2000; Sato *et al.*, 2001). A putative tyrosine hydroxylase activity from betacyanin-producing callus cultures from *P. grandiflora* was separated from a polyphenol oxidase activity (Yamamoto *et al.*, 2001). The purified enzyme catalysed the formation of DOPA from tyrosine and was activated by Fe^{2+} and Mn^{2+}, and inhibited by metal

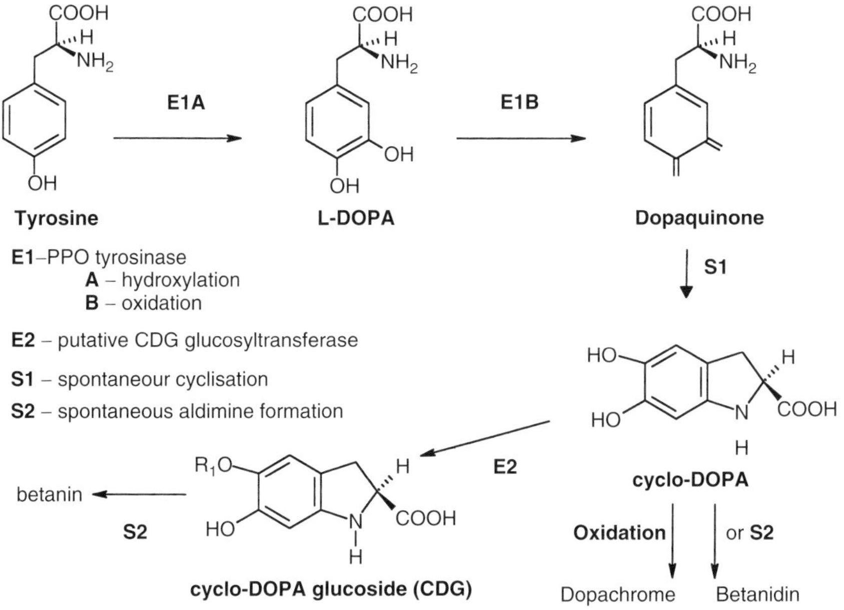

Fig. 6.7 The PPO reactions and the possible fates of the unstable *cyclo*-DOPA.

chelating agents. Unfortunately the authors did not give details about the substrate specificity of this enzyme and therefore the question of the existence of a plant tyrosine hydroxylase similar to the mammalian enzyme remains open.

6.5.2 The ring-opening reaction

6.5.2.1 The fungal (Amanita muscaria) enzyme

Betalamic acid is formed through an enzymatic cleavage of the DOPA aromatic ring at the position 4,5; this produces an unstable *seco*-DOPA intermediate (Fischer & Dreiding, 1972; Terradas & Wyler, 1991b), which spontaneously closes on itself (Schliemann *et al.*, 1998). In *A. muscaria* an active 3,4 DOPA extradiol-dioxygenase catalyses this extradiol cleavage both in the 4,5 and 2,3 positions (Girod & Zrÿd, 1991a; Terradas & Wyler, 1991a). Depending of the pH, part of the substrate undergoes a 2,3 cleavage leading to the formation of the fungal pigment muscaflavin (see Fig. 6.8). The purified enzyme is an iron-containing homo-multimeric protein with a 22 000 kDa subunit. This enzyme does not show a strict specificity toward DOPA (Km = 3.9 mM) and is able to use caffeic acid (Km = 0.9 mM), dopamine (Km = 6.3 mM) or catechol as substrates. The gene for this enzyme has been isolated (Hinz *et al.*, 1997); this *DODA* gene codes for a 228 amino acid protein that shows no homology with any sequence yet published. The closest matches at the protein level are sequences

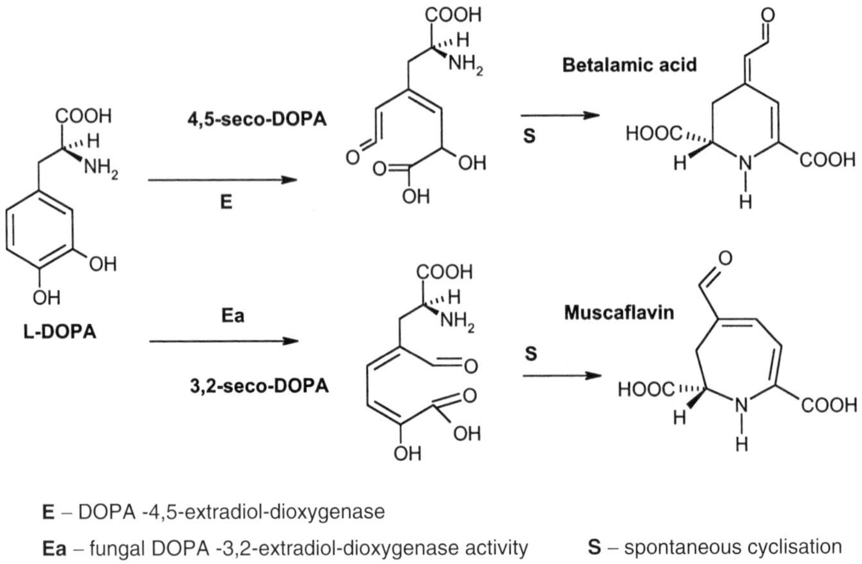

Fig. 6.8 The DOPA 4,5 and 3,2 extradiol ring-opening reactions in betalain-containing plants and in the fungus *Amanita muscaria* and the spontaneous cyclisation products.

from *Burkholderia fungorum*, *Botrytis cinerea* (ESTs) and *Nostoc* species (complete sequencing). The recombinant enzyme displays Km and Vmax values close to those of the native enzyme and is also able to produce both betalamic acid and muscaflavin (Mueller *et al.*, 1997b).

Interestingly, the fungal enzyme is able to complement the betalain pathway in *P. grandiflora cc* genotypes deficient in the gene *C*, which was postulated to code for a plant ring-opening dioxygenase (Mueller *et al.*, 1997a). This successful complementation was obtained by biolistic transformation of petals using a *pNco DODA* vector; the construct being the *DODA* cDNA under the control of a *CaMV 35S* promoter and terminator cloned into a *pUC18* vector. The main pigments found in transformed, coloured petal cells were dopaxanthin, betanin and miraxanthin V, depending on the genetic background; muscaflavin, which is specific to fungi and normally absent from plants (Barth *et al.*, 1979), was also found, showing that the fungal enzyme was working properly in a plant cell environment retaining its full catalytic properties.

6.5.2.2 The plant (Portulaca grandiflora) enzyme
The complementation experiments performed on *P. grandiflora* with the *Amanita* enzyme confirmed simultaneously that the plant dioxygenase could be expected to be different from the fungal enzyme and probably totally unrelated to it. The plant enzyme escaped all isolation attempts; despite the effort of many research groups no enzymatic activity was ever detected in plants.

Based on the availability of *P. grandiflora* inbred lines (Trezzini & Zrÿd, 1990), subtractive libraries where constructed in order to isolate directly plant genes associated with betalain biosynthesis (Zaiko, 2000). After PCR-Select® cDNA subtraction, betalain-specific transcripts present in yellow or violet-flowered (*C*-) genotypes and not in white (*cc*) -flowered genotypes where obtained. Putative interesting transcripts were selected for their strong expression during the early stages of flower development when betalain rate of synthesis is at its maximum. Corresponding full-length cDNAs were obtained using the RACE amplification strategy. A potentially interesting candidate showing strong homology at the level of translated protein with the LigB domain present in the bacteria *Sphingomonas paucimobilis* protocatechuate extradiol 4,5-dioxygenase was identified (Christinet *et al.*, 2004). This single-copy gene was expressed only in coloured flower petals and stem epidermal cells and not in non-pigmented plant tissues. The function of this gene in the betalain biosynthetic pathway was confirmed by using biolistic genetic complementation of *P. grandiflora* petals (Mueller *et al.*, 1997a). The recombinant iron protein has yet to be produced.

This gene named *DODA* is the first characterised member of a novel family of plant non-haem dioxygenases phylogenetically distinct from *Amanita* DOPA-dioxygenase; not only does the DODA protein share no homology with the fungal protein but its catalytic activity is that of a genuine 4,5-extradiol dioxygenase with no 2,3- activity. Homologues of DODA are present not only in betalain-producing plants but also, albeit with some changes near the catalytic site, in other angiosperms and in the bryophyte *Physcomitrella patens*. These homologues are part of a novel conserved plant gene family that could be involved in aromatic compound metabolism. Due to the broad distribution of sequences similar to the *P. grandiflora* DODA across terrestrial plants, it can be hypothesised that the betalain pathway originated by recruitment of an existing metabolic pathway compensating for loss(es) in the capacity of synthesising coloured anthocyanins in plants of the order Caryophyllales. A conserved amino acid motif (HNL-R/G) part of the catalytic site is present in the DODA-like protein of all organisms tested except in plants synthesising betalains. In these plants, a completely different conserved motif is present: the highly conserved catalytic amino acid His177 is followed by the pattern P-(S,A)-(N,D)-x-T-P. Both motifs begin with the strictly conserved His177, which is essential for the catalytic activity of extradiol dioxygenase class III enzyme like *P. grandiflora* DODA (Sugimoto *et al.*, 1999). 3D modelling of the enzyme, based on the crystal structure from the protocatechuate 4,5-dioxygenase LigAB, revealed that the amino acids following His177 control the access of the substrate. The *Sphingomonas* enzyme involved in lignin degradation uses a substrate with a lateral chain shorter than that of the DOPA molecule specific for DODA enzyme from *P. grandiflora* (Fig. 6.9).

It is worth remembering here that, in Fabaceae, where DOPA is present at high concentration (5% in the seeds from the genera *Mucuna* and *Stizolobium* (Yang *et al.*, 2001)) another DOPA ring–cleavage activity is present; those plants have been shown to contain stizolobic or stizolobinic acid (Saito & Komamine, 1978).

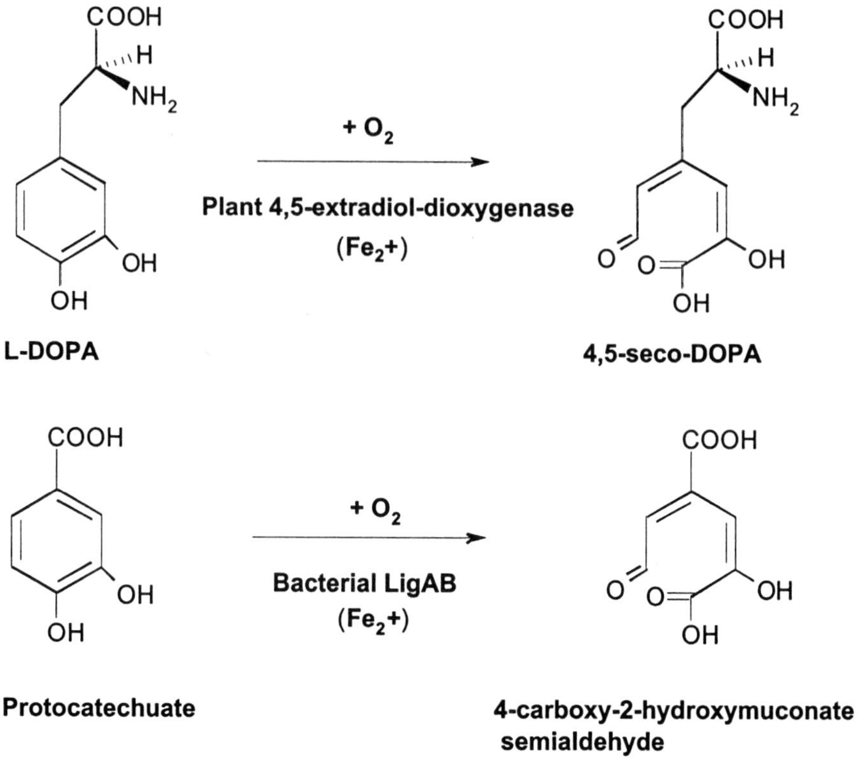

Fig. 6.9 The two homologue reactions catalysed by the plant DOPA 4,5-extradiol-dioxygenase and by the protocatechuate 4,5-extradiol-dioxygenase from the bacteria *Sphingomonas paucimobilis*.

The synthesis of those compounds needs a supplementary reduction step that a classical DOPA-dioxygenase does not catalyse alone. Both compounds are naturally present in *Amanita* (Saito & Komamine, 1978).

6.5.3 *Glycosylation and other modifications*

Most betacyanins are obtained by a final step involving the 5- or 6-*O*-glucosylation of betanidin or isobetanidin (Kobayashi *et al.*, 2001) (Fig. 6.10). Two distinct enzymes have been isolated from *Dorotheanthus bellidiformis* cell suspensions (Heuer *et al.*, 1996; Vogt *et al.*, 1997). Those enzymes catalyse the indiscriminate transfer of glucose from UDP-glucose to hydroxyl groups of betanidin, flavonols, anthocyanidins and flavones. Sequence comparison shows that these enzymes have only 19% amino acid sequence identity, suggesting a paraphyletic origin of these two glucosyltansferases (GT). Their kinetic properties are similar to other known GTs related to flavonoid biosynthesis (Vogt *et al.*, 1997).

Betanidin   **Betacyanins**

E – betanidin 5-*O*-glucosyltransferase: R1=glucosyl   R2=H
    – betanidin 6-*O*-glucosyltransferase: R1=H   R2=glucosyl

Fig. 6.10 The formation of betanins from betanidin through the action of 5-*O*- and 6-*O*-glucosyltransferases.

It was suggested that *cyclo*-DOPA glycoside, observed at trace level only, is an artefact that originates from the hydrolysis of betanin (Strack *et al.*, 2003), and that glycosylation occurs at a later step on betanidin itself by means of specific GTs. Nevertheless, labelled *cyclo*-DOPA glycoside is formed in red beet cell culture after addition of radiolabelled tyrosine or DOPA to the culture medium (Zrÿd *et al.*, 1982; Bauer, 2001); this would indicate that glucosylation occurs before the condensation step with betalamic acid. Careful compartmental kinetic analysis using H$_3$-labelled tyrosine feeding of *B. vulgaris* cell suspensions showed that glucosylated *cyclo*-DOPA accumulates before its incorporation into betanin. Rate constant for the formation of *cyclo*-DOPA glucoside from DOPA is 0.005/s whereas the rate constant for the formation of betanin is only 0.0029/s (Bauer, 2001). A fast and early glycosylation of *cyclo*-DOPA would make sense in view of the well-known instability of this molecule. We already know that DOPA is oxidised to DOPA-quinone, which itself spontaneously rearranges into 5,6-dihydroxyindole-2-carboxylate (*cyclo*-DOPA); this unstable compound could either be further oxidised or stabilised by glycosylation (see Fig. 6.7). In intact, not-wounded, plant cells neither *cyclo*-DOPA nor pigmented (black) oxidation compounds have ever been detected. The violet phenotypes of *B. vulgaris* cells cultivated *in vitro* are very sensitive to mechanical stress and cell disruption and become rapidly black if wounded, this is not the case with the yellow phenotypes that do not synthesise *cyclo*-DOPA. It seems therefore that rapid glucosylation of *cyclo*-DOPA to CDG could protect cells against damage from oxidation products. The question of the place of the glucosylation step(s) in betanin biosynthesis remains open to further investigations (Sasaki *et al.*, 2004).

Celosia argentea variety plumosa contains, in addition to the known compounds amaranthin and betalamic acid, three yellow pigments that are immonium conjugates of betalamic acid with dopamine, 3-methoxytyramine and (S)-tryptophan and 2-descarboxy-betanidin, a dopamine-derived betacyanin (Schliemann *et al.*, 2001)

Betanins are, like flavonoids, frequently acylated via an ester linkage to the sugar moiety. An acyltransferase from cell cultures of *Chenopodium rubrum* was purified (Bokern *et al.*, 1992). This enzyme catalyses the transfer of hydroxycinnamic acids from 1-*O*-hydroxycinnamoyl-β-glucose to the C-2 hydroxy group of glucuronic acid of amaranthin (betanidin 5-*O*-glucuronosylglucose). The *in vivo* products formed are celosianin I (4-coumaroylamaranthin) and celosianin II (feruloylamaranthin).

6.5.4 DOPA decarboxylation

Dopamine-derived betalains like miraxanthin V or the 2-descarboxy-betanin need a supplementary decarboxylation step of the DOPA into dopamine (Dunkelblum *et al.*, 1972; Kobayashi *et al.*, 2001). Protein extract from red beet revealed the presence of a decarboxylase, transforming DOPA into dopamine, thus supporting the synthesis of dopamine-derived betacyanins as observed previously (Terradas, 1989).

6.5.5 Degradation

In vivo betalain degradation, albeit an important factor in the maintenance of betalain steady state and in the recycling of nitrogen, has been somehow a neglected field. Some of our knowledge comes from *in vitro* or post-harvesting studies (Cai & Corke, 2001). It is known that betanin can be degraded spontaneously into betalamic acid and *cyclo*-DOPA-5-*O*-β-*D*-glucoside (Schwartz & von Elbe, 1983); this could happen when cells are damaged but also under normal condition. Degradation of betacyanin is reversible and could be accompanied by synthesis of betaxanthins in presence of the proper amine. Degradation of betanin to betanidin is probably catalysed by a β-glucosidase (Zakharova & Petrova, 2000).

An enzyme catalysing the discolouration and breakdown of betacyanins was isolated and purified from beet hypocotyls *B. vulgaris* (Soboleva *et al.*, 1976). The enzyme activity induced the oxidative discolouration of betanin and betanidin. The enzyme is membrane-associated, degrades pigments in presence of oxygen and is inhibited by chelating agents and could well be a PPO. Peroxidases (EC 1.11.1.7) have been suspected to be involved in betalain degradation (Martinez-Parra & Munoz, 2001). A protein fraction with peroxidase activity against guaiacol from *B. vulgaris* hypocotyls oxidised both betanidin and betanin. Betanidin quinone was formed as the only product in the course of enzymatic betanidin oxidation, whereas betalamic acid and several oxidised *cyclo*-DOPA-5-*O*-β-*D*-glucoside polymers were generated during the oxidation of betanin; the apparent Km for the reaction was 0.46 mM.

6.6   Evolution of betalains

Betalains replace anthocyanins in most families of the order Caryophyllales (Clement & Mabry, 1996), except in the families Caryophyllaceae and Mollugi-naceae where anthocyanins are still exclusively present (Table 6.4). Mutual exclusion of betalain and anthocyanin pigments (Wyler & Dreiding, 1961) and their respective biosynthetic pathways (Kimler *et al.*, 1971) provides the main chemotaxonomic criterion to differentiate Chenopodiaceae from Caryophyllaceae. Betalains have also been identified in the pilei from a very restricted number of basidiomycetes species belonging to *Amanita* and *Hygrocybe* genera (Döpp & Musso, 1973b; von Ardenne *et al.*, 1974).

Betalains and anthocyanins have a similar localisation in plant and both use a large spectra of colour, which is part of the attraction syndrome toward insects, mammals and birds involved in pollination and seed dispersal (Clement & Mabry, 1996). The origin of betalains in plant is a fascinating question. The biological role of betalain in fungal species is unknown.

Ehrendorfer (1976) was the first to propose a general hypothesis for the presence of betalains in Caryophyllales. Ancestral group of plants may have evolved under environmental conditions (drought) that could have induced a preference for anemochory (wind pollination), and at the same time the loss of insect attracting capability (pigmentation). It is remarkable that the majority of the present betalain-containing plants are mostly drought- and salt-resistant. Subsequently the acquisi-

Table 6.4 Caryophyllales and the distribution of betalain and anthocyanin-producing taxa

Sub-order	Family	Examples of genus
	Achatocarpaceae	*Achatocarpus*
	Aizoaceae	*Dorotheanthus, Mesembryanthemum, Carpobrotus*
	Amaranthaceae	*Amaranthus, Iresine, Gomphrena*
Chenopodiineae	Basellaceae	*Basella*
Betalain-producing	Cactaceae	*Mammillaria, Opuntia, Pereskia*
Anthocyanin-free taxa	Chenopodiaceae	*Beta, Chenopodium, Spinacia*
	Didiereaceae	*Didierea*
	Halophytaceae	*Halophytum*
	Hectorellaceae	*Hectorella*
	Nyctaginaceae	*Bougainvillea, Mirabilis*
	Phytolaccaceae	*Phytolacca, Gisekia*
	Portulacaceae	*Portulaca, Claytonia*
	Stegnospermataceae	*Stegnosperma*
Caryophyllineae		
Betalain-free	Caryophyllaceae	*Dianthus, Silene*
Anthocyanin-producing taxa	Molluginaceae	*Mollugo, Limeum*

tion of betalain-type pigments would have permitted a new evolution back to zoochory (animal-linked pollination). In recent phylogenetic studies of the Caryo-phyllales (Cuenoud *et al.*, 2002), it was shown that the Caryophyllaceae that possess anthocyanins are closely related to the Amaranthaceae (Chenopodiaceae species strain and Amaranthaceae species strain) which synthesise betalains but are also known for their inconspicuous flowers and predominant anemophily.

Evolution of new types of highly coloured flowers (in Portulacaceae, Cactaceae and others) from almost flowerless plants is also coherent with this hypothesis. Amaranthaceae have small sepals but petals are absent; the flowers of Cactaceae have numerous separate large and colourful petals, the lower-most of which may appear sepal-like. Throughout most of the order Caryophyllales, flowers have true sepals but no true petals. Some families that display what appear to be petals, such as Portulacaceae and Nyctaginaceae, are thought to have greatly modified sepals, in which case the sepal-like appendages are interpreted as bracts. For a description of floral ontogeny in *Portulaca* see papers by Soetiart (Soetiart & Ball, 1968, 1969).

Substrate specificity for DOPA of the specific ring-opening dioxygenase DODA should have occurred during the evolution of early betalain-synthesising plants. Only an extensive analysis across the phylogenetic tree will answer the question of the point of divergence. We could expect to find a few plants with intermediate se-quences, at the interface of betalain-producing and-non-producing plants, which have escaped current investigations; those plants could help to solve some critical aspects of the molecular phylogeny of Caryophyllales (Cuenoud *et al.*, 2002). Exam-ination of the phylogenetic tree of DODA-type proteins (Christinet *et al.*, 2004) puts Chenopodiaceae (*B. vulgaris*) at the base of the branch of betalain-producing plants. DODA-like proteins are present not only in plants but also in eubacteria and archea; they are notably absent from the fungal and animal kingdoms (Fig. 6.11).

6.7   Social and economic value

Due to their colourant properties, betalains can be used in various applications in the food and agriculture industries. People are more often in contact with betalains than thought, while eating red beet or strawberry yoghourt coloured with beet juice, or admiring cactus or bougainvillea flowers.

Betacyanin-producing plants, mainly in the Chenopodiaceae family (beet) and in the Amaranthaceae family (*A. paniculatus, A. hypochondriacus*), have been used traditionally as food, medicine and cult items in sacred ceremonies. The Aztec have used red-violet amaranths leaves or seed heads (*huauhtli* in classic nahuatl) to symbolise human blood during sacrificial ceremonies or to make coloured (bleeding) idols either as a substitute for human sacrifice or simply as representation of harvest gods; in an apparently paradoxical way this could be related with the use of red wine (anthocyanidins) as representation of Christ's blood in the Christian Eucharist. In fact the Catholic Church eventually forbade the use of *Amaranthus* as food crop, and it is only recently that cultivation and use

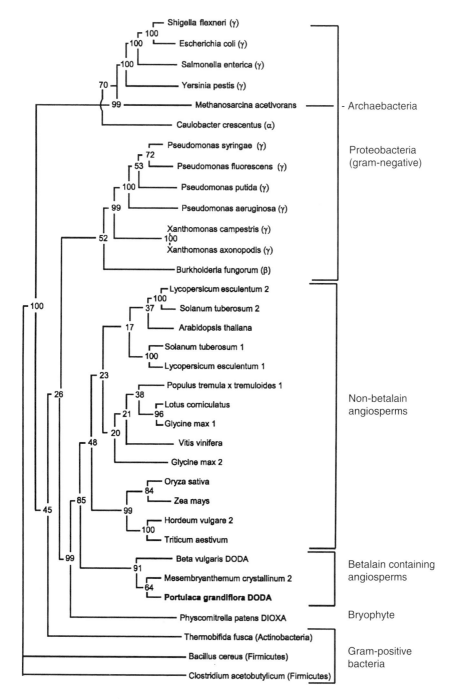

Fig. 6.11 Phylogeny analysis of *Portulaca grandiflora* DOPA-dioxygenase. This strict consensus phylogenetic tree has been built across all prokaryotic and eukaryotic genomes in which sequence homology was found (note the absence of animals and fungi). The bootstrap values at the forks indicate the number of times out of 100 trees that this grouping occurred.

of this plant has resumed in Central America. In Eastern Europe red beet species were not related directly to specific religious ceremony; nevertheless red beet soup (borsch) fermented or not is often offered on special occasions (Christmas, Jewish celebrations – Pesach etc.). On the negative side, it has been reported in an ethnological study in Sri Lanka, that red-coloured food, including red beet, is considered inappropriate for pregnant women (De Silva, 1996). At the beginning of the twentieth century, red beet hypocotyl (root) was highly valued as a remedy against cancer, and those beliefs have been partly confirmed recently (Kapadia et al., 1996, 1998; Wettasinghe et al., 2002).

The betaxanthin-coloured fungus A. muscaria is used by Siberian shaman as a potent healing drug, which is not given to the patient but used by the medium (Eliade, 1968). The hallucinogenic properties of A. muscaria are not related to its betalain content but to its ibotenic acid and muscimol content. It is possible nevertheless that its attractive, unique pattern of colour was a major factor in its use as a revered and popular medicine.

In Europe, Russia and Northern America the main betalain-containing plant with an economic value is red beet. Beet has been cultivated in Europe since the antiquity. Today it is used either as a salad (cooked or raw) in Western Europe or as a basic soup ingredient (borsch) in Poland and Russia; in the USA red beet is sold mainly canned. The use of raw or refined beet extract as a colouring food additive represents, in the EU (with France as the major producer), a market of more than 10 m. for a production of 4000 tons/year. Red beet extract (N° CEE: E162) is used mainly in dairy food products, confectionery, jams, sausages, pâté and cannery.

Among other interesting food products, the juices from cacti (O. ficus-indica, Myrtillocactus geometrizans, Hylocereus polyrhizus) are a readily available source of various betalain pigments (Stintzing et al., 2001). The fact that those plants are adapted to semi-arid regions adds a further value to this source of betalains in the perspective of sustainable development. It has been suggested that, through the application of the relatively simple technique of semi-synthesis, betaxanthins can be used as a means to introduce essential amino acids into food giving rise to 'essential dietary colourant' (Leathers et al., 1992).

In horticulture there is no betalain-containing plant in the top cultivated species; nevertheless, there is a large amateur market for Cactaceae, and various species of Bougainvillea, Mirabilis, Phytolacca, Portulaca, Mesembryanthemum, Amaranthus, Iresine (Celosia) and Beta are widely used as ornamentals.

6.8 Prospectives

Considering the actual state of the art, it is now possible to use recombinant DODAs (either from Amanita or from Portulaca) to boost the level of betalamic acid in order to increase pigment content either for ornamental or nutritional purposes. The limiting factor will probably be DOPA. The general availability of amino acids and other amines will potentially allow flux toward the synthesis of betaxanthins.

The limiting factor in this case would probably be the vacuolar transport system. For betacyanin synthesis there is a strong requirement for *cyclo*-DOPA (or *cyclo*-DOPA-glucoside). We should realise that unfortunately we are still far from a clear understanding of those specific steps (corresponding to the *R* and *I* gene loci in *P. grandiflora*); nevertheless transformation experiments can and should be done with the DODA-containing vectors, as we know already that their products are correctly expressed in competent cells. As long as DOPA is present, DODA could be used as coloured reporter molecule in transgenic plants (either as a transient marker or as an integrated one) and other organisms. It has the advantage of being non-toxic and detectable *in vivo* by simple spectroscopic means or microscopic multispectral analysis (Fernandez *et al.*, 1995). It could be used for the in situ quantitative analysis of DOPA, not only in plant cells but also in animal cells and specifically in the nervous system. DODA genes and DODA-like genes of the same family are now at the reach of interested investigators; studies on the characteristics of those genes and their evolutionary relationship will certainly help us in designing new strategies in the field of biotechnology of aromatic compounds.

Among other betalain-specific enzymes, GTs have been characterised enough to be potentially useful (Vogt *et al.*, 1999a). If a suitable PPO can be engineered to produce DOPA, the whole betalain pathway will be available for optimisation and expression in new plants.

To complete the picture of the present situation, we could add that *P. grandiflora* transformation is possible with some further optimisation (Rossi-Hassani *et al.*, 1995) and that transgenic beet has already been produced (Lindsey & Gallois, 1990; Dhalluin *et al.*, 1992; Jacq *et al.*, 1993) and tested in the field (Meier & Wackernagel, 2003). Large collections of ESTs are available for *B. vulgaris* (Bellin *et al.*, 2002) and *M. crystallinum* (Andolfatto *et al.*, 1994) for gene identification and isolation; in the case of *B. vulgaris*, the Gabi-beet programme *http://www.mips.biochem.mpg.de/proj/gabi/projects/gabibeet.htm* has put among other important genes to be mapped the *R* gene for red hypocotyl.

Acknowledgements

We would like to dedicate this review to all previous members of the Laboratory of Plant Cell Genetics of the University of Lausanne, whose dedication, hard work and enthusiasm have significantly contributed to our understanding of the betalain biosynthetic pathway. Some of their work has been published in peer-reviewed journals, some in PhD thesis, some in difficult-to-access congress reports; in any case, all pieces of this puzzle were needed to get a better picture of this fascinating biological enigma.

References

Adachi, T. (1972) Chemogenetic studies on flower color in the genus *Portulaca* in relation to breeding. *Bull Lab Plant Breed Miyazaki University, Japan*, **4**, 1–98.

Andolfatto, P., Bornhouser, A., Bohnert, H. J. & Thomas, J. C. (1994) Transformed hairy roots of *Mesembry-anthemum crystalinum*: gene expression patterns upon salt stress. *Physiol. Plant.*, **90**, 708–714.

Austin, D. & Goldman, I. L. (2001) Transmission ratio distortion due to the *bl* gene in table beet. *J. Am. Soc. Hort. Sci.*, **126**, 340–343.

Barth, H., Kobayashi, M. & Musso, H. (1979) Pigments of *Amanita muscaria* 5. Synthesis of muscaflavin. *Helv. Chim. Acta*, **62**, 1231–1235.

Bauer, J. A. (2001) *Etude des voies métaboliques conduisant à la bétanine dans des cultures de 'Beta vulgaris L.' par incorporation de tyrosine et de dihydroxyphenylalanine*. PhD Thesis. Université de Lausanne, Switzerland, p. 198.

Bellin, D., Werber, M., Theis, T., Schulz, B., Weisshaar, B. & Schneider, K. (2002) EST sequencing, annotation and macroarray transcriptome analysis identify preferentially root-expressed genes in sugarbeet. *Plant Biol.*, **4**, 700–710.

Biddington, N. & Thomas, T. H. (1973) A modified *Amaranthus betacyanin* bioassay for the rapid determination of cytokinins in plant extracts. *Planta*, **111**, 183–186.

Bigot, C. (1968) Action des adénines subsituées sur la synthèse des bétacyanines dans les plantules d' *Amarantus caudatus* L. Possibilité d'un test biologique de la distribution des cytokinines. *Comptes rendus hebdomadaires des séances de l'Académie des Sciences Série D*, **266**, 349–352.

Bokern, M., Heuer, S. & Strack, D. (1992) Hydroxycinnamic acid transferases in the biosynthesis of acylated betacyanins: purification and characterization from cell cultures of *Chenopodium rubrum* and occurrence in some other members of the Caryophyllales. *Bot. Acta*, **105**, 146–151.

Cai, Y. Z. & Corke, H. (2001) Effect of postharvest treatments on *Amaranthus* betacyanin degradation evaluated by visible/near-infrared spectroscopy. *J. Food Sci.*, **66**, 1112–1118.

Cai, Y., Sun, M. & Corke, H. (2001) Identification and distribution of simple and acylated betacyanins in the Amaranthaceae. *J. Agric. Food Chem.*, **49**, 1971–1978.

Christinet, L., Burdet, F., Zaiko, M., Hinz, U. & Zrÿd, J.-P. (2004) Characterization and functional identification of a novel plant 4,5-extradiol dioxygenase involved in betalain pigment biosynthesis in *Portulaca grandiflora*. *Plant Physiol.*, **134**, 265–274.

Clement, J. S. & Mabry, T. J. (1996) Pigment evolution in the Caryophyllales: a systematic overview. *Bot. Acta*, **109**, 360–367.

Cuenoud, P., Savolainen, V., Chatrou, L. W., Powell, M., Grayer, R. J. & Chase, M. W. (2002) Molecular phylogenetics of Caryophyllales based on nuclear 18S rDNA and plastid *rbcL*, *atpB* and *matK* DNA sequences. *Am. J. Bot.*, **89**, 132–144.

De Silva, W. I. (1996) Towards safe motherhood in Sri Lanka: knowledge, attitudes and practices during the period of maternity. *J. Fam. Welfare*, **41**, 18–26.

D'halluin, K., Bossut, M., Bonne, E., Mazur, B., Leemans, J. & Botterman, J. (1992) Transformation of sugarbeet (*Beta vulgaris L*) and evaluation of herbicide resistance in transgenic plants. *Bio-Technology*, **10**, 309–314.

Döpp, H. & Musso, H. (1973a) Die konstitution dess muscaflavins aus *Amanita muscaria* und über betalaminsäure. *Naturwissenschaften*, **60**, 477–478.

Döpp, H. & Musso, H. (1973b) Isolierung und chromophore der farbstoffe aus *Amanita muscaria*. *Chemische Berichte*, **106**, 3473–3482.

Dunkelblum, E., Miller, H. E. & Dreiding, A. S. (1972) On the mechanism of decarboxylation of betanidine. A contribution to the interpretation of the biosynthesis of betalaines. *Helv. Chim. Acta*, **55**, 642–648.

Eagen, K. A. & Goldman, I. L. (1996) Assessment of RAPD marker frequencies over cycles of recurrent selection for pigment concentration and percent solids in red beet (*Beta vulgaris* L.). *Mol. Breed.*, **2**, 107–115.

Ehrendorfer, F. (1976) Closing remarks: evolution and systematics of centrospermous families. *Plant. Syst. Evol.*, **126**, 99–105.

Eliade, M. (1968) *Le chamanisme et les techniques archaïques de l'extase*, Payot, Paris.

Endress, R., Jager, A. & Kreis, W. (1984) Catecholamine biosynthesis dependent on the dark in betacyanin-forming *Portulaca callus*. *J. Plant Physiol.*, **115**, 291–295.

Fernandez, G., Kunt, M. & Zrÿd, J.-P. (1995) Multi-spectral based cell segmentation and analysis, in *Workshop on Physics-Based Modeling in Computer Vision*, IEEE Computer Society Press, Cambridge, USA, pp. 166–172.

Fischer, N. & Dreiding, A. S. (1972) Biosynthesis of betalains. On the cleavage of the aromatic ring during the enzymatic transformation of DOPA into betalamic acid. *Helv. Chim. Acta*, **55**, 649–658.

Girod, P. A. & Zrÿd, J.-P. (1987) Clonal variability and light induction of betalain synthesis in red beet cell cultures. *Plant Cell Rep.*, **6**, 27–30.

Girod, P. A. & Zrÿd, J.-P. (1991a) Biogenesis of betalains: purification and partial characterization of DOPA 4,5-dioxygenase from *Amanita muscaria*. *Phytochemistry*, **30**, 169–174.

Girod, P. A. & Zrÿd, J.-P. (1991b) Secondary metabolism in cultured red beet (*Beta vulgaris* L.) cells: differential regulation of betaxanthin and betacyanin biosynthesis. *Plant Cell Tiss. Organ Cult.*, **25**, 1–12.

Giudici De Nicola, M., Amico, V., Sciuto, S. & Piattelli, M. (1975) Light control of amaranthin synthesis in isolated *Amaranthus* cotyledons. *Phytochemistry*, **14**, 479–481.

Goldman, I. L. & Austin, D. (2000) Linkage among the *R, Y* and *Bl* loci in table beet. *Theor. Appl. Genet.*, **100**, 337–343.

Guggenheim, M. (1913) Dioxyphenylalanine, eine neue Aminosaure aus *Vicia faba. Z. Physiol. Chem.*, **88**, 276–284.

Heuer, S., Wray, V., Metzger, J. W. & Strack, D. (1992) Betacyanins from flowers of *Gomphrena globosa*. *Phytochemistry*, **31**, 1801–1807.

Heuer, S., Richter, S., Metzger, J. W., Wray, V., Nimtz, M. & Strack, D. (1994) Betacyanins from bracts of *Bougainvillea glabra. Phytochemistry*, **37**, 761–767.

Heuer, S., Vogt, T., Bohm, H. & Strack, D. (1996) Partial purification and characterization of UDP-glucose: betanidin 5-*O*- and 6-*O*-glucosyltransferases from cell suspension cultures of *Dorotheanthus bellidiformis* (Burm. f.) N.E.Br. *Planta*, **199**, 244–250.

Hinz, U. G., Fivaz, J., Girod, P. A. & Zrÿd, J.-P. (1997) The gene coding for the DOPA dioxygenase involved in betalain biosynthesis in *Amanita muscaria* and its regulation. *Mol. Gen. Genet.*, **256**, 1–6.

Hooker, W. J. (1829) *Portulaca grandiflora*. Large flowered purslane. *Curti's Botanical Magazine*, Vol. III, 2885.

Ibdah, M., Krins, A., Seidlitz, H. K., Heller, W., Strack, D. & Vogt, T. (2002) Spectral dependence of flavonol and betacyanin accumulation in *Mesembryanthemum crystallinum* under enhanced ultraviolet radiation. *Plant Cell Environ.*, **25**, 1145–1154.

Ikeno, S. (1921) Studies on the genetics of flower-colours in *Portulaca grandiflora. Journal of the College of Agriculture, Imperial University Tokyo*, **8**, 93–133.

Impellizzeri, G., Piattelli, M. & Sciuto, S. (1973) A new betaxanthin from *Glottiphyllum longum. Phytochemistry*, **12**, 2293–2294.

Jacq, B., Lesobre, O., Sangwan, R. S. & Sangwannorreel, B. (1993) Factors influencing T-DNA transfer in *Agrobacterium*-mediated transformation of sugarbeet. *Plant Cell Rep.*, **12**, 621–624.

Joy, R. W., Sugiyama, M., Fukuda, H. & Komamine, A. (1995) Cloning and characterization of polyphenol oxidase cDNAs of *Phytolacca americana. Plant Physiol.*, **107**, 1083–1089.

Kapadia, G. J., Tokuda, H., Konoshima, T. & Nishino, H. (1996) Chemoprevention of lung and skin cancer by *Beta vulgaris* (beet) root extract. *Cancer Lett.*, **100**, 211–214.

Kapadia, G. J., Tokuda, H., Sridhar, R., Balasubramanian, V., Takayasu, J., Bu, P., Enjo, F., Takasaki, M., Konoshima, T. & Nishino, H. (1998) Cancer chemopreventive activity of synthetic colorants used in foods, pharmaceuticals and cosmetic preparations. *Cancer Lett.*, **129**, 87–95.

Kimler, L., Larson, R. A., Messenge, L., Moore, J. B. & Mabry, T. J. (1971) Betalamic acid, a new naturally occurring pigment *J. Chem. Soc. D- Chem. Comm.*, 1329–1330.

Kishima, Y., Shimaya, A. & Adachi, T. (1995) Evidence that blue light induces betalain pigmentation in *Portulaca* callus. *Plant Cell Tiss. Organ Cult.*, **43**, 67–70.

Kobayashi, N., Schmidt, J., Nimtz, M., Wray, V. & Schliemann, W. (2000) Betalains from Christmas cactus. *Phytochemistry*, **54**, 419–426.

Kobayashi, N., Schmidt, J., Wray, V. & Schliemann, W. (2001) Formation and occurrence of dopamine-derived betacyanins. *Phytochemistry*, **56**, 429–436.

Kubota, S., Imamura, H., Hisamatsu, T. & Koshioka, M. (1999) An enhanced *Amaranthus* betacyanin bioassay for detection of cytokinins. *J. Plant Physiol.*, **155**, 133–135.

Leathers, R. R., Davin, C. & Zrÿd, J.-P. (1992) Betalain producing cell cultures of *Beta vulgaris* L. var. bikores monogerm (red beet). *In Vitro Cell. Dev. Biol.- Plant*, 28P, 39–45.

Liebisch, H. W. & Bohm, H. (1981) Studies on the physiology of betalain formation in cell-cultures from *Portulaca grandiflora*. *Pharmazie*, **36**, 218.

Lindsey, K. & Gallois, P. (1990) Transformation of sugarbeet (*Beta vulgaris*) by *Agrobacterium tumefaciens*. *J. Exp. Bot.*, **41**, 529–536.

Mabry, T. J. (2001) Selected topics from forty years of natural products research: betalains to flavonoids, antiviral proteins, and neurotoxic nonprotein amino acids. *J. Nat. Prod.*, **64**, 1596–1604.

Martinez-Parra, J. & Munoz, R. (2001) Characterization of betacyanin oxidation catalyzed by a peroxidase from *Beta vulgaris* L. roots. *J. Agric. Food Chem.*, **49**, 4064–4068.

Meier, P. & Wackernagel, W. (2003) Monitoring the spread of recombinant DNA from field plots with transgenic sugar beet plants by PCR and natural transformation of *Pseudomonas stutzeri*. *Trans. Res.*, **12**, 293–304.

Miller, H. E., Rosler, H., Wohlpart, A., Wyler, H., Wilcox, M. E., Frohofer, H., Mabry, T. J. & Dreiding, A. S. (1968) Biogenesis of betalain. Biotransformation of dopa and tyrosine in betalaminic acid part of the betanins. *Helv. Chim. Acta*, **51**, 1470–1474.

Minale, L., Piattelli, M., De Stefano, S., 1967. Pigments of centrospermae – VII. Betacyanins from *Gomphrena globosa* L. *Phytochemistry*, **6**, 703–709.

Mueller, L. A., Hinz, U. & Zrÿd, J.-P. (1996) Characterization of a tyrosinase from *Amanita muscaria* involved in betalain biosynthesis. *Phytochemistry*, **42**, 1511–1515.

Mueller, L. A., Hinz, U., Uze, M., Sautter, C. & Zrÿd, J.-P. (1997a) Biochemical complementation of the betalain biosynthetic pathway in *Portulaca grandiflora* by a fungal 3,4-dihydroxyphenylalanine dioxygenase. *Planta*, **203**, 260–263.

Mueller, L. A., Hinz, U. & Zrÿd, J.-P. (1997b) The formation of betalamic acid and muscaflavin by recombinant DOPA-dioxygenase from *Amanita*. *Phytochemistry*, **44**, 567–569.

Nakayama, T., Yonekura-Sakakibara, K., Sato, T., Kikuchi, S., Fukui, Y., Fukuchi-Mizutani, M., Ueda, T., Nakao, M., Tanaka, Y., Kusumi, T. & Nishino, T. (2000) Aureusidin synthase: a polyphenol oxidase homolog responsible for flower coloration. *Science*, **290**, 1163–1166.

Piattelli, M. & Minale, L. (1964) Pigments of centrospermae – I: betacyanins from *Phyllocactus hybridus* Hort. and *Opuntia ficus-indica* Mill.. *Phytochemistry*, **3**, 307–311.

Piattelli, M., Minale, L. & Prota, G. (1964) Isolation, structure and absolute configuration of indicaxanthin. *Tetrahedron*, **20**, 2325–2329.

Piattelli, M., Minale, L. & Nicolaus, R. A. (1965a) Pigments of centrospermae – V : betaxanthins from *Mirabilis jalapa* L.. *Phytochemistry*, **4**, 817–823.

Piattelli, M., Minale, L. & Nicolaus, R. A. (1965b) Ulteriori ricerche sulle betaxantine. *Rend. Accad. Sci. Fis. Mat. Napoli*, **32**, 55–56.

Piattelli, M., Minale, L. & Prota, G. (1965c) Pigments of centrospermae – III: betaxanthins from *Beta vulgaris* L. *Phytochemistry*, **4**, 121–125.

Piattelli, M., Giudici De Nicola, M. & Castrogiovanni, V. (1969) Photocontrol of amaranthin synthesis in *Amaranthus tricolor*. *Phytochemistry*, **8**, 731–736.

Piattelli, M. & Imperato, F. (1970) Pigments of centrospermae – 13: pigments of *Bougainvillea glabra*. *Phytochemistry*, **9**, 2557–2560.

Rast, D., Skrivanova, R. & Wohlpart, A. (1972) Betalain biosynthesis in Cetrospermae seedlings. The action of light on betacyanin formation. *Ber. Schweiz. Bot. Ges.*, **82**, 213–222.

Romanov, G. A., Getman, I. A. & Schmulling, T. (2000) Investigation of early cytokinin effects in a rapid *Amaranthus* seedling test. *Plant Growth Reg.*, **32**, 337–344.

Rossi-Hassani, B. D. & Zrÿd, J.-P. (1994) Genetic instability in *Portulaca grandiflora* (Hook). *Annales Génétique*, **37**, 53–59.

Rossi-Hassani, B. D. & Zrÿd, J.-P. (1995) Demonstration of transposition in *Portulaca grandiflora* (Hook). *Annales Génétique*, **38**, 90–96.

Rossi-Hassani, B. D., Bennani, F. & Zrÿd, J.-P. (1995) *Agrobacterium*-mediated transformation of large-flowered purslane (*Portulaca grandiflora* H). *Genome*, **38**, 752–756.

Saito, K. & Komamine, A. (1978) Biosynthesis of stizolobinic acid and stizolobic acid in higher plants. *Eur. J. Biochem.*, **82**, 385–392.

Sakuta, M., Hirano, H., Kakegawa, K., Suda, J., Hirose, M., Joy, R. W., Sugiyama, M. & Komamine, A. (1994) Regulatory mechanisms of biosynthesis of betacyanin and anthocyanin in relation to cell division activity in suspension cultures. *Plant Cell Tiss. Organ Cult.*, **38**, 167–169.

Sasaki, N., Adachi, T., Koda, T. & Ozeki, Y. (2004) Detection of UDP-glucose: cyclo-DOPA 5-*O*-glucosyltransferase activity in four o'clocks (*Mirabilis jalapa* L.). *FEBS Letters*, **568**, 159–162.

Sato, T., Nakayama, T., Kikuchi, S., Fukui, Y., Yonekura-Sakakibara, K., Ueda, T., Nishino, T., Tanaka, Y. & Kusumi, T. (2001) Enzymatic formation of aurones in the extracts of yellow snapdragon flowers. *Plant Sci.*, **160**, 229–236.

Schliemann, W., Joy, R. W., Komamine, A., Metzger, J. W., Nimtz, M., Wray, V. & Strack, D. (1996) Betacyanins from plants and cell cultures of *Phytolacca americana*. *Phytochemistry*, **42**, 1039–1046.

Schliemann, W., Steiner, U. & Strack, D. (1998) Betanidin formation from dihydroxyphenylalanine in a model system. *Phytochemistry*, **49**, 1593–1598.

Schliemann, W., Kobayashi, N. & Strack, D. (1999) The decisive step in betaxanthin biosynthesis is a spontaneous reaction. *Plant Physiol.*, **119**, 1217–1232.

Schliemann, W., Cai, Y., Degenkolb, T., Schmidt, J. & Corke, H. (2001) Betalains of *Celosia argentea*. *Phytochemistry*, **58**, 159–165.

Schudel, G. (1919) *Über Alkalischmelzen*. Thesis. Eidgenössische Technische Hochschule Zürich, p. 48.

Schwartz, S. J. & von Elbe, J.H. (1983) Identification of betanin degradation products. *Z. Lebensm. Unters. Forsch.*, **176**, 448–453.

Sciuto, S., Oriente, G., Piattelli, M., Impellizzeri, G. & Amico, V. (1974) Pigments of centrospermae – 19: biosynthesis of amaranthin in *Celosia plumosa*. *Phytochemistry*, **13**, 947–951.

Sepúlveda Jiménez, G., Rueda Benítez, P., Porta, H. & Rocha Sosa, M. (2003) Wounding and infiltration with *Pseudomonas syringae* pv. tabaci or *Agrobacterium tumefaciens* induced the synthesis of betacyanins in red beet (Beta vulgaris var. Crosby Egyptian) leaves. Participation of oxygen reactive species, in *7th International Congress of Plant Molecular Biology Book of Abstracts*, June 23–28, Barcelona, Spain.

Shin, K. S., Murthy, H. N., Ko, J. Y. & Paek, K. Y. (2002) Growth and betacyanin production by hairy roots of *Beta vulgaris* in airlift bioreactors. *Biotech. Lett.*, **24**, 2067–2069.

Soboleva, G. A., Ulyianova, M. S., Zakharova, N. S. & Bokuchava, M. A. (1976) Study of betacyanin-discoloring enzyme. *Biokhimiia*, **41**, 968–974.

Soetiart, S. & Ball, E. (1968) Origin of perigyny in *Portulaca grandiflora*. *Am. J. Bot.*, **55**, 715.

Soetiart, S. & Ball, E. (1969) Ontogenetical and experimental studies of floral apex of *Portulaca grandiflora* 1. Histology of transformation of shoot apex into floral apex. *Can. J. Bot.*, **47**, 133–140.

Steddom, K., Heidel, G., Jones, D. & Rush, C. M. (2003) Remote detection of rhizomania in sugar beets. *Phytopathology*, **93**, 720–726.

Steiner, U., Schliemann, W. & Strack, D. (1996) Assay for tyrosine hydroxylation activity of tyrosinase from betalain-forming plants and cell cultures. *Anal. Biochem.*, **238**,72–75.

Steiner, U., Schliemann, W., Bohm, H. & Strack, D. (1999) Tyrosinase involved in betalain biosynthesis of higher plants. *Planta*, **208**, 114–124.

Stintzing, F. C., Schieber, A. & Carle, R. (2001) Phytochemical and nutritional significance of cactus pear. *Eur. Food Res. Tech.*, **212**, 396–407.

Stintzing, F. C., Schieber, A. & Carle, R. (2002) Identification of betalains from yellow beet (*Beta vulgaris* L.) and cactus pear [*Opuntia ficus-indica* (L.) Mill.] by high-performance liquid chromatography-electrospray ionization mass spectrometry. *J. Agric. Food Chem.*, **50**, 2302–2307.

Stobart, A. K. & Kinsman, L. T. (1977) The hormonal control of betacyanin synthesis in *Amaranthus caudatus*. *Phytochemistry*, **16**, 1137–1142.

Strack, D., Schmitt, D., Reznik, H., Boland, W., Grotjahn, L. & Wray, V. (1987) Humilixanthin a new betaxanthin from *Rivina humilis*. *Phytochemistry*, **26**, 2285–2287.

Strack, D., Vogt, T. & Schliemann, W. (2003) Recent advances in betalain research. *Phytochemistry*, **62**, 247–269.

Sugimoto, K., Senda, T., Aoshima, H., Masai, E., Fukuda, M. & Mitsui, Y. (1999) Crystal structure of an aromatic ring opening dioxygenase LigAB, a protocatechuate 4,5-dioxygenase, under aerobic conditions. *Struc. Fold. Des.*, **7**, 953–965.

Suzuki, H., Nakayama, T., Yonekura-Sakakibara, K., Fukui, Y., Nakamura, N., Yamaguchi, M., Tanaka, Y., Kusumi, T. & Nishino, T. (2002) cDNA cloning, heterologous expressions, and functional characterization of malonyl-coenzyme A: anthocyanidin 3-*O*-glucoside-6''-*O*-malonyltransferase from dahlia flowers. *Plant Physiol.*, **130**, 2142–2151.

Terradas, F. (1989) *Etude par voie enzymatique des métabolites des bétalaines dans Beta vulgaris et Amanita muscaria*. Thesis. Université de Lausanne, Switzerland, p. 147.

Terradas, F. & Wyler, H. (1991a) 2,3- and 4,5-Secodopa, the biosynthetic intermediates generated from L-DOPA by an enzyme system extracted from the fly agaric, *Amanita muscaria* L., and their spontaneous conversion to muscaflavin and betalamic acid, respectively, and betalains. *Helv. Chim. Acta*, **74**, 124–140.

Terradas, F. & Wyler, H. (1991b) The secodopas, natural pigments in *Hygrocybe conica* and *Amanita muscaria*. *Phytochemistry*, **30**, 3251–3253.

Trezzini, G. F. (1990) *Génétique des bétalaïnes chez Portulaca grandiflora Hook*. Thesis. Université de Lausanne, Switzerland, p. 135.

Trezzini, G. F. & Zrÿd, J.-P. (1990) *Portulaca grandiflora*: a model system for the study of the biochemistry and genetics of betalain synthesis. *Acta Hort.*, **280**, 581–585.

Trezzini, G. F. & Zrÿd, J.-P. (1991a) Two betalains from *Portulaca grandiflora*. *Phytochemistry*, **30**, 1897–1899.

Trezzini, G. F. & Zrÿd, J.-P. (1991b) Characterization of some natural and semi-synthetic betaxanthins. *Phytochemistry*, **30**, 1901–1904.

Vogt, T., Zimmermann, E., Grimm, R., Meyer, M. & Strack, D. (1997) Are the characteristics of betanidin glucosyltransferases from cell-suspension cultures of *Dorotheanthus bellidiformis* indicative of their phylogenetic relationship with flavonoid glucosyltransferases? *Planta*, **203**, 349–361.

Vogt, T., Grimm, R. & Strack, D. (1999a) Cloning and expression of a cDNA encoding betanidin 5-O-glucosyltransferase, a betanidin- and flavonoid-specific enzyme with high homology to inducible glucosyltransferases from the Solanaceae. *Plant J.*, **19**, 509–519.

Vogt, T., Ibdah, M., Schmidt, J., Wray, V., Nimtz, M. & Strack, D. (1999b) Light-induced betacyanin and flavonol accumulation in bladder cells of *Mesembryanthemum crystallinum*. *Phytochemistry*, **52**, 583–592.

von Ardenne, R., Döpp, H., Musso, H. & Steiglich, W. (1974) Über das vorkommen von muscaflavin bei Hygrocyben (Agaricales) und seine dihydroazepin-struktur. *Z. Naturforsch.*, **29c**, 637–639.

Wettasinghe, M., Bolling, B., Plhak, L., Xiao, H. & Parkin, K. (2002) Phase II enzyme-inducing and antioxidant activities of beetroot (*Beta vulgaris* L.) extracts from phenotypes of different pigmentation. *J. Agric. Food Chem.*, **50b** 6704–6709.

Wilcox, M. E., Wyler, H. & Dreiding, A. S. (1965) Stereochemie von betanidin und isobetanidin 8. Zur konstitution des randenfarbstoffes betanin. *Helv. Chim. Acta*, **48**, 1134–47.

Wohlpart, A. & Mabry, T. J. (1966) Biogenesis of betacyanins. *Plant Physiol.*, R72.

Wohlpart, A. & Mabry, T. J. (1968) On light requirement for betalain biogenesis. *Plant Physiol.*, **43**, 457–459.

Wolyn, D. J. & Gabelman, W. H. (1989) Inheritance of root and petiole pigmentation in red table beet. *J. Hered.*, **80**, 33–38.

Wybraniec, S., Platzner, I., Geresh, S., Gottlieb, H. E., Haimberg, M., Mogilnitzki, M. & Mizrahi, Y. (2001) Betacyanins from vine cactus *Hylocereus polyrhizus*. *Phytochemistry*, **58**, 1209–1212.

Wyler, H. & Dreiding, A. S. (1961) On betacyanins, the nitrogen containing pigment of centrospermae. Preliminary report. *Experientia*, **17**, 23–25.

Wyler, H., Mabry, T. J. & Dreiding, A. S. (1963) Zur struktur des betanidins 6. Über die konstitution des randenfarbstoffes Betanin. *Helv. Chim. Acta*, **46**, 1745–1748.

Yamamoto, K., Kobayashi, N., Yoshitama, K., Teramoto, S. & Komamine, A. (2001) Isolation and purification of tyrosine hydroxylase from callus cultures of *Portulaca grandiflora*. *Plant Cell Physiol.*, **42**, 969–975.

Yang, X. H., Zhang, X. T. & Zhou, R. H. (2001) Determination of L-DOPA content and other significant nitrogenous compounds in the seeds of seven *Mucuna* and *Stizolobium* species in China. *Pharma. Biol.*, **39**, 312–316.

Yashi, K. (1920) Genetical studies in *Portulaca grandiflora. Bot. Mag. Tokyo*, **34**, 55–65.

Zaiko, M. (2000) *Colour-specific genes from betalain producing plants*. PhD Thesis. Université de Lausanne, Switzerland, p. 129.

Zakharova, N. S. & Petrova, T. A. (2000) Beta-glucosidases from leaves and roots of the common beet, *Beta vulgaris*. Appl. *App. Biochem. Microbiol.*, **36** 394–397.

Zrÿd, J.-P., Bauer, J., Wyler, H. & Lavanchy, P. (1982) Pigment biosynthesis and precursor metabolism in red beet semi-continuous cell suspension cultures, in *Proceedings of the 5th International Congress of Plant Tissue & Cell Culture* (ed. A. Fujiwara), Tokyo, Japan, pp. 387–388.

7 Important rare plant pigments

Kevin M. Davies

7.1 Introduction

Most reviews on plant pigments concentrate on the major groups – chlorophylls, betalains, carotenoids and flavonoids. Of course there is much justification for this, as these pigments are of almost universal distribution in the plant kingdom and of key importance to many aspects of plant survival. However, there are a large number of other plant pigments that are also of interest. In particular, several of the textile dyes, cosmetics, tattooing agents and food colourants used throughout history have their origin outside the major pigment groups. Most of these pigments are of rare or sporadic occurrence, compared to the major pigment groups; others may be widespread, but found in plant tissues less studied for pigmentation, such as roots and bark. Most of these 'unusual' biosynthetic pathways are poorly characterised but, for a few, the biosynthetic enzymes and genes have been determined and genetic modification approaches taken to alter pigment production. Many of these rare pigments have also been the subject of studies for their health-promoting properties, in a similar manner to the more extensive research carried out on flavonoids and carotenoids. Some of these 'rare' plant pigments are collected together for review in this chapter, with an emphasis on those of particular commercial significance or for which there is good knowledge of the biosynthetic pathways.

Some compounds used extensively as pigments for textile dyes or cosmetics by humans for many centuries are weakly pigmented or not pigmented in the intact plant tissues – most notably indigo and weld (Gilbert & Cooke, 2001). Indigo is a blue-coloured dimer formed from colourless precursors under oxidative conditions, such as those in wounded plant tissues. Weld is a strong yellow dye commonly produced from *Reseda* species, such as *R. luteola* (weld) and *R. lutea* (dyer's greenweed or dyer's mignonette), and is based on luteolin-derived flavonoids (Cerrato *et al.*, 2002; Cristea *et al.*, 2003). Other yellow dyes may be produced from closely related flavonoids, such as quercetin extracted from the inner bark of *Quercus tinctoria* trees (a type of oak) (Gilbert & Cooke, 2001). Weld is not covered further in this review, as the compounds in intact plants are generally weakly coloured, and also the biosynthesis of the flavonoids falls outside of the scope of this chapter. However, indigo is featured, as it has been the target of genetic modification strategies to enable production in intact plant tissues.

7.2   Quinones

The quinones are compounds with either a 1,4-diketocyclohexa-2,5-denoid or a
1,2-diketocyclohexa-3,5-denoid moiety (in essence, a phenolic-derived ring modified with carbonyl groups). The former types are *p*-quinones and the latter
o-quinones (Fig. 7.1). The pigments discussed in this review are all *p*-quinones.
Under some conditions quinones will be in a state of equilibrium with the
corresponding hydroquinone, and such compounds may be important biosynthetic
intermediates. There are widespread examples of pigmented quinones in plants, in
particular in the Rubiaceae. They may occur with one, two or three rings (benzo-,
naphtho- and anthraquinones, respectively; Fig. 7.1), or as larger polycyclic
quinones (Thomson, 1976a; Leistner, 1980; 1981). A large variety of quinone
structures are known, formed by secondary modifications such as hydroxylation
and glycosylation. Furthermore, dimeric or high multimer forms of quinones can
occur in plants (Leistner, 1980).

A striking feature of quinones is that they can be derived from several biosynthetic routes, including a number of alternative pathways in plants. Leistner (1980)
identified at least six possible biosynthetic routes to benzoquinones (BQs). For
naphthoquinones (NQs), two main routes have been characterised, involving
contribution of a phenolic ring (chorismate) from the shikimate pathway and
formation of the quinone ring with a mechanism involving either α-ketoglutarate

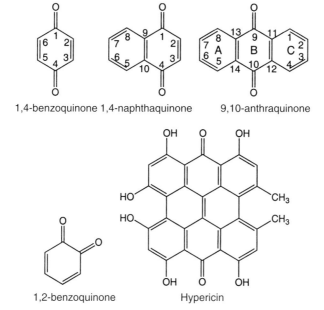

1,4-benzoquinone 1,4-naphthaquinone   9,10-anthraquinone

1,2-benzoquinone   Hypericin

Fig. 7.1 The basis of benzo-, naphtho- and anthraquinones with regard to the ring structure and carbon
numbering, and an example of a highly polycyclic quinone, hypericin, found in *Hypericum* species.

or isopentenyl diphosphate (IPP) derivatives. At least two biosynthetic routes have also been suggested for the anthraquinones (AQs), either starting from a similar pathway to that for NQs (from chorismate/IPP) or arising from acetyl-CoA and malonyl-CoA in the polyketide pathway. The enzymology for NQs has been well defined, for at least one biosynthetic route, and there is increasing information on the enzymes involved in biosynthesis of AQs.

Although they occur commonly in plants, compared to the major pigment groups, quinones make only a minor contribution to general plant pigmentation. One reason for this is that many forms are weakly pigmented or colourless, but another reason for their less notable presence is that they are predominantly in not readily observable plant parts, such as the roots and wood (Thomson, 1976a). Most quinone pigments used as dyes or colourants are of fungal origin. There are, however, some notable examples of plant quinone pigments with long histories of human use, including madder (comprising AQs), shikonin, alkannin and henna (all NQs). Some of these have been the traditional mainstay of textile dying, with long histories of the cultivation of the source species solely for extraction of the pigment. For example, the AQ pigments found in madder were used for dying mummy wraps (Smith, 1999) and have been identified in the fourth and twelfth century AD Egyptian textiles (Orska-Gawrys *et al.*, 2003) and the AQs extracted from *Morinda citrifolia* (Noni) may have been used by Polynesian people for dying Tapa cloth for over 2000 years, as the species was carried with them on their Pacific migrations (Wang *et al.*, 2002).

The best-characterised plant quinone pigments belong to the NQ and AQ types, and these are discussed in more detail in the following sections. The simpler BQs are generally colourless, except for some highly substituted examples. The highly polymeric quinones are often coloured, and can form dark brown or black pigments that are melanin-like in appearance. Of particular note are hypericin and pseudohypericin that colour the red and black glands of *Hypericum* species (Piovan *et al.*, 2004). There are also reports of various yellow and red pigments sharing features with the well-defined BQs, NQs and AQs but which are probably of unrelated origin (Thomson, 1976b; Eugster, 1980), including the quinone-methides that are discussed in a later section.

7.2.1 Naphthoquinones

Two NQ plant pigments are of particular note – shikonin, and related structures, and lawsone. Lawsone is a 2-hydroxy-1,4-NQ that is the main pigment in henna (Fig. 7.2). Henna is traditionally prepared from the leaves of *Lawsonia inermis*, but is also present in other members of the Balsaminaceae, in particular the aerial parts of *Impatiens balsamina*. There is a long tradition of henna use for staining skin and hair by people in India, the Middle East and North Africa. Little is known on the biosynthetic pathway for lawsone, but radiolabelling feeding experiments have identified potential precursors (Thomson, 1976a; Leistner, 1981; De-Eknamkul, 1998). It is thought that the benzenoid ring is derived from the

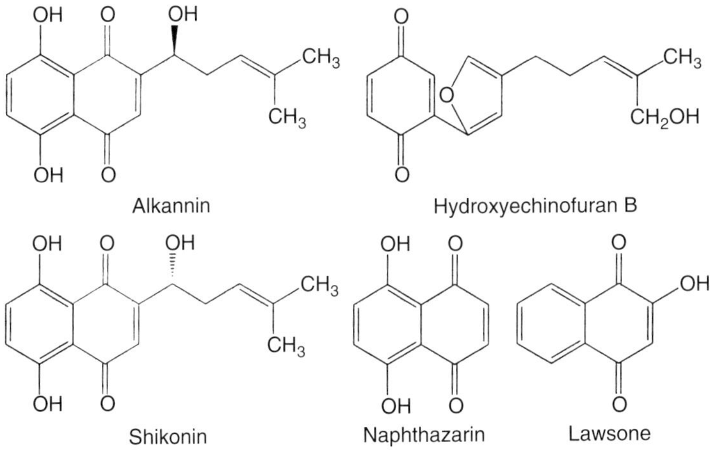

Fig. 7.2 The structures of the naphthoquinone plant pigments alkannin, shikonin, lawsone and naphthazarin, and the benzoquinone, hydroxyechinofuran B.

shikimate pathway, potentially from chorismate, and that formation of the quinone ring proceeds via addition of α-ketoglutarate to form an *O*-succinylbenzoic acid (OSB) intermediate (Fig. 7.3). This is similar to the proposed formation of intermediates in the biosynthesis of AQs in the Rubiaceae (Han *et al.*, 2002).

Much more is known on the biosynthesis of shikonin, which is derived by a distinct biosynthetic route from that proposed for lawsone. Shikonin (*S*-enantiomer) and alkannin (*R*-enantiomer) are two examples of a large range of red NQ pigments produced in the roots of various Boraginaceae species (Fig. 7.2). They have long, parallel histories of use as textile dyes: shikonin extracted from *Lithospermum erythrorhizon* in China and alkannin extracted from *Alkanna tinctoria* in Europe, with recorded use dating back to the fourth century BC (Papageorgiou *et al.*, 1999). Today they are still used as textile dyes, and additionally as food colourants, cosmetics and for medical purposes (Papageorgiou *et al.*, 1999). The high value of shikonin, and the ability to produce it at up to 4 g/l in culture, led to it being the first plant secondary metabolite to be commercially produced by plant cell culture (Papageorgiou *et al.*, 1999).

Shikonin and alkannin are related chemically to a widely studied purple dye, naphthazarin (Fig. 7.2), which occurs naturally in the bark or seed husks of some plant species and can also be easily synthesised chemically (Papageorgiou *et al.*, 1999). In addition to the NQs, some Boraginaceae species produce related pigments in their roots, which may share a common biosynthetic pathway. For example, *Agrobacterium rhizogenes* transgenic cultures of *L. erthrorhizon* have been produced that synthesise a brown BQ derivative (hydroxyechinofuran B; Fig. 7.2) (Fukui *et al.*, 1998).

The biosynthesis of shikonin has been extensively studied, allowing manipulation of shikonin production by altering cell culture conditions and through genetic

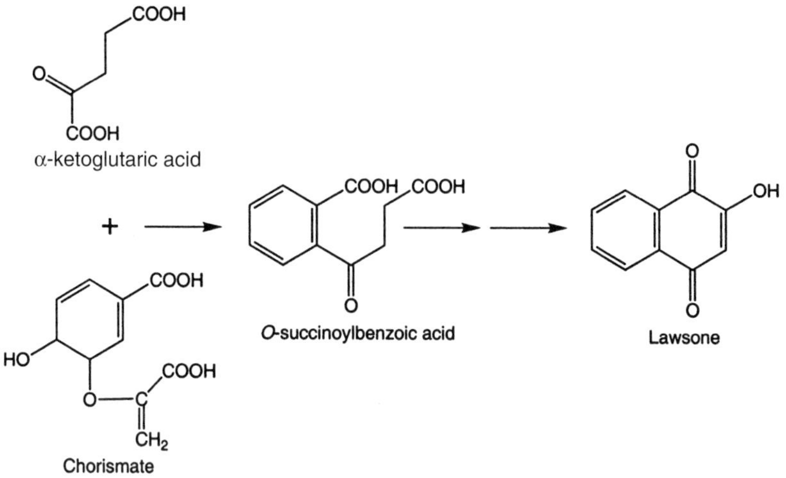

Fig. 7.3 A currently proposed biosynthetic pathway for the naphthoquinone lawsone.

modification in transgenic plants and cell lines. The currently proposed biosynthetic pathway for shikonin is presented in Fig. 7.4. Shikonin is derived from two precursors, 4-hydroxybenzoate (4HB), produced by the phenylpropanoid pathway, and geranyl diphosphate (GPP), probably produced by the cytosolic mevalonic acid (MVA) isoprenoid pathway. 4HB is derived from a series of reactions from shikimate, ending with the well-characterised enzymatic reactions of phenylalanine ammonia lyase (PAL), cinnamate 4-hydroxylase (C4H) and 4-coumaroyl CoA-ligase (4CL), and a partially characterised cleavage to 4HB and acetyl-CoA (Löscher & Heide, 1994). A bacterial enzyme that can convert 4-coumaroyl CoA to a compound closely related to 4-HB, 4-hydroxybenzaldehyde, has been characterised, including expression in transgenic plants (Mayer et al., 2001), but there are no published reports on plant genes for the cleavage enzyme. In non-shikonin producing cells, 4HB has been shown to accumulate as a glucoside in the vacuole, perhaps as a storage form. Three *Arabidopsis thaliana* (arabidopsis) cDNAs have been shown to encode 4HB-glycosyltransferase activities (Lim et al., 2002), but it is not clear how these may relate to the endogenous *L. erythrorhizon* activity identified by Li et al. (1997a).

Transcript abundance for PAL (Yazaki et al., 1997), C4H (Yamamura et al., 2001) and 4CL (Yazaki et al., 1995) has been measured in relation to shikonin formation in *L. erythrorhizon*, using the homologous cDNA clones. PAL, C4H and 4CL transcript abundance was constitutive during increases in shikonin production, and showed no difference between shikonin-producing and -non-producing cell lines, even though PAL transcript levels and enzyme activity showed variation with regard to other aspects of the cell culturing period (Sommer et al., 1999). Thus, the formation of the immediate 4HB precursors does not appear to be a key regulatory point in shikonin biosynthesis in *L. erythrorhizon*.

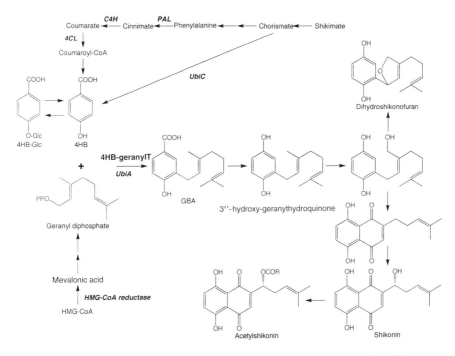

Fig. 7.4 A currently proposed biosynthetic pathway for shikonin. Compound and enzyme abbreviations are as given in the text, and included are points of action of the two bacterial enzymes used to modify shikonin production in transgenic plants (UBIA and UBIC).

Several well-defined enzymes are involved in the conversion of 3-hydroxy-3-methyl glutaryl-coenzyme A (HMG-CoA) to GPP via MVA and IPP (Li *et al.*, 1998). The activity of one of these, HMG-CoA reductase (HMGR), that converts HMG-CoA to MVA has been studied in *L. erythrorhizon* (Lange *et al.*, 1998). HMGR transcript levels and enzyme activity closely correlated with shikonin production in cell cultures. This and other evidence suggest that GPP for NQ formation is produced from IPP sourced from the cytosolic MVA pathway. However, a second route to IPP is active in plant plastids starting with pyruvate and glyceraldehyde, the methylerythritol 4-phosphate (MEP) pathway (see Lichtenthaler, 1999; Cunningham & Gantt, 2002 for reviews; and Chapter 3), and this can supply IPP for formation of some quinones (Li *et al.*, 1997b; Han *et al.*, 2002).

GPP and 4HB are brought together by the enzyme GPP: 4HB 3-geranyltransferase, forming geranyl-hydroxybenzoate (GBA). Two cDNA clones encoding this enzyme have been isolated from *L. erythrorhizon*, and their encoded protein activities studied by expression in yeast (Yazaki *et al.*, 2002). They show strict substrate specificity for GPP as the prenyl donor, and their transcript abundance follows shikonin production closely in cell cultures. Subsequent biosynthetic steps from GBA to shikonin have been shown biochemically to involve decarboxylation, oxidation and hydroxylation, then cyclisation to yield the naphthalene ring system of shikonin, although the specific enzymatic reactions are not well characterised. However, a cytochrome P450 hydroxylase has been characterised, geranyl hydroquinone $3''$-hydroxylase, that may carry out the final hydroxylation prior to cyclisation (Yamamoto *et al.*, 2000). Furthermore, two cDNA clones, *LeDI-2* (Yazaki *et al.*, 2001) and *LEPS-2* (Yamamura *et al.*, 2003), have been isolated that have spatial and temporal expression patterns coincident with shikonin accumulation. The role of *LeDI-2* in shikonin biosynthesis is further supported by the reduction in shikonin levels in *L. erythrorhizon* roots expressing a *LeDI-2* antisense RNA construct (Yazaki *et al.*, 2001). However, neither cDNA appears to encode one of the known biosynthetic activities. *LeDI-2* encodes a small, highly hydrophobic polypeptide that may be membrane-associated, and *LEPS-2* encodes an apoplastic cell wall protein. It is possible that the proteins have a role in the subcellular localisation of shikonin biosynthesis. The biosynthetic steps from the formation of GBA onwards occur in membrane vesicles derived from the endoplasmic reticulum, with a final secretion of shikonin into the cell wall after fusion of the vesicles with the plasma membrane (Tsukata & Tabata, 1984).

Shikonin production has been the target of a number of genetic modification projects, in addition to the inhibition of *LeDI-2* activity. In particular, two *Escherichia coli* genes under the control of strong plant gene promoters have been introduced into transgenic hairy root cultures of *L. erythrorhizon* (using *A. rhizogenes*) and transgenic tobacco plants (using *A. tumefaciens*). The *E. coli* genes are *ubiA*, which encodes 4HB-3-poly-prenyltransferase, and *ubiC*, which encodes chorismate pyruvate lyase. UBIA forms GBA from 4HB and GPP, while UBIC shortcuts the normal route for formation of 4HB in plants by converting chorismate directly to 4HB (Fig. 7.4). The *ubiA* product was targeted to the endoplasmic reticulum and the *ubiC* product to the chloroplast.

Introduction of the *ubiA* construct increased levels of GBA in transgenic lines compared to control lines by factors up to $50\times$ (Boehm *et al.*, 2000). However, the *ubiA* lines had only small increases in shikonin production, and no correlation was found between UBIA enzyme activity and shikonin levels, suggesting formation of GBA is not a rate-determining step of shikonin formation in these cell culture conditions.

Introduction of the *ubiC* construct also failed to significantly increase shikonin levels above those of the controls (Sommer *et al.*, 1999). However, it was shown that the UBIC enzyme was forming 4HB *in planta*, and that 4HB formed by the UBIC enzyme was used by the cultures to make shikonin. In addition to shikonin levels not being higher in the cultures, 4HB or 4HB-glucoside accumulation was not seen. This is in contrast to expression of *ubiC* constructs in transgenic tobacco plants and cell cultures, which led to accumulation of two 4HB-glucosides at up to 0.5% dry weight in the plant leaves (Siebert *et al.*, 1996; Li *et al.*, 1997b). Even more dramatic results in potato and tobacco are cited in Köhle *et al.* (2002), with 4HB-glucoside accumulation at up to 5% dry weight. In *L. erythrorhizon* cell lines expressing the *ubiC* construct it is possible that 4HB catalysis occurred, as levels of a presumed aromatic amino acid derivative, menisdaurin, increased fivefold. Köhle *et al.* (2002) followed up these experiments by using the (*ocs*)₃*mas* promoter to drive the *ubiC* construct in *L. erythrorhizon*, which is known to generate markedly higher transcription rates than the previously used *35SCaMV* promoter. Again, this did not result in significantly increased shikonin levels.

High-level accumulation of 4HB derivatives has also been reported for transgenic tobacco in which the normal phenylpropanoid pathway has been diverted using a *Pseudomonas* gene encoding an enoyl-CoA hydratase/lyase (HCHL) (Mayer *et al.*, 2001). This enzyme can convert 4-coumaroyl-CoA, caffeoyl-CoA or feruloyl-CoA to the corresponding hydroxybenzaldehydes. Transgenic tobacco expressing *35SCaMV::HCHL* had reduced levels of phenylpropanoids and accumulated novel glucosides and glucose esters of the acid and alcohol derivatives of the hydroxybenzaldehydes at levels up to 0.45% fresh weight in seeds. These compounds included 4HB glucosides.

Köhle *et al.* (2002) introduced *A. thaliana HMGR* cDNA constructs under the (*ocs*)₃*mas* promoter into *L. erythrorhizon* cell lines. While this approach resulted in an increase of up to a 30-fold of HMGR activity, it did not lead to higher shikonin levels. This is in spite of HMGR transcript levels and enzyme activity correlating with shikonin production in non-transgenics (Lange *et al.*, 1998), and HMGR overexpression being successful in increasing production of isoprenoids in other plant transgenics (Chappell *et al.*, 1995; Schaller *et al.*, 1995). For shikonin, it may be that combinations of biosynthetic transgenes are required – an approach successful for manipulating flavonoid levels (Davies *et al.*, 2003). However, the failure of increased GBA production to raise shikonin levels suggests that later biosynthetic steps for which genes have yet to be isolated may also be rate-limiting.

There is little published on the genetic modification of the production of other NQs. Generation of transgenic cell cultures of *L. inermis* accumulating lawsone has been reported (Bakkali *et al.*, 1997), using *A. rhizogenes*, but no attempt to modify the pigment biosynthesis was made.

7.2.2   Anthraquinones

The AQs contain the largest range of plant quinones, occurring in several families, and are also widespread in fungi, lichens, bacteria and insects (Thomson, 1976a; Han *et al.*, 2001). One of the major commercial food colourants, cochineal, is an AQ extracted from the bodies of female coccid insects. The most studied plant pigment examples are those of the Rubiaceae, in particular the dye madder from the roots of dyer's madder (*Rubia tinctorum*), Indian madder (*Oldenlandia umbellata*) and closely related plants. Madder extract consists of a range of AQs, most notably from the perspective of the red dye colour, alizarin, pseudopurpurin and purpurin. *In planta*, *R. tinctorum* produces at least 36 AQs, with the major ones being the glycosides lucidin primeveroside and ruberythric acid and the carboxylic AQs pseudopurpurin and munjstin, with minor amounts of the aglycones of alizarin and purpurin also present (Fig. 7.5; Derksen *et al.*, 2002). During dye

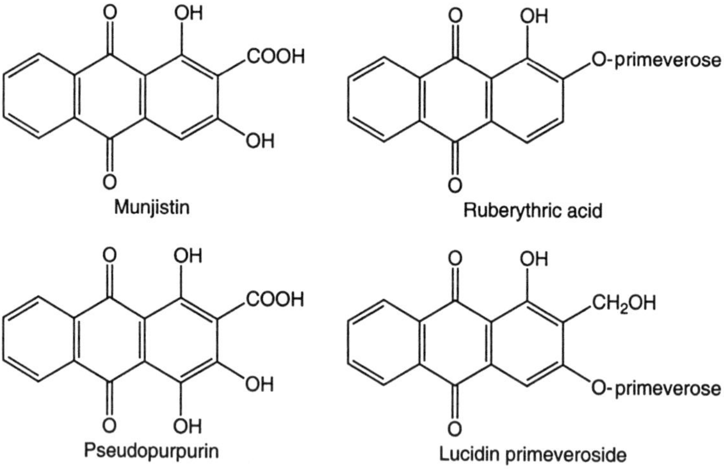

Fig. 7.5 Structures of major anthraquinones of *Rubia tinctorum* (dyer's madder): the glycosides, lucidin primeveroside and ruberythric acid; and the carboxylic anthraquinones, pseudopurpurin and munjistin.

extract preparation the glycosides may commonly be converted to the aglycones (Derksen *et al.*, 2003). In *Cinchona* species up to 25 different AQs have been reported (Schripsema *et al.*, 1999). Typically, AQs are substituted only in the C-ring. However, there are reports of AQs from plants substituted in both the A- and C-rings, particularly for *Cinchona* species.

As mentioned earlier, there are two major routes by which AQs are formed in plants (Leistner, 1981; Han *et al.*, 2001). The polyketide pathway that is common in fungi may also operate in Leguminosae, Rhamnaceae and Polygonaceae plant species. It involves one acetyl-CoA unit being extended by seven malonyl-CoA units, by undefined enzyme systems. However, the alternative chorismate/OSB pathway forms the important AQ pigments from Rubiaceae species, and this is the focus of the remainder of this section. In general terms, the pathway to the AQ pigments is similar to that for some of the NQs, i.e. the joining together of a phenolic ring from the shikimate pathway with IPP derivatives (van Tegelen *et al.*, 1999; Han *et al.*, 2001, 2002), although the specific enzymatic steps may differ.

There have been several studies on the enzymes involved in formation of AQs in cell cultures of Rubiaceae species such as *Cinchona* 'Robusta', *M. citrifolia* and *R. tinctorum* (Ramos-Valdivia *et al.*, 1997a, 1998; van Tegelen *et al.*, 1999; Han *et al.*, 2001, 2002; Stalman *et al.*, 2003). The A- and B-rings of AQs are formed in a similar manner to the path proposed for lawsone (Figs 7.3 and 7.6), i.e. joining of isochorismate with α-ketoglutarate in the presence of thiamine diphosphate to form an OSB intermediate, catalysed by the enzyme OSB synthase. Isochorismate is formed from chorismate by the enzyme isochorismate synthase (ICS). Elicitation of AQ formation in cell cultures causes an attendant marked increase in ICS activity. In *M. citrifolia* cultures this increase in ICS activity was not accompanied by induction of other enzymes of the shikimate pathway that direct chorismate into other biosynthetic pathways, specifically deoxy-D-arabinoheptulosante 7-phosphate synthase and chorismate mutase (Stalman *et al.*, 2003). Thus, ICS may be a key regulatory target of AQ biosynthesis. This is further supported by transgenic experiments using a bacterial *ICS* gene (Lodhi *et al.*, 1996). The recent isolation of the *ICS* gene from *M. citrifolia* (cited in Han *et al.*, 2001) should allow genetic approaches to test the role of the endogenous *ICS* gene(s) in controlling the rate of AQ biosynthesis. OSB is activated at the aliphatic carboxyl group to produce OSB-CoA ester, a reaction carried out by the enzyme OSB:CoA-ligase (Simantiras & Leistner, 1992). Ring closure then produces the A- and B-rings as the intermediate 1,4-dihydroxynaphthalene-2-caroxylic acid (DHNA). Subsequent prenylation of DHNA yields a prenylated NQ intermediate and finally the formation of the C-ring (Fig. 7.6). The source of the prenyl groups is either IPP or 3,3-dimethylallyl diphosphate (DMAPP), which are interconverted by the enzyme IPP isomerase (Ramos-Valdivia *et al.*, 1997a). Changes in IPP isomerase activity

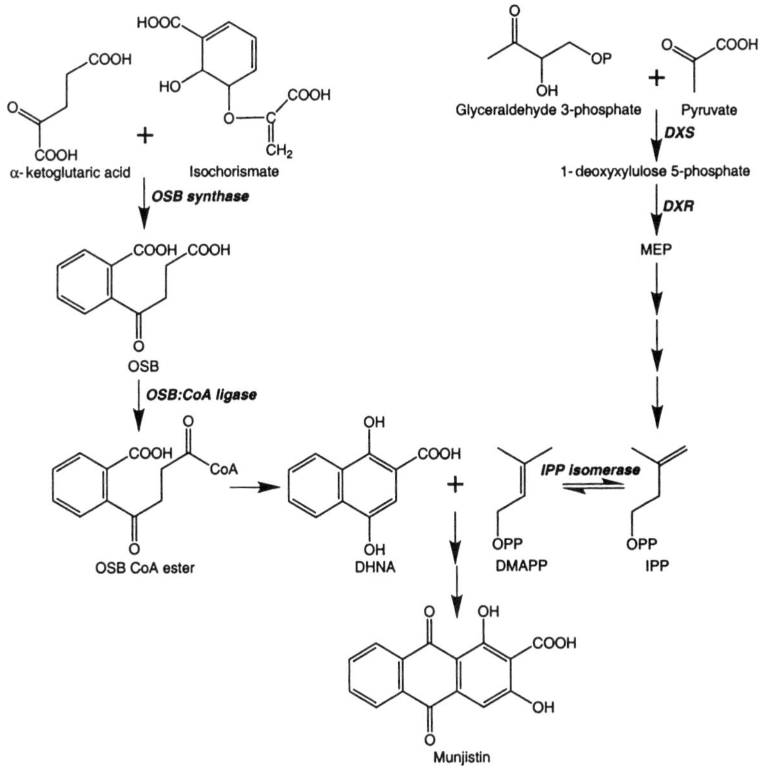

Fig. 7.6 A currently proposed biosynthetic pathway for the anthraquinone pigments of the Rubiaceae, such as munjistin. Compound and enzyme abbreviations are as given in the text.

accompany induction of coloured AQ production in Rubiaceae cell cultures (Ramos-Valdivia *et al.*, 1997b, 1998). Induction of IPP isomerase activity at the same time as a reduction in the activity of the enzyme farnesyl diphosphate synthase (which converts IPP and DMAPP into farnesyl diphosphate) may assist in channelling more IPP into AQ biosynthesis. However, the presence of isozymes of IPP isomerase, and possible variations in their subcellular compartmentation, complicates the analysis of the specific induction process. In *Cinchona* 'Robusta' and *M. citrifolia* cultures it is likely that a specific isoform of IPP isomerase is induced in conditions favouring AQ formation.

Identifying the source of IPP for formation of the C-ring has been the target of recent research. Early feeding experiments with radiolabelled MVA suggested that IPP for AQ formation came from the MVA pathway (Leistner, 1981). However, recent results have shown that IPP for AQ biosynthesis in the species *Cinchona* 'Robusta' (Han *et al.*, 2002) and *R. tinctorum* (Eichinger *et al.*, 1999) is derived from the MEP pathway, and this is likely the case for other Rubiaceae species (Han *et al.*, 2001). The MEP pathway is shown briefly in Fig. 7.6 and is discussed in detail in Lichtenthaler (1999) and Han *et al.* (2001). 1-deoxy-D-xylulose 5-phosphate synthase (DXS) is the first committed enzyme step of the MEP pathway, and has been suggested as a key regulatory step. Han *et al.* (2003) cloned a cDNA for DXS from *M. citrifolia*, and showed that transcript abundance correlated with AQ biosynthesis in cell cultures.

Divergence into the wide variety of AQ structures observed in plants is likely to occur late in the biosynthetic pathway by specific secondary modification enzymes, rather than by incorporating different phenylpropanoid precursors (Leistner, 1981; Han *et al.*, 2001). These later reactions would include methylation, hydroxylation and *C*- and *O*-glycosylation, although appropriate enzymes have not been characterised for the coloured AQs. However, Schripsema *et al.*, (1999) proposed a possible order of biosynthesis for formation of the highly substituted AQs of *Cinchona* 'Robusta'. The final AQ products are vacuolar-located, requiring a yet-to-be-described transport mechanism.

With regard to genes encoding the biosynthetic enzymes, or pathway regulatory factors, little is published. Furthermore, it is not known which signals regulate the AQ biosynthetic genes, although fungal elicitors, methyl jasmonate, salicylic acid and the protein phosphatase inhibitor cantharidin, all induce AQ accumulation in transgenic cell cultures. On the other hand, light generally inhibits AQ accumulation and the impact of auxins is variable (Mantrova *et al.*, 1999; Han *et al.*, 2001; Bulgakov *et al.*, 2002, 2003a). These gaps in the knowledge of AQ biosynthesis may start to be filled as data emerges from proteomic and gene studies in progress for *M. citrifolia* (Han *et al.*, 2001)

There have been two notable successes in modifying AQ production by genetic engineering in cell cultures of *Rubia* species. Lodhi & Charlwood (1996) developed an *A. rhizogenes*–mediated system for generating transgenic root culture lines of *R. peregrina*, and this was used to introduce the *E. coli* gene for ICS (Lodhi *et al.*, 1996). Cell lines expressing the gene had up to twice as much ICS activity as control lines, were visibly more pigmented, and had significantly higher levels of total AQs and alizarin in particular. In the second set of experiments, Bulgakov *et al.* (2002) used *A. rhizogenes* to introduce *35SCaMV::rolB* or *35SCaMV::rolC* into *R. cordifolia* cell lines. Levels of total AQs, comprised mainly of munjistin and purpurin, were significantly higher in transgenic compared to non-transgenic cultures. Pigment contents of almost 5% dry weight were reached, with stable production occurring over 3 years for some lines (Bulgakov *et al.*, 2002). The transgenic cultures have also been characterised with regard to the mechanism by which the transgenes induce AQ production (Bulgakov *et al.* 2003a, 2003b).

7.3 Indigo

The blue compound indigo is one of the oldest dyes known to humans. There is written evidence of use of indigo dating back 4600 years (Gilbert & Cooke, 2001), and it was a tattooing agent in Britain when Roman colonisation occurred (Smith, 1999). Until the development of synthetic indigo at the end of the nineteenth century, major plant-based indigo production industries were present in Europe and the tropics (Kokubun *et al.*, 1998). Indigo can be derived from a wide range of plants, with the major traditional and commercial sources being *Isatis tinctoria* (dyer's wold) in temperate climates, *Indigofera* species in tropical climates, and various other species (e.g. *Polygonum tinctorum*) in the Far East. Today it is still a dye of major commercial importance, principally for the dying of denim, but is often now obtained from non-plant source material.

Indigo is an aromatic compound derived from the precursor indole and, along with the betalains, is one of the few notable nitrogen-containing pigments of plants (Fig. 7.7). Indole can be converted to indoxyl and this can spontaneously dimerise to the blue-coloured indigo under oxidative conditions. In addition, related compounds isatin and, in small amounts, dioxindole can be formed. These can dimerise with each other or combine with indoxyl to form the red-coloured compounds indirubin (*trans*-indirubin) and, in minor amounts, isoindirubin (*cis*-indirubin). *Indigofera* species and *P. tinctorum* produce indoxyl as the major precursor, but for *I. tinctoria* isatin is most abundant (Kokubun *et al.*, 1998; Maugard *et al.*, 2001). Small quantities of related dimeric compounds can form in some plant tissues, including *cis*-indigo, the brown pigment isoindigo, and tryptanthrin, which is derived from an unusual indole glycoside named calanthoside (Murakami *et al.*, 2001). Along with co-extracted flavonoids these minor compounds can contribute to the distinctive hue of natural indigo dye (Maugard *et al.*, 2001).

Although all the precursors and biosynthetic activities necessary may be present in the plant, the formation of the coloured dimers does not occur in intact plant tissues. The precursors accumulate as colourless glycosides in the plant vacuole – in some cases at up to a significant per cent of wet weight of the leaves – and it is

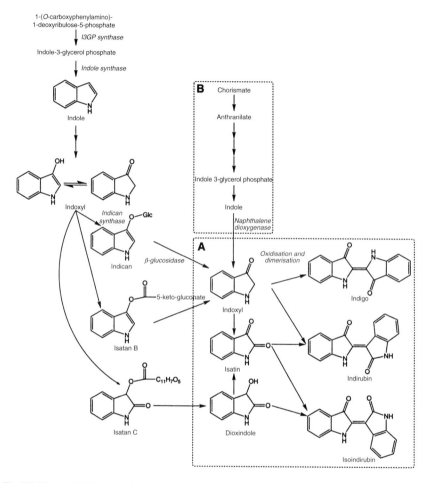

Fig. 7.7 The possible biosynthetic pathway to indigo and related pigments in plants. The final formation of the pigmented compounds occurs naturally only upon loss of cellular compartmentation and exposure to air (reactions in box A). Elements of the pathway are based on the results of Maugard *et al.* (2001), who identified a new precursor compound, isatan C, but did not solve the full structure. *E. coli* have been modified to produce indigo via the pathway shown in box B.

only upon cellular damage, allowing deglycosylation by chloroplastic enzymes, and exposure to air that the pigments are formed (Minami *et al.*, 1996, 1997, 2000; Gilbert & Cooke, 2001). The dimer formation is spontaneous following the oxidation upon exposure to air.

Bacteria produce a range of enzymes that will act on the indigo biosynthetic intermediates, and indeed, the pathway is much better defined in bacteria than it is in plants. An entire indoxyl biosynthetic pathway has been assembled in *E. coli* (Ensley *et al.*, 1983), and optimised for commercial use by modifying the endogenous *E. coli* tryptophan pathway to cause production of indole, combined with *Pseudomonas putida* genes for naphthalene dioxygenase to convert this to indoxyl (Berry *et al.*, 2002). The pathway has been further modified in *E. coli* by reducing isatin levels with a transgene for isatin hydrolase (Berry *et al.*, 2002). Bacterial mono-oxygenases and mammalian cytochrome P450 mono-oxygenases will also trigger indigo production in *E. coli*, through conversion of indole to indoxyl, and subsequently to isatin, indigo and indirubin (Mermod *et al.*, 1986; Gillam *et al.*, 1999, 2000; Gillam & Guengerich, 2001). However, in plants there is no evidence as yet for presence of naphthalene dioxygenase or other specific mono-oxygenases described in bacteria and mammals. Plant cytochrome P450s analogous to those of bacteria may be involved in the formation of indoxyl (probably a single-step 3-hydroxylation of indole) and dioxindole, although as mentioned, these enzymes have yet to be characterised.

The biosynthetic route to indole in plants is better characterised than the subsequent conversion to indoxyl. It is likely to be derived from indole-3-glycerol phosphate (IGP), a product in the pathway from chorismate to tryptophan (Xia & Zenk, 1992). IGP is formed by IGP synthase, and its biosynthesis has been studied extensively in relation to the formation of the phytohormone indole-3-acetic acid, particularly in *A. thaliana* (Ouyang *et al.*, 2000). Indole can be formed from IGP by the action of tryptophan synthase α (TSA) (Kramer & Koziel, 1995; Melanson *et al.*, 1997). Whether this is the pathway for indole formation for indigo is uncertain, however, as it is thought that TSA operates as a complex with trypto-phan synthase β, so that tryptophan is formed directly from IGP with indole as an ephemeral intermediate that is passed between the active sites of the two sub-units (Radwanski *et al.*, 1995). Thus, tryptophan would need to be converted back to indole for indoxyl formation. Furthermore, C_{13} feeding experiments have shown that tryptophan is not a precursor of indoxyl (Xia & Zenk, 1992). However, a second TSA-like enzyme, termed the indole synthase, can also operate in plants and may be the enzyme that forms indole for indole-derived secondary metabol-ites (Ouyang *et al.*, 2000). Evidence in support of this has come from analysis of maize plants in which the indole synthase gene is mutated (Melanson *et al.*, 1997). These plants still make tryptophan but lack the usual accumulation of indole-derived metabolites, which thus seem to rely on a tryptophan-independent biosyn-thetic route from IGP via the indole synthase. Transgenic plant studies have been undertaken on whether the indole synthase genes of maize influence indigo production, but with the results unpublished to date (NCBI GenBank entries AY254103 and AY254104).

The formation, and subsequent utilisation, of the stable glycosylated storage intermediate of indoxyl is also well defined for plants. Most commonly, indoxyl is glucosylated with UDP-glucose to form indoxyl β-D-glucoside, called indican. Some species may also add 5-ketogluconate to indoxyl to form isatin B, or organic acids to dioxindole to form isatin C (Maugard *et al.*, 2001). The enzyme adding UDP-glucose to indoxyl has been termed the indican synthase, and has been purified to homogenity from *P. tinctorium* (Minami *et al.*, 2000) and *Baphica-canthus cusia* (Marcinek *et al.*, 2000). When tested with over 60 substrates the enzyme used only indoxyl (3-OH-indole) or closely related artificial substrates (4-OH-, 5-OH-, 6-OH- and 7-OH-indole) (Marcinek *et al.*, 2000). The cloning of the associated gene has yet to be reported.

The hydrolysis of indican to indoxyl and glucose, allowing subsequent oxidation and dimerisation to indigo, is carried out by a β-glucosidase. Minami *et al.* (1996) purified such a β-glucosidase enzyme from *P. tinctorium*, and used the partial amino acid sequence to isolate the corresponding cDNA clone (Minami *et al.*, 1999). The deduced amino acid sequence included a putative plastid transport peptide at the N-terminal end, which fits with the proposed location of the indican glycosidase activity. The expressed protein in *E. coli* had activity against indican and related compounds. To date, this is the only plant cDNA published for a specific indigo biosynthetic enzyme. However, other molecular studies are in progress (Gilbert & Cooke, 2001).

To date there are no journal reports on the modification of indigo biosynthesis in plants, either altering production in indican-producing species or introducing the biosynthetic pathway into species that normally lack it. This is surprising, given the availability of indigo biosynthetic cDNAs from plants, bacteria and mammals. It has certainly been raised as a possible route for novel colours in plant tissues from biotechnology (Gillam & Guengerich, 2001), and patents related to indigo production in cotton (McBride *et al.*, 1996) and indigo or indirubin production in plant tissues (Notley *et al.*, 2001) have been filed.

In the patent of McBride *et al.* (1996), an *E. coli* gene for tryptophanase, which converts tryptophan to indole, and a *Rhodococcus* gene for an indole oxygenase, which converts indole to indigo and indirubin (Hart *et al.*, 1990, 1992), were targeted to the plastid of cotton fibre cells. Transgenics expressing both genes showed a weak colour phenotype of pale blue fibres. Based on these results, it may be necessary to increase levels of the precursor compounds to indole for generating strong blue colours in plant tissues. Furthermore, the localisation of the indigo produced, for example to the vacuole, may be important to generating successful transgenic plants.

In the patent of Notley *et al.* (2001) a range of human cytochrome P450 cDNAs (Gillam *et al.*, 1999; Gillam & Guengerich, 2001) formed the basis of a scheme to use the cytochrome P450s for indigo and indirubin production, including in plant tissues. Production of the pigments was demonstrated in *E. coli*, but no examples of plant tissues transformed to produce the pigments were presented, although various *A. tumefaciens* plant transformations and particle bombardment of white petunia petals with the gene constructs were described.

7.4 Quinochalcones

There are few quinoflavonoids reported from plants, but this group of compounds does include one notable group of flower pigments, from *Carthamus tinctorius* (dyer's saffron, safflower). *C. tinctorius* is an annual Compositae whose florets produce yellow- and red-coloured quinochalcones, first described by Kametaka and Perkin (1910). Extracts from the florets have been used for centuries as dyes, and more recently for cosmetics and natural food colourants. Several different compounds have been described, with the major ones being the red carthamin and the yellow precarthamin, safflower yellow A and safflower yellow B (Fig. 7.8). Kazuma *et al.* (2000) identified six quinochalcones and 11 other flavonoids in florets of various *C. tinctorius* cultivars.

The flower colour of *C. tinctorius* changes from yellow to red as the flower matures, and it is thought this is due to the conversion of precarthamin to carthamin (Saito *et al.*, 1983; Cho & Hahn, 2000; Cho *et al.*, 2000). The biosynthetic route to precarthamin and the safflower yellow-type pigments is unknown. However, it is likely that it proceeds via a chalcone precursor derived from the flavonoid pathway. C_{14}-phenylalanine fed to intact flowers of *C. tinctorius* is incorporated into carthamin, supporting the flavonoid origin (Saito *et al.*, 1993). This is further indicated by the studies of Kazuma *et al.* (2000) on the types of flavonoid and quinochalcones that occur in cultivars of *C. tinctorius* with orange, yellow or white petals. Some of the flavonols and quinochalcones have a hydroxyl group in the equivalent position (6-hydroxyl flavonols), which is a rare pattern of substitution for flavonols, and suggesting to Kazuma *et al.* (2000) a common biosynthetic intermediate. From their results they propose a biosynthetic pathway for precarthamin involving the joining of two intermediates derived from penta-

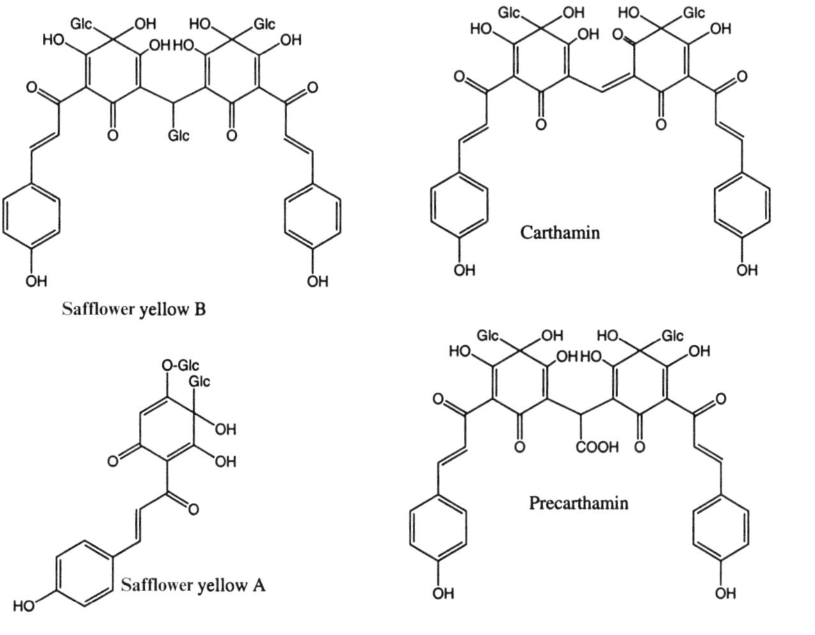

Fig. 7.8 The structures of the pigments precarthamin, carthamin, safflower yellow A and safflower yellow B from *Carthamus tinctorius*.

hydroxychalcone. *C*-glycosylation of the chalcone may prevent ring closure and formation of the common flavonoids, encouraging formation of the quinochalcones. However, confirmation of the chalcone-based biosynthetic route awaits further radiolabelled substrate or enzymology studies.

The conversion of precarthamin to carthamin by oxidative decarboxylation is better understood. An enzyme named precarthamin decarboxylase has been purified to homogeneity and characterised (Cho & Hahn, 2000; Cho *et al.*, 2000). However, the enzyme activity was found in all vegetative tissues, rather than showing activity related to the change in colour in floret development (Saito *et al.*, 1983). To date, there are no cDNA cloning studies reported for the enzymes forming the pigments of *C. tinctorius*, and no published attempts to modify the pigment biosynthesis by genetic modification approaches. Cell cultures from *C. tinctorius* that accumulate the pigments have been established (Gao *et al.*, 2000).

7.5   Apocarotenoids

Plant carotenoids are based on a common C_{40} linear terpene, which may be formed into cyclic or other types of group at one or both ends, and are reviewed in detail in Chapter 3. However, there are some vividly coloured pigments in plants, known as apocarotenoids, which are formed by the removal of an end group from the normal C_{40} structure. Some apocarotenoids are of economic importance as food colourants, in particular providing the colours of annatto and saffron.

Saffron is an expensive spice used to flavour and colour food and as a herbal medicine (Winterhalter & Straubinger, 2000). It is made up of the dried red styles of *Crocus sativus*, and contains three major carotenoid derivatives, crocetin glycosides (Fig. 7.9), picrocrocin and safranal, that are key to imparting the distinctive colour, bitter taste and aroma of the spice, respectively. It is the C_{20} apocarotenoid crocetin glycosides that are responsible for the bright red colour. They are probably formed from the cleavage of zeaxanthin, initially by the zeaxanthin $7,8(7'8')$-cleavage dioxygenase, a plastid-localised enzyme that removes the cyclic rings from both ends (Bouvier *et al.*, 2003a). It is proposed that the crocetin dialdehyde product is subsequently acted upon by an aldehyde oxydoreductase and a UDPG-glucosyltransferase to produce water-soluble crocetin glycosides that are transferred from the plastid to the vacuole via direct interactions between the two organelles (Bouvier *et al.*, 2003a).

Annatto extracts have long been used as dyes, in particular by early societies in tropical America, and are still in use today as a food colourant and ingredient in cosmetics. The pigments are extracted from the resinous coating on the seeds of the tropical bush *Bixa orellana*, and can reach up to 10% of the seed coat dry weight (Britton, 1996). The main pigment is the yellow-orange apocarotenoid bixin, accompanied by lesser quantities of norbixin (Fig. 7.10). Gene expression related to bixin biosynthesis during development of the seed has been investigated by examining known isoprenoid biosynthetic genes (Narvaez *et al.*, 2001) and by

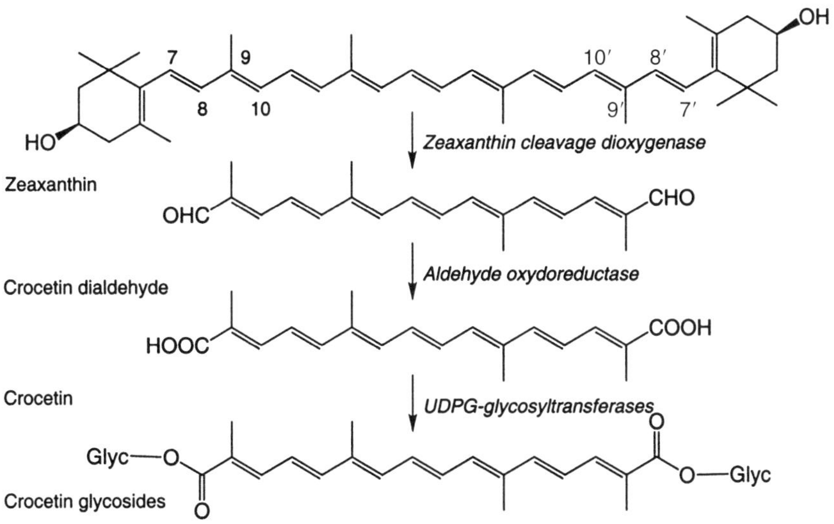

Fig. 7.9 The biosynthetic pathway to crocetin glycosides from zeaxanthin in *Crocus sativus*.

analysis of expressed sequence tags (ESTs) (Jako *et al.*, 2002). The enzyme HMGR, which is key to the production of some isoprenoids, shows enhanced transcript levels and enzyme activity in the seeds, particularly in the developing seeds (Narvaez *et al.*, 2001). Jako *et al.* (2002) proposed a biosynthetic pathway to bixin from lycopene, which involved dioxygenases, aldehyde dehydrogenases and carboxyl methyltransferases, and looked for sequences encoding proteins that might correspond to these activities in an EST database made from a cDNA library enriched for seed coat–specific cDNAs. Out of 870 ESTs analysed, 15 different sequences were identified as corresponding to known carotenoid and putative bixin biosynthetic enzymes.

Bouvier *et al.* (2003b) confirmed the bixin biosynthetic pathway in *B. orellana* by isolating putative cDNAs for the biosynthetic enzymes, using PCR and primers to conserve sequences of related enzymes, and analysis of the encoded activities in *E. coli*. The initial biosynthetic step from the carotenoid precursor is analogous to that for formation of the related saffron compounds, specifically, cleavage of lycopene by a dioxygenase to yield bixin aldehyde (Fig. 7.10). This compound is subsequently acted upon by bixin aldehyde dehydrogenase to produce the pigment norbixin. Conversion of norbixin to bixin (and a bixin dimethylester) is achieved by the action of norbixin carboxyl methyltransferase. *B. orellana* is currently the only commercial source of annatto. By introducing the three bixin biosynthetic cDNAs into *E. coli* that had previously been modified to accumulate lycopene, production of bixin in the bacteria was achieved. It is proposed to transfer the same genes into tomato, which accumulates large amounts of lyco-pene in its fruit, to enable bixin production in this crop species.

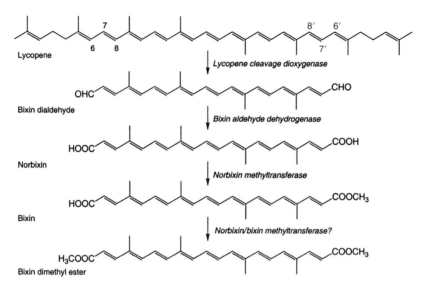

Fig. 7.10 The biosynthetic pathway to bixin, and related pigments, from lycopene in *Bixa orellana*.

Apocarotenoids also occur as yellow pigments in roots in a wide range of plant species following colonisation by arbuscular mycorrhizal fungi (Fester *et al.*, 2002a). The yellow pigmentation is likely to consist of a mix of different ester-ification products of mycorradicin (an acyclic C_{14} polyene) and blumenol C cellobioside (a C_{13} cyclohexenone). The pigment accumulates as hydrophobic bodies in the vacuole of root cortical cells. It is likely that both components of the pigment are formed via cleavage of a C_{40} precursor, as reported for formation of crocetin glycosides, picrocrocin and safranal in saffron. Carotenoid biosynthesis is induced upon colonisation, probably to provide more of the proposed C_{40} precursor compound, and the carotenoid biosynthesis gene phytoene desaturase and genes for two key enzymes of the MEP pathway, DXS and 1-deoxyxylulose 5-phosphate reductoisomerase (DXR) (Fig. 7.6), show increased transcript levels following colonisation (Walter *et al.*, 2000; Fester *et al.*, 2002b). The compounds may have a functional role in the colonisation process, as maize mutants deficient in carotenoid biosynthesis are impaired in colonisation (Fester *et al.*, 2002b).

7.6   Phenylphenalenones and related pigments

Phenylphenalenones are plant pigments thought to be derived from cinnamic acids, perhaps via diarylheptanoid intermediates. As the principal colourants in turmeric are diarylheptanoids, and as they may share a similar biosynthetic mechanism to phenylphenalenones, they are considered together here.

Phenylphenalenones were first identified in the Haemodoraceae family, especially the genus *Anigozanthos* (kangaroo paw) (Thomson, 1976b). They have
since been found in other monocotyledonous families, the Musaceae, Ponteder-
iaceae and Strelitziaceae. They have most commonly been observed as red root
pigments, particularly in the root tip. However, yellow, orange and purple variants
have been reported for the species *Lachnanthes tinctoria* (red root), including
occurrence of the yellow and orange pigments in the flowers (Thomson, 1976b).
Opitz and Schneider (2003) measured levels of phenylphenalenone-related compound in various tissues of *Xiphidium caeruleum*, and found their presence in most
samples but the highest levels in the root tip and stamens.

Phenylphenalenones are comprised of a series of phenolic rings that may be
altered by methylation, glycosylation or other modifications; some representative
structures are shown in Fig. 7.11. Indeed, although they occur in a limited group of
plants, they show significant structural variation. The related phenylbenzoisochromenones that may be derived from phenylphenalenones can also occur (Opitz &
Schneider, 2003; Opitz *et al.*, 2003), and dimeric phenylphenalenones have been
identified (Otálvaro *et al.*, 2002). The function of phenylphenalenones is not clear,
but their prominent occurrence in root tips and induction following wounding or
infection of *Musa* plants suggests a role as phytoalexins.

Although no specific biosynthetic activities have been characterised or genes
identified for phenylphenalenone biosynthesis, significant progress in elucidating
their possible biosynthetic pathway has been made. Feeding of radiolabelled
precursors into phenylphenalenone-accumulating root cultures of *A. preissii*

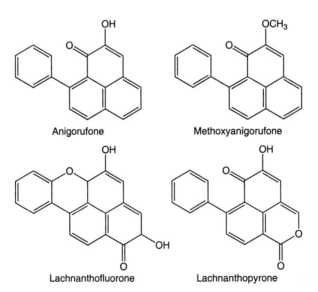

Anigorufone

Methoxyanigorufone

Lachnanthofluorone

Lachnanthopyrone

Fig. 7.11 Examples of structures of three phenylphenalenone pigments and a phenylbenzoisochromenone
(lachnanthopyrone).

showed that the compounds are derived from phenylalanine, with a likelihood of diarylheptanoids as intermediates (Schmitt & Schneider, 1999; Schmitt *et al.*, 2000). The central one-carbon linking unit is likely derived from an acetyl residue. Cinnamic acid, coumaric acid, caffeic acid or ferulic acid could be incorporated as part of their biosynthesis; however, there was variability of incorporation of the different substrates that suggested coumaric acid is the preferred substrate for the formation of the initial diarylheptanoid. Also, methylation (for methoanigorufone) was unlikely to arise via use of the methylated precursor ferulic acid, but rather likely to be the product of subsequent methylation of anigorufone. Feeding experiments also support the oxidative formation of phenylbenzoisochromenones as a late biosynthetic branch from phenylphenalenones (Opitz *et al.*, 2003). For *X. caeruleum*, the phenylbenzoisochromenones were found in aerial parts and phenylphenalenones in the roots, and it was suggested that phenylphenalenones biosynthesis may occur in the roots with translocation to the aerial parts and formation of phenylbenzoisochromenones there (Opitz & Schneider, 2003).

The chalcone synthase/stilbene synthase family of plant-specific polyketide synthases can use a range of phenylpropanoid starter molecules, including 4-coumaroyl-CoA, cinnamoyl-CoA, caffeoyl-CoA, dihydrocoumaroyl-CoA and dihydrocinnamoyl-CoA, that then typically react with two or three units of malonyl-CoA to form a range of products depending on the specific enzyme (Schröder, 1997; Austin & Noel, 2003). It is possible that similar enzymes are involved in the formation of the diarylheptanoid intermediate of phenylphenalenone biosynthesis (Schröder, 1997, 1999). Such a polyketide synthase may carry out a condensing reaction between the first phenylpropanoid unit and malonyl-CoA, with the reaction also yielding the one-carbon unit for the central link (Schmitt *et al.*, 2000).

Turmeric has been used as a spice since the times of early civilisations in Asia, and is prepared from rhizomes of species of the *Curcuma* genus (Zingiberaceae), usually *Curcuma longa*. It contains the yellow-orange diarylheptanoid curcumin as the principle pigment, with smaller quantities of the related pigments demethoxycurcumin and bis-demethoxycurcumin (Fig. 7.12). However, the balance of the three pigments can vary depending on the species and geographical location (Francis, 1996). Despite a large research effort investigating the medicinal properties of curcumin, there is little information on its biosynthetic pathway in plants. It has been suggested that it is derived from the phenylpropanoid pathway, possibly based on a diketide intermediate by a mechanism similar to that of phenylphenalenones (Schröder, 1997, 1999). This is supported by feeding experiments looking at formation of the related structure [6]-gingerol from *Zingiber officinale* (ginger), which showed use of ferulic acid, malonate and hexanoate substrates (Denniff *et al.*, 1980). The formation of the diketide, and subsequently curcumin, may share features with the reaction of benzalacetone synthase that forms phenylbutanones in *Rheum palmatum* (Rhubarb) (Abe *et al.*, 2001), a member of the chalcone synthase/stilbene synthase family of enzymes (Schröder, 1997). Schmitt *et al.* (2000) suggest that the structure of one of the phenolic rings

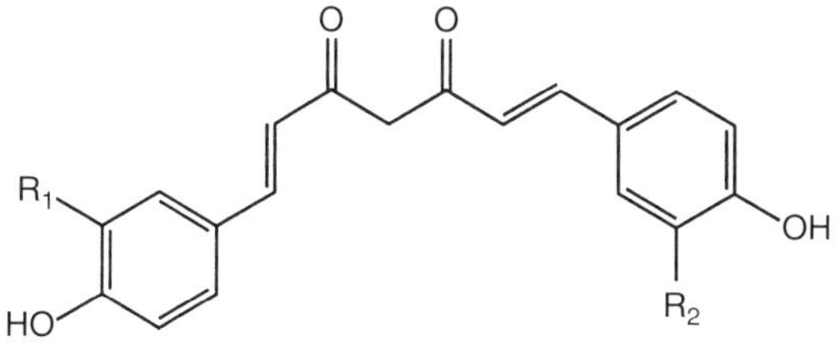

Curcumin; R$_1$ and R$_2$=OCH
Demethoxycurcumin; R$_1$=H and R$_2$=OCH$_3$
Bis-demethoxycurcumin; R$_1$ and R$_2$=H

Fig. 7.12 The structures of the pigments curcumin, demethoxycurcumin and bis-demethoxycurcumin, which form the main colourants of the spice turmeric.

of the diarylheptanoids may influence whether it remains as a diarylheptanoid, as in *Curcuma*, or becomes an intermediate in phenylphenalenone synthesis.

7.7 Miscellaneous plant pigments

There are numerous individual plant pigments for which there are few reports in the literature on their occurrence or structure, and generally little detail of their possible biosynthetic routes. In other cases, the pigmented product may be an offshoot of a pathway much studied for other interests, such as plant defence or medicinal applications. Of particular significance are a large range of alkaloids, terpenoids and flavonoid-related pigments found in specific plant tissues such as secretory glands on leaves, root and stem bark and the heartwood of trees and shrubs. Eugster (1980) reviewed the terpenoid-derived plant pigments (with the exception of the common carotenoids), and was able to report on over 100 different diterpenoid pigments alone. Some of these rarely described pigments still have long histories of human use. For example, huangbo, extracted from the bark of *Phellodendron amurense*, has been used as a dye for many centuries. There are extensive studies on several of these pigments as potential medicinal compounds. However, there are few details in the literature on them as plant pigments or on their biosynthesis. Some examples, with representatives from the different pigment groups and plant tissues, are featured here, and the reader is referred to Eugster (1980) and Thomson (1976b) for comprehensive reviews on the chemistry and occurrence.

The dye huangbo contains a mix of related benzylisoquinoline alkaloids (BIAs), including berberine, palmatine and jatrorrhizine (Bell *et al.*, 2000). Berberine (Fig. 7.13) can also be extracted from the roots of *Coptis japonica* or the xylem of *Berberis* species (Francis, 1996; Facchini, 2001). It is also present in many organs of the medicinal plant *Thalictrum flavum* (meadow rue), with highest levels in the roots, and provides a conspicuous yellow colour to the rhizome and older petioles (Samanani *et al.*, 2002). Related BIAs, produced by branches off the berberine biosynthetic pathway, may also provide colour to plant organs. In particular, the root exudates of many Papaveraceae species are intensely red due to the accumulation of sanguinarine (Fig. 7.13) and related compounds (e.g. Bloodroot plant, *Sanguinaria canadensis*). BIAs have received little attention as pigments but have been the subject of much study in relation to plant defence and medicinal activities. Indeed, berberine was the first alkaloid to have its biosynthetic pathway defined from the primary metabolite precursor (tyrosine) through to the final product (Hashimoto & Yamada, 1994; Facchini, 2001). The pathway involves 13 enzymatic steps, and may involve the transport of intermediate compounds between different sites in the plant and also between different cellular compartments. Several genes involved in the biosynthesis of BIAs have been cloned, allowing genetic modification approaches. However, while there has been some success in the metabolic engineering of alkaloid biosynthesis, with over a dozen examples of transgenic plants or cell lines with altered alkaloid production listed in Facchini (2001), to date these have been principally for terpenoid indole alkaloids (TIAs) and not BIAs. Most experiments have focused on the early stages of TIA biosynthesis, in particular the enzymes tryptophan decarboxylase and strictosidine synthase. However, genetic transformation systems are now available for several BIA-producing species, and this has allowed the first examples of modifying BIA production (Sato *et al.*, 2001; Park *et al.*, 2002a, 2002b). In particular, antisense RNA inhibition of the production of one of the biosynthetic enzymes, berberine bridge enzyme, prevented formation of the red BIA pigment of *Eschscholzia californica* (California poppy) cell cultures (Park *et al.*, 2002b).

Quinonemethides are commonly part of the coloured resins of the heartwood of trees and shrubs, and are also found in root bark and leaf secretary glands. They are unrelated biosynthetically to the true quinones and lack a quinone ring. Most commonly reported ones are diterpene and triterpene compounds, but there are

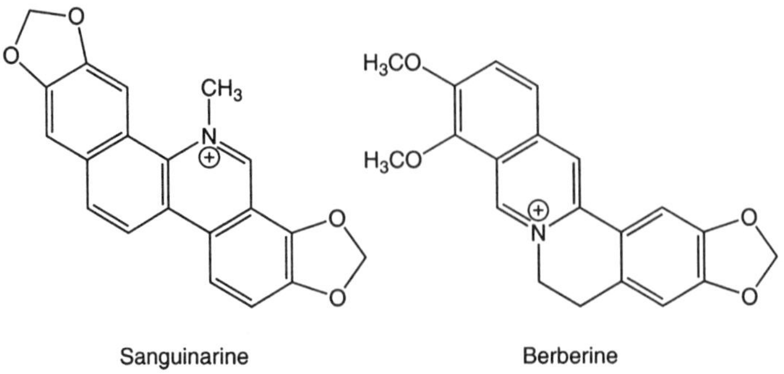

Sanguinarine   Berberine

Fig. 7.13 The structures of the benzylisoquinoline alkaloid pigments berberine and sanguinarine.

others related structurally to flavonoids, and which may be flavonoid- or isofla-
vonoid-derived (Thomson, 1976b; Eugster, 1980). Well-known examples of the
flavonoid type from wood resins are the red dye santalin (also known as red
sandalwood pigment) (Arnone *et al.*, 1972; Mathieson *et al.*, 1973) and dracorubin
from the 'dragon's blood' of the dragon tree *Dracaena* (Thomson, 1976b; Arnone
et al., 1997) (Fig. 7.14). Czakó & Márton (2001) established pigmented cell
culture lines from many shrub or tree species and looked for those that produced

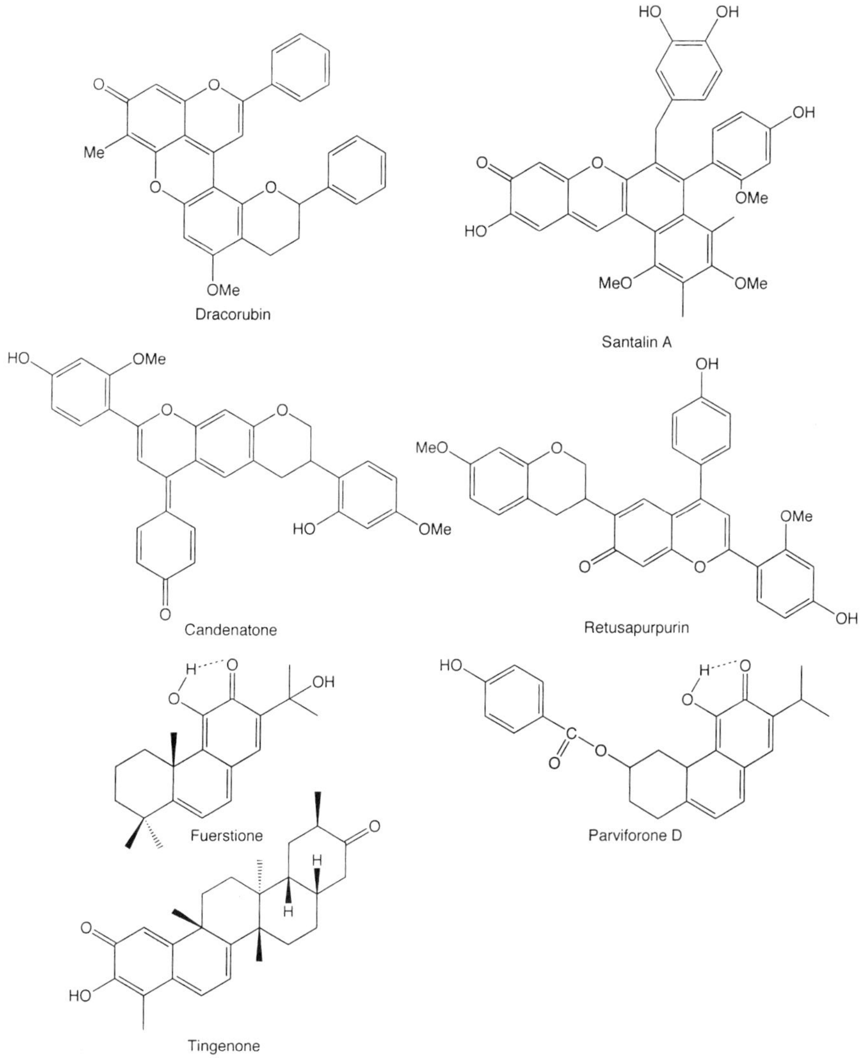

Fig. 7.14 The structures of sample quinonemethide pigments: candenatone, retusapurpurin A (both from
Dalbergia species), santalin A, dracorubin, fuerstione, parviforone D and tingenone.

heartwood-associated secondary metabolites in culture. Of particular interest were several *Dalbergia* (rosewood) species that produced brilliantly coloured quinone-methides in culture. One of the main purple pigments in *D. retusa* was identified as retusapurpurin A, and other *Dalbergia* species produced different quinone-methide pigments, such as candenatone (Fig. 7.14) and obtusaquinone.

There are a range of terpenoid quinonemethides occurring in root bark and leaf glands. The notable ones in leaf glands are the large group of coleone pigments, found in various genera of Lamiaceae (Labiatae), in particular *Solenostemon* and *Plectranthus*. For example, fuerstione (in *Fuerstia africana*) and the various parviflorone pigments (in *Plectranthus parviflorus*) (Fig. 7.14) accumulate in sufficient quantities to provide deep red colour to leaf glands and visible pigmentation to the overall leaf (Eugster, 1980). The root bark quinonemethide pigments are prominent in the Celastraceae as well as the Lamiaceae. In addition to diterpenes, they include a range of triterpenoids, such as tingenone (Fig. 7.14).

To date there is little published information on the biosynthesis of the quino-nemethides and similar pigments of heartwood. It has been suggested that some quinonemethides may arise from oxidation of a C_{15} flavonoid followed by coupling to a second C_{15} flavonoid unit (Mathieson *et al.*, 1973; Arnone *et al.*, 1975, 1997); however, there are few enzymatic or genetic studies. This may change with the data emerging from several major genomic studies in progress around the world on heartwood formation in timber species.

The xanthone group of flavonoid-related compounds includes tricyclic pigments that may be found in roots and heartwood, specifically in the Gentianaceae and Hypericaceae (Guttiferae), and that may also accumulate in cell cultures of such species (Thomson, 1976b; Bennett & Lee, 1989). Xanthone pigments also occur in fungi and lichens, and are typically yellow to yellow-orange in colour. Well-characterised examples from plants are mangostin (from *Garcinia mangostana*) and gentisin (from *Gentiana lutea*) (Fig. 7.15). Their induction in cell cultures of *Centaurium* species turn the cultures yellow or brownish, dependent on the elicitor used and the mix of compounds induced (Beerhues & Berger, 1995). Xanthones are likely to be formed from a polyketide synthase enzyme reaction, similar to the chalcone synthase reaction of flavonoid biosynthesis. Indeed, a polyketide synthase has been cloned from *Hypericum androsaemum*, and termed benzophenone synthase, that uses benzoyl-CoA and 3X malonyl-CoA to produce putative xanthone precursors (Liu *et al.*, 2003). The benzoyl-CoA is formed from the shikimate pathway by reactions including the enzyme 3-hydroxybenzoate:CoA ligase (Barillas & Beerhues, 2000; Wang *et al.*, 2003). The formation of the various xanthones from benzophenones may involve many reactions, including cyclisation, methylation, hydroxylation, prenylation (especially in the Hypericaceae) and glycosylation (especially in the Gentianaceae).

Gossypol is a yellow sesquiterpenoid pigment found in the Malvaceae. Related compounds may be bright yellow and red pigments, but there is limited information available on them (Eugster, 1980). The best-characterised system for gossypol biosynthesis is the pigment glands of seeds and leaves of *Gossypium*

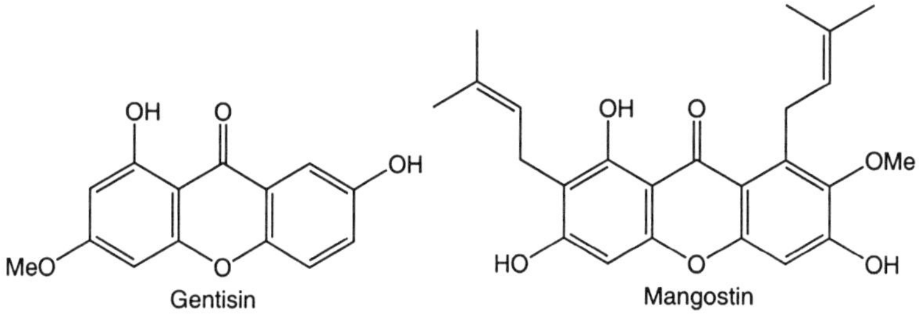

Fig. 7.15 The structures of the xanthone pigments mangostin and gentisin.

species (cotton), in which they have been studied in detail with regard to their role in pathogen and insect resistance. Furthermore, gossypol is toxic, and thus needs to be removed from cottonseed and cottonseed oil for consumption, which has also lead to studies on controlling its biosynthesis (Martin *et al.*, 2003). Gossypol is a dimer of hemigossypol, which itself is formed from deoxyhemigossypol. In cotton, deoxyhemigossypol is also the precursor of a range of related sesterterpe-noid phytoalexin compounds. Deoxyhemigossypol is derived from the MVA isoprenoid pathway, via the intermediate (+)-δ-cadinene (Fig. 7.16) (Martin *et al.*, 2003). The formation of (+)-δ-cadinene from farnesyl diphosphate is catalysed by the (+)-δ-cadinene synthase (CDN synthase). Four cDNAs for CDN synthase have been cloned from cotton (Chen *et al.*, 1995, 1996), and

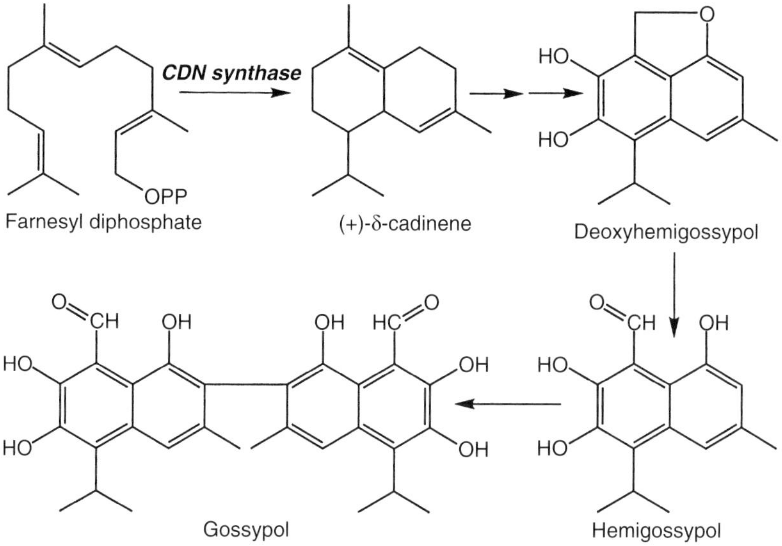

Fig. 7.16 A currently proposed biosynthetic pathway for gossypol in *Gossypium* species.

transgenic cotton plants have been generated containing an antisense RNA construct using one of the cDNAs (Martin *et al.*, 2003). In some of the transgenics, gossypol levels were reduced up to 70% in cottonseed and up to 92% in the leaves. Levels of the related phytoalexin compounds were also markedly lower.

Melanins are one of the most prominent and important pigments of animals (Riley, 1997). However, their role and structure in plants is less well defined. A general definition often used for melanins is that they are high-molecular weight polymers formed by the oxidation of phenols. However, this is too broad a definition, as some of the brown/black polymers that appear as melanins contain nitrogenous compounds derived from tyrosine. These nitrogenous polymers are referred to as eumelanins (Thomson, 1976b). They are widespread in animals and may also occur in plants, although little is reported on their structure or biosynthesis. The high-molecular weight polymers of simple phenolics are referred to as allomelanins. Allomelanins may occur in oxidative environments in plants, such as in seed coats or following cell damage (Riley, 1997). However, some of the black pigment layers found in plants, such as the phytomelanin of the Compositae (Pandey & Dhakal, 2001), are unlikely to be made up of pigments related to the true melanins.

There are many reports in the literature, most particularly the patent literature, on a range of compounds extracted from plants that may be used as colouring agents for food directly or following derivatisation. These include a blue colourant from callus generated from *Clerodendrum trichotomum*, a yellow pigment from the roots of *Glycyrrhiza* (liquorice), and brown and red polyphenols from *Phaseolus* species (bean) seed coats (Francis, 1996). The compound azulene can form orange, red, blue, violet and green pigments upon extraction from some plant species, and it is possible that orange-red azulene-derived compounds exist *in planta* (Eugster, 1980). There are extensive details in numerous patents of pigments for food use developed from iridoid compounds of plants, particularly those found in fruits of *Gardenia* species, which contain a mix of flavonoids, apocarotenoids and iridoids (Francis, 1996; Park *et al.*, 2002c). The endogenous iridoid structures identified to date are glycosides, and many of the patented processes for generating colourants involve treatment with glycosidases followed by reaction of the aglycones with molecules such as amino acids (Francis, 1996; Park *et al.*, 2002c). By varying the reaction substrates and conditions, compounds can be produced covering much of the colour spectrum including red, yellow, green, blue and violet.

7.8 Concluding comments

Plants produce a large array of different pigment structures. As will be evident from the other chapters of this volume, the biosynthesis of the major plant pigment groups – chlorophylls, carotenoids, flavonoids and betalains – is now well defined, and there have been notable successes in manipulating their production in transgenic plants. The biosynthesis of the pigment types of rare occurrence is, however, often poorly understood. Nevertheless, there has been success in elucidating and modifying the production of some of the colourants of economic importance, such as shikonin and annatto, and similar success with other pigment types would be expected in the next few years.

Acknowledgements

It is my pleasure to thank Dr Nigel Perry for commenting on the manuscript, and generally encouraging my interest in compounds other than flavonoids.

References

Abe, I., Takahashi, Y., Morita, H. & Noguchi, H. (2001) Benzalacetone synthase: a novel polyketide synthase that plays a crucial role in the biosynthesis of phenylbutanones in *Rheum palmatum*. *Eur. J. Biochem.*, **268**, 3354–3359.

Arnone, A., Merlini, L. & Nasini, G. (1972) Santalin pigments: on the structure of santalin A (from *Pterocarpus santalinus*). *Tetrahedron Lett.*, **33**, 3503–3506.

Arnone, A., Camarda, L., Merlini, L. & Nasini, G. (1975) Structures of the Red Sandalwood (*Pterocarpus santalinus*) pigments santalins A and B. *J. Chem. Soc. Perkin Trans. I*, 186–194.

Arnone, A., Nasini, G., Vajna de Pava, O. & Merlini, L. (1997) Constituents of Dragon's Blood. 5. Dracoflavans B1, B2, C1, C2, D1, and D2, new A-type deoxyproanthocyanidins. *J. Nat. Prod.*, **60**, 971–975.

Austin, M. B. & Noel, J. P. (2003) The chalcone synthase superfamily of type III polyketide synthases. *Nat. Prod. Rep.*, **20**, 79–110.

Bakkali, A. T., Jaziri, M., Foriers, A., Vander Heyden, Y., Vanhaelen, N. & Homes, J. (1997) Lawsone accumulation in normal and transformed cultures of henna, *Lawsonia inermis*. *Plant Cell Tissue Organ Cult.*, **51**, 83—87.

Barillas, W. & Beerhues, L. (2000) 3-Hydroxybenzoate:coenzyme A ligase from cell cultures of *Centaurium erythraea*: isolation and characterization. *Biol. Chem.*, **381**, 155–160.

Beerhues, L. & Berger, U. (1995) Differential accumulation of xanthones in methyl-jasmonate- and yeast-extract-treated cell cultures of *Centaurium erythraea* and *Centaurium littorale*. *Planta*, **197**, 608–612.

Bell, S. E. J., Bourguignon, E. S. O., Dennis, A. C., Fields, J. A., McGarvey, J. J. & Seddon, K. R. (2000) Identification of dyes on ancient Chinese paper samples using the subtracted shifted Raman spectroscopy method. *Anal. Chem.*, **72**, 234–239.

Bennett, G. J. & Lee, H. H. (1989) Xanthones from Guttiferae. *Phytochemistry*, **28**, 967–998.

Berry, A., Dodge, T. C., Pepsin, M. & Weyler, W. (2002) Application of metabolic engineering to improve both the production and use of biotech indigo. *J. Ind. Microbiol. Biotech.*, **28**, 127–133.

Boehm, R., Sommer, S., Li, S. M. & Heide, L. (2000) Genetic engineering on shikonin biosynthesis: expression of the bacterial *ubiA* gene in *Lithospermum erythrorhizon*. *Plant Cell Physiol.*, **41**, 911–919.

Bouvier, F., Suire, C., Mutterer, J. & Camara, B. (2003a) Oxidative remodeling of chromoplast carotenoids: identification of the carotenoid dioxygenase CsCCD and CsZCD genes involved in crocus secondary metabolite biogenesis. *Plant Cell*, **15**, 47–62.

Bouvier, F., Dogbo, O. & Camara, B. (2003b) Biosynthesis of the food and cosmetic plant pigment bixin (Annatto). *Science*, **300**, 2089–2092.

Britton, G. (1996) Carotenoids, in *Natural Food Colorants* (eds G. A. F. Hendry & J. D. Houghton), Chapman & Hall, London, UK, pp. 197–243.

Bulgakov, V. P., Tchernoded, G. K., Mischenko, N. P., Khodakovskaya, M., Glazunov, V. P., Radchenko, S. V., Zvereva, E. V., Fedoreyev, S. A. & Zhuravlev, Y. N. (2002) Effect of salicylic acid, methyl jasmonate, ethephon and cantharidin on anthraquinone production by *Rubia cordifolia* callus cultures transformed with the *rolB* and *rolC* genes. *J. Biotech.*, **97**, 213–221.

Bulgakov, V. P., Tchernoded, G. K., Mischenko, N. P., Shkryl, Y. N., Glazunov, V. P., Fedoreyev, S. A. & Zhuravlev, Y. N. (2003a) Effects of Ca^{2+} channel blockers and protein kinase/phosphatase inhibitors on growth and anthraquinone production in *Rubia cordifolia* callus cultures transformed by the *rolB* and *rolC* genes. *Planta*, **217**, 349–355.

Bulgakov, V. P., Tchernoded, G. K., Mischenko, N. P., Shkryl, Y. N., Glazunov, V. P., Fedoreyev, S. A., Zhuravlev, Y. N. (2003b) Increase in anthraquinone content in *Rubia cordifolia* cells transformed by *rol* genes does not involve activation of the NADPH oxidase signaling pathway. *Biochemistry (Moscow)*, **68**, 795–801.

Cerrato, A., De Santis, D. & Moresi, M. (2002) Production of luteolin extracts from *Reseda luteola* and assessment of their dying properties. *J. Sci. Food Ag.*, **82**, 1189–1199.

Chappell, J., Wolf, F., Proulx, J., Cuellar, R. & Saunders, C. (1995) Is the reaction catalyzed by 3-hydroxy-3-methylglutaryl-coenzyme A reductase a rate limiting step for isoprenoid biosynthesis in plants? *Plant Physiol.*, **109**, 1337–1343.

Chen, X.-Y., Chen, Y., Heinstein, P. & Davisson, V. J. (1995) Cloning, expression, and characterization of (+)-δ-cadinene synthase: a catalyst for cotton phytoalexin biosynthesis. *Arch. Biochem. Biophys.*, **324**, 255–266.

Chen, X.-Y., Wang, M., Chen, Y., Davisson, V. J. & Heinstein, P. (1996) Cloning and heterologous expression of a second (+)-δ-cadinene synthase from *Gossypium arboreum*. *J. Nat. Prod.*, **59**, 944–951.

Cho, M. H. & Hahn, T. R. (2000) Purification and characterization of precarthamin decarboxylase from the yellow petals of *Carthamus tinctorius* L. *Arch. Biochem. Biophys.*, **382**, 238–244.

Cho, M. H., Paik, Y.-S. & Hahn, T. R. (2000) Enzymatic conversion of precarthamin to carthamin by a purified enzyme from the yellow petals of safflower. *J. Agric. Food Chem.*, **48**, 3917–3921.

Cristea, D., Bareu, I. & Vilarem, G. (2003) Identification and quantitative HPLC analysis of the main flavonoids present in weld (*Reseda luteola* L.). *Dyes Pigments*, **57**, 267–272.

Cunningham, F. X. & Gantt, E. (2002) Molecular control of floral pigmentation: carotenoids, in *Breeding for Ornamentals: Classical and Molecular Approaches* (ed. A. Vainstein), Kluwer Academic Publishers, Dordrecht, The Netherlands, pp. 273–293.

Czakó, M. & Márton, L. (2001) A heartwood pigment in *Dalbergia* cell cultures. *Phytochemistry*, **57**, 1013–1022.

Davies, K. M., Schwinn, K. E., Deroles, S. C., Manson, D. G., Lewis, D. H., Bloor, S. J & Bradley, J. M. (2003) Enhancing anthocyanin production by altering competition for substrate between flavonol synthase and dihydroflavonol 4-reductase. *Euphytica*, **131**, 259–268.

De-Eknamkul, W. (1998) Chasing the key enzymes of secondary metabolite-biosynthesis from Thai medicinal plants. *Pure Appl. Chem.*, **70**, 2107.

Denniff, P., Macleod, I. & Whiting, D. A. (1980) Studies in the biosynthesis of [6]-gingerol, pungent principle of ginger (*Zingiber officinale*). *J. Chem. Soc. Perkin Trans. 1*, 2637–2644.

Derksen, G. C. H., Niederlander, H. A. G. & van Beek, T. A. (2002) Analysis of anthraquinones in *Rubia tinctorum* L. by liquid chromatography coupled with diode-array UV and mass spectrometric detection. *J. Chromatogr. A*, **978**, 119–127.

Derksen, G. C. H., Naayer, M., van Beek, T. A., Capelle, A., Haaksman, I. K., van Doren, H. A. & de Groot, A. E. (2003) Chemical and enzymatic hydrolysis of anthraquinone glycosides from madder roots. *Phytochem. Anal.*, **14**, 137–144.

Eichinger, W., Bacher, A., Zenk, M. H. & Eisenreich, W. (1999) Quantitative assessment of metabolic flux by ^{13}C NMR analysis. Biosynthesis of anthraquinones in *Rubia tinctorum*. *J. Am. Chem. Soc.*, **121**, 7469–7475.

Ensley, B. D., Ratzkin, B. J., Osslund, T. D., Simon, M. J., Wackett, L. P. & Gibson, D. T. (1983) Expression of naphthalene oxidation genes in *Escherichia coli* results in biosynthesis of indigo. *Science*, **222**, 167–169.

Eugster, C. H. (1980) Terpenoid, especially diterpenoid pigments, in *Pigments in Plants* (ed. F.-C. Czygan), GustavFischer, Stuttgart, Germany, pp. 149–186.

Facchini, P. J. (2001) Alkaloid biosynthesis in plants: biochemistry, cell biology, molecular regulation, and metabolic engineering applications. *Annu. Rev. Plant Physiol. Plant Mol. Biol.*, **52**, 29–66.

Fester, T., Hause, B., Schmidt, D., Halfmann, K., Schmidt, J., Wray, V., Hause, G. & Strack, D. (2002a) Occurrence and localization of apocarotenoids in arbuscular mycorrhizal plant roots. *Plant Cell Physiol.*, **43**, 256–265.

Fester, T., Schmidt, D., Lohse, S., Walter, M. H., Giuliano, G., Bramley, P. M., Fraser, P. D., Hause, B. & Strack, D. (2002b) Stimulation of carotenoid metabolism in arbuscular mycorrhizal roots. *Planta*, **216**, 148–154.

Francis, F. J. (1996) Less common natural colourants, in *Natural Food Colorants* (eds G. A. F. Hendry & J. D. Houghton), Chapman & Hall, London, UK, pp. 112–132.

Fukui, H., Hasan, A. F. M. F., Ueoka, T. & Kyo, M. (1998) Formation and secretion of a new brown benzoquinone by hairy root cultures of *Lithospermum erythrorhizon*. *Phytochemistry*, **47**, 1037–1039.

Gao, W. F., Fan, L. & Paek, K. Y. (2000) Yellow and red pigment production by cell cultures of *Carthamus tinctorius* in a bioreactor. *Plant Cell Tiss. Org. Cult.*, **60**, 95–100.

Gilbert, K. G. & Cooke, D. T. (2001) Dyes from plants: past usage, present understanding and potential. *Plant Growth Reg.*, **34**, 57–69.

Gillam, E. M. J., Aguinaldo, A. M., Notley, L. M., Kim, D., Mundkowski, R. G., Volkov, A. A., Arnold, F. H., Soucek, P., DeVoss, J. J. & Guengerich, F. P. (1999) Formation of indigo by recombinant mammalian cytochrome P450s. *Biochem. Biophys. Res. Commun.*, **265**, 469–472.

Gillam, E. M. J. & Guengerich, F. P. (2001) Exploiting the versatility of human cytochrome P450 enzymes: the promise of blue roses from biotechnology. *IUBMB Life*, **52**, 271–277.

Gillam, E. M. J., Notley, L. M., Cai, H. L., De Voss, J. J. & Guengerich, F. P. (2000) Oxidation of indole by cytochrome P450 enzymes. *Biochem.*, **39**, 13817–13824.

Han, Y. S., Van der Heijden, R. & Verpoorte, R. (2001) Biosynthesis of anthraquinones in cell cultures of the Rubiaceae. *Plant Cell Tissue Organ Cult.*, **67**, 201–220.

Han, Y. S., van der Heijden, R., Lefeber, A. W. M., Erkelens, C. & Verpoorte, R. (2002) Biosynthesis of anthraquinones in cell cultures of *Cinchona* 'Robusta' proceeds via the methylerythritol 4-phosphate pathway. *Phytochemistry*, **59**, 45–55.

Han, Y. S., Roytrakul, S., Verberne, M. C., van der Heijden, R., Linthorst, H. J. M. & Verpoorte, R. (2003) Cloning of a cDNA encoding 1-deoxy-D-xylulose 5-phosphate synthase from *Morinda citrifolia* and analysis of its expression in relation to anthraquinone accumulation. *Plant Sci.*, **164**, 911–917.

Hart, S., Koch, K. R. & Woods, D. R. (1992) Identification of indigo-related pigments produced by *Escherichia coli* containing a cloned *Rhodococcus* gene. *J. Gen. Microbiol.*, **138**, 211–216.

Hart, S., Kirby, R. & Woods, D. R. (1990) Structure of a *Rhodococcus* gene encoding pigment production in *Escherichia coli*. *J. Gen. Microbiol.*, **136**, 1357–1363.

Hashimoto, T. & Yamada, Y. (1994) Alkaloid biogenesis: molecular aspects. *Annu. Rev. Plant Physiol. Plant Mol. Biol.*, **45**, 257–285.

Jako, C., Coutu, C., Roewer, I., Reed, D. W., Pelcher, L. E. & Covello, P. S. (2002) Probing carotenoid biosynthesis in developing seed coats of *Bixa orellana* (Bixaceae) through expressed sequence tag analysis. *Plant Sci.*, **163**, 141–145.

Kametaka, T. & Perkin, A. G. (1910) Cartamine, I. *J. Chem. Soc.*, **97**, 1415–1427.

Kazuma, K., Takahashi, T., Sato, K., Takeuchi, H., Matsumoto, T. & Okuno, T. (2000) Quinochalcones and flavonoids from fresh florets in different cultivars of *Carthamus tinctorius* L.. *Biosci. Biotech. Biochem.*, **64**, 1588–1599.

Köhle, A., Sommer, S., Yazaki, K., Ferrer, A., Boronat, A., Li, S. M. & Heide, L. (2002) High level expression of chorismate pyruvate-lyase (*UbiC*) and HMG-CoA reductase in hairy root cultures of *Lithospermum erythrorhizon*. *Plant Cell Physiol.*, **43**, 894–902.

Kokubun, T., Edmonds, J. & John, P. (1998) Indoxyl derivatives in woad in relation to medieval indigo production. *Phytochemistry*, **49**, 79–87.

Kramer, V. C. & Koziel, M. G. (1995) Structure of a maize tryptophan synthase alpha subunit gene with pith enhanced expression. *Plant Mol. Biol.*, **27**, 1183–1188.

Lange, B. M., Severin, K., Bechthold, A. & Heide, L. (1998) Regulatory role of microsomal 3-hydroxy-3-methylglutaryl-coenzyme A reductase for shikonin biosynthesis in *Lithospermum erythrorhizon* cell suspension cultures. *Planta*, **204**, 234–241.

Leistner, E. (1980) Quinonoid pigments, in *Pigments in Plants* (ed. F.-C. Czygan), GustavFischer, Stuttgart, Germany, pp. 352–368.

Leistner, E. (1981) Biosynthesis of plant quinones, in *The Biochemistry of Plants,* Vol. 7 (ed. E. E. Conn), Academic Press, London, UK, pp. 403–423.

Li, S.-M., Wang, Z.-X. & Heide, L. (1997a) Purification of UDP-glucose: 4-hydroxybenzoate glucosyltrans-ferase from cell cultures of *Lithospermum erythrorhizon*. *Phytochemistry*, **46**, 27–32.

Li, S.-M., Wang, Z.-X., Wemakor, E. & Heide, L. (1997b) Metabolization of the artificial secondary metabolite 4-hydroxybenzoate in *ubiC*-transformed tobacco. *Plant Cell Physiol.*, **38**, 844–850.

Li, S.-M., Hennig, S. & Heide, L. (1998) Shikonin: a geranyl diphosphate-derived plant hemiterpenoid formed via the mevalonate pathway. *Tetrahedron Lett.*, **39**, 2721–2724.

Lichtenthaler, H. K. (1999) The 1-deoxy-D-xylulose-5-phosphate pathway of isoprenoid biosynthesis in plants. *Annu. Rev. Plant Physiol. Plant Mol. Biol.*, **50**, 47–65.

Lim, E. K., Doucet, C. J., Li, Y., Elias, L., Worrall, D., Spencer, S. P., Ross, J. & Bowles, D. J. (2002) The activity of *Arabidopsis* glycosyltransferases toward salicylic acid, 4-hydroxybenzoic acid, and other benzoates. *J. Biol. Chem.*, **277**, 586–592.

Liu, B. Y., Falkenstein-Paul, H., Schmidt, W. & Beerhues, L. (2003) Benzophenone synthase and chalcone synthase from *Hypericum androsaemum* cell cultures: cDNA cloning, functional expression, and site-directed mutagenesis of two polyketide synthases. *Plant J.*, **34**, 847–855.

Lodhi, A. H. & Charlwood, B. V. (1996) *Agrobacterium rhizogenes*-mediated transformation of *Rubia peregrina* L.: *in vitro* accumulation of anthraquinones. *Plant Cell Tissue Organ Cult.*, **46**, 103–108.

Lodhi, A. H., Bongaerts, R. J. M., Verpoorte, R., Coomber, S. A. & Charlwood, B. V. (1996) Expression of bacterial isochorismate synthase (EC 5.4.99.6) in transgenic root cultures of *Rubia peregrina*. *Plant Cell Rep.*, **16**, 54–57.

Löscher, R. & Heide, L. (1994) Biosynthesis of *p*-hydroxybenzoate from *p*-coumarate and *p*-coumaroyl-coenzyme A in cell-free extracts of *Lithospermum erythrorhizon* cell cultures. *Plant Physiol.*, **106**, 271–279.

Mantrova, O. V., Dunaeva, M. V., Kuzovkina, I. N., Schneider, B. & Muller-Uri, F. (1999) Effect of methyl jasmonate on anthraquinone biosynthesis in transformed madder roots. *Russ. J. Plant Physiol.*, **46**, 248–251.

Marcinek, H., Weyler, W., Deus-Neumann, B. & Zenk, M. H. (2000) Indoxyl-UDPG-glucosyltransferase from *Baphicacanthus cusia*. *Phytochemistry*, **53**, 201–207.

Martin, G. S., Liu, J., Benedict, C. R., Stipanovic, R. D. & Magill, C. W. (2003) Reduced levels of cadinane sesquiterpenoids in cotton plants expressing antisense (+)-δ-cadinene synthase. *Phytochemistry*, **62**, 31–38.

Mathieson, D. W., Millward, B. J., Powell, J. W. & Whalley, W. B. (1973) The chemistry of the 'insoluble red' woods. Part XI. Revised structure of santalin and santarubin. *J. Chem. Soc. Perkin Trans. I*, 184–188.

Maugard, T., Enaud, E., Choisy, P. & Legoy, M. D. (2001) Identification of an indigo precursor from leaves of *Isatis tinctoria* (Woad). *Phytochemistry*, **58**, 897–904.

Mayer, M. J., Narbad, A., Parr, A. J., Parker, M. L., Walton, N. J., Mellon, F. A. & Michael, A. J. (2001) Rerouting the plant phenylpropanoid pathway by expression of a novel bacterial enoyl-CoA hydratase/lyase enzyme function. *Plant Cell*, **13**, 1669–1682.

Melanson, D., Chilton, M.-D., Masters-Moore, D. & Chilton, W. S. (1997) A deletion in an indole synthase gene is responsible for the DIMBOA-deficient phenotype of *bxbx* maize. *Proc. Natl. Acad. Sci. USA*, **94**, 13345–13350.

Mermod, N., Harayama, S. & Timmis, K. N. (1986) New route to bacterial production of indigo. *BioTechnology*, **4**, 321–326.

McBride, K., Pear, J. R., Perez-Grau, L. & Stalker, D. M. (1996) Cotton fiber transcriptional factors. International application published under the Patent Cooperation Treaty, WO 96/40924.

Minami, Y., Kanafuji, T. & Miura, K. (1996) Purification and characterization of a β-glucosidase from *Polygonum tinctorium* which catalyzes preferentially the hydrolysis of indican. *Biosci. Biotech. Biochem.*, **60**, 147–149.

Minami, Y., Takao, H., Kanafuji, T., Miura, K., Kondo, M., Hara-Nishimura, I., Nishimura, M. & Matsubara, H. (1997) Glucosidase in the indigo plant: intracellular localisation and tissue specific expression in leaves. *Plant Cell Physiol.*, **38**, 1069–1074.

Minami, Y., Shigeta, Y., Tokumoto, U., Tanaka, Y., Yonekura-Sakakibara, K., Oh-oka, H. & Matsubara, H. (1999) Cloning, sequencing, characterization, and expression of a β-glucosidase cDNA from the indigo plant. *Plant Sci.*, **142**, 219–226.

Minami, Y., Nishimura, O., Hara-Nishimura, I., Nishimura, M. & Matsubara, H. (2000) Tissue and intracellular localization of indican and the purification and characterization of indican synthase from indigo plants. *Plant Cell Physiol.*, **41**, 218–225.

Murakami, T., Kishi, A., Sakurama, T., Matsuda, H. & Yoshikawa, M. (2001) Chemical constituents of two oriental orchids, *Calanthe discolor* and *C. liukiuensis*: precursor indole glycoside of tryptanthrin and indirubin. *Heterocycles*, **54**, 957–966.

Narvaez, J. A., Canche, B. B. C., Perez, P. F. & Madrid, R. R. (2001) Differential expression of 3-hydroxy-3-methylglutaryl-CoA reductase (HMGR) during flower and fruit development of *Bixa orellana*. *J. Plant Physiol.*, **158**, 1471–1477.

Notley, L., Devoss, J., Gillam, E. J., Guengerich, F. P. & Volkov, A. (2001) Pigment production by cells having introduced cytochrome P450 sequences. International application published under the Patent Cooperation Treaty, WO 01/14565.

Opitz, S. & Schneider, B. (2003) Oxidative biosynthesis of phenylbenzoisochromenones from phenylphenalenones. *Phytochemistry*, **62**, 307–312.

Opitz, S., Schnitzler, J.-P., Hause, B. & Schneider, B. (2003) Histochemical analysis of phenylphenalenone-related compounds in *Xiphidium caeruleum* (Haemodoraceae). *Planta*, **216**, 881–889.

Orska-Gawrys, J., Surowiec, I., Kehl, J., Rejniak, H., Urbaniak-Walczak, K. & Trojanowitcz, M. (2003) Identification of natural dyes in archeological Coptic textiles by liquid chromatography with diode array detection. *J. Chromatogr. A.*, **898**, 239–248.

Otálvaro, F., Görls, H., Hölscher, D., Schmitt, B., Echeverri, F., Quiñones, W. & Schneider, B. (2002) Dimeric phenylphenalenones from *Musa acuminata* and various Haemodoraceae species. Crystal structure of anigorootin. *Phytochemistry*, **60**, 61–66.

Ouyang, J., Shao, X. & Li, J. Y. (2000) Indole-3-glycerol phosphate, a branchpoint of indole-3-acetic acid biosynthesis from the tryptophan biosynthetic pathway in *Arabidopsis thaliana*. *Plant J.*, **24**, 327–333.

Pandey, A. K. & Dhakal, M. R. (2001) Phytomelanin in compositae. *Curr. Sci.*, **80**, 933–940.

Papageorgiou, V. P., Assimopoulou, A. N., Couladouros, E. A., Hepworth, D. & Nicolaou, K. C. (1999) The chemistry and biology of alkannin, shikonin, and related naphthazarin natural products. *Angew. Chemie Int. Ed.*, **38**, 271–300.

Park, J. E., Lee, J. Y., Kim, H. G., Hahn, T. R. & Paik, Y. S. (2002c) Isolation and characterization of water-soluble intermediates of blue pigments transformed from geniposide of *Gardenia jasminoides*. *J. Agric. Food Chem.*, **50**, 6511–6514.

Park, S. U., Yu, M. & Facchini, P. J. (2002a) Modulation of berberine bridge enzyme levels in transgenic root cultures of California poppy alters the accumulation of benzophenanthridine alkaloids. *Plant Mol. Biol.*, **51**, 153–164.

Park, S. U., Yu, M. & Facchini, P. J. (2002b) Antisense RNA-mediated suppression of benzophenanthridine alkaloid biosynthesis in transgenic cell cultures of California poppy. *Plant Physiol.*, **128**, 696–706.

Piovan, A., Filippini, R., Caniato, R., Borsarini, A., Maleci, L. B. & Cappelletti, E. M. (2004) Detection of hypericins in the ''red glands'' of *Hypericum elodes* by ESI-MS/MS. *Phytochemistry*, **65**, 411–414.

Radwanski, E. R., Zhao J.-M. & Last, R. L. (1995) *Arabidopsis thaliana* tryptophan synthase alpha: gene cloning, expression, and subunit interaction. *Mol. Gen. Genet.*, **248**, 657–667.

Ramos-Valdivia, A. C., van der Heijden, R. & Verpoorte, R. (1997a) Isopentenyl diphosphate isomerase: a core enzyme in isoprenoid biosynthesis. A review of its biochemistry and function. *Nat. Prod. Rep.*, **14**, 591–603.

Ramos-Valdivia, A. C., van der Heijden, R. & Verpoorte, R. (1997b) Elicitor-mediated induction of anthraquinone biosynthesis and regulation of isopentenyl diphosphate isomerase and farnesyl diphosphate synthase activities in cell suspension cultures of *Cinchona robusta* How.. *Planta*, **203**, 155–161.

Ramos-Valdivia, A. C., van der Heijden, R. & Verpoorte, R. (1998) Isopentenyl diphosphate isomerase and prenyltransferase activities in rubiaceous and apocynaceous cultures. *Phytochemistry*, **48**, 961–969.

Riley, P. A. (1997) Melanin. *Int. J. Biochem. Cell Biol.*, **29**, 1235–1239.

Saito, K., Takahashi, Y. & Wada, M. (1983) Enzymic synthesis of carthamin in safflower. *Biochim. Biophys. Acta.*, **8**, 127–133.

Saito, K., Kanehira, T., Horimoto, M., Moritome, N. & Komamine, A. (1993) Biosynthesis of carthamin in florets and cultured cells of *Carthamus tinctorius*. *Biologia Plant.*, **35**, 537–546.

Samanani, N., Yeung, E. C. & Facchini, P. J. (2002) Cell type-specific protoberberine alkaloid accumulation in *Thalictrum flavum*. *J. Plant Physiol.*, **159**, 1189–1196.

Sato, F., Hashimoto, T., Hachiya, A., Tamura, K., Choi, K. B., Morishige, T., Fujimoto, H. & Yamada, Y. (2001) Metabolic engineering of plant alkaloid biosynthesis. *Proc. Natl. Acad. Sci. USA*, **98**, 367–372.

Schaller, H., Grausem, B., Benveniste, P., Chye, M. L., Tan, C. T., Song, Y. H. & Chua, N. H. (1995) Expression of the *Hevea brasiliensis* (H.B.K.) Müll. Arg. 3-hydroxy-3-methylglutaryl-coenzyme A reductase 1 in tobacco results in sterol overproduction. *Plant Physiol.*, **109**, 761–770.

Schmitt, B. & Schneider, B. (1999) Dihydrocinnamic acids are involved in the biosynthesis of phenylphena-lenones in *Anigozanthos preissii*. *Phytochemistry*, **52**, 45–53.

Schmitt, B., Holscher, D. & Schneider, B. (2000) Variability of phenylpropanoid precursors in the biosyn-thesis of *Anigozanthos preissii*. *Phytochemistry*, **53**, 331–337.

Schripsema, J., Ramos-Valdivia, A. & Verpoorte, R. (1999) Robustaquinones, novel anthraquinones from an elicited *Cinchona robusta* suspension culture. *Phytochemistry*, **51**, 55–60.

Schröder, J. (1997) A family of plant-specific polyketide synthases: facts and predictions. *Trends Plant Sci.*, **2**, 373–378.

Schröder, J. (1999) The chalcone/stilbene synthase-type family of condensing enzymes, in *Comprehensive Natural Products Chemistry*, Vol. 1: *Polyketides and Other Secondary Metabolites Including Fatty Acids and their Derivatives* (ed. U. Sankawa), Elsevier, Oxford, UK, pp. 749–771.

Siebert, M., Sommer, S., Li, S.-M., Wang, Z.-W., Severin, K. & Heide, L. (1996) Genetic engineering of plant secondary metabolism: accumulation of 4-hydroxybenzoate glucosides as a result of the expression of the bacterial *ubiC* gene in tobacco. *Plant Physiol.*, **112**, 811–819.

Simantiras, M. & Leistner, E. (1992) *O*-succinylbenzoate: coenzyme A ligase from anthraquinone producing cell suspension cultures of *Galium mollugo*. *Phytochemistry*, **31**, 2329–2335.

Smith, H. B. (1999) Photosynthetic pigmentation – variegations on a theme. *Plant Cell*, **11**, 1–3.

Sommer, S., Kohle, A., Yazaki, K., Shimomura, K., Bechthold, A. & Heide, L. (1999) Genetic engineering of shikonin biosynthesis hairy root cultures of *Lithospermum erythrorhizon* transformed with the bacterial *ubiC* gene. *Plant Mol. Biol.*, **39**, 683–693.

Stalman, M., Koskamp, A. M., Luderer, R., Vernooy, J. H. J., Wind, J. C., Wullems, G. J. & Croes, A. F. (2003) Regulation of anthraquinone biosynthesis in cell cultures of *Morinda citrifolia*. *J. Plant Physiol.*, **160**, 607–614.

Thomson, R. H. (1976a) Quinones: nature, distribution and biosynthesis, in *Chemistry and Biochemistry of Plant Pigments,* 2nd edn, Vol. 1 (ed. T. W. Goodwin), Academic Press, London, UK, pp. 527–559.

Thomson, R. H. (1976b) Miscellaneous pigments, in *Chemistry and Biochemistry of Plant Pigments,* 2nd edn, Vol. 1 (ed. T. W. Goodwin), Academic Press, London, UK, pp. 597–623.

Tsukata, M. & Tabata, M. (1984) Intracellular localization and secretion of naphthoquinone pigments in cell cultures of *Lithospermum erythrorhizon*. *Planta Med.*, **50**, 338–340.

van Tegelen, L. J. P., Bongaerts, R. J. M., Croes, A. F., Verpoorte, R. & Wullems, G. J. (1999) Isochorismate synthase isoforms from elicited cell cultures of *Rubia tinctorum*. *Phytochemistry*, **51**, 263–269.

Wang, M.-Y., West, B. J., Jensen, C. J., Nowicki, D., Su, C., Palu, A. K. & Anderson, G. (2002) *Morinda citrifolia* (Noni): a literature review and recent advances in Noni research. *Acta Pharmacol. Sin.*, **23**, 1127–1141.

Wang, C. Z., Maier, U. H., Keil, M., Zenk, M. H., Bacher, A., Rohdich, F. & Eisenreich, W. (2003) Phenylalanine-independent biosynthesis of 1,3,5,8-tetrahydroxyxanthone – a retrobiosynthetic NMR study with root cultures of *Swertia chirata*. *Eur. J. Biochem.*, **270**, 2950–2958.

Walter, M. H., Fester, T. & Strack, D. (2000) Arbuscular mycorrhizal fungi induce the non-mevalonate methylerythritol phosphate pathway of isoprenoid biosynthesis correlated with accumulation of the 'yellow pigment' and other apocarotenoids. *Plant J.*, **21**, 571–578.

Winterhalter, P. & Straubinger, M. (2000) Saffron: renewed interest in an ancient spice. *Food Rev. Int.*, **16**, 39–59.

Xia, Z.-Q. & Zenk M. H. (1992) Biosynthesis of indigo precursors in higher plants. *Phytochemistry*, **31**, 2695–2697.

Yamamoto, H., Inoue, K., Li, S. M. & Heide, L. (2000) Geranylhydroquinone 3″-hydroxylase, a cytochrome P450 monooxygenase from *Lithospermum erythrorhizon* cell suspension cultures. *Planta*, **210**, 312–317.

Yamamura, Y., Ogihara, Y. & Mizukami, H. (2001) Cinnamic acid 4-hydroxylase from *Lithospermum erythrorhizon*: cDNA cloning and gene expression. *Plant Cell Rep.*, **20**, 655–662.

Yamamura, Y., Sahin, F. P., Nagatsu, A. & Mizukami, H. (2003) Molecular cloning and characterization of a cDNA encoding a novel apoplastic protein preferentially expressed in a shikonin-producing callus strain of *Lithospermum erythrorhizon*. *Plant Cell Physiol.*, **44**, 437–446.

Yazaki, K., Ogawa, A. & Tabata, M. (1995) Isolation and characterization of two cDNAs encoding 4-coumarate:CoA ligase in *Lithospermum* cell cultures. *Plant Cell Physiol.*, **36**, 1319–1329.

Yazaki, K., Kataoka, M., Honda, G., Severin, K. & Heide, L. (1997) cDNA cloning and gene expression of phenylalanine ammonia-lyase in *Lithospermum erythrorhizon*. *Biosci. Biotech. Biochem.*, **61**, 1995–2003.

Yazaki, K., Matsuoka, H., Shimomura, K., Bechthold, A. & Sato, F. (2001) A novel dark-inducible protein, LeDI-2, and its involvement in root-specific secondary metabolism in *Lithospermum erythrorhizon*. *Plant Physiol.*, **125**, 1831–1841.

Yazaki, K., Kunihisa, M., Fujisaki, T. & Sato, F. (2002) Geranyl diphosphate: 4-hydroxybenzoate geranyl-transferase from *Lithospermum erythrorhizon* – cloning and characterization of a key enzyme in shikonin biosynthesis. *J. Biol. Chem.*, **277**, 6240–6246.

8 Plant pigments and human health

Mary Ann Lila

8.1 Introduction/Overview

While all color additives to foods are strictly regulated (Feord, 2003), naturally pigmented foods, and food products enhanced with pigments extracted from plants, are overwhelmingly preferred by consumers, especially in light of adverse publicity surrounding the consumption of some synthetic colorants. Natural pigments not only circumvent the putative health hazards posed by synthetic chemical colorants but also confer substantive and multifaceted health benefits (nutraceutical value) to the diet. The natural plant pigments are considered to have bioactive (or functional food) value because, unlike the traditional classes of nutrients (proteins, vitamins, fat, minerals), they are not generally considered vital to survival, but instead, in a wide range of ways, have properties that can promote optimal human health. The nutraceutical value of natural plant pigments such as carotenoids, anthocyanins, chlorophylls, and betalains is based on their recognized biologically active properties. These pigments have been implicated in regimes to maintain human health, to protect against chronic disease incidence, or to restore wellness by repairing tissues after disease has been established. Levels of specific natural pigments vary with plant species and variety, and they are a natural target for manipulation and augmentation using conventional and biotechnological breeding tactics. Given that the biosynthetic pathways of the pigments have been studied extensively, it is now feasible to enhance or alter pigment content in a food crop using either strategy. As examples, the United States Department of Agriculture/Agricultural Research Service (USDA ARS) bred a tomato with ten times the lycopene level of its original parent genotype; 'golden rice' was genetically engineered to produce a carotenoid precursor to vitamin A and a source of this deficient vitamin in a grain staple which normally would not carry it (Potrykus, 2003); and anthocyanin regulatory genes have been manipulated to produce unique color combinations in various floral crops that have no counterpart in nature.

A wide range of bioassays and tests have been forwarded to establish the biological efficacy of natural pigments in human health intervention, including *in vitro* (laboratory bench top) bioassays, *in vivo* (animal trials), epidemiological (population surveys) and more rarely, clinical (human intervention) trials (Table 8.1). Almost all early support for the roles of pigments in biological systems was based on an eclectic set of anecdotal and/or epidemiological evidence, not always well documented; however, increasingly, with the surge of interest in functional/natural foods for health, the evidence is shifting towards replicated *in vivo* animal

Table 8.1 Levels of research inquiry applied to natural plant pigments and human health interventions

Pigment	Disease condition or health target	Selected research evidence supporting the role of pigment in chronic disease prevention or therapy			
		In vitro	In vivo	Epidemiological	Clinical
Anthocyanins	**Vision disorders**	—	Mastsumoto et al., 2001	Morazzoni & Bombardelli, 1996	Mastsumoto et al., 2001
	Neuroprotection	Youdim et al., 2000c	Joseph et al., 1999 Youdim et al., 2000b	—	—
	Cardiovascular disease	Oak et al., 2003 Serraino et al., 2003	Nielsen et al., 2003	Havsteen, 2002 Duthie et al., 2000	Shanmuganayagam et al., 2002
	Cancer	Kang et al., 2003 Naasani et al., 2003	Kang et al., 2003 Naasani et al., 2003	Horvathova et al., 2001 Duthie et al., 2000	—
Betalains	**Cancer**	Wettasinghe et al., 2002	Kapadia et al., 1996	—	—
Carotenoids	**Photooxidative damage**	Sies & Stahl, 2003	—	Khachik et al., 2002	Snodderly, 1995
	Immune system	Jyonouchi et al., 1996	—	—	Hughes et al., 1997
	Cardiovascular disease	Fuhrman et al., 1997	Fuhrman et al., 1997	Delgado-Vargas et al., 2000 Platz et al., 2003	Platz et al., 2003
	Cancer	Kotake-Nara et al., 2001 Liu et al., 2003	Wang et al., 1999 Jain et al., 1999 Teplizky et al., 2001	Goodman et al., 2003 Toniolo et al., 2001 Nyberg et al., 2003	Baron et al., 2003 Deming & Erdman, 1999 Kucuk et al., 2002
Chlorophylls	**Cancer**	Chung et al., 1999	Park & Surh, 1996 Rebeiz et al., 1996	—	—

models and clinical trials, especially for the most widely researched pigments like anthocyanins and some carotenoids.

The natural pigments and related phytochemicals are integral to the plant's strategies for survival, serving protective roles (as shields against abiotic stresses like temperature, UV-B radiation, low water potential, or mineral stress); active defensive roles (against pathogen or insect attack, or against herbivory and inter-plant competition); attractant roles (inviting pollinators and seed dispersors), or roles in cell-level regulation (as gene expression modulators). The pigments contribute to a wealth of survival strategies and competitive advantages for the host plants. The ecological benefits to the plant have at least some bearing on the potential medicinal benefits for humans. For example, cytotoxic properties, which serve the defense/protective strategies of the plant, also relate to antimicrobial properties benefiting human consumers, and may also relate to cytotoxicity to malignant cells. Pigment molecules that have evolved to mimic hormones or other metabolites can achieve medicinal effects by interfacing with the same potential target sites (endocrine system, nervous system).

Pigments as health supplements or ingredients can be extracted from crop plants and by- products, and several laboratories have explored bioreactor-based produc-tion of the same. For example, betalain pigments have been synthesized actively in hairy root cultures (Pavlov *et al.*, 2002) as well as in suspension cultures (Akita *et al.*, 2000) and large batch-type bioreactor cultures (Leathers *et al.*, 1992), under sponsorship of pharmaceutical industries interested in expanding their product portfolios to naturally occurring plant pigments with biomedicinal value. Simi-larly, anthocyanins have been accumulated in a wide range of *in vitro*-cultivated callus and cell cultures from species including *Ajuga*, *Daucus*, *Rudbeckia*, *Vacci-nium*, and many others (Pépin *et al.*, 1999; Kandil *et al.*, 2000; Bourgaud *et al.*, 2001; Hirner *et al.*, 2001; Luczkiewcz & Cisowski, 2001; Terahara *et al.*, 2001). Through the technology of large-scale plant *in vitro* culture, an alternative means of producing natural color additives under a controlled industrial regime is available (Sahai, 1994). Still, relatively few plant pigments have yet reached commercial scale production using plant cell culture strategies.

While the nutritional and nutraceutical value of a well-balanced 'colorful' diet has been widely advocated, individual contributions of plant pigments to human health have been difficult to precisely document. Four major challenges to health investigations involving natural pigments can be cited. These versatile compounds have unfortunately been extremely difficult to study due to their (1) often large, complex structures, (2) ephemeral nature (tendency to degrade during chemical separation), and (3) the fact that pigments can be highly metabolized after inges-tion, making it very difficult to assign the contribution of particular constituents to human health intervention, or assess their absorption and bioavailability. In addition, (4) natural plant pigments often exert biological activity in concert with other co-occurring phytochemicals in a food or in the food matrix. These potentiating interactions are responsible for observed interventions in human health status, and it is difficult to assign contributions of single components.

Other factors that complicate the task of pinpointing the bioactivity of individual natural pigments include the multitude of structures and isomers involved, the analytical limits imposed by the complexity of the chemical structures (Santos-Buelga & Scalbert, 2000), the losses in efficacy that occur when interacting components are isolated and purified for chemical characterization (Smith, 2000), and the lack of adequate means to follow the uptake, metabolism, and metabolic fate of pigments after ingestion (Deprez *et al.*, 1999). In part, the discrepancies between different reports concerning the bioactivity and efficacy of certain pigments against chronic disease are due to current limitations in testing techniques and instrumentation. New paradigms and more robust guidelines about recommended intakes will follow as research continues to elucidate the contribution of natural pigments in human health maintenance (Fox, 2003).

Despite the intense research efforts surrounding natural plant pigments and their manipulation (especially in light of health benefits), a number of key questions remain to be answered. At present, neither scientists nor medical professionals can establish optimal levels for human intake. At least in the case of some carotenoids (see below), excessive levels of intake have proven deleterious to human health. Requirements are likely to vary between individuals depending on their health status, gender, ethnicity, age, body mass index, and other factors. Given the potentiation effects of interacting phytochemicals, it is evident in many cases that the pigments intervene in human health maintenance most effectively when delivered as a component of the food matrix, rather than in supplement form. In addition, various forms of processing or cooking can impinge on the subsequent absorption and bioavailability of pigments and related phytochemicals. While in some cases, thermal processing, light, or oxygen may degrade these natural pigmented compounds, in others, the bioavailability may actually be enhanced.

The chemical structures and nature of each of the pigments have been covered in previous chapters. In this chapter, for each of the major plant pigment groups, the evidence for biological activity against human chronic disease conditions is reviewed, along with evidence for interactions between pigments and other phytochemical compounds leading to more potent bioactivity (potentiation). Finally, recent progress towards determining the metabolic fate of pigments after ingestion into the body, using tracking mechanisms, is described.

8.2   Anthocyanins (and associated flavonoids)

8.2.1   Diverse roles of anthocyanin pigments in human health maintenance

The flavonoid group, which includes anthocyanin pigments, has been linked to an incredibly diverse range of biological functions in human metabolism. Recently the USDA ARS constructed an extensive database for flavonoid content of selected foods www.nal.usda.gov/fnic/foodcomp/index.html. It is hoped that the provision of detailed compositional data will allow the research community to assess dietary intakes and identify relationships between those intakes and chronic disease risk factors. Because the flavonoid biosynthetic pathway is probably the most extensively researched natural product pathway in plants, it has been suggested as an excellent target for metabolic engineering, using sense or antisense manipulation of pathway genes, modification of regulatory gene expression, or generation of novel enzymatic specificities (Dixon & Steele, 1999; Parr & Bowell, 2000; Davies et al., 2003).

Anthocyanins exert significant antimicrobial properties, and (in association with other flavonoids) have demonstrated quite effective inhibition of aflatoxin biosynthesis (Norton, 1999). Folk medicine has relied on anthocyanins taken from *Hibiscus sabdariffa* L. as a remedy for liver dysfunction and hypertension; the antioxidant capacity of the pigments is considered integral to efficacy in these cases (Wang et al., 2000). Particularly complex anthocyanin profiles in members of the genus *Vaccinium* including bilberry (in Europe) and blueberry (in North America) have been credited with superb antioxidant capacities, especially in acylated anthocyanin mixtures (Smith et al., 2000). Anthocyanins, depending on the concentrations ingested, confer protection to DNA by preventing radical hydroxyl attack. In these reactions, cyanidin-DNA co-pigmentation is suggested as a possible defense mechanism against oxidative damage of DNA; once an anthocyanin (e.g. cyanidin) complexes with DNA, it is no longer vulnerable to nucleophilic attack by an OH group (Sarma & Sharma, 1999). In studies with chokeberry, honeysuckle, and sloe fruits, the antioxidative potency of anthocyanin extracts was shown to be concentration-dependent when end products of lipid membrane oxidation (induced by UV radiation) were evaluated using HPLC (Garielska et al., 1999). Anthocyanin aglycones were shown to have significant estrogenic activity, like other bioflavonoids in the metabolic pathway, and thus these natural pigments may play a role in altering the development of hormone-dependent conditions (Schmitt & Stopper, 2001). An anthocyanin's structure is intrinsic to its antioxidant capacity, appearing to be related to patterns of hydroxylation and glycosylation (Wang et al., 1997). Hydroxyl B-ring substituents induce high antioxidant activity in glycosylated anthocyanins, and the presence of saturated 2,3-double bonds and 3'- and 4'-OH in the B-ring structure are all positively correlated with antioxidant capacity (Delgado-Vargas & Paredes-López, 2003). Antioxidant capacity of anthocyanin pigments (including ability to inhibit oxidative enzymes) is by far the most widely publicized mechanism, although it is clear that these pigments benefit human health maintenance via other potential mechanisms as well. In a survey of high- and lowbush blueberry genotypes analyzed for antioxidant potential, there was an excellent direct correlation with anthocyanin as well as total polyphenolic content (Kalt et al., 2001), whereas in a survey of over 100 berry fruits, total polyphenolic content was found to be more highly correlated to antioxidant capacity than just the content of anthocyanins in the berries (Moyer et al., 2002). Structure/function relationships in anthocyanin molecules and derivatives, related to the presence of a sugar moiety, influence their ability to protect against DNA damage in smooth muscle and hepatoma cells in rodent models (Lazze et al., 2003).

Anthocyanins are credited with ability to enhance immune function by boosting the production of cytokines, which help regulate immune responses. Anthocyanins/anthocyanidins and anthocyanin-rich extracts induced tumor necrosis factor α (TNF-α) production in macrophages that had been gamma-activated – an induction that was requisite for the macrophages to engage in their scavenging function, and these flavonoids also acted as modulators of the immune response in activated macrophages (Wang & Mazza, 2002).

Although strong research progress has been advanced recently, absorption and bioavailability issues with regard to anthocyanin pigments are still largely uncertain; metabolites are unknown and unaccounted for. An increase in antioxidant capacity of plasma after consumption of flavonoid-rich foods gives indirect evidence for the absorption and bioavailability (Scalbert & Williamson, 2000), as well as measurable increases in flavonoid concentrations in plasma and urine. Only a small proportion of flavonoids can be detected in serum or urine, which either implies poor bioavailability or that the metabolites have not been detected and isolated. Whereas earlier studies assumed that only aglycones could be effectively absorbed from foods (that glycosides must be hydrolyzed by bacteria in the colon before absorption), more recently it has been established that glycosides were well absorbed in humans without prior hydrolysis by microorganisms (Hollman & Katan, 1998), although extensive modification occurs during metabolism (by liver or colonic flora). Anthocyanidins are much more limitedly metabolized than many other associated flavonoids (Hollman & Katan, 1998). Twenty percent of bilberry (*Vaccinium myrtillus*) anthocyanins administered intravenously to rodents were excreted intact in urine (Lietti & Forni, 1976).

The propensity of a flavonoid to inhibit free radical-mediated events is governed by its chemical structure, in particular the number, positions, and types of substitutions that influence radical scavenging and chelating activity (Heim *et al.*, 2002). Multiple hydroxyl groups typically confer antioxidant, chelating, and pro-oxidant activity, whereas methoxyl groups introduce unfavorable steric effects and increase lipophilicity and membrane partitioning. Some researchers have noted that, given their increasing levels of consumption especially in dietary supplements, some caution needs to be taken with regard to potentially toxic effects of high doses of flavonoids, which can act as pro-oxidants and mutagens (Skibola & Smith, 2000).

8.2.2 Anthocyanins and treatment of vision disorders

The most extensively documented phytomedicinal role of anthocyanin pigments is in improving eyesight, including night vision. Black currant (*Ribes nigrum* L.) anthocyanins concentrated from juice, when administered to human subjects, markedly improved visual adaptation to darkness. Anthocyanin pigments have been administered to remedy vision disorders, enhance visual acuity, and increase capillary resistance. After oral administration of three anthocyanin species to rats, intact pigments can be found in plasma; similar results were achieved subsequently in human subjects (Matsumoto *et al.*, 2001).

Retinopathy and cataracts, serious consequences of diabetes mellitus, can be combated using plant-derived anthocyanin pigments, because the pigments inhibit the α-glucosidase enzyme (AGH) located in the epithelium of the small intestine. AGH normally catalyzes cleavage of glucose from disaccharides, leading to excess glucose absorption and aggravation of diabetes mellitus. Anthocyanin extracts from a wide range of plant species including *Ipomoea batatas* and *Brassica oleracea* have demonstrated ability to inhibit AGH, especially the acylated forms of the anthocyanin pigments.

8.2.3   Anthocyanins and neuroprotection

Significant new research has revealed a role for anthocyanin pigments in protection against neurological disorders, especially declines related to aging. Originally, this association was based on *in vivo* evidence demonstrating that rats administered an anthocyanin-rich diet showed significant improvements in motor and cognitive function, and protection against age-associated impairments. In fact, the research demonstrated not only protection but also reversal of age-related decrements when artificially aged rats were fed a diet that was comprised of 2% blueberry extract (high in anthocyanin content) (Joseph *et al.*, 1999; Youdim *et al.*, 2000b). These *in vivo* feeding studies led to further research to determine how bioavailable the anthocyanin pigments were. Anthocyanins were found to confer significant protection from oxidative stress and to be highly bioavailable in endothelial cells, which has direct relevance to atherosclerosis and neurodegenerative disorders. Both *in vivo* and *in vitro* research probes demonstrated that these polyphenolic compounds enhanced red blood cell resistance to oxidative stress (Youdim *et al.*, 2000c). Cyanidin-3-sambubioside-5-glucoside and cyanidin-3,5-diglucoside are two of the anthocyanin species responsible for the protective effects in these studies (Youdim *et al.*, 2000a, 2002).

8.2.4   Anthocyanins and inhibition of cardiovascular disease

Intake of flavonoids, including tannins and anthocyanins, is associated with reduced incidence of coronary heart disease (Peterson & Dwyer, 1998; Havsteen, 2002). A typical daily dietary intake (25–215 mg) without supplementation is sufficient to provide pharmacological benefits. The rich anthocyanin pigment content of red wines, as well as associated flavonoids, are purported to be responsible for the well-publicized correlation between red wine consumption and reduced cardiovascular mortality (the French paradox). Because anthocyanins are potent inhibitors of nitrated tyrosine formation *in vitro*, they may have a role in prevention of atherosclerotic lesions in human coronary arteries (Tsuda *et al.*, 2000). Bioflavonoid mixtures (including the anthocyanin pigments cyanidin, cyanidin-3-glucoside, and cyanidin-3,5-diglucoside) derived from fruit sources

(*Rubus occidentalis*, *Sambucus nigra*, bilberry, and *Aronia melanocarpa*) have demonstrated ability to accelerate ethanol metabolism, reduce inflammatory and edematic symptoms, and improve permeability and strength of capillaries (Francis, 1989). Pure anthocyanins are up to seven times more effective as antioxidants inhibiting lipid peroxidation than α-tocopherol. Anthocyanins in red wines and grape juice contributed to the ability of these dietary sources to inhibit heart attacks, by virtue of their antithrombotic potential, and to inhibit platelet function and enhance nitric oxide release (Folts, 1998; Stein *et al.*, 1999; Freedman *et al.*, 2001). The ability of anthocyanins and related polyphenolics in red wines to prevent the development of atherosclerotic lesions (in part, by inhibiting vascular endothelial growth factor expression in vascular smooth muscle cells) was demonstrated recently by Oak *et al.* (2003). Development of peroxynitrite-induced vascular dysfunction could be ameliorated by blackberry juice (active ingredient: cyanidin-3-*O*-glucoside) (Serraino *et al.*, 2003).

8.2.5   Anthocyanins and cancer chemoprevention

Anthocyanins (and the aglycone cyanidin) were noted to inhibit cycloxygenase enzymes, which can be one marker for the initiation stage of carcinogenesis. Recently, both the anthocyanins and cyanidin aglycone from tart cherries reduced cell growth of human colon cancer cell lines (Kang *et al.*, 2003). As part of the same study, mice consuming a cherry diet, or a diet enriched with the anthocyanin or aglycone exclusively, exhibited significantly fewer and smaller adenomas. Anthocyanins and related flavonoids from berries including cranberry, lingonberry, bilberry and blueberry have demonstrated the ability to inhibit not only the initiation stages of chemically induced carcinogenesis but also the later promotion and proliferation stages (Bomser *et al.*, 1996; Kandil *et al.*, 2000, 2002; Smith *et al.*, 2000; Schmidt *et al.*, 2003).

8.2.6   Potentiation of anthocyanin bioactivity

Bioflavonoids like anthocyanins occur in mixtures within edible foods, and are ingested in mixtures. It is difficult to distinguish the biological activity of the red to blue anthocyanin pigments from other associated flavonoids, because these flavonoid compounds invariably function interactively, and biological activity is potentiated with additive or synergistic interactions between co-occurring compounds in a fruit or vegetable (Santos-Buelga & Scalbert, 2000). In a comparison of uptake and excretion of anthocyanins in humans and in a rabbit model, a significant food matrix effect was detected in rabbits, resulting in the absorption of a higher proportion of anthocyanins from black currant than from an aqueous citric acid matrix (Nielsen *et al.*, 2003). However, in humans the absorption and urinary excretion were proportional with dose and not influenced by the co-ingestion of a rice cake. High recoveries were achieved in both urine and plasma samples.

Different flavonoids present in grape skin extracts versus grape seed extracts apparently interact together to potentiate the antiplatelet activity in human and animal subjects (Shanmuganayagam et al., 2002). Components from both sources are present in wines or grape juice, but may not be able to act additively or synergistically in a supplement from only one source. Other investigators concur that synergism among flavonoids is responsible for antiplatelet activity of red wine and grape juice. Concentrations for any one anthocyanin or other flavonoid component are too low in the human bloodstream to effectively inhibit platelet aggregation, and the investigators were able to demonstrate that catechin monomers mixed with quercetin flavonoids, at concentrations comparable to those expected after grape juice consumption, were synergistic in their activity (Violi et al., 2002). Similar effects are noted in co-occurring flavonoids, which interact synergistically to inhibit platelet function by antagonizing hydrogen peroxide generation in the cell (Pignatelli et al., 2000). Synergism in certain flavonoid combinations can result in simultaneous efficacy in radical scavenging and in transition–metal ion chelation properties, depending on the components in the mixtures (Shafiee et al., 2002).

8.3 Betalains

8.3.1 Diverse roles of betalain pigments in human health maintenance

Betalains have not been as extensively researched in terms of health benefits as the other major plant-derived pigments; however, antiviral and antimicrobial properties are well documented (Strack et al., 2003). Commercially, betaxanthin pigments have been introduced as a food supplement in order to fortify processed food products with a nutraceutical natural colorant containing essential amino acids (Lee & Min, 1990). The very potent antiradical scavenging activity of betalains has been demonstrated repeatedly in a wide range of assays (Escribano et al., 1998; Zakharova & Petrova, 1998; Kanner et al., 2001; Kujala et al., 2001; Pavlov et al., 2002; Wettasinghe et al., 2002; Cai et al., 2003). Recently, betalain enrichment of human low-density lipoproteins effectively increased their resistance to oxidation (Tesoriere et al., 2003). A striking characteristic of betalain pigments is that they can be naturally absorbed by the body; at least 15% of consumers absorb large amounts resulting in temporary red urinary discharge, which indicates minimal transformation during metabolism (Wettasinghe et al., 2002).

8.3.2 Betalains and cancer chemoprevention

Preliminary cancer-chemopreventive studies with crude beetroot extract have shown that the protection against skin and lung tumors in animal (mice) models is slightly higher than that afforded by anthocyanin pigments (from red onion skin) or carotenoids (from red bell peppers). In particular, one pigment in the betalain group, betanin, appears to be a potent anticancer compound. Beetroot extract, rich in betalain pigments, has demonstrated chemopreventive properties against both lung and skin cancers (Kapadia *et al.*, 1996). In a study of different beetroot phenotypes, it was the most highly pigmented (betacyanin-rich) beets that were most efficient at free-radical scavenging and inducing quinine reductase (a phase II enzyme important as a marker of the initiation stage of carcinogenesis) in murine hepatoma cells *in vitro* (Wettasinghe *et al.*, 2002). Although earlier screening trials had not indicated strong chemopreventive properties for betalains, Wettasinghe *et al.* demonstrated that the use of acetonitrile as the extractant was likely responsible for the failure to demonstrate enzyme induction.

8.4   Carotenoids

8.4.1   *Diverse roles of carotenoid pigments in human health maintenance*

In terms of nutraceutical contribution, carotenoids are best recognized for their antioxidant capacity, especially in the membranes, since they are pigments located within membranes. Hydrogen peroxide, singlet oxygen, nitrogen oxides, super-oxide anion, and other reactive oxygen species insulting the body from either endogenous or exogenous routes (and thus aggravating to a plethora of chronic human disease conditions) can be inactivated by carotenoids; in fact, carotenoids are considered the most potent of biological quenchers of singlet oxygen (Boileau *et al.*, 1999; Paiva & Russell, 1999), and can react with virtually any radical species likely to be encountered in a biological system. In the majority of these reactions, the carotenoids break down to biologically active degradation products (Krinsky & Yeum, 2003). The same molecular characteristic that is responsible for absorbing visible light to provide the characteristic orange-yellow and red appearance of carotenoids – the system of nine conjugated double bonds – is also a feature intrinsic to all carotenoids found in human serum and tissues. A second feature common to all dietary carotenoids that are absorbed by humans (and can be found in human tissues and serum) is an unsubstituted β-ionone ring at the end of their conjugated double-bond chain (Boileau *et al.*, 1999).

About 50 carotenoids have been identified in the human diet, and 34 have been identified in human serum and/or breast milk (13 as geometrical isomers of their *all-trans* parent structures and 8 as metabolites). β-carotene, lycopene, and lutein are the most prevalent carotenoids in the diet, and are also most frequently detected in tissues and serum, although concentrations vary widely (Khachik *et al.*, 1997). β-carotene is the best studied, and in most countries the most prevalent carotenoid in the diet; however, lycopene has approached the same level in the USA (Paiva & Russell, 1999). The USDA recently brought on-line a carotenoid database for US Foods www.nal.usda.gov/fnic/foodcomp/index.html in collaboration with the Nutrition Coordinating Center (NCC) at the University of Minnesota. The broad variation in serum carotenoid concentrations that have been reported (in μmol/l: lycopene 0.13–0.82; lutein + zeaxanthin, 0.16–0.72; β-carotene, 0.09–0.91; β-cryptoxanthin 0.05–0.38; α-carotene 0.02–0.22) may be associated with gender, lifestyle, and physiological factors (Boileau et al., 1999).

Chemically, the polyene chain length of carotenoid molecules is correlated with efficiency in quenching singlet oxygen, with lycopene being the most efficient type of carotenoid. Factors including its isolated double bonds, open chain, and lack of oxygen substituents increase its quenching activity (Di Mascio et al., 1989). Carotenoids can also protect cells from oxidative damage using alternative mechanisms. β-carotene may terminate lipid oxidation as a chain-breaking antioxidant; both β-carotene and lutein protect cells from lipid peroxidation and membrane damage as measured by decreased cellular release of lactate dehydrogenase (Martin et al., 1996). Carotenoid cis-isomers are superior to all-trans forms as radical chain reaction terminators in both in vitro and in vivo experiments (Levin & Mokady, 1994). An advantageous association of carotenoid intake and protection from type 2 diabetes was recently investigated with glucose metabolism in men at high risk (Ylönen et al., 2003); dietary carotenoids were inversely associated with fasting plasma glucose concentrations, and plasma β-carotene concentrations were inversely associated with insulin resistance. Baseline serum concentrations of non-esterified fatty acids were directly related to dietary lycopene.

The carotenoids are released from the food matrix by mastication, gastric action, and digestive enzymes (Deming & Erdman, 1999). The carotenoids are nonpolar lipids that cannot dissolve in aqueous solutions (as lipid-soluble molecules, they follow the absorption pathway of dietary fat), so absorption in the human intestinal tract occurs following the formation of bile acid micelles, and micellar solubilization facilitates the diffusion of lipids across the unstirred water layer. Because fat in the small intestines stimulates secretion of bile from the gall bladder (increasing the size and stability of micelles), dietary fat consumption improves the solubilization and uptake of carotenoids. Subsequent transport of carotenoids in serum occurs in lipoproteins, with more polar carotenoids like xanthophylls close to the surface, and hydrocarbon carotenoids like β-carotene in the hydrophobic core (Parker, 1996).

Since they can be complexed with other components such as protein, their bioavailability is variable. Relative bioavailability has been estimated at less than 10% (raw uncooked vegetables) to over 50% when prepared in an oil matrix. The carotenoids exist in the all-trans configuration in raw fruits and vegetables, but processing can promote isomerization, and various isomers appear in metabolites recovered after ingestion. Lycopene is consumed primarily as all-trans in food sources, but is detected almost exclusively in cis-isomers in human tissues (Boileau et al., 2002; Platz et al., 2003). Specific isomers are apparently involved in different biological reactions, which opens up the possibility that patterns of isomers may provide insight into the risk or pathogenesis of disease processes. Current data support the hypothesis that cis-isomers (shorter in length, less prone to aggregation, and more soluble in mixed micelles) are substantially more bioavailable. Although heating and processing food products clearly tends to increase the bioavailability of lycopene, most evidence suggests that only very little shift to *cis*-isomers occurs during even rigorous thermal processing or dehydration; other physiological processes must be responsible for the large differences in *cis/trans* ratios observed between foods and tissues. *In vivo* studies (using ferret and gerbil animal models, because they absorb carotenoids in a similar manner to humans) show that some isomerization of lycopene does occur in the gastric system, and in addition, it seems that *cis*-isomers are preferentially absorbed by humans (Boileau *et al.*, 2002).

Carotenoids are accumulated in the human body in various tissues including liver, adipose, serum, breast milk, adrenal, prostate, macula, kidney, lung, brain, and skin (Deming & Erdman, 1999). Carotenoid bioavailability is considered to be the ability to accumulate in dynamic or static pools in the human body, and no standard methods are validated for quantitative assessment in humans. All measurements in humans have been confined to plasma or serum concentrations. Because carotenoids impact on a wide variety of age-related ailments, the relationship between lutein status and the age of healthy subjects was investigated; however, there was no major age effect on the levels of lutein in serum, adipose tissue, or other sites monitored (Cardinault *et al.*, 2003).

Previous reports linking β-carotene intake to high lung cancer mortality in smokers created unwarranted adverse publicity regarding carotenoid intake and human health. While a large body of epidemiological evidence supported the hypothesis that dietary carotenoids reduce cancer risk, three randomized lung cancer chemoprevention trials using β-carotene supplementation contradicted this epidemiological evidence. However, the unnaturally high levels of carotenoids tested in these experiments were not representative of actual dietary hazards, and would only be a consideration if excessive carotenoid supplements were consumed (Ziegler, 1993). In the supplementation studies, β-carotene, due to its ability to generate oxidative stress, exerted co-carcinogenic activity and predisposed individuals to cancer risk from bioactivated tobacco smoke procarcinogens. Similar increases in cancer and angina cases with high-dose β-carotene supplementation were noted for asbestos workers, considered a high-risk population (Paiva & Russell, 1999).

More recently, it has been demonstrated that mid-level dietary carotenoid intake is correlated with the lowest incidence of hypoxanthine-guanine phosphoribosyl transferase (HPRT) gene mutant frequency; however, very high or very low intakes resulted in higher mutation frequencies. This study provides support for observed increases in lung cancer risk in smokers under heavy β-carotene supplementation (Nyberg *et al.*, 2003). Similarly, β-carotene was clearly associated with a marked decrease in risk for colorectal adenoma recurrence (44%) in nonsmoking patients with low or moderate alcohol intakes (Baron *et al.*, 2003). In the same study, this antioxidant carotenoid doubled the risk of adenoma recurrence in patients who smoked and also drank more than one alcoholic drink per day. In current smokers, a statistically significant lower mean level of zeaxanthin, β-cryptoxanthin, α-carotene, α-tocopherol, retinol, and retinyl palmitate was detected, compared to former smokers (Goodman *et al.*, 2003). Both low and high doses of lycopene were able to inhibit smoke-induced squamous metaplasia in ferrets, and smoke exposure was found to markedly lower the levels of available lycopene in lycopene-supplemented lab animals (Liu *et al.*, 2003). The importance of lifestyle factors in recommendations for carotenoid intake for health and wellness cannot be overestimated, given these results. Potential dangers of administering high doses of single 'active ingredients' as supplements, without the protection afforded by the balanced food matrix, are accentuated (Paolini *et al.*, 2003).

8.4.2   Carotenoids and photooxidative damage

Also like the flavonoids, carotenoids provide protection to vision and eye function, and against macular degeneration and cataracts. Two carotenoids present within the entire retina and macula, lutein and zeaxanthin (classified together as xanthophylls), account for the yellowish cast in the macular region of the eye. Because of their antioxidant capacity, these carotenoids are assumed to protect against light-mediated damage – a hypothesis that is well supported by epidemiological research on intake of vegetables with content of the two carotenoid types, and risk of macular degeneration in a population (Snodderly, 1995). Macular degeneration, caused by blue light–mediated free radical damage to the retina, is the leading cause of irreversible blindness in aging adults over 65. Lutein has been loosely referred to as 'the eye-protective nutrient' and its intake has been shown in epidemiological trials to be inversely associated with ocular diseases such as age-related macular degeneration and cataracts (Khachik *et al.*, 2002).

In part, the benefits of carotenoids for vision health are tied to the vitamin A activity of a subset of carotenoid pigments, which makes them retinoid precursors. For example, the most well recognized in the carotenoid group of pigments, β-carotene, can be converted to two molecules of retinal (vitamin A) within the intestinal mucosal cell by the enzyme β-carotene 15,15′-dioxygenase. Retinoic acids are capable of effectively reducing swelling and destruction of collagen, and hence have been integral to some arthritis therapy regimes. However, of the over 600 naturally occurring carotenoid species, most have no vitamin A activity (Boileau *et al.*, 1999).

Carotenoids have a photoprotective function in human metabolism, and hence have been used in therapies for photosensitivity diseases (Zeigler, 1993), although as in many other disease conditions, β-carotene must be administered in conjunction with vitamin E to ensure beneficial effects. Although they do not contribute to vitamin A synthesis like β-carotene, the carotenoids lycopene, lutein, and zeaxanthin are excellent antioxidants and oral sun protectants that protect exposed tissues from light-induced damages. Mechanisms suspected to be involved in the protective effects of these macula carotenoids are filtering blue light, and scavenging reactive intermediates generated in photooxidation (Sies & Stahl, 2003).

8.4.3 Carotenoids and modulation of the immune system response

Carotenoids are credited with biological promotion of immune system response. β-carotene may influence immune cells to act more efficiently via a mechanism that increases expression of adhesion molecules by monocytes, increased secretion of TNF-α, and increased percentage of monocytes expressing a cell surface molecule responsible for presenting antigen to T-helper cells (Hughes *et al.*, 1997).

A broad range of other studies point to multiple potential mechanisms by which carotenoids may bolster the immune system (Boileau *et al.*, 1999) including their roles in increasing lymphocyte response to mitogens, protecting immune cells from the reactive bactericidal species they produce, increasing natural killer cell activity in aging subjects, and increasing total white blood cells and CD4/CD8 ratios; the latter in HIV-infected humans.

8.4.4 Carotenoids and inhibition of cardiovascular disease

A series of studies have now demonstrated a correlation between high intakes or high serum concentrations of carotenes and lowered risk of coronary artery disease. Inhibition of low-density lipoprotein (LDL) oxidation has been demonstrated, especially with low levels of carotene, although interestingly, the carotenoid lycopene did not protect against oxidation in these tests (Dugas *et al.*, 1999). Clinical studies show that a 2-week dosage of tomato juice, rich in carotenoids, effectively lowers LDL in male subjects (Bub *et al.*, 2000).

Work by Fuhrman *et al.* (1997) suggests that the antioxidant mechanism is not solely responsible for the ability of β-carotene or lycopene to lower LDL cholesterol. In this work, cholesterol synthesis in a macrophage cell line was inhibited by $10\,\mu\text{M}$ concentrations of either carotenoid. Another potential mechanism for carotenoids to inhibit cardiovascular disease may be through a feedback mechanism to inhibit HMG-CoA reductase in humans – a hypothesis that was supported by 14% decrease in plasma LDL cholesterol with supplementation of 60 mg/d lycopene for 3 months.

8.4.5 Carotenoids and cancer chemoprevention

One critical mechanism of cell-to-cell communication is through gap junction communication, and alterations in this means of communication can lead to unrestrained cell proliferation, as in cancer. Carotenoids enhance cell-to-cell communication because they can upregulate gap junctions by increasing expression of connexin-43, a six-subunit protein formed in cell membranes (Yamasaki, 1990). Both natural (six-membered rings) and synthetic (five-membered rings) carotenoids are able to induce gap junctional communication, although the natural carotenoids are more efficient (Stahl *et al.*, 1997).

Carotenoids are associated with inhibition of several types of cancers including cervical, esophageal, pancreatic, lung, prostate, colorectal and stomach.

Less-compelling evidence has been shown for a link between carotenoids and breast cancer; however, observation indicates that very low carotenoid intakes may be associated with increased risk of breast cancer (Parker, 1989; Toniolo et al., 2001). In particular, the carotenoid lycopene has been extensively researched for its role in cancer inhibition. Lycopene's role in colon carcinogenesis has been documented in rat models, where selective uptake in key tissues (prostate, testes and adrenals) was noted (Jain et al., 1999). Lutein and zeaxanthin intake has been linked to reduced risk of prostate cancer (Cohen et al., 2000). An epidemiological study found an inverse relationship between consumption of lycopene-rich foods (pizza, tomato, tomato paste) and prostate cancer risk, which may be related to the tissue-specific accumulation of lycopene in the human prostate (Deming & Erdman, 1999). A strong association between levels of serum carotenoids and lung cancer in females was also detected, but no association was found with prostate cancer (Goodman et al., 2003). Trials conducted with both β-carotene and vitamin A supplementation indicate that both substances inhibit cell proliferation in carcinogenesis (Moreno et al., 2002). In another study, carotenoids demonstrated potential to reduce risk of prostate cancer in that proliferation of human prostate cancer cell lines were inhibited (Kotake-Nara et al., 2001). Of 15 carotenoids examined, lycopene was the most efficacious, at low levels. Lycopene's inhibitory effect on prostate cancer cells has been linked to antioxidant mechanisms although other mechanisms may also be operative (Kim et al., 2002).

Hormone-responsive cancers in particular are inhibited by retinoids, as shown by reduced growth in rat mammary tumors or inhibition of human breast cancer cell lines; however, doses required to achieve effective results experimentally are toxic (Teplizky et al., 2001). Another mechanism for cancer chemoprevention that involves the carotenoids is their ability to modulate or change the expression of xenobiotic enzymes. Phase I xenobiotic metabolizing enzymes can activate a procarcinogen into a carcinogen, and phase II enzymes make xenobiotics unreactive and more easily excreted. Therefore, any substance that can induce phase II enzymes, while decreasing phase I enzyme activity, can fall into the category of cancer chemoprevention agent, as xenobiotics can be cleared from the body without damage by potential carcinogens. While many experimental trials have produced equivocal results in this arena, Wang et al. (1995) have demonstrated that β-carotene induces quinine reductase (a phase II enzyme), although the effect was purported to be mediated by retinoids rather than β-carotene alone.

8.4.6 Potentiation of carotenoid bioactivity

Interactions between carotenoids and other components in the food matrix (dietary fats, dietary fiber) as well as between different carotenoid species are evident during the absorptive process and in potentiation of biological effects (Boileau et al., 1999). Mixtures of carotenoids or associations with other antioxidants (such as vitamin E) can intensify their antioxidant capacity (Paiva & Russell, 1999). Given the widespread public acceptance of β-carotene supplements, the significance of these interactions is underscored. On the other hand, a modified fat diet and a high-carbohydrate low-fat diet were compared to determine effects on lycopene and other carotenoids, with no statistically significant differences observed (Ahuja *et al.*, 2003). *In vivo* experiments in a rodent model demonstrated that the presence of citric pectin in the diet diminished the bioavailability of synthetic β-carotene, and resulted in reduced reserves of vitamin A and β-carotene in rat liver (Zanutto *et al.*, 2002).

Null results in trials of β-carotene and coronary artery disease have refocused attention on the interactions of other dietary carotenoids in prevention activity. Both β-carotene and α-carotene intakes are associated with reduction in heart disease, but no significant relation with intakes of lutein, zeaxanthin, or lycopene were noted (Osganian *et al.*, 2003). Only a portion of the full range of dietary carotenoids has actually been detected after ingestion. For prostate cancer interventions as well, a large body of nutritional scientists are now turning away from a pure pharmacologic approach with pure lycopene, and instead turning attention to the complex mixture of biologically active phytochemicals and mix of carotenoids that have antiprostate cancer efficacy beyond any one component (Hadley *et al.*, 2002).

For the carotenoids, the structure of individual pigments has an enormous effect on their biological activity. Canthaxanthin and β-carotene have very different antioxidant activities than zeaxanthin due to different polar environments; as a result, the latter pigment is ineffective against macrophage formation from human monocytes, whereas the former two are inhibitory. Diets that incorporate diverse mixtures of different carotenoids are recommended due to this wide range of activities associated with different individual carotenoid compounds (Carpenter *et al.*, 1997). Other compounds such as ascorbic acid and α-tocopherol potentiate the antimutagenic activity of carotenoids.

8.5 Chlorophylls

8.5.1 Diverse roles of chlorophyll pigments in human health maintenance

Most plant pigments comprise only a small percentage of common foods; thus it may require ingestion of concentrated supplements to achieve biologically active levels for some human health therapies. Chlorophylls, in contrast, are typically consumed in much higher doses in a diet that incorporates green and leafy vegetables (Delgado-Vargas & Paredes-López, 2003). The same photodynamic properties of metabolic tetrapyrroles that have led to development of valuable herbicidal and insecticidal applications are also instrumental in their role as therapeutic agents (Rebeiz *et al.*, 2002).

Chlorophylls in green foods are credited with numerous interventions in human health maintenance including reduction of blood pressure, blood sugar control, brain activation, antimutagenic and anticlastogenic effects, and more (Bailey,

2003). In Asian countries, the dietary sources for chlorophyll-rich foods are far more extensive than in Western diets, as an abundance of sea vegetables are common in prepared foods. The dark green vegetables are simultaneously excellent sources of carotenoids, in particular lutein, which contribute with the chlorophylls to the health-protective properties.

8.5.2 Chlorophylls and cancer chemoprevention

The antimutagenic properties of chlorophylls have been demonstrated in various assays, and clearly chlorophyll has potential to act as a chemopreventive compound in humans. However, much of the research is equivocal. Chlorophyllin, a water-soluble, food-grade, stable sodium and copper salt of chlorophyll, has shown both anticancer properties and cancer-promoting activity (Magnuson et al., 1998). Dietary consumption of green vegetables has been associated with the protection against mutagenic and clastogenic activity of genotoxicants, with chlorophyll as the principal protective agent involved (Sarkar et al., 1996). After administration of either a crude Indian spinach leaf extract containing chlorophyll, purified extract, purified isolated chlorophyll, or chlorophyllin for 7 days to mice, only the crude leaf extract or the chlorophyllin (but not purified chlorophyll) proved to reduce clastogenic effects of potassium dichromate treatment.

Research supporting chlorophyll's roles in cancer chemoprevention indicates that these pigments are capable of deactivating potential carcinogens in the digestive tract (e.g. heterocyclic amines which provoke colorectal cancer), and preventing absorption.

When chlorophyllin was administered by gavage to female mice prior to topical treatment with a tumor-inducing electrophilic and carcinogenic metabolite (benzo(α) pyrene, 8-dihydrodiol-9,10-epoxide), a significant reduction in both incidence and multiplicity of skin tumors resulted (Park & Surh, 1996). Similarly, chlorophyllin decreased papillomagenesis in mice treated during the promotion stage of carcinogenesis, indicating that this form of chlorophyll can inhibit both tumor promotion and the progression of papillomagenesis in the two-stage mouse skin carcinogenesis (Chung et al., 1999).

Aminolevulinic acid (ALA) is a building block of tetrapyrroles that is synthesized in green plants as part of the chlorophyll biosynthetic pathway. The porphyrin-inducing properties of ALA have recently been adapted for photodynamic destruction of cancer cells, following the successful application of these properties as porphyric insecticides and photodynamic herbicides (Rebeiz et al., 2002). ALA has been administered in topical formulations, ingested, or delivered systemically in the course of photodynamic therapy. A response rate of 90% was achieved after topical treatment of basal cell carcinomas with ALA, with less dramatic response achieved for superficial squamous cell carcinomas and actinic keratoses. Because ALA is a small water-soluble molecule, it can easily penetrate and be taken up by transformed cells, and is rapidly cleared from the circulatory stream within 48 h of treatment (Rebeiz et al., 2002).

8.5.3   Potentiation of chlorophyll bioactivity

The ability of a crude chlorophyll-containing leaf extract to protect against mutagenic and clastogenic effects was suggested to be due to the interaction between chlorophyll pigments and other co-occurring phytochemicals in the leaf, because chlorophyll alone, whether extracted from the leaves or administered in a commercial purified formulation, proved to be clastogenic (Sarkar *et al.*, 1996). The efficacy of ALA in cancer therapy is significantly enhanced by addition of the tetrapyrrole modulator Oph to ALA formulations (Rebeiz *et al.*, 1992). In rapidly proliferating mammalian cancer cell lines, treatment with 1.0 mM ALA induced significant Proto accumulation in the dark, followed by a triggered destruction of treated cells only 30 min after illumination. Both Proto accumulation and specific cell lysis was significantly enhanced by addition of 0.75 mM Oph to the treatment along with the ALA formulation (Rebeiz *et al.*, 1994). In separate experiments, Oph synergized with ALA to significantly enhance indication of Proto in tumors (in mice), and phototreatment resulted in tumor necrosis as determined by significant reduction in both size and histopathology, with little damage to healthy surrounding tissues (Rebeiz *et al.*, 2002).

8.6   Other pigments and human health intervention

Iridoids, found in certain plants including the groundcover *Ajuga* and fruits of *Gardenia*, have demonstrated efficacy against mouse skin tumorigenesis, and have shown antiviral properties (Konoshima *et al.*, 2000). Curcumin (as turmeric, curry, and mustard) is widely consumed, especially in middle-eastern diets, and has a broader range of pharmacological applications. It acts as an anti-inflammatory agent. Low levels of curcumin in the diet of laboratory animals resulted in significant lowering of blood cholesterol levels, and improvement in the LDL/HDL ratios (Suresh-Babu & Srinivasan, 1997). As such, a therapeutic role for turmeric for individuals with diabetes as well as predisposition to coronary heart disease is suggested. Similarly, turmeric has demonstrated antimutagenic and anticarcinogenic properties in both *in vitro* and *in vivo* chemopreventive assays. Curcumin has also been shown to potentiate the beneficial effects of drugs and vitamins.

8.7   Conclusions and prospects for future research on pigments and human health

Plant pigments have multiple roles in human health protection, maintenance and therapy through a diverse array of biological activities, evidenced through *in vitro*, *in vivo*, and clinical trials, as well as through extensive epidemiological evidence (Delgado-Vargas & Paredes-López, 2003). As a body, the combined evidence of numerous research trials indicates that variety of different pigments in a well-balanced diet is key to prevention and therapy for a range of chronic human disease conditions.

One of the biggest challenges to conclusively determining the biological properties of natural plant pigments in human health maintenance has been the inability to track the fate of these pigments after ingestion. Certainly, as repeatedly noted in this chapter, epidemiological studies have suggested that reduction in cardiovascular disease, cancer risk and a range of other chronic disease conditions is linked to consumption in the diet of anthocyanins, betalains, chlorophylls, and carotenoids. *In vitro* studies have provided further insight into potential modes of action for various pigments (Codorniu-Hernández *et al.*, 2003; Sadik *et al.*, 2003), and *in vivo* experiments have further corroborated the roles of these pigments as antioxidants, enzyme inhibitors or activators, anti-inflammatory agents, anti-adhesins, antimicrobials, etc. Still, very little is known about accumulation of pigments or pigment metabolites in various organs of the body, or absorption after digestion. Knowledge is hindered because of the inherent problems associated with selective monitoring of ingested phytochemicals from a food source in a complete, complex diet, and discerning newly absorbed compounds of interest from the background levels already in the system. Lack of understanding concerning mechanisms of action, absorption, bioavailability, and metabolic fate hinders the comprehensive demonstration of health benefits (Grusak, 1997).

For example, the bioavailability of flavonoids such as anthocyanin is largely unknown. Anthocyanins are absorbed into plasma mainly in intact form, unlike tannins, which are highly derivatized. Anthocyanins bind to proteins and certain lipids, and have pH-dependent ionic forms. The presence of sugars and carbohydrates in the diet has an impact on anthocyanin absorption, because of competition for time on the glucose transporters. Dietary proteins also influence the amount of anthocyanin absorption. Contradictory research has reported that anthocyanins are bioactive only as aglycones, whereas other more recent research demonstrates that the native glycosidic forms are actually the most active in various assays. Some researchers have speculated that anthocyanin absorption and bioavailability is probably quite low (less than 1%), based solely on measurable levels detected in the serum. However, this estimate is inadequate, as it does not begin to account for all the potential pools for anthocyanin (and metabolite) accumulation, and it has not been feasible to measure accumulation in organs or fully monitor presence in the waste stream.

Isotopic labeling of natural pigment molecules, in order to track their progress during metabolism, is a potential solution to this challenge, as it can provide a traceable tool for following the absorption and metabolic fate of natural ingested pigments (Grusak, 1997). Oral and intravenous administration of simple radiolabeled compounds such as drugs and minerals has been a powerful tactic for monitoring distribution and excretion patterns (Weaver, 1985). Mullen *et al.* (2002) were able to examine flavonol metabolites in plasma and tissues in rodents following oral ingestion of labeled quercetin-4'-glucoside, using HPLC radiocounting and tandem mass spectrometry. The highest proportion of the ingested label was detected in the gastrointestinal tract, with relatively low levels in plasma and body tissues, indicative of potentially poor absorption. The use of the traceable label allowed conclusive determination of phytochemical bioavailability in this case.

Given the relative complexity of the natural plant pigments, these molecules for the most part cannot be synthesized in the laboratory, but must be produced by plant cells. Deprez *et al.* (1999) incorporated C_{14} phenylalanine into catechins and proanthocyanidin oligomers using acetate as an elicitor in willow tree cuttings. The use of plant cell cultures to produce pigments and associated phytochemicals for subsequent use in feeding trials is advantageous in that additions can be made to the plant cell culture broth (precursors, elicitors, or labeled carbohydrate sources) that can influence the composition of the pigmented compounds accumulated by the plant cells. Anthocyanin pigments and associated flavonoid and other phytochemicals accumulated by grape cell suspension cultures have been successfully labeled both with stable isotope (C_{13}) as well as radiolabel (C_{14}). Labeled flavonoid precursors (C_{13} or C_{14} phenylalanine) were provided to metabolizing cell cultures of *Vitis vinifera* in order to harvest biolabeled wine polyphenols, including resveratrol and anthocyanins. The levels of enrichment were investigated by means of proton NMR spectrum (Krisa *et al.*, 1999; Vitrac *et al.*, 2003).

Vitrac *et al.* (2003) were able to track the distribution of the hydroxyl stilbene resveratrol in a mouse model following oral intubation. Distribution of radioactivity was subsequently evaluated using whole-body autoradiography and microautoradiography; quantitative measurements in key organs were also made. Both intact C_{14} *trans*-resveratrol molecules, and glucurono- and sulfoconjugates were detected in tissues, demonstrating that *trans*-resveratrol was bioavailable and remained largely in the intact form. Grusak *et al.* (2004) recently custom constructed an enclosed chamber system to facilitate the safe introduction of a radiolabeled precursor (including sucrose, a typical carbon source for metabolizing cultures) into plant cell suspension cultures, for recovery of a wide range of secondary metabolites. Initial trials resulted in successful recovery of enriched fractions of anthocyanins and other flavonoids from both grape and berry cell cultures. Differential distribution of the C_{14}-labeled carotenoid lycopene and its polar metabolic products was reported by Zaripheh *et al.* (2003), who showed that the majority of labeled lycopene accumulated in liver, and that the majority of label in the prostate was in the form of C_{14} polar products, suggesting differential metabolism in various tissue types. Emerging studies of this type, using labeled pigments in feeding or intubation experiments, are capable of providing a much clearer view of how these natural phytochemicals are able to intervene to promote human health, and give stronger clues as to mechanisms of action, which are currently unknown in most cases.

References

Ahuja, D., Ashton, E. & Ball, J. (2003) Effects of a monounsaturated fat, tomato-rich diet on serum levels of lycopene. *Eur. J. Clin. Nutr.*, **57**, 832–841.

Akita, T., Hina, Y. & Nishi, T. (2000) Production of betacyanins by a cell suspension culture of table beet (*Beta vulgaris* L.). *Biosci. Biotech. Biochem.*, **64**, 1807–1812.

Bailey, R. (2003) Green foods in Japan. *Nutraceuticals World*, **6**, 22–24.

Baron, J., Cole, B. F., Mott, L., Haile, R., Grau, M., Church, T. R., Beck, G. J. & Greenberg, E. R. (2003) Neoplastic and antineoplastic effects of β-carotene on colorectal adenoma recurrence: results of a randomized trial. *J. Natl. Cancer I.*, **95**, 717–722.

Boileau, T. W. M., Boileau, A. C. & Erdman, J. W. Jr (2002) Bioavailability of *all-trans-* and *cis-*isomers of lycopene. *Exp. Biol. Med.*, **227**, 914–919.

Boileau, T. W. M., Moore, A. C. & Erdman, J. W. Jr (1999) Carotenoids and vitamin A, in *Antioxidant Status, Diet, Nutrition, and Health* (ed. A. M. Papas), CRC Press, Boca Raton, Florida, USA, pp. 133–158.

Bomser, J., Madhavi, D., Singletary, K. & Smith, M. A. (1996) *In vitro* anticancer activity of fruit extracts from *Vaccinium* species. *Planta Medica*, **62**, 212–216.

Bourgaud, F., Gravot, A., Milesi, S. & Gontier, E. (2001) Production of plant secondary metabolites: a historical perspective. *Plant Sci.*, **161**, 839–851.

Bub, A., Watzi, B., Abrahamse, L., Delincée, H., Adam, S., Wever, J., Müller, H. & Rechkemmer, G. (2000) Moderate intervention with carotenoid-rich vegetable products reduces lipid peroxidation in men. *J. Nutr.*, **130**, 2200–2206.

Cai, Y., Sun, M. & Corke, H. (2003) Antioxidant activity of betalains from plants of the Amaranthaceae. *J. Agr. Food Chem.*, **51**, 2288–2294.

Cardinault, N., Gorrand, J., Tyssandier, V., Grolier, P., Rock, E. & Borel, P. (2003) Short-term supplementation with lutein affects biomarkers of lutein status similarly in young and elderly subjects. *Exp. Gerontol.*, **38**, 573–582.

Carpenter, K., Vander veen, C., Hird, R., Dennis, I., Ding, T. & Mitchinson, M. (1997) The carotenoid beta-carotene, canthaxanthin and zeaxanthin inhibit macrophage-mediated LDL-oxidation. *FEBS Lett.*, **401**, 262–266.

Chung, W., Lee, J., Park, M., Yook, J., Kim, J., Chung, A., Surh, Y. & Park, K. (1999) Inhibitory effects of chlorophyllin on 7,12-dimethylbenz α-anthracene-induced bacterial mutagenesis and mouse skin carcinogenesis. *Cancer Lett.*, **145**, 57–64.

Cohen, J. H., Kristal, A. R. & Stanford J. L. (2000) Fruit and vegetable intakes and prostate cancer risk. *J. Natl. Cancer I.*, **92**, 61–67.

Codorniu-Hernández, E., Mesa-Ibirico, A., Montero-Cabrera, L. A., Martínez-Luzardo, F., Borrmann, T. & Stohrer, W.-D. (2003) Theoretical study of flavonoids and proline interactions. Aqueous and gas phases. *J. Mol. Struc. (Theochem)*, **623**, 63–73.

Davies, K., Schwinn, K., Deroles, S., Manson, D., Lewis, D., Bloor, S & Bradley, J. (2003) Enhancing anthocyanin production by altering competition for substrate between flavonol synthase and dihydroflavonol 4-reductase. *Euphytica*, **131**, 259–268.

Delgado-Vargas, F. & Paredes-López, O. (eds.) (2003) *Natural Colorants for Food and Nutraceutical Uses*. CRC Press, Boca Raton, Florida, USA.

Delgado-Vargas, F., Jimenez, A. & Paredes-López, O. (2000) Natural pigments: carotenoids, anthocyanins, and betalains - characteristics, biosynthesis, processing, and stability. *Crit. Rev. Food Sci. Nutr.*, **40**, 173–289.

Deming, D. & Erdman, J. W. Jr (1999) Mammalian carotenoid absorption and metabolism. *Pure Appl. Chem.*, **71**, 2213–2223.

Deprez, S., Mila, I. & Scalbert, A. (1999) Carbon-14 biolabeling of (+) -catechin and proanthocyanidin oligomers in willow tree cuttings. *J. Agr. Food Chem.*, **47**, 4219–4230.

Di Mascio, P., Kaiser S. & Sies, H. (1989) Lycopene as the most efficient biological carotenoid singlet oxygen quencher. *Arch. Biochem. Biophys.*, **274**, 532–538.

Dixon, R. & Steele, C. (1999) Flavonoids and isoflavonoids – a gold mine for metabolic engineering. *Trends Plant Sci.*, **4**, 394–400.

Dugas, T. R., Morel, D. W. & Harrison, E. H. (1999) Dietary supplementation with β-carotene, but not with lycopene, inhibits endothelial cell-mediated oxidation of low-density lipoprotein. *Free Radical Biol. Med.*, **26**, 1238–1244.

Duthie, G., Duthie, S. & Kyle, J. (2000) Plant polyphenols in cancer and heart disease: implications as nutritional antioxidants. *Nutr. Res. Rev.*, **13**, 79–106.

Escribano, J., Pedreno, M., Garcia-Carmona, F. & Munoz, R. (1998) Characterization of the antiradical activity of betalains from *Beta vulgaris* L. roots. *Phytochem. Anal.*, **9**, 124–127.

Feord, J. (2003) Food colours and the law. *International Food Ingredients*, April/May 2003, 6–10.

Folts, J. (1998) Antithrombotic potential of grape juice and red wine for preventing heart attacks. *Pharma. Biol.*, **36**, S21–S27.

Fox, B. (2003) Separating fact from fiction: evaluating the medical literature. *J. Am. Nutr. Assoc.*, **6**, 3–16.

Francis, F. J. (1989) Food colorants: anthocyanins. *Crit. Rev. Food Sci. Nutr.*, **28**, 273–314.

Freedman, J., Parker, C., Li, L., Perlman, J., Frei, B., Ivanov, V., Deak, L., Iafrati, M. & Folts, J. (2001) Select flavonoids and whole juice from purple grapes inhibit platelet function and enhance nitric oxide release. *Circulation*, **103**, 2792–2798.

Fuhrman, B., Elis, A. & Aviram, M. (1997) Hypocholesterolemic effect of lycopene and β-carotene is related to suppression of cholesterol synthesis and augmentation of LDL receptor activity in macrophages. *Biochem. Bioph. Res. Co.*, **233**, 658–662.

Gabrielska, J., Oszmianski, J., Komorowska, M. & Langner, M. (1999) Anthocyanin extracts with antioxidant and radical scavenging effect. *Z. Naturforsch.*, **54**, 319–324.

Gaziano, F. M., Johnson, E. J., Russell, R. M., Manson, J. E., Stampfer, M. J. & Ridker, P. M. (1995) Discrimination in absorption or transport of β-carotene isomers after oral supplementation with either *all-trans* or 9-*cis* β-carotene. *Am. J. Clin. Nutr.*, **61**, 1248–1252.

Goodman, G., Schaeffer, S., Omenn, G., Chen, C. & King, I. (2003) The association between lung and prostate cancer risk, and serum micronutrients: results and lessons learned from β-carotene and retinol efficacy trial. *Cancer Epidem. Biomark. Preven.*, **12**, 518–526.

Grusak, M. (1997) Intrinsic stable isotope labeling of plants for nutritional investigations in humans. *J. Nutr. Biochem.*, **8**, 164–171.

Grusak, M. A., Rogers, R. B, Yousef, G. G., Erdman, J. W., Jr & Lila, M. A. (2004) An enclosed-chamber labeling system for the safe ^{14}C-enrichment of phytochemicals in plant cell suspension cultures. *In Vitro Cell. Dev. Biol. – Plant*, **40**, 80–85.

Hadley, C., Miller, E., Schwartz, S. & Clinton, S. (2002) Tomatoes, lycopene, and prostate cancer: progress and promise. *Exp. Biol. Med.*, **227**, 869–880.

Havsteen, B. (2002) The biochemistry and medical significance of the flavonoids. *Pharma. Therapeut.*, **96**, 67–202.

Heim, K., Tagliaferro, A. & Bobilya, D. (2002) Flavonoid antioxidants: chemistry, metabolism and structure-activity relationships. *J. Nutr. Biochem.*, **13**, 572–584.

Hirner, A., Veit, S. & Seitz, H. (2001) Regulation of anthocyanin biosynthesis in UV-A-irradiated cell cultures of carrot and in organs of intact carrot plants. *Plant Sci.*, **161**, 315–322.

Hollman, P. & Katan, M. (1998) Absorption, metabolism, and bioavailability of flavonoids, in *Flavonoids in Health and Disease* (eds C. Rice-Evans & L. Packer), Marcel Dekker, New York, USA, pp. 483–522.

Horvathova, K., Vachalkova, A. & Novotny, L. (2001) Flavonoids as chemoprotective agents in civilization diseases. *Neoplasma*, **48**, 435–441.

Hughes, D. A., Wright, A. J., Finglas, P. M., Peerless, A. C., Bailey, A. L., Astley, S. B., Pinder, A. C. & Southon, S. (1997) The effect of β-carotene supplementation on the immune function of blood monocytes from healthy male nonsmokers. *J. Lab. Clin. Med.*, **129**, 309–317.

Jain, C. K., Agarwal, S. & Rao, A. V. (1999) The effect of dietary lycopene on bioavailability, tissue distribution, and *in vivo* antioxidant properties and colonic preneoplasia in rats. *Nutr. Res.*, **19**, 1383–1391.

Joseph, J. A., Shukitt-Hale, B., Denisova, N., Bielinski, D., Martin, A., McEwen, J. J. & Bickford, P. (1999). Reversals of age-related declines in neuronal signal transduction, cognitive and motor behavioral deficits with blueberry, spinach or strawberry dietary supplementation. *J. Neurosci.*, **19**, 8114–8121.

Jyonouchi, H., Sun, S., Mizokami, M. & Gross, M. (1996) Effects of various carotenoids on cloned effector-stage T-helper cell activity. *Nutr. Cancer*, **26**, 313–324.

Kandil, F, Song, L, Pezzuto, J., Seigler, D. & Smith, M. A. (2000) Isolation of oligomeric proanthocyanidins from flavonoid-producing cell cultures. *In Vitro Cell. Dev. Biol. – Plant*, **36**, 492–500.

Kalt, W., Ryan, D., Duy, J., Prior, R., Ehlenfeldt, M. & Vander Kloet, S. (2001) Interspecific variation in anthocyanins, phenolics and antioxidant capacity amng genotypes of highbush and lowbush blueberries (*Vaccinium* Section *cyanococcus* spp.). *J. Agr. Food Chem.*, **49**, 4761–4767.

Kang, S., Seeram, N., Nair, M. & Bourquin, L. (2003) Tart cherry anthocyanins inhibit tumor development in Apc(Min) mice and reduce proliferation of human colon cancer cells, *Cancer Lett.*, **194**, 13–19.

Kanner, J., Harel, S. & Granit, R. (2001). Betalains – a new class of dietary cationized antioxidants. *J. Agr. Food Chem.*, **49**, 5178–5185.

Kapadia, G., Tokuda, H., Konoshima, T. & Nishino, H. (1996) Chemoprevention of lung and skin cancer by *Beta vulgaris* (beet) root extract. *Cancer Lett.*, **100**, 211–214.

Khachik, F., Carvalho, L., Bernstein, P., Muir, G., Zhao, D. & Katz, N. (2002) Chemistry, distribution, and metabolism of tomato carotenoids and their impact on human health. *Exp. Biol. Med.*, **227**, 845–851.

Khachik F., Spangler C. J., Cecil J. & Smith J. (1997) Identification, quantification, and relative concentrations of carotenoids and their metabolites in human milk and serum. *Anal. Chem.*, **69**, 1873–1881.

Kim, L., Rao, V. & Rao, L. (2002) Effect of lycopene on prostate LNCaP cancer cells in culture. *J. Med. Food*, **5**, 181–187.

Konoshima, T., Takasaki, M., Tokuda, H. & Nishino, H. (2000) Cancer chemopreventive activity of an iridoid glycoside, 8-acetylharpagide, from *Ajuga decumbens*. *Cancer Lett.*, **158**, 87–92.

Kotake-Nara, E., Kushiro, M., Zhang, H., Sugawara, T., Miyashita, K. & Nagao, A. (2001) Carotenoids affect proliferation of human prostate cancer cells. *J. Nutr.*, **131**, 3303–3306.

Krinsky, N. & Yeum, K. (2003) Carotenoid-radical interactions. *Biochem. Bioph. Res. Co.*, **305**, 754–760.

Kujala, T., Loponen, J. & Pihlaja, K. (2001). Betalains and phenolics in red beetroot (*Beta vulgaris*) peel extracts: extraction and characterization. *Z. Naturforsch.*, **56c**, 343–348.

Kucuk, O., Sarkar, F., Djuric, Z., Sakr, W., Pollak, M., Khachik, F., Banerjee, M., Bertram, J. & Wood, D. (2002) Effects of lycopene supplementation in patients with localized prostate cancer. *Exp. Biol. Med.*, **227**, 881–885.

Lazze, M., Pizzala, R., Savio, M., Stivala, L., Prosperi, E. & Bianchi, L. (2003) Anthocyanins protect against DNA damage induced by tert-butyl-hydroperoxide in rat smooth muscle and hepatoma cells. *Mutat. Res.— Gen. Tox. En.*, **535**, 103–115.

Leathers, R., Davin, C. & Zrÿd, J.-P. (1992) Betalain producing cell cultures of *Beta vulgaris* L. var. bikores monogerm (red beet). *In Vitro Cell. Dev. Biol. – Plant*, **28**, 39–45.

Lee, S. H. & Min, D. B. (1990) Effects, quenching mechanisms, and kinetics of carotenoids in chlorophyll-sensitized photooxidation of soybean oil. *J. Agr. Food Chem.*, **38**, 1630–1634.

Levin, G. & Mokady S. (1994) Antioxidant activity 9-*cis* compared to *all-trans* β-carotene *in vitro*. *Free Radicals Biol. Med.*, **17**, 77–82.

Lietti, A. & Forni, G. (1976) Studies on *Vaccinium myrtillus* anthocyanosides. II. Aspects of anthocyanins pharmacokinetics in the rat. *Arzneim. Forsch.*, **26**, 832–835.

Liu, C., Lian, F., Smith, D., Russell, R. & Wang, X. (2003) Lycopene supplementation inhibits lung squamous metaplasia and induces apoptosis via upregulating insulin-like growth factor-binding protein 3 in cigarette smoke-exposed ferrets. *Cancer Res.*, **63**, 3138–3144.

Luczkiewcz, M. & Cisowski, W., (2001) Optimisation of the second phase of a two-phase growth system for anthocyanin accumulation in callus cultures of *Rudbeckia hirta*. *Plant Cell Tiss. Org. Cult.*, **65**, 57–68.

Magnuson, B. A., Exon, J. H., South, E. H. & Hendrix, K. (1998). Effects of various phytochemicals on colonic cancer biomarkers, in *Functional Foods for Disease Prevention II. Medicinal Plants and Other Foods*, Vol. 2 (eds T. Shibamoto, J. Terao, & T. Osawa), American Chemical Society, Washington, DC, USA, pp. 231–243.

Martin, K. R., Failla, M. L. & Smith, J. C. (1996) β-carotene and lutein protect HepG2 human liver cells against oxidant-induced damage. *J. Nutr.*, **126**, 2098–2106.

Matsumoto, H., Inaba, H., Kishi, M., Tominanga, S., Hirayama, M. & Tusda, T. (2001) Orally administrated delphinidin 3-rutinoside and cyanidin 3-rutinoside are directly absorbed in rats and humans and appear in the blood as the intact forms. *J. Agr. Food Chem.*, **49**, 1546–1551.

Moreno, F., S-Wu, T., Naves, M., Silveira, E., Oloris, S., da Costa, M., Dagli, M. & Ong, T. (2002) Inhibitory effects of beta-carotene and vitamin A during the progression phase of hepatocarcinogenesis involve inhibition of cell proliferation but not alternations in DNA methylation. *Nutr. Cancer*, **44**, 80–88.

Morazzoni, P. & Bombardelli, E. (1996) *Vaccinium myrtillus* L. Fitoterapia, **47**, 3–29.

Moyer, R., Hummer, K., Finn, C., Frei, B. & Wrolstad, R. (2002) Anthocyanins, phenolics, and antioxidant capacity in diverse small fruits: *Vaccinium, Rubus*, and *Ribes*. *J. Agr. Food Chem.*, **50**, 519–525.

Mullen, W., Graf, B., Caldwell, S., Hartley, R., Duthie, G, Edwards, C., Lean, M. & Crozier, A. (2002) Determination of flavonol metabolites in plasma and tissues of rats by HPLC-radiocounting and tandem mass spectrometry following oral ingestion of [2-^{14}C] quercetin-4'-glucoside. *J. Agr. Food Chem.*, **50**, 6902–6909.

Naasani, I., Oh-hashi, F., Oh-hara, T., Feng, W., Johnston, J., Chan, K. & Tsuruo, T. (2003) Blocking telomerase by dietary polyphenols is a major mechanism for limiting the growth of human cancer cells *in vitro* and *in vivo*. *Cancer Res.*, **63**, 824–830.

Nielsen, I. L., Dragsted, L. O., Ravn-Haren, G., Freese, R. & Rasmussen, S. E. (2003) Absorption and excretion of black currant anthocyanins in humans and Watanabe heritable hyperlipidemic rabbits. *J. Agr. Food Chem.*, **51**, 2813–2820.

Norton, R. A. (1999). Inhibition of aflatoxin B1 biosynthesis in *Aspergillus favus* by anthocyanidins and related flavonoids. *J. Agr. Food Chem.*, **47**, 1230–1235.

Nyberg, F., Hous, S., Pershagen, G. & Lambert, B. (2003) Dietary fruit and vegetables protect against somatic mutation *in vivo*, but low or high intake of carotenoids does not. *Carcinogenesis*, **24**, 689–696.

Oak, M., Chataigneau, M., Keravis, T., Chataigneau, T., Beretz, A., Andriantsitohaina, R., Stoclet, J., Chang, S. & Schini-Kerth, V. (2003) Red wine polyphenolic compounds inhibit vascular endothelial growth factor expression in vascular smooth muscle cells by preventing the activation of the p38 mitogen-activated protein kinase pathway. *Arterioscl. Throm. Vas. Biol.*, **23**, 1001–1007.

Osganian, S. K., Stampfer, M. J., Rimm, E., Spiegelman, D., Manson, J. E., & Willett, W. C. (2003) Dietary carotenoids and risk of coronary artery disease in women. *Am. J. Clin. Nutr.*, **77**, 1390–1339.

Paiva, S. & Russell, R. (1999) Beta-carotene and other carotenoids as antioxidants. *J. Am. Coll. Nutr.*, **18**, 426–433.

Paolini, M., Abdel-Rahman, S., Sapone, A., Pedulli, G., Perocco, P., Cantelli-Forti, G. & Legator, M., (2003) Beta-carotene: a cancer chemopreventive agent or a co-carcinogen? *Mutat. Res. – Rev. Mutat. Res.*, **543**, 195–200.

Park, K. K., & Surh, Y. J. (1996) Chemopreventive activity of chlorophyllin against mouse skin carcinogenesis by benzo(a)pyrene and benzo(a)pyrene-7,8-dihydrodiol-9,10-epoxide. *Cancer Lett.*, **102**, 143–149.

Parker, R. S. (1989) Carotenoids in human blood and tissues. *J. Nutr.*, **119**, 101–104.

Parker, R. S. (1996) Absorption, metabolism, and transport of carotenoids. *FASEB J.*, **10**, 542–551.

Parr, A. J. & Bolwell, G. P. (2000) Phenols in the plant and in man. The potential for possible nutritional enhancement of the diet by modifying the phenols content or profile. *J. Sci. Food Ag.*, **80**, 985–1012.

Pavlov, A., Kovatcheva, P., Georgiev, V., Koleva, I. & Ilieva, M. (2002) Biosynthesis and radical scavenging activity of betalains during the cultivation of red beet (*Beta vulgaris*) hairy root cultures. *Z. Naturforsch.*, **57c**, 640–644.

Pépin, M.-F., Smith, M. A. L. & Reid, J. (1999) Application of imaging tools to plant cell culture: relationship between plant cell aggregation and flavonoid production. *In Vitro Cell. Dev. Biol. – Plant*, **35**, 290–295.

Peterson, J. & Dwyer, J. (1998) Flavonoids: dietary occurrence and biochemical activity. *Nutr. Res.*, **18**, 1995–2018.

Pignatelli, P., Pulcinelli, F., Celestini, A., Lenti, L., Ghiselli, A., Gazzaniga, P. & Violi, F. (2000) The flavonoids quercetin and catechin synergistically inhibit platelet function by antagonizing the intracellular production of hydrogen peroxide. *Am. J. Clin. Nutr.*, **72**, 1150–1155.

Platz, E. A., Clinton, S., Erdman, J. W. Jr, Ferruzzi, M. G., Willett, M. G. & Giovannucci, E. L. (2003) Variations in plasma lycopene and specific isomers over time in a cohort of U.S. men. *J. Nutr.*, **133**, 1930–1936.

Potrykus, I. (2003) Nutritional improvement of rice to reduce malnutrition in developing countries, in *Plant Biotechnology 2002 and Beyond* (ed. I. Vasil), Kluwer Academic Publishers, Dordrecht, The Netherlands, pp. 401–406.

Rebeiz, C. A., Kolossov, V. L., Briskin, D. & Gawienowski, M. (2002) Chloroplast biogenesis: chlorophyll biosynthetic heterogeneity, multiple biosynthetic routes and biological spin-offs, in *Handbook of Photochemistry and Photobiology*, Vol. IV (ed. H. S. Nalwa; Foreword by Professor Jean-Marie Lehn, Nobel Prize Laureate in chemistry), American Scientific Publishers, Los Angeles, USA, pp. 183–268.

Rebeiz, N., Arkins, Rebeiz, C. A., Simon, J., Zakary, J. & Kelley, K. (1996) Induction of tumor necrosis by delta-aminolevulinic acid and 1,10-phenanthroline photodynamic therapy, *Cancer Res.*, **56**, 339–344.

Rebeiz, N., Kelley, K. & Rebeiz, C. (1994) Porphorins as chemotherapeutic agents: biochemistry of protoporphyrin 9 accumulation in mammalian cells, in *Porphyric Pesticides; Chemistry, Toxicology, and Pharmaceutical Applications* (ed. S. Duke & C. Rebeiz), American Chemistry Society Symposium 559, Washington, DC, USA, pp. 233–246.

Rebeiz, N., Rebeiz, C. C., Arkins, S., Kelley, K. W. & Rebeiz, C. A. (1992) Photodestruction of tumor cells by induction of endogenous accumulation of photoporphyrin IX: Enhancement by 1,10-phenanthroline. *Photochem. Photobiol.*, **55**, 431–445.

Sadik, C. D., Sies, H. & Schewe, T. (2003) Inhibition of 15-lipoxygenases by flavonoids: structure-activity relations and mode of action. *Biochem. Pharma.*, **65**, 773–781.

Sahai, O. (1994) Plant tissue culture, in, *Bioprocess Production Flavor, of Fragrance, and Color Ingredients* (ed. A. Gabelman), John Wiley, New York, USA, pp. 239–268.

Santos-Buelga, C. & Scalbert, A. (2000) Proanthocyanidins and tannin-like compounds – nature, occurrence, dietary intake, and effects on nutrition and health. *J. Sci. Food Ag.*, **80**, 1094–1117.

Sarkar, D., Sharma, A. & Talukder, G. (1996) Clastogenic activity of pure chlorophyll and anticlastogenic effects of equivalent amounts of crude extract of Indian spinach leaf and chlorophyllin following dietary supplementation to mice. *Environ. Mol. Mut.*, **28**, 121–126.

Sarma, A. D. & Sharma, R. (1999) Anthocyanin-DNA copigmentation complex: mutual protection against oxidative damage. *Phytochemistry*, **52**, 1313–1318.

Scalbert, A. & Williamson, G. (2000) Dietary intake and bioavailability of polyphenols. *J. Nutr.*, **130**, 2073S–2085S.

Schmidt, B., Kraft, T. Seigler, D., Erdman, J. W. Jr & Lila, M. A. (2003) Antiproliferation activities of wild blueberry (*Vaccinium angustifolium*) fruit extracts. *FASEB J.*, Experimental Biology Meetings, San Diego, CA. Abstract #B487.

Schmitt, E. & Stopper, H. (2001) Estrogenic activity of naturally occurring anthocyanidins. *Nutr. Cancer*, **41**, 145–149.

Serraino, I., Dugo, L., Dugo, P., Mondello, L., Mazzon, E., Dugo, G., Caputi, A. & Cuzzocrea, S. (2003) Protective effects of cyanidin-3-*O*-glucoside from blackberry extract against peroxynitrite-induced endothelial dysfunction and vascular failure, *Life Sci.*, **73**, 1097–1114.

Shafiee, M., Carbonneau, M., Huart, J., Descomps, B. & Leger, C. (2002) Synergistic antioxidative properties of phenolics from natural origin toward low-density lipoproteins depend on the oxidation system. *J. Med. Food*, **5**, 69–78.

Shanmuganayagam, D., Beahm, M., Osman, H., Krueger, C., Reed, J. & Folts, J. (2002) Grape seed and grape skin extracts elicit a greater antiplatelet effect when used in combination than when used individually in dogs and humans. *J. Nutr.*, **132**, 3592–3598.

Sies, H. & Stahl, W. (2003) Non-nutritive bioactive constituents of plants: lycopene, lutein, and zeaxanthin. *Int. J. Vit. Nutr. Res.*, **73**, 95–100.

Skibola, C. & Smith, M. (2000) Potential health impacts of excessive flavonoid intake. *Free Radical Biol. Med.*, **29**, 375–383.

Smith, M. (2000) Secondary product expression in vitro, in *Plant Tissue Culture Concepts and Laboratory Exercises*, 2nd edn (eds R. Trigiano & D. Gray), CRC Press, Boca Raton, Florida, USA, pp. 355–359.

Smith, M. A. L., Marley, K. A., Seigler, D., Singletary K. W. & Meline B. (2000) Bioactive properties of wild blueberry fruits. *J. Food Sci.*, **65**, 352–356.

Snodderly, D. M. (1995) Evidence for protection against age-related macular degeneration by carotenoids and antioxidant vitamins. *Am. J. Clin. Nutr.*, **62S**, 1448S–1461S.

Stahl, W., Nicolai, S., Briviba K. & Hanusch, M. (1997) Biological activities of natural and synthetic carotenoids: induction of gap junctional communication and singlet oxygen quenching. *Carcinogenesis*, **18**, 89–92.

Stahl, W., Schwarz, W., Sundquist, A. R. & Sies, H. (1992) *Cis-trans* isomers of lycopene and β-carotene in human serum and tissues. *Arch. Biochem. Biophys.*, **294**, 173–177.

Stein, J., Keevil, J., Wiebe, D., Aeschlimann, D. & Folts, J. (1999) Purple grape juice improves endothelial function and reduces the susceptibility of LDL cholesterol to oxidation in patients with coronary artery disease. *Circulation*, **100**, 1050–1055.

Strack, D., Vogt, T. & Schliemann, W. (2003) Recent advances in betalain research, *Phytochemistry*, **62**, 247–269.

Suresh-Babu, P. & Srinivasan, K. (1997) Hypolipidemic action of curcumin, the active principle of turmeric (*Curcuma longa*) in streptozotocin induced diabetic rats. *Mol. Cell Biol.*, **166**, 169–175.

Teplizky, S. R., Kiefer, T. L., Cheng, Q., Dwivedi, P. D., Moroz, K., Myers, L., Anderson, M. B., Collins, A., Dai, j., Yuan, L., Spriggs, L. L., Blask, D. E. & Hill, S. M. (2001) Chemoprevention of NMU-induced rat mammary carcinoma with the combination of melatonin and 9-*cis*-retinoic acid. *Cancer Lett.*, **168**, 155–163.

Terahara, N., Callebaut, A., Ohba, R., Nagata, R., Ohnishi-Kameyama, M. & Suzuki, M. (2001) Acylated anthocyanidin 3-sophoroside-5-glucosides from *Ajuga reptans* flowers and the corresponding cell cultures. *Phytochemistry*, **58**, 493–500.

Tesoriere, L., Butera, D., D'Arpa, D., Di Gaudio, F., Allegra, M., Gentile, C. & Livrea, M. (2003). Increased resistance to oxidation of betalain-enriched human low density lipoproteins. *Free Radical Res.*, **37**, 689–696.

Toniolo, P., van Kappel., A., Akhmedkhanov, A., Ferrari, P., Kato, I., Shore, R. & Riboli, E. (2001) Serum carotenoids and breast cancer. *Am. J. Epidemiol.*, **153**, 1142–1147.

Tsuda, T., Kato, Y. & Osawa, T. (2000) Mechanism for the peroxynitrite scavenging activity by anthocyanins. *FEBS Lett.*, **484**, 207–210.

Violi, F., Pignatelli, P. & Pulcinelli, F. M. (2002) Synergism among flavonoids in inhibiting platelet aggregation and H_2O_2 production. *Circulation*, **105**, 53.

Vitrac, X., Desmouliere, A., Brouillaud, B., Krisa, S., Deffieux, G., Barthe, N., Rosenbaum, J. & Merillon, J. M. (2003) Distribution of [C-14] – *trans*-resveratrol, a cancer chemopreventive polyphenol, in mouse tissues after oral administration. *Life Sci.*, **72**, 2219–2233.

Wang, H., Cao, G. & Prior, R. (1997) Oxygen radical absorbing capacity of anthocyanins. *J. Agr. Food Chem.*, **45**, 304–309.

Wang, J. & Mazza, G. (2002) Effects of anthocyanins and other phenolic compounds on the production of tumor necrosis factor alpha in LPS/IFN-gamma-activated RAW 264.7 macrophages. *J. Agr. Food Chem.*, **50**, 4183–4189.

Wang, W. & Higuchi, C. M. (1995) Induction of NAD(P)H: quinine reductase by vitamins A, E, and C in Colo205 colon cancer cells. *Cancer Lett.*, **98**, 63–69.

Wang, C. J., Wang, J. M., Lin, W. L., Chu, C. Y., Chou, F. P. & Tseng, T. H. (2000). Protective effect of *Hibiscus* anthocyanins against *tert*-butyl hydroperoxide-induced hepatic toxicity in rats. *Food Chem. Tox.*, **38**, 411–416.

Wang, X., Liu, C., Bronson, R., Smith, D., Krinsky, N. & Russell, R. (1999) Retinoid signaling and activator protein-1 expression in ferrets given β-carotene supplements and exposed to tobacco smoke. *J. Natl. Cancer I.*, **91**, 60–66.

Weaver, C. (1985) Intrinsic mineral labeling of edible plants: methods and uses. *CRC Crit. Rev. Food Sci. Nutr.*, **23**, 75–101.

Wettasinghe, M., Bolling, B., Plhak, L., Xiao, H. & Parkin, K. (2002) Phase II enzyme-inducing and antioxidant activities of beetroot (*Beta vulgaris* L.) extracts from phenotypes of different pigmentation. *J. Agr. Food Chem.*, **50**, 6704–6709.

Yamasaki, H. (1990) Gap junctional intercellular communication and carcinogenesis. *Carcinogenesis*, **11**, 1051–1058.

Ylönen, K., Alfthan, G., Groop, L., Saloranta, C., Aro, A. & Virtanen, S. (2003) Dietary intakes and plasma concentrations of carotenoids and tocopherols in relation to glucose metabolism in subjects at high risk of type 2 diabetes: The Botnia Dietary Study. *Am. J. Clin. Nutr.*, **77**, 1434–1441.

Youdim, K. A., Martin, A. & Joseph, J. A. (2000a) Incorporation of the elderberry anthocyanins by endothelial cells increases protection against oxidative stress. *Free Radical Biol. Med.*, **29**, 51–60.

Youdim, K. A., Shukitt-Hale, B., Martin, A., Wang, H., Denisova, N. & Joseph, J. (2000b) Short-term dietary supplementation of blueberry polyphenolics: beneficial effects on aging brain performance and peripheral tissue function. *Nutr. Neurosci.*, **3**, 383–397.

Youdim, K. A., Shukitt-Hale, B., MacKinnon, S., Kalt, W. & Joseph, J. A. (2000c) Polyphenolics enhance red blood cell resistance to oxidative stress: *in vitro* and *in vivo*. *Biochim. Biophys. Acta.* **1519**, 117–122.

Youdim, K. A., McDonald, J., Kalt, W. & Joseph, J. A. (2002) Potential role of dietary flavonoids in reducing microvascular endothelium vulnerability to oxidative and inflammatory insults. *J. Nutr. Biochem.*, **13**, 282–288.

Zakharova, N., Petrova, T. (1998) Relationships between the structure and antioxidant activity of certain betalains. *Appl. Biochem. Microbiol.*, **34**, 182–185.

Zanutto, M., Junior, A., Meirelles, M., Favaro, R. & Vannucchi, H. (2002) Effect of citric pectin on beta-carotene bioavailability in rats. *Int. J. Vit. Nutr. Res.*, **72**, 199–203.

Zaripheh, S., Boileau, T., Lila, M. A. & Erdman, J. W. Jr (2003) [^{14}C] – *lycopeneand*[^{14}C]-polar products are differentially distributed in tissues of F344 rats prefed lycopene. *J. Nutr.*, **133**, 4189–4195.

Zeigler, R. (1993) Carotenoids, cancer, and clinical trials, in *Carotenoids in Human Health* (eds I. Canfield, N. Drinski & J. Olson), Annals of the New York Academy of Sciences, New York, USA, pp. 110–119.

9   Plant pigments and protection against UV-B radiation

Brian R. Jordan

9.1   Introduction

Light, including visible, ultraviolet and far-red, is a major determinant of plant growth and development. It provides energy for photosynthesis and information that influences a wide variety of characteristics, including development, morphology and reproductive status. To bring about these responses, light must be absorbed, and plants have many light-absorbing pigments. These include light-harvesting chlorophyll protein complexes as part of the photosynthetic apparatus and the photoreceptor molecules such as phytochrome, cryptochrome and phototropin. Plant pigments play another very important function, which is to protect the complex cellular processes from the damaging consequences of light – either excessive amounts or specific wavelengths that create damage. Ultraviolet (UV) radiation is considered to be a particular hazard to living cells. The UV spectrum can be divided into three regions: UV-A (320–380 nm), UV-B (280–315 nm) and UV-C (< 280 nm). The shorter the wavelength, the more damaging is the radiation. UV-A passes through the protective ozone layer above the earth's surface while UV-B is absorbed to some extent and shorter wavelengths of UV-B (<290 nm) and UV-C are completely excluded. The recent deterioration in the ozone layer, however, has led to increasing amounts of UV-B radiation reaching the earth's surface (McKenzie et al., 1999). Because plants are sessile they are unable to avoid the damaging consequences of UV radiation and any increase is potentially a problem.

A large number of studies have been carried out to investigate the impact of UV-B radiation on plants (Jordan, 1996, and references therein). These studies show a very large variation in response and many factors that can determine these responses (Table 9.1). Most notably, plants vary in sensitivity between species and even between varieties of the same species. The response is also dependent on other environmental parameters such as the intensity of visible irradiance, water status and temperature. This variability has provided an incentive to understand the underlying defence mechanisms of the plant and how they provide protection against UV-B. Three cellular defence mechanisms have been clearly identified; UV-B absorption by protective pigments, DNA repair and antioxidants (Jordan, 1996, 2002). UV-B reflectance by surface waxes, etc. plays only a minor role in protection (Barnes et al., 1996; Jordan, 1996). In addition, the morphology and anatomy of the plant can contribute to the overall protection (Holmes & Keiller, 2002). It is the purpose of this review to concentrate on the role of protective

Table 9.1 Effects of UV-B radiation (280 nm–315 nm) on plants

Effects vary between species and within varieties of the same species
Reduction in photosynthesis and metabolic activity
Downregulation of some genes for photosynthesis and metabolism
Increase in gene expression to produce enzymes involved in UV-B defence
Damage to DNA and activation of DNA repair mechanisms
Changes to plant development and morphology
Complex interactions with other environmental parameters

pigments in UV-B defence. It is, however, particularly important to understand their role in the context of the other defence mechanisms (Fig. 9.1) and the general metabolism of the cell. In addition, this review will discuss the biosynthesis of plant pigments and the signal transduction processes that are involved. A number of reviews on molecular aspects of UV-B response will provide valuable insight to the reader (Strid *et al.*, 1994; Jordan, 1996, 2002; Jansen *et al.*, 1998; Mackerness & Jordan 1999; Mackerness, 2000; Jenkins *et al.*, 2001; Brosché & Strid, 2003).

9.2 The role of plant pigments

The role of protective pigments in UV-B protection has been postulated to be a major determinant for plants to evolve and inhabit the terrestrial environment (Jorgensen, 1994). Aquatic species can be protected to some extent from UV-B by the attenuation of water. This is, however, not possible on the land. Thus land-based autotrophs relying on photosynthesis must evolve protective pigments to allow establishment. To provide protection against UV-B radiation, pigments must absorb at the appropriate wavelengths. They must also allow other wavelengths of light to pass into the tissue, e.g. for photosynthesis. These characteristics are provided mainly by phenylpropanoid pigments that absorb light between 230 nm and 380 nm (Markham, 1982; Cerovic *et al.*, 2002). These compounds are largely water-soluble flavonoids including flavones, flavonols, isoflavonoids and synapate esters, usually located in the vacuoles of cells. The predominant location of these compounds is in the epidermal cells as determined by a variety of techniques such as methanolic extraction (Caldwell *et al.*, 1983; Kalbin *et al.*, 2001), in situ hybridisation (Schmelzer *et al.*, 1988), fibre optics (Day *et al.*, 1993; Vogelmann, 1993) and fluorescence (Kolb *et al.*, 2001). A common characteristic of these compounds is that they are known to increase in amount on exposure to UV-B radiation (Fig. 9.2 and see below). Some polyphenolic compounds, however, are present in other tissue layers and may not increase in amounts on exposure to UV-B (Bornmann *et al.*, 1997). Polyphenolic compounds can also be hydrophobic and may be co-polymerised with cutin and lignin in epidermal cell

Protective mechanisms against UV-B

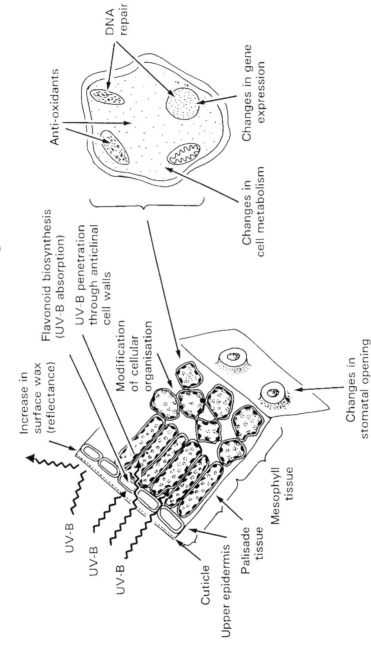

Fig. 9.1 Representation of a leaf cross-section and a mesophyll cell to illustrate potential defense mechanisms against UV-B radiation (from Jordan, 1996).

walls. Studies using fibre-optic microprobes have shown that these compounds prevent UV-B radiation penetrating through anticlinal cell walls to damage underlying mesophyll tissue (Fig. 9.1; Day *et al.*, 1993; Vogelmann, 1993). In a study of over 20 species, conifer needles were the most effective at preventing UV-B transmittance with herbaceous plants being least effective (Jordan, 1996). The spatial transmittance strongly suggested that this was due to UV-B passage through the anticlinal cell walls in herbaceous plants.

In some species there appears to be a constitutive level of protective pigment biosynthesis. For example, in the primary leaf of rye, epidermal hydroxycinnamic acid esters do not respond to UV-B, whereas flavonoids increase twofold. In mesophyll tissue no UV-B response is found in light-exposed green tissue. Etiolated tissue does, however, show an increase in UV-B-absorbing compounds (Bornmann *et al.*, 1997). Furthermore, mRNA levels for genes of the phenylpropanoid pathway vary between leaf epidermal and mesophyll tissues (Kalbin *et al.*, 2001). Chalcone synthase (CHS) mRNA accumulation was similar between the tissues, whereas phenylalanine ammonia lyase (PAL) and chalcone isomerase (CHI) showed substantial differences. No correlation was found between gene expression for the protective pigments and DNA damage. In many species the chemically related anthocyanins readily accumulate on exposure to UV-B. These compounds are coloured and absorb maximally at 530 nm. Consequently, they do not provide protection per se (Solovchenko & Schmitz-Eiberger, 2003), but may be esterified to cinnamic acid to modify their absorption spectrum (Jordan, 1996).

The role of phenylpropanoids in UV-B protection has largely been implied due to their increase on exposure to UV-B (Fig. 9.2). More recently, however, a clear requirement has been confirmed by a number of approaches. Using *transparent testa* mutants of *Arabidopsis thaliana* that have reduced polyphenolic levels (Li *et al.*, 1993, Jordan *et al.*, 1998), flavonoids and synapate esters were shown to provide UV-B protection. Jordan *et al.* (1998) used *tt* mutants (*tt4* and *tt5*, which are deficient in CHS and CHI respectively) to demonstrate that these plants were more susceptible to UV-B as indicated by reduced RNA transcripts. Nuclear- and chloroplast-encoded gene expression for photosynthetic proteins were studied in *tt* mutants exposed to UV-B. Both nuclear- and chloroplast-encoded genes were more severely downregulated in *tt* mutants than in wild-type *A. thaliana*. The nuclear-encoded genes were, however, more severely inhibited by UV-B as is found in wild-type plants. In addition, protective pigment mRNA transcripts were not induced by UV-B to the same extent in *tt* mutants as in wild-type plants. Landry *et al.* (1995) studied the UV-B response of the *A. thaliana* mutant *fah1* that is deficient in the synthesis of synapic acid esters. The *fah1* mutant was shown to be more sensitive than the *tt* mutants suggesting an important role for compounds related to hydroxycinnamic acid esters. Recently a knockout mutant has been identified for an *A. thaliana* gene *AtMYB4*. The mutant plant is more tolerant than wild-type to UV-B and accumulates synapate esters as UV-B-protective pigments. The MYB-related transcription factor AtMYB4 is downregulated by UV-B radi-

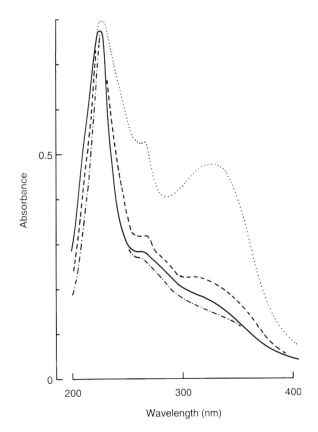

Fig. 9.2 Increases in acidic methanol-extracted flavonoids in response to UV-B radiation (control: 8 h_._._, 3 days _ _ _; UV-B-treated: 8 h ___, 3 days ....).

ation. From these studies, AtMYB4 normally represses the gene for cinnamate 4-hydroxylase and reduces its expression. UV-B exposure lowers AtMYB4 levels and allows enhanced expression of an important rate-limiting enzyme for synapate ester formation (Jin *et al.*, 2000). Lois and Buchanan (1994) used *A. thaliana* mutants produced by chemical mutagenesis to study the role of protective pigments. Plants that had lost the potential to produce kaempferol appeared susceptible to UV-B damage, suggesting that a few specific compounds may be mainly important in UV-B protection (see below). In similar experiments in maize and rice, results suggested an important role for flavonoids (Stapleton & Walbot, 1994; Maekawa *et al.*, 2001). In maize, isogenic lines lacking regulatory genes for anthocyanin biosynthesis were more susceptible to UV-B damage. Both cyclobutane pyrimidine dimers (CPDs) and 6,4 photoproducts were increased in the isogenic lines. Interestingly, plants in natural field-grown conditions showed no overall damage, reflecting the very efficient photorepair mechanisms present in plants (Stapleton & Walbot, 1994). In rice, three near-isogenic lines (NILs) for purple leaf genes *Pl*, *Pl*[W] and *Pl*[i] were tested for anthocyanin induction by UV-B (Maekawa *et al.*, 2001). The NILs showed differential induction and accumulation of anthocyanins. In addition, the total biomass of the NILs decreased with increasing dose of irradiation. This suggests that the anthocyanins may not be fully protective or that there is some growth compensation to produce the protective pigments.

In addition to UV-B absorption as a protective mechanism, polyphenolic compounds may also function as antioxidants. The actual antioxidant capacity will vary according to the chemical structure. For instance, flavonoids with additional hydroxyl groups on the B-ring, such as quercetin, are stronger antioxidants than those without, e.g. monohydroxylated compounds such as kaempferol. There is now accumulating evidence that UV-B exposure favours the formation of these dihydroxylated flavonoids presumably to enhance the antioxidant potential. In *Brassica napus* and *Hordeum vulgare*, quercetin-glycoside and lutonarin respectively increase substantially compared to kaempferol-glycoside and saponarin (Reuber *et al.*, 1996; Bornman *et al.*, 1997). Similar results have been found in rice (Markham *et al.*, 1998a), *Brassica* (Olsson *et al.*, 1998), birch (Lavola, 1997), *Marchantia polymorpha* (Markham *et al.*, 1998b) and *Trifolium repens* (Hofmann *et al.*, 2000, 2003). In the study of *T. repens* an extended range of populations of the leguminous species was investigated. The relative preference for quercetin biosynthesis was consistently observed; however, importantly there was an inverse relationship between plant productivity and UV-B-generated accumulation of quercetin.

The response of *T. repens* to UV-B was also investigated during water stress (Hofmann *et al.*, 2003). Nine populations of *T. repens* were exposed to UV-B supplements of $13.3\,kJ/m^2/day$ for 12 weeks with or without 4 weeks of drought. The expected increase in protective pigments was enhanced by the drought conditions and favoured the accumulation of ortho-dihydroxylated quercetin rather than monohydroxylated kaempferol glycosides (Table 9.2). Intraspecific comparisons linked increased quercetin glycosides to lower plant productivity and higher UV-B tolerance under well-watered conditions. In addition to these studies, transgenic plants with modified flavonoid biochemistry have been used recently to investigate quercetin/kaempferol ratios (Ryan *et al.*, 1998, 2002). In these studies, comparison was made between wild-type, transgenic and mutant *F3'H* petunia. Wild-type petunia increased flavonoid content on exposure to UV-B and the quercetin (Q)/kaempferol (K) ratio increased. In the mutant *F3'H*, overall flavonoid levels increased, but the Q:K ratio decreased with a rise in kaempferol as the biosynthesis of quercetin was blocked by the ineffective F3'H activity. The growth rate of the *F3'H* mutant was significantly reduced compared to the wild-type. Thus despite the increase in flavonoids by UV-B, there is a clear requirement for quercetin biosynthesis. Transgenic lines with antisense *F3'H* produced varying Q : K ratios though in most cases this did not change and did not respond to UV-B. These results suggest that the antisense construct did have some effect on the plants' ability to respond to UV-B radiation. The transgenic plants had similar

Table 9.2 Changes in protective pigment composition in response to UV-B and water stress

Attribute	Water	UV−	UV+	P
UV-B-absorbing compounds (A_{300}/mg)	WW	0.994 ± 0.019	1.027 ± 0.023	NS
	DR	1.026 ± 0.020	1.146 ± 0.022	***
Total flavonols (mg/g)	WW	0.559 ± 0.045	1.245 ± 0.078	***
	DR	0.477 ± 0.041	1.700 ± 0.073	***
Quercetin (mg/g)	WW	0.262 ± 0.031	0.771 ± 0.049	***
	DR	0.125 ± 0.019	1.049 ± 0.048	***
Kaempferol (mg/g)	WW	0.297 ± 0.028	0.474 ± 0.035	***
	DR	0.352 ± 0.034	0.651 ± 0.040	***
Chlorophyll (mg/g)	WW	15.36 ± 0.225	15.98 ± 0.306	+
	DR	13.04 ± 0.242	13.32 ± 0.191	NS

Biochemical and physiological attributes averaged across nine populations of *T. repens* (mean $\pm$ 1 SE), grown with (UV+) and without (UV-) supplementation of 13.3 kJ/m^2/d UV-B, under well-watered (WW) or droughted (DR) conditions. (Probabilities reflect significance of the UV-B effect under the two water regimes. ***, $P < 0.001$; +, $P < 0.10$; NS, $P \geq 0.10$).
Adapted from Table 1, Hofmann *et al.*, 2003.

growth rates to the wild-type, but reduced overall flavonoids due to inhibition of a number of key biosynthetic genes.

In addition to increasing antioxidant potential, dihydroxylated flavonoids may enhance energy dissipation (Smith & Markham, 1998). Anthocyanin has also been shown to contribute substantially to the antioxidant potential of plants. The role of anthocyanins as antioxidants has been studied using electron spin resonance in *A. thaliana* plants (Nagata *et al.*, 2003). From these studies, anthocyanin levels appear to control the cellular reactive oxygen species in plants subject to irradiation (in addition to ascorbic acid). High levels of anthocyanin also maintained the growth and flowering potential of the plants. Differences in anthocyanin accumulation were identified between the two ecotypes *Ler* and *Col*, indicating how different cultivars or even species may vary in their sensitivity to UV-B. In studies on red versus green leaves of *Elatostema rugosum*, red leaves had higher levels of antioxidants (Neill *et al.*, 2002). These antioxidants included anthocyanin, superoxide dismutase, catalase and hydroxycinnamic acid. Of these compounds, anthocyanins contributed to the antioxidant potential more than any other phenolic compound. In agreement with other studies, UV-B radiation increased the biosynthesis of the anthocyanins, but their content declined with leaf age as did the overall antioxidant potential of the leaf. A similar decline has been found in flavonoid levels. For instance, in *Silene vulgaris* the concentration at day 18 of UV-B exposure was only 1% of day 1 (van de Staaij *et al.*, 1995).

In addition to the phenylpropanoids, carotenoids are important in protecting plants from light-mediated damage. Carotenoids function as antioxidants and are closely associated with the photosynthetic apparatus. There have not been many detailed studies of carotenoids in relation to UV-B damage. Some studies have shown increased levels of carotenoids (Steel & Keller, 2000; Xiong & Day, 2001) while others have found reductions (Strid *et al.*, 1990; Musil *et al.*, 1999) or no change (Kirchgeβner *et al.*, 2003). Recently Götz *et al.* (2002) used transgenic plants transformed with β-carotene hydroxylase to study the role of carotenoids in UV-B protection. UV-B- irradiated transgenic plants retained a higher biomass, higher levels of photosynthetic pigments and less impairment of photosynthetic function than controls. In addition, they had greater resistance to lipid peroxidation. The xanthophyll cycle plays an important role in this protection, and enzymes of this cycle can be inhibited by UV-B, reducing photosynthetic efficiency (Pfündel *et al.*, 1992). These results suggest that under normal physiological conditions carotenoids provide UV-B protection to the photosynthetic system. Furthermore, they may contribute to the 'chloroplast signal' that is involved in signal transduction (Taylor, 1989; Jordan *et al.*, 1998).

9.3   Biosynthesis of UV-B-protective pigments

UV-B-induced changes in the biosynthesis of protective pigments is a well-characterised and a widespread response (Jordan, 1996). Levels of phenylpropanoids increase rapidly in UV-B-irradiated tissue (Fig. 9.2), as do other protective compounds such as antioxidants. RNA levels for enzymes such as PAL, CHS and dihydroflavonol 4-reductase increase within a few hours of UV-B exposure and then decline to low levels. The changes in mRNAs reflect a sequential modification of enzyme activity in order of the enzymes of the phenylpropanoid pathway (Kubasek *et al.*, 1992). Using 'run-off' transcription assays on isolated nuclei these changes have been shown to involve increased transcriptional activity (Chappell & Hahlbrock, 1984; Feinbaum & Ausubel, 1988). Initially, using *in vivo* DNA footprinting, two *cis*-acting units were identified in the parsley *chs* promoter (Schulze-Lefert *et al.*, 1989a, 1989b; Weisshaar *et al.*, 1991a, 1991b). These promoter elements showed differential reactivity to dimethyl sulphate methylation in control and UV-treated tissue. In addition, this alteration in methylation correlated to the transcription kinetics of *chs*. The two UV-inducible *cis*-acting elements are now called ACE (ACGT-containing elements that recognise common plant regulatory factors) and MRE (Myb-recognition elements). Two bZip transcription factors bind to the ACE (Feldbrügge *et al.*, 1994; Sprenger-Gaussels & Weisshaar, 2000) and two MYB-proteins to the MRE (Feldbrügge *et al.*, 1997). These interactions are involved in the UV-inducible transcription activation of *chs*. Other defence-related genes are also activated, such as those associated with antioxidants and DNA repair. However, many other genes are downregulated in response to UV-B (Fig. 9.3), most notably the genes that code for photosynthetic proteins (Jordan *et al.*, 1991, 1992, 1994; Casati & Walbot, 2003; Izaguirre *et al.*, 2003). These results suggest that there is a UV-B-induced switch from normal metabolism and photosynthesis to the biosynthesis of defence-related compounds. This response is determined by a number of factors, including photoperception, signal transduction (see below), developmental stage, biotic and abiotic factors (Table 9.3). The UV-B-induced response is particularly dependent on developmental stage (Jordan *et al.*, 1994, 1998; Day *et al.*, 1996; Mackerness *et al.*, 1998a). Thus, etiolated tissue can show a limited response to UV-B despite having apparently less protective pigments. Similarly, young leaf tissue appears less susceptible than older tissue as reflected by changes in gene expression for photosynthetic genes (Fig. 9.4). Gene expression for protective pigment synthesis is also differentially activated as illustrated by increased expression of *chs* in the outer (older) leaves of *A. thaliana* (Fig. 9.4). Thus older leaves are more susceptible to UV-B damage than younger leaves. There is an apparent 'accelerated senescence' during UV-B exposure and senescence-related genes are also activated (John *et al.*, 2001). In field-grown pea plants subject to supplementary UV-B radiation over a period of time, leaf age was the major determinant of the response (Day *et al.*, 1996). Although the levels of protective pigments per se did not change, there was a redistribution of these compounds to the more adaxial epidermal tissue. Recently, near isogenic lines of maize varying in flavonoid content have been used to study UV-B responses by microarray analysis (Casati & Walbot, 2003). Two thousand five hundred expression sequence tags (ESTs) were tested and 355 responded to UV-B. Of these, 232 ESTs have been assigned a possible function. UV-B increased the expression of stress-related genes, ribosomal protein genes and downregulated a limited number, notably photosynthetic genes (these field studies are consistent with more extreme controlled environment studies, suggesting that the latter are indicative of molecular changes that take place in the natural environment). In near isogenic lines with reduced protective pigments, the UV-B-induced response was more prominent, again indicating the important protective role of these compounds.

Other environmental parameters can influence the induction of protective pigment biosynthesis by UV-B, such as the different quality and quantity of light. For instance, in parsley tissue culture, *chs* is strongly induced after a 2 h lag phase by UV light. Pretreatment of the tissue culture removes the lag phase

Table 9.3 Factors affecting UV-B-induced changes in protective plant pigment composition

Penetration of UV-B into the tissue
Perception of the light environment
Signal transduction pathways and crosstalk between them
Damage to DNA and efficiency of repair mechanisms
Indirect regulation of gene expression through metabolic feedback, oxidation, etc.
Developmental stage of the tissue
Tissue-specific gene expression
Differential response of gene family members
Metabolic channelling
Interaction of other environmental parameters

UV-B exposure

C   4h   8h

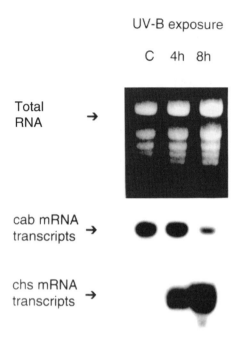

Total
RNA →

cab mRNA
transcripts →

chs mRNA
transcripts →

Fig. 9.3 Changes in gene expression in response to UV-B radiation to illustrate downregulation of the *cab* (*chlorophyll a/b-binding protein*) and the induction of *chs* (from Jordan, 1996).

UV-B exposure

0   8h   1d   3d   7d

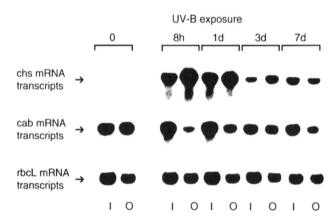

chs mRNA
transcripts →

cab mRNA
transcripts →

rbcL mRNA
transcripts →

I   O   I   O   I   O   I   O   I   O

Fig. 9.4 Changes in gene expression in young (I: inner) and older (O: outer) leaves of *Arabidopsis thaliana* exposed to UV-B radiation for various time periods (*chs* represents gene expression for protective pigments; *cab* represents nuclear-encoded gene expression and *rbcL* represents chloroplast-encoded gene expression).

and red/far-red irradiation treatments alter both the magnitude and extent of the *chs* induction (Jordan, 1996, and references therein). These responses work through different photoreceptors, such as phytochrome, cryptochrome and photo-tropin. A major determinant of UV-B regulation of gene expression is the level of photosynthetically active radiation (PAR). High PAR is known to protect the plant from UV-B damage (Jordan, 1996). Although this amelioration response has been well established at the physiological level, Jordan *et al.* (1991, 1992) first documented the effect of PAR on UV-B-induced gene expression. Thus, genes for photosynthetic proteins, such as *rubisco* and *chlorophyll a/b-binding protein* (*cab*), are not as dramatically downregulated by UV-B if the PAR is high. In a similar though reverse response, the induction of *chs* is not as strongly upregulated in high PAR compared to low PAR. A number of studies have been carried out to investigate this amelioration of the UV-B response by high PAR (Mackerness *et al.*, 1996, 1998a; Strid *et al.*, 1996). These studies clearly indicate that photosynthesis per se is controlling the response (and not increased photolyase activity etc.), possibly through the generation of ATP, but certainly through electron transport activity. This is consistent with the view that the balance between photosystem I and II can act as a photosensor in mature green tissue and modify gene expression (Chow *et al.*, 1990). Another potential regulation of UV-B-induced biosynthesis is metabolic feedback, particularly changes in carbohydrate levels. Recent studies, however, indicate that feedback regulation from carbohydrates is not a significant factor in UV-B-induced changes in gene expression (Mackerness *et al.*, 1997). Tissue-specific expression will also be important in determining the response and this may involve multigene families (Jordan, 1996, and references therein). In petunia the nine members of the *chs* gene family are expressed in different tissues and with different photoreceptors during development (Koes *et al.*, 1989). Similar responses can be found in the multigene *cab* family of pea (Mackerness *et al.*, 1998b). These different responses are particularly interesting in the context of metabolic channelling (Rasmussen & Dixon, 1999). In studies on tobacco cell cultures, Rasmussen and Dixon showed that specific cellular compartments exist to channel metabolites through linked enzyme systems. They used transgenic plants and elicitors to perturb the biosynthesis of phenolic compounds and concluded that specific forms of enzymes were closely aligned on microsomal membranes. This channelling of resources must play an important role in the specificity of the plants' response.

9.4 Signal transduction involved in the biosynthesis of protective pigments

The synthesis of protective pigments is very rapid in response to UV-B radiation. This response requires the perception of the UV-B radiation followed by a signal transduction pathway leading to changes in gene expression. Recently, substantial research has been focused in the area of signal transduction (Jordan, 2002; Casati & Walbot, 2003; Holley *et al.*, 2003; Izaguirre *et al.*, 2003). At present no UV-B photoreceptor has been identified, although some evidence does indicate a degree of specificity (Ballaré *et al.*, 1995; Boccalandro *et al.*, 2001). After UV-B perception there are a number of separate pathways with different components. These pathways can overlap and 'crosstalk' exists between the different pathways (Logemann & Halhbrock, 2002). In addition, signal transduction pathways are formed from other stimuli and they interact with those established by UV-B. These stimuli include herbivory, pathogens, systemin and oligosaccharide elicitors (OEs). OEs and their respective signal transduction components have been compared to UV-B. These comparisons show both clear similarities and differences between pathways. For instance, systemin, OEs and UV-B all induce mitogen-activated protein kinase (MAPK) activity. However, two homologous MAPK were activated in response to systemin, four different OEs and UV-B. In contrast, a third MAPK was only activated by UV-B (Holley *et al.*, 2003). These differences may account, to some extent, for the specificity of UV-B responses. Furthermore, a hierarchy of stress responses seems to be present with pathogen response overriding UV-B through modification of transcription (Logemann & Halhbrock, 2002). One of the first components of the UV-B-induced signal pathway is reactive oxygen species (ROS). These ROS can be generated through a number of reactions and in particular through NADPH oxidase and peroxidase(s) (Mackerness *et al.*, 2001). It seems likely that the systemin receptor may also be involved (Yalamanchili & Stratmann, 2002). This receptor is known to be involved in a variety of hormonal signals. Hydrogen peroxide is generated via NADPH in response to systemin and activates the plant defence systems. An increase in NADPH oxidase has been detected (Rao *et al.*, 1996; Mackerness *et al.*, 2001), and Casati and Walbot (2003) using microarray analysis found NADPH gene expression was upregulated by UV-B in field-grown maize. From the ROS a number of pathways seem to be involved with components such as salicylic acid, jasmonic acid and ethylene (Mackerness *et al.*, 1999). These particular signal transduction pathways lead to changes in gene expression for pathogen defence–related gene expression and photosynthetic genes. Of particular interest, however, is that expression for genes of the phenylpropanoid pathway does not seem to follow these pathways. CHS, which is the first committed enzyme of the flavonoid pathway, has been extensively investigated and involves phytochrome, cryptochrome and UV-B signal transduction pathways (Wade *et al.*, 2001; Davies & Schwinn, 2003). Pharmacological approaches have been used to develop an understanding of signal transduction for *chs* genes. Christie and Jenkins (1996) demonstrated that the UV-B pathways were distinct from UV-A/blue and phytochrome. UV-B and UV-A/blue share common pathways involving Ca^{2+} and protein phosphorylation. They differed, however, in that the UV-B pathway to *chs* was inhibited by the calmodulin antagonist W-7 while UV-A/blue was not. Both UV-A/blue through cry1 and UV-B probably involve redox changes associated with the plasma membranes (Long & Jenkins, 1998). This cry1 could interact with the plasma membrane and generate a redox potential that subsequently stimulates a Ca^{2+} release from an internal compartment of the cells. Changes in both Ca^{2+} influx and efflux regulate the Ca^{2+} concentration. Downstream of these changes very little is known and this is a major area for further investigation. Another significant difference is that *chs* is not induced through changes in ROS

(Jenkins *et al.*, 2001; Mackerness *et al.*, 2001). The normal UV-B-induced upregulation response is not affected by ROS scavengers, but rather by inhibitors of nitric oxide synthase or scavengers of nitric oxide (Mackerness *et al.*, 2001). This signal pathway then involves both calcium and calmodulin. From these studies, the UV-B-induced phenylpropanoid pathway may be induced separately to a number of other pathways and work in parallel to them. It is apparent from many of these studies that UV-B shares components with many other signal transduction pathways, including defence, herbivory and environmental stress. There are, however, discrete pathways that relate specifically to UV-B that have been identified (Brosché & Strid, 2002, 2003; Casati & Walbot, 2003).

9.5 The cost and consequences of protection

It is clear from this review that plant pigments, mainly polyphenolics, play a significant role in protecting plants from UV-B damage. It is also apparent that UV-B-induced signal transduction and biosynthesis of protective pigments is extremely complex. The changes induced by UV-B instigate a switch from general metabolism to the production of protective compounds (Logemann *et al.*, 2000). This involves a sophisticated regulatory network to distribute the response along the appropriate biochemical pathways. This response is relatively broad (Logemann *et al.*, 2000) with factors such as gene expression being extensively modified (Casati & Walbot, 2003). To add to the complexity there is a hierarchy of responses to other external stimuli (Logemann & Hahlbrock, 2002). Overall, this sophisticated switch from primary to secondary metabolism is likely to impart a 'cost' to the plant. Secondary metabolites can constitute a substantial percentage of a plant; for instance in buckwheat herbs, between 2% and 10% of dry weight of rutin can be found in leaves and flowers (Kreft *et al.*, 2002). If these levels are then substantially increased by UV-B there is a potential extra cost to the plant (Seigler, 2002). This cost will undoubtedly be reflected in loss of productivity as has been found by Hofmann *et al.* (2000, 2003). Other environmental parameters will influence the UV-B-induced loss of productivity, particularly light (Jordan, 1996), water (Hofmann *et al.*, 2003) and nitrogen availability (Hunt & McNeil, 1998). In addition, there is significant potential for the nutritional value of the plant to be changed as many of these secondary compounds have important physiological properties. It is therefore important that the consequences of UV-B-induced changes to the phenylpropanoid pathway are understood at all levels from molecular to dietary consequences for both humans and animals.

References

Ballaré, C. L., Barnes, P. W. & Flint, S. D. (1995) Inhibition of hypocotyl elongation by ultraviolet-B radiation in de-etiolating tomato seedlings. I. The photoreceptor. *Physiol. Plant.*, **93**, 584–592.

Barnes, J. D., Percy, K. E., Paul, N. D., Jones, P., McLaughlin, C. K., Mullineaux, P. M. Creissen, G. & Wellburn, A. R. (1996) The influence of UV-B radiation on the physicochemical nature of tobacco (*Nicotiana tabacum* L.) leaf surfaces. *J. Exp. Bot.*, **47**, 99–109.

Boccalandro, H. E., Mazza, C. A., Mazzella, M. A., Casal, J. J. & Ballaré, C. L. (2001) Ultraviolet-B radiation enhances a phytochrome-B-mediated photomorphogenic response in *Arabidopsis*. *Plant Physiol.*, **126**, 780–788.

Bornman, J. F., Reuber, S., Cen Y.-P. & Weissenböck, G. (1997) Ultraviolet radiation as a stress factor and the role of protective pigments, in *Plants and UV-B. Responses to Environmental Change*. Society for Experimental Biology Seminar Series 64 (ed. P. Lumsden), Cambridge University Press, Cambridge, UK, pp. 157–168

Brosché, M. & Strid, Å (2003) Molecular events following perception of ultraviolet-B radiation by plants. *Physiol. Plant.*, **117**, 1–10.

Brosché, M., Schuler, M. A., Kalbina, I., Conner, L. & Strid, Å. (2002) Gene regulation by low-level UV-B radiation: identification by DNA array analysis. *Photochem. Photobiol. Sci.*, **1**, 656–664.

Caldwell, M. M., Robberecht, R. & Flint, S. D. (1983) Internal filters: prospects for UV-acclimation in higher plants. *Physiol. Plant.*, **58**, 445–450.

Casati, P. & Walbot, V. (2003) Gene expression profiling in response to ultraviolet radiation in maize genotypes with varying flavonoid content. *Plant Physiol.*, **132**, 1739–1754.

Cerovic, Z. G., Ounis, A., Cartelat, A., Latouche, G., Goulas, Y., Meyer, S. & Moya, I. (2002) The use of chlorophyll fluorescence excitation spectra for the non-destructive *in situ* assessment of UV-absorbing compounds in leaves. *Plant Cell Environ.*, **25**, 1663–1676.

Chappell, J. & Hahlbrock, K. (1984) Transcription of plant defence genes in response to UV light or fungal elicitor. *Nature*, **311**, 76–78.

Chow, W. S., Goodchild, D. J., Miller, C. & Anderson, J. M. (1990) The influence of high levels of brief or prolonged supplementary far-red illumination during growth on the photosynthetic characteristics, composition and morphology of *Pisum sativum* chloroplasts. *Plant Cell Environ.*, **13**, 135–145.

Christie, J. M. & Jenkins, G. I. (1996) Distinct UV-B and UV-A/blue light signal transduction pathways induce chalcone synthase gene expression in arabidopsis cells. *Plant Cell*, **8**, 1555–1567.

Davies, K. M. & Schwinn, K. E. (2003) Transcriptional regulation of secondary metabolism. *Fun. Plant Biol.*, **30**, 913–925.

Day, T. A., Howells, B. W. & Ruhland, C. T. (1996) Changes in growth and pigment concentrations with leaf age in pea under modulated UV-B radiation field treatments. *Plant Cell Environ.*, **19**, 101–108.

Day, T. A., Martin, G. & Vogelmann, T. C. (1993) Penetration of UV-B radiation in foliage: evidence that the epidermis behaves as a non-uniform filter. *Plant Cell Environ.*, **16**, 735–741.

Feinbaum, R. L. & Ausubel, F. M. (1988) Transcriptional regulation of the *Arabiodpsis thaliana* chalcone synthase gene. *Mol. Cell. Biol.*, **8**, 1985–1992.

Feldbrügge, M., Sprenger, M., Dinkelbach, M., Yazaki, K., Harter, K. & Weisshaar, B. (1994) Functional-analysis of a light-responsive plant bZIP transcriptional regulator. *Plant Cell*, **6**, 1607–1621.

Feldbrügge, M., Sprenger, M., Hahlbrock, K. & Weisshaar, B. (1997) PcMYB1, a novel plant protein containing a DNA-binding domain with one MYB repeat, interacts *in vivo* with a light-regulatory promoter unit. *Plant J.*, **11**, 1079–1093.

Götz, T., Sandmann, G. & Römer, S. (2002) Expression of a bacterial carotene hydroxylase gene (*crtZ*) enhances UV tolerance in tobacco. *Plant Mol. Biol.*, **50**, 129–142.

Hofmann, R. W., Campbell, B. D., Bloor, S. J., Swinny, E. E., Markham, K. R., Ryan, K. G. & Fountain, D. W. (2003) Responses to UV-B radiation in *Trifolium repens* L. – physiological links to plant productivity and water availability. *Plant Cell Environ.*, **26**, 603–612.

Hofmann, R. W., Swinny, E. E., Bloor, S. J., Markham, K. R., Ryan, K. G., Campbell, B. D., Jordan, B. R. & Fountain, D. W. (2000) Response of nine *Trifolium repens* L. populations to ultraviolet-B radiation: differential flavonol glycoside accumulation and biomass production. *Ann. Bot.*, **86**, 527–537.

Holley, S. R., Yalamanchili, R. D., Moura, D. S., Ryan, C. A. & Stratmann, J. W. (2003) Convergence of signalling pathways induced by systemin, oligosaccharide elicitors, and ultraviolet-B radiation at the level of mitogen-activated protein kinases in *Lycopersicon peruvianum* suspension-cultured cells. *Plant Physiol.*, **132**, 1728–1738.

Holmes, M. G. & Keiller, D. R. (2002) Effects of pubescence and waxes on the reflectance of leaves in the ultraviolet and photosynthetic wavebands: a comparison of a range of species. *Plant Cell Environ.*, **25**, 85–93.

Hunt, J. E. & McNeil, D. L. (1998) Nitrogen status affects UV-B sensitivity of cucumber. *Aust. J. Plant Physiol.*, **25**, 79–86.

Izaguirre, M. M., Scopel, A. L., Baldwin, I. T. & Ballaré, C. L. (2003) Convergent responses to stress. Solar ultraviolet-B radiation and *Manduca Sexta* herbivory elicit overlapping transcriptional responses in field-grown plants of *Nicotiana longiflora. Plant Physiol.*, **132**, 1755–1767.

Jansen, M. A. K., Gaba, V. & Greenberg, B. M. (1998) Higher plants and UV-B radiation: balancing damage, repair and acclimation. *Trends Plant Sci.*, **3**, 131–135.

Jenkins, G. I., Long, J. C., Wade, H. K., Shenton, M. R. & Bibikova, T. N. (2001) Research review: UV and blue light signalling: pathways regulating chalcone synthase gene expression in *Arabidopsis. New Phytol.*, **151**, 121–131.

Jin, H., Cominelli, E., Bailey, P., Parr, A., Mehrtens, F., Jones, J., Tonelli, C., Weisshaar, B. & Martin, C. (2000) Transcriptional repression by At-MYB4 controls production of UV-protecting sunscreens in *Arabidopsis. EMBO J.*, **19**, 6150–6161.

John, C. F., Morris, K., Jordan, B. R., Thomas, B. & Mackerness, S. A.-H. (2001) Ultraviolet-B exposure leads to up-regulation of senescence-associated genes in *Arabidopsis thaliana. J. of Exp. Bot.*, **52**, 1367–1373.

Jordan, B. R. (1996) The effects of ultraviolet-B radiation on plants: a molecular perspective, in *Advances in Botanical Research*, Vol. 22 (ed. J. A. Callow), Academic Press, London, UK, pp. 97–162.

Jordan, B. R. (2002) Review: molecular response of plant cells to UV-B stress. *Fun. Plant Biol.*, **29**, 909–916.

Jordan, B. R., Chow, W. S., Strid, Å. & Anderson, J. M. (1991) Reduction in *cab* and *psbA* RNA transcripts in response to supplementary ultraviolet-B radiation. *FEBS Lett.*, **284**, 5–8.

Jordan, B. R., He, J., Chow, W. S. & Anderson, J. M. (1992) Changes in mRNA levels and polypeptide subunits of ribulose 1,5-bisphosphate carboxylase in response to supplementary ultraviolet-B radiation. *Plant Cell Environ.*, **15**, 91–98.

Jordan, B. R., James, P., Strid, Å. & Anthony, R. (1994) The effect of ultraviolet-B radiation on gene expression and pigment composition in etiolated and green pea leaf tissue: UV-B induced changes are gene-specific and dependent upon the developmental stage. *Plant Cell Environ.*, **17**, 45–54.

Jordan, B. R., James, P. & Mackerness, S. A.-H. (1998) Factors affecting UV-B induced changes in *Arabidopsis thaliana* gene expression: role of development, protective pigments and the chloroplast signal. *Plant Cell Physiol.*, **39**, 769–778.

Jorgensen, R. (1994) The genetic origins of biosynthesis and light-responsive control of the chemical UV screen of land plants, in *Genetic Engineering of Plant Secondary Metabolism* (ed. R. E. Ellis, G. W. Kuroki & H. A. Stafford), Plenum, New York, USA, pp. 179–192.

Kalbin, G., Hidema, J., Brosché, M., Kumagai, T., Bornman, J. F., Strid, Å. (2001) UV-B-induced DNA damage and expression of defence genes under UV-B stress: tissue-specific molecular marker analysis in leaves. *Plant Cell Environ.*, **24**, 983–990.

Kirchgeßner, H.-D., Reichert, K., Hauff, K., Steinbrecher, R., Schnitzler, J.-P. & Pfündel, E. E. (2003) Light and temperature, but not UV radiation, affect chlorophylls and carotenoids in Norway spruce needles (*Picea abies* (L.) Karst.). *Plant Cell Environ.*, **26**, 1169–1179.

Koes, R. E., Spelt, C. E. & Mol, J. N. M. (1989) The chalcone synthase multigene family of *Petunia hybrida* (V30): differential light-regulated expression during flower development and UV light induction. *Plant Mol. Biol.*, **12**, 213–225.

Kolb, C. A., Kaser, M. A., Kopecky, J., Zotz, G., Riederer, M. & Pfündel, E. E. (2001) Effects of natural intensities of visible and ultraviolet radiation on epidermal ultraviolet screening and photosynthesis in grape leaves. *Plant Physiol.*, **127**, 863–875.

Kreft, S., Strukelj, B., Gaberscik, A. & Kreft, I. (2002) Rutin in buckwheat herbs grown at different UV-B radiation levels: comparison of two UV spectrophotometric and an HPLC method. *J. Exp. Bot.*, **53**, 1801–1804.

Kubasek, W. L., Shirley, B. W., McKillop, A., Goudman, H. M., Briggs, W. R. & Ausubel, F. M. (1992) Regulation of flavonoid biosynthetic genes in germinating *Arabidopsis* seedings. *Plant Cell*, **4**, 1229–1236.

Landry, L. G., Chapple, C. C. S. & Last, R. L. (1995) *Arabidopsis* mutants lacking phenolic sunscreens exhibit enhanced ultraviolet-B injury and oxidative damage. *Plant Physiol.*, **109**, 1159–1166.

Lavola, A., Julkunen-Tiitto, R., Aphalo, P., de la Rosa, T. & Lehto, T. (1997) The effect of UV-B radiation on UV-absorbing secondary metabolites in birch seedlings grown under simulated forest soil conditions. *New Phytol.*, **137**, 617–621.

Li, J., Ou-Lee, T.-M., Raba, R., Amundson, R. G. & Last, R. L. (1993) *Arabidopsis* flavonoid mutants are hypersensitive to UV-B irradiation. *Plant Cell.*, **5**, 171–179.

Logemann, E. & Halhbrock, K. (2002) Crosstalk among stress responses in plants: pathogen defense overrides UV protection through an inversely regulated ACE/ACE type of light-responsive gene promoter unit. *Proc. Nat. Acad. Sci. USA*, **99**, 2428–2432.

Logemann, E., Tavernaro, A., Schulz, W., Somssich, I. E. & Hahlbrock, K (2000) UV light selectivity coinduces supply pathways from primary metabolism and flavonoid secondary product formation in parsley. *Proc. Nat. Acad. Sci. USA*, **97**, 1903–1907.

Long, J. C. & Jenkins, G. I. (1998) Involvement of plasma membrane redox activity and calcium homeostasis in the UV-B and UV-A/blue light induction of gene expression in *Arabidopsis. Plant Cell*, **10**, 2077–2086.

Lois, R. & Buchanan, B. B. (1994) Severe sensitivity to ultraviolet radiation in an *Arabidopsis* mutant deficient in flavonoid accumulation. II. Mechanisms of UV-resistance in *Arabidopsis. Planta*, **194**, 504–509.

Mackerness, S. A.-H. (2000) Plant responses to ultraviolet-B (UV-B: 280—320 nm) stress: what are the key regulators? *Plant Growth Reg.*, **32**, 27–29.

Mackerness, S. A.-H. & Jordan, B. (1999) Change in gene expression in response to UV-B induced stress, in *Handbook of Plant and Crop Stress* (ed. M. Pessaraki), Marcel Dekker, New York, USA, pp. 749–768.

Mackerness, S. A.-H., Butt, J. P., Jordan, B. R. & Thomas, B. (1996) Amelioration of ultraviolet-B-induced down-regulation of mRNA transcripts for chloroplast proteins, by high irradiance, is mediated by photosynthesis. *J. Plant Physiol.*, **148**, 100–106.

Mackerness, S., Surplus, S. Jordan, B. R. & Thomas, B. (1997) Ultraviolet-B effects on transcript levels for photosynthetic genes are not mediated through carbohydrate metabolism. *Plant Cell Environ.*, **20**, 1431–1437.

Mackerness, S. A.-H., Surplus, S. L., Jordan, B. R. & Thomas, B. (1998a) Effects of supplementary ultraviolet-B radiation on photosynthetic transcripts at different stages of leaf development and light levels in pea (*Pisum sativum* L.): role of active oxygen species and antioxidant enzymes. *Photochem. Photobiol.*, **68**, 88–96.

Mackerness, S. A.-H., Liu, Liansen, Thomas, B., Thompson, W. F., Jordan, B. R. & White, M. J. (1998b) Individual members of the light-harvesting complex II chlorophyll *a*/*b*-binding protein gene family in pea (*Pisum sativum*) show differential responses to ultraviolet-B radiation. *Physiol. Plant.*, **103**, 377–384.

Mackerness, S. A.-H., Surplus, S. L., Blake, P., John, C. F., Buchanan-Wollaston, V., Jordan, B. R. & Thomas, B. (1999) Ultraviolet-B-induced stress and changes in gene expression in *Arabidopsis thaliana*: role of signalling pathways controlled by jasmonic acid, ethylene and reactive oxygen species. *Plant Cell Environ.*, **22**, 1413–1423.

Mackerness, S. A.-H., John, C. F., Jordan, B. & Thomas, B. (2001) Early signalling components in ultraviolet-B responses: distinct roles for different reactive oxygen species and nitric oxide. *FEBS Lett.*, **489**, 237–242.

McKenzie, R., Conner, B. & Bodeker, G (1999) Increased summer time UV radiation in New Zealand in response to ozone loss. *Science*, **285**, 1709–1711.

Maekawa, M., Sato, T., Kumagai, T. & Noda, K. (2001) Differential responses to UV-B irradiation of three near isogenic lines carrying different purple leaf genes for anthocyanin accumulation in rice (*Oryza sativa* L.). *Breed. Sci.*, **51**, 27–32.

Markham, K. R. (1982) *Techniques of Flavonoid Identification*, Academic Press (Biological Techniques Series), London. UK.

Markham, K. R., Ryan, K. G., Bloor, S. J. & Mitchell, K. A. (1998b) An increase in luteolin: apigenin ratio in *Marchantia polymorpha* on UV-B enhancement. *Phytochemistry*, **48**, 791–94.

Markham, K. R., Tanner, G., Caasi-Lit, M., Whitecross, M., Nayudu, M. & Mitchell, K. (1998a) Protectant 3′,4′-dihydroxyflavones induced by enhanced UV-B in a UV tolerant rice cultivar. *Phytochemistry*, **49**, 1913–1919.

Musil, C. F., Midgley, G. F. & Wand, S. J. E. (1999) Carry-over of enhanced ultraviolet-B exposure effects to successive generations of a desert annual: interaction with atmospheric CO_2 and nutrient supply. *Global Change Biol.*, **5**, 311–329.

Nagata, T., Todoriki, S., Masumizu, T., Suda, I., Furuta, S., Du, Z. & Kikuchi, S. (2003) Levels of active oxygen species are controlled by ascorbic acid and anthocyanin in *Arabidopsis*. *J. Ag. Food Chem.*, **51**, 2992–2999.

Neill, S. O., Gould, K. S., Kilmartin, P. A., Mitchell, K. A. & Markham, K. R. (2002) Antioxidant activities of red versus green leaves in *Elatostema rugosum*. *Plant Cell Environ.*, **25**, 539–547.

Olsson, L. C., Veit, M., Weissenböck, G. & Bornman, J. F. (1998) Differential flavonoid response to enhanced UV-B radiation in *Brassica napus*. *Phytochemistry*, **49**, 1021–1028.

Pfündel, E. E., Pan, R. S. & Dilley, R. A. (1992) Inhibition of violaxanthin deepoxidation by ultraviolet-B radiation in isolated chloroplasts and intact leaves. *Plant Physiol.*, **98**, 1372–1380.

Rasmussen, S. & Dixon, R. A. (1999) Transgene-mediated and elicitor-induced perturbation of metabolic channelling at the entry point into the phenylpropanoid pathway. *Plant Cell*, **11**, 1537–1551.

Rao, M. V., Paliyath, G. & Ormrod, D. P. (1996) Ultraviolet-B- and ozone-induced biochemical changes in antioxidant enzymes of *Arabidopsis thaliana*. *Plant Physiol.*, **110**, 125–136.

Reuber, S., Bornman, J. F. & Weissenböck, G. (1996) A flavonoid mutant of barley (*Hordeum vulgare* L.) exhibits increased sensitivity to UV radiation in the primary leaf. *Plant Cell Environ.*, **19**, 593–601.

Ryan, K. G., Markham, K. R., Bloor, S. J., Bradley, J. M. Mitchell, K. A. & Jordan, B. R. (1998) UV-B radiation induced increase in quercetin: kaempferol ratio in normal and transgenic lines of *Petunia*. *Photochem. Photobiol.*, **68**, 323–330.

Ryan, K. G., Swinny, E. E., Markham, K. R. & Winefield, C. (2002) Flavonoid gene expression and UV photoprotection in transgenic and mutant petunia leaves. *Phytochemistry*, **59**, 23–32.

Schmelzer, E., Jahnen, W. & Hahlbrock, K. (1988) *In situ* localisation of light-induced chalcone synthase and flavonoid end products in epidermal cells of parsley leaves. *Proc. Nat. Acad. Sci. USA*, **85**, 2989–2993.

Schulze-Lefert, P., Dangl, J. L., Becker-André, M., Hahlbrock, K & Schulz, W. (1989a) Inducible *in vivo* DNA footprints define sequences necessary for UV light activation of the parsley chalcone synthase gene. *EMBO J.*, **8**, 651–656.

Schulze-Lefert, P., Becker-André, M., Schulz W., Hahlbrock, K. & Dangl, J. L. (1989b) Functional architecture of the light-responsive chalcone synthase promoter from parsley. *Plant Cell*, **1**, 707–714.

Seigler, D. S. (2002) *Plant Secondary Metabolism*, Kluwer Academic Publishers, Boston, USA.

Smith, G. J.& Markham, K. R. (1998) Tautomerism of flavonol glucosides: relevance to plant UV protection and flower colour. *J. Photochem. Photobiol.*, **A118**, 99–105.

Solovchenko, A. & Schmitz-Eiberger, M. (2003) Significance of skin flavonoids for UV-B-protection in apple fruits. *J. Exp. Bot.*, **54**, 1977–1984.

Sprenger-Haussels, M. & Weisshaar, B. (2000) Transactivation properties of parsley proline-rich bZIP transcription factors. *Plant J.*, **22**, 1–8.

Stapleton, A. E. & Walbot, V. (1994) Flavonoids can protect maize DNA from the induction of ultraviolet radiation damage. *Plant Physiol.*, **105**, 881–889.

Steel, C. C. & Keller, M. (2000) Influence of UV-B irradiation on the carotenoid content of *Vitis vinifera* tissues. *Biochem. Soc. Trans.*, **28**, 883–885.

Strid, Å., Chow, W. S. & Anderson, J. M. (1990) Effects of supplementary ultraviolet-B radiation on photosynthesis in *Pisum sativum*. *Biochim. Biophys. Acta*, **1020**, 260–268.

Strid, Å., Chow, W. S. & Anderson, J. M. (1994) UV-B damage and protection at the molecular level in plants. *Photosynth. Res.*, **39**, 475–489.

Strid, Å., Chow, W. S. & Anderson, J. M. (1996) Changes in the relaxation of electrochromic shifts of photosynthetic pigments and in the levels of mRNA transcripts in leaves of *Pisum sativum* as a result of exposure to supplementary UV-B radiation. The dependency on the intensity of the photosynthetically active radiation. *Plant Cell Physiol.*, **37**, 61–67.

Taylor, W. C. (1989) Regulatory interactions between nuclear and plastid genomes. *Ann. Rev. Plant Physiol. Plant Mol. Biol.*, **40**, 211–233.

Van de Staaij, J. W. M., Ernst, W. H. O., Hakvoort, H. W. J. & Rozema, J. (1995) Ultraviolet-B (280–320 nm) absorbing pigments in the leaves of *Silene vulgaris*: their role in UV-B tolerance. *J. Plant Physiol.*, **147**, 75–80.

Vogelmann, T. C. (1993) Plant tissue optics. *Ann. Rev. Plant Physiol. Plant Mol. Biol.*, **44**, 231–251.

Wade, H. K., Bibikova, T. N., Valentine, W. J. & Jenkins, G. I. (2001) Interactions within a network of phytochrome, cryptochrome and UV-B phototransduction pathways regulate chalcone synthase gene expression in *Arabidopsis* leaf tissue. *Plant J.*, **25**, 675–685.

Weisshaar, B., Armstrong, G. A., Block, A., da Costa e Silva, O. & Hahlbrock, K. (1991a) Light inducible and constitutively expressed DNA-binding proteins recognizing a plant promoter element with functional relevance in light responsiveness. *EMBO J.*, **10**, 1777–1786.

Weisshaar, B., Block, A., Armstrong, G. A Herrmann, A., Schulze-Lefert, P. & Hahlbrock, K. (1991b) Regulatory elements required for light-mediated expression of the *Petroselinum crispum* chalcone synthase gene, in *Society for Experimental Biology, SEB Symposia Series 45* (eds G. I. Jenkins & W. Schuch), The Company of Biologists Limited, Cambridge, UK, pp. 191–210.

Xiong, F. S. & Day, T. A. (2001) Effect of solar ultraviolet-B radiation during springtime ozone depletion on photosynthesis and biomass production of Antarctic vascular plants. *Plant Physiol.*, **125**, 738–751.

Yalamancila, R. D. & Stratmann, J. W. (2002) Ultraviolet-B activates components of the system in signalling pathway in *Lycopersicon peruvianum* suspension-culture cells. *J. Biol. Chem.*, **277**, 28424–28430.

10   Techniques of pigment identification

Øyvind M. Andersen and George W. Francis

This chapter aims to provide an overview of the modern and most recent techniques used for extraction, separation, identification and quantification of the major pigment classes in plants. Limited space restricts comprehensive treatment of the various techniques involved; however, an effort has been made to give enough information (and references) to carry out the isolation and structure elucidation of even a novel compound.

10.1   Introduction

The importance of plant pigments to human society is difficult to exaggerate. This may primarily be due to the fact that man, like all other animals with polychromatic vision, uses perceived colour to judge the ripeness and quality of food plants. Plant pigments have, moreover, been utilised in a wide variety of other roles as dyes for cloth and colouring materials for all sorts of industrial products from food to paint (De Jong, 1997). The desire to understand and identify the source of these colours has fascinated the practising chemist and his predecessors since before the Renaissance. Newer knowledge has shown that pigments and other secondary metabolites are both environmentally and genetically determined, and this has led to increasing interest in the pigments themselves and their manipulation (Harborne & Turner, 1984; Amaya & Valpuesta, 2002; Bjoern, 2002).

Plants exhibit two fundamentally different types of colour, which are due respectively to structural effects of the tissue and to the presence of pigments. These two forms of colouration may, and often do, co-exist in the same tissues, and the perceived colour will thus depend on the joint effect of two distinct phenomena. Many pigments have quite different colour characteristics depending on the environment in which they find themselves: in fat globules, aqueous suspension, pH, protein or other complexes. While the investigation of the pigments present may proceed along chemical lines, a complete appreciation of the observed colour should take into account all factors affecting colour (Schoefs, 2002). Gonnet (1998) has for instance specified that an adequate description of anthocyanin colour variation caused by pH differences requires that spectral variations considered should be those affecting the entire spectral curve (not only its visible λ_{max}), that three colour attributes (hue, saturation and lightness) should be used to describe colour (e.g. CIELAB parameters), and that these should refer to light source and observer condition.

The above need not deter investigation of colour in plants, and in fact preliminary conclusions about the classes of pigment present can often be immediately implied. The reality is that the number of pigment classes to be commonly found in plants has proved to be strictly limited (Goodwin, 1976). This is not to say that these main classes answer for all natural plant pigments, but they do include the majority of all known pigments of general occurrence. The observed colours of plant tissues are widely recognised by scientists from many disciplines, and are indeed known also to an increasing number of non-specialists. Chlorophylls are immediately recognised by their characteristic green colours, as are many carotenoids due to their deep yellow to orange-red hues. Less blatant are the majority of flavonoid pigments, which provide delicate white to yellow colours in a large number of flowers. The anthocyanins, a special class of flavonoids, are responsible for bright, and often intense, red to blue colours in a variety of flowers, leaves and fruits. The betacyanins with restricted distribution mainly in Caryophyalles show superficial colour similarities to anthocyanins, although their occurrence seems to be mutually exclusive. The quinonoid pigments vary somewhat more in both structure and colour, but often provide colours in yellow-orange-red hues. Skeletal variation within each class is limited and preliminary conclusions about the class involved can often be implied even from the *in vivo* colour of the tissue. The degree of oxidation and the presence, absence and type of substituents are largely responsible for variations within each class.

10.1.1 Sample preparation and extraction

Plant pigments are of greatly differing polarities, and solvent extractions must thus be tailored specifically for each class. This, together with the fact that many of the pigments are sensitive to acids and/or bases, oxygen and light, means that it is always advisable to carry out preliminary tests prior to extraction. All that is required is to take a small sample of the tissue to be investigated, cut it into small pieces and then attempt extraction with a series of organic solvents of increasing polarity and eventually with water. If extraction is unsatisfactory in neutral conditions, acidified methanol can be utilised. This usually provides an indication of solvent choice and a small sample of extracted pigment. Together with an absorption spectrum of the sample, it will usually be possible to identify the main pigment classes present (Figs 10.1, 10.12 and 10.14), and thus to decide on an initial extraction method. It should be kept in mind that several pigments of the same or different classes may well be present, and it is far from certain that a single method of extraction will be satisfactory for all of the compounds present. While it is true that extraction processes should always be carried out with due care, it is still important to work rapidly and carefully, rather than overcarefully.

Any extract, or indeed any single pigment that appears to be pure, should always be investigated for the presence of colourless impurities. These are

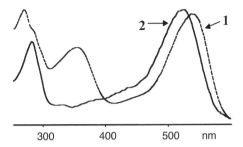

Fig. 10.1 The UV–Vis absorption spectra of the covalent anthocyanin–flavonol complex, (cyanidin 3-*O*-β-glucoside)(kaempferol 3-*O*-(2-*O*-β-glucosyl-β-glucoside)-7-*O*-β-glucosiduronic acid) malonate (**1**), and the anthocyanin cyanidin 3-*O*-β-glucoside (**2**) (Fossen *et al.*, 2000). The bathochromic shift (16 nm) of the visible absorption maximum of **1** compared to **2** indicates intramolecular association between the anthocyanin and flavonol units of **1**.

normally detected by spraying a trial chromatogram with a strong oxidising agent (Stahl, 1969; Wagner *et al.*, 1984). The presence of such impurities is particularly undesirable during chromatography as these greatly reduce adsorbent efficiency by blocking active sites. The most common clean-up treatment is the extraction of polar impurities from less polar extracts, or the reverse where waxes and the like are removed either prior to or after the extraction process. Afterwards, it is important to ensure the removal of any residual solvents, which are a common source of problems during the following analysis. In particular, the presence of even minor traces of water can often spoil the separation of hydrophobic pigments, affect spectroscopic measurements and interfere with chemical reactions.

10.1.2 *Purification and separation*

Once a clean extract has been obtained, it becomes necessary to isolate the individual pigments. The methods used will depend on the pigment class involved, but wet chromatographic methods are the most common. Solvent partition, acid and/or base extraction, zone refining, sublimation and fractional crystallisation can occasionally be utilised. The most usual chromatographic methods in current practice are high-performance liquid chromatography (HPLC) (Figs 10.2 and 10.16) and thin-layer chromatography (TLC) (Tables 10.1, 10.3, 10.4 and 10.5), although open column and flash chromatography (FC) are also often used in initial separations. A variety of other methods, including electrophoresis and countercurrent chromatographic techniques (Fig. 10.9), find application in specific cases. Where a single separation does not provide single pigments, the same method may be repeated or combined with one or more additional chromatographic steps (Fig. 10.3). Isolated pigments can be crystallised to obtain analytical samples for final characterisation.

10.1.3   Characterisation and identification

The amounts of pigments present in most plant tissues are relatively modest even when the visual impression is quite striking. The total carotenoid content of ripe red tomatoes is for instance about 100 µg/g wet weight, while the content of chlorophylls in spinach leaves is about 1000 µg/g wet weight (Gross, 1991). Methods for the characterisation of individual pigments have thus traditionally reflected the lack of available material, and sensitive chromatographic and spectroscopic techniques have achieved a prominent place in pigment characterisation. The introduction of hyphenated methods, in particular HPLC coupled to a diode-array detector, a mass spectrometer or both, has, if anything, given added emphasis to this situation. Microchemical methods provide additional corroboration in many cases. While visible absorption spectroscopy will normally form part of the identification of any particular pigment, the other methods applied will depend on the class of compound being examined.

Recent developments have made nuclear magnetic resonance (NMR) techniques the most important tools for complete structure elucidation for most pigments when isolated on a moderate scale (mg quantities) (Andersen & Fossen, 2003). We have therefore included here some information regarding the potential of these modern NMR techniques for pigment analysis. The purpose of the standard 1-D ^{1}H NMR experiment is to record chemical shifts, spin–spin couplings, and integration data telling the relative number of hydrogen atoms. Applied for instance to a flavonoid, this information may help to identify the aglycone, the number of monosaccharides present and the anomeric configuration of the monosaccharides. However, for most pigments this information is insufficient for a complete structure elucidation, and the 1-D ^{13}C NMR experiments (spin–echo Fourier transform – SEFT, compensated attached proton test – CAPT) combined with 2-D NMR experiments are used to identify the aglycone, the number and identities of sugar and potential acyl moieties.

The 2-D ^{1}H-^{1}H-correlated NMR experiments (double quantum filtered correlation spectroscopy – DQF-COSY, total correlation spectroscopy – TOCSY) (Andersen & Fossen, 2003) generate NMR spectra in which ^{1}H chemical shifts along two axes are correlated with each other. The DQF-COSY spectrum shows couplings between neighbouring protons ($^2J_{HH}$, $^3J_{HH}$ and $^4J_{HH}$), while the TOCSY experiment identifies protons belonging to the same spin system. Since each sugar ring contains one discrete spin system separated by oxygen, this latter experiment is especially useful for assignments of the sugar protons belonging to each sugar unit. After assignments of the ^{1}H signals, the one-bond ^{1}H-^{13}C correlations ($^1J_{CH}$) observed in the heteronuclear spectrum allow assignment of the corresponding sugar ^{13}C signals (Fig. 10.10). Assignment of quaternary ^{13}C signals as well as determination of the linkages between the building blocks (aglycone, sugar units, acyl moieties) may afterwards be established by the heteronuclear multiple bond correlation (HMBC) experiment. This experiment shows couplings between proton and carbon atoms, which are mainly three ($^3J_{CH}$)

and two ($^2J_{CH}$) bonds away. Long-range ^{1}H-^{13}C correlations are also accessible in heteronuclear single quantum coherence (HSQC)–TOCSY spectra (Fig. 10.5). Alternatively, intra- and intermolecular neighbourship can be determined by through-space interactions (nuclear Overhaüser enhancement spectroscopy – NOESY, rotating frame Overhaüser enhancement spectroscopy – ROESY) and distance geometry calculations (Nerdal & Andersen, 1992; Gakh *et al.*, 1998).

10.1.4   Quantification

Quantitative work with pigments nearly always involves spectroscopic methods, and with well-chosen standards nearly any spectroscopic measuring techniques can be utilised. In mixtures, and when degradation products and other interfering substances are present, chromatography (HPLC, TLC) combined with spectroscopic detection is the method of choice for quantitative analysis of both individual pigments and total amounts. Other methods will be discussed where they are relevant.

10.2   Flavonoids

10.2.1   Introduction

More than 7000 flavonoid structures divided into more than 15 classes have been reported in the literature (Bohm, 1998; Harborne & Baxter, 1999). Most of them occur *in vivo* as glycosides, which may have acyl substituents on their sugar moieties (see also Chapter 4). From an analytical point of view this multitude of structures may be grouped into anthocyanins, flavonoid glycosides, non-polar flavonoids (aglycones, *C*-alkylated flavonoids or flavonoids with high levels of *O*-methylation), and proanthocyanidins. The anthocyanins are treated separately in Section 10.3. The proanthocyanidins are not discussed at all in this chapter, but an example of mass spectrometric (MS) analysis of proanthocyanidins is presented in Chapter 5.

Before a species is analysed with respect to its flavonoid content, knowledge about previous reports on chemistry and flavonoid distribution within the genus and related species may be valuable. The most exhaustive source for such information is *Chemical Abstracts*, and reviews on flavonoids have appeared regularly (Harborne *et al.*, 1975; Harborne & Mabry, 1982; Harborne, 1988, 1994; Harborne & Williams, 1998, 2001). Several excellent reviews have recently addressed the field of flavonoid analysis more specifically (Bohm, 1998; Markham & Bloor, 1998; Sivam, 2002; Santos-Buelga & Williamson, 2003). A limited number of flavonoids are commercially available from Carl Roth *http://www.carl-roth.de/jsp/reddot/index.jsp*, Extrasynthese http://www.extrasynthese.com, Indofine Chemical Company *http://www.indofinechemical.com*, Polyphenols Laboratories *http://www.polyphenols.com*, although some other companies sell a few of the more common flavonoids.

10.2.2 Sample preparation and extraction

There is a wide variety of methods to prepare flavonoid samples for analysis (Waterman & Mole, 1994; Escribano-Bailon & Santos-Buelga, 2003) depending on location of flavonoid in tissue, type of flavonoid, and whether the extraction is for qualitative or quantitative purposes. Plant tissue can be extracted fresh, dried or freeze-dried and often ground to a fine powder. The solvents of choice are combinations of water with methanol, ethanol or acetone. The use of more than 50% alcohol restricts unwanted enzyme activity. Non-polar flavonoids located for instance on leaf surface can be extracted by just soaking the fresh material in ether or dichloromethane. Extractions are, almost always, repeated two to three times.

10.2.3 Purification and separation

None of the extraction solvents is specific for flavonoids. If appreciable quantities of lipids or chlorophylls are suspected to be present in the aqueous extract after concentration under reduced pressure, these compounds may be removed by washing with petroleum ether or diethyl ether. However, one should be aware that non-polar flavonoids might be partitioned into water-immiscible solvents such as ethyl acetate.

10.2.3.1 Column chromatography

Unwanted compounds in the extract (e.g. sugars) can be removed by passing aqueous samples through a solid phase extraction column (for instance C_{18} Sep-Pak). The cartridge is then washed with water, and the retained flavonoids are eluted with methanol. Care must be taken to ensure that conjugated flavonoids carrying a negative charge are also retained on the column by, for instance, addition of acid to the solvent in order to provide a counter-ion.

The separation of large quantities of flavonoids can be carried out by column chromatography using a variety of packing materials such as Sephadex LH-20, Amberlite XAD-7, reversed phase (C_{18} or C_8), silica, cellulose and polyamide (Markham & Bloor, 1998). In our labs we have achieved efficient resolution of a variety of flavonoid mixtures with step elution from 20–60% MeOH in H_2O (0.1% TFA) on Sephadex LH-20 columns (100×5.0 cm). For semi-preparative isolation of pure flavonoids, we have used a reversed-phase C_{18} HPLC column (250×22 mm, $10 \mu m$) with formic acid in water–methanol mixtures as mobile phase.

10.2.3.2 Countercurrent chromatography

Countercurrent chromatography (CCC) includes techniques that are based on partitioning of compounds between two immiscible liquid phases. In particular, HSCCC has in recent years been shown to be a powerful tool for preparative separation and purification of flavonoids. HSCCC has a high sample-load capacity, and has been considered to be superior in some cases to column chromatography, which also suffers from undesirable adsorption problems. For instance, a crude sample (20 mg) has been separated within 2 h into five flavonoids (purity between 85–98%) by a solvent system composed of chloroform–methanol–water (8 : 10 : 5 – all ratios given in this chapter are v/v unless otherwise stated) (Chen *et al.*, 2003).

10.2.4   Characterisation and identification

During analysis of flavonoids, much effort normally goes into the isolation and identification of known compounds. The first step in structure elucidation may thus be co-chromatography (TLC and HPLC) of isolated flavonoids with authentic standards. When HPLC is coupled with diode array UV-visible detection or MS, much information about individual flavonoids is achieved even in crude or partly purified extracts. However, structure elucidation of novel flavonoids demands normally both NMR and MS analysis involving a few milligrams of pure compounds.

10.2.4.1   Thin-layer chromatography

TLC is a technique that is applicable to all classes of flavonoids (Markham & Bloor, 1998; Francis & Andersen, 2003). The solvent ethyl acetate–formic acid–acetic acid–water (100 : 11 : 11 : 27) on silica support can be used for separation of a wide range of flavonoids. Polyamide layers are often used for separation of medium polar to apolar flavonoids, and a solvent containing methanol–formic acid–water (58 : 10 : 16) seems to be ideal for the separation of flavonoid aglycones on reversed-phase layers (Table 10.1). Paper chromatography (PC) was the original method of chromatographic analysis for flavonoids; however, TLC on cellulose is a useful alternative. The solvents 15% acetic acid, TBA (*t*-butanol–acetic acid–water, 3 : 1 : 1) or BAW (butanol–acetic acid–water, 4 : 1 : 5, upper layer) on cellulose (or PC) have routinely been used for flavonoid analysis (Harborne, 1967; Markham & Wilson, 1988).

Being colourless or having yellow colours, the flavonoids may be difficult to detect in visible light. By using TLC layers containing a fluorescent indicator, it is possible to detect most flavonoids as quenching bands or spots. Alternatively, a number of spray reagents may be used to enhance spot detection (Table 10.1). The strength of, for instance, Naturstoffreagenz A (1% diphenyl–boric acid–ethanolamine complex in methanol) as a spray reagent lies in its ability to distinguish between different substitution patterns. Inspection under UV light with or without spray reagents will normally improve detection.

10.2.4.2   High-performance liquid chromatography

HPLC is the method of choice for the accurate determination of both the composition and the absolute concentrations of the flavonoids in a given sample (Merken & Beecher, 2000). Most of the HPLC analyses are run using C_{18} reverse phase columns ranging from 10 cm to 30 cm in length usually with 2.1–4.6 mm internal diameter and particle size of 3–5 μm. Elution systems are usually binary, with an aqueous acidified solvent (A) and a less polar organic solvent (B) such as methanol or acetonitrile, possibly acidified (Fig. 10.2). The advent of photodiode

Table 10.1 Colours and R_F values of flavone and flavonol aglycones on silica gel, polyamide and reversed-phase TLC layers

Pigment, spot colour[*]	Substituent position		System[†]			
	OH	OCH$_3$	1	2	3	4
Flavones:						
Flavone b			0.50	0.60	0.86	0.29
5-Hydroxyflavone br	5		0.63	0.74	0.88	0.22
6-Hydroxyflavone gr	6		0.31	0.50	0.65	0.39
7-Hydroxyflavone b	7		0.29	0.48	0.51	0.41
7,8-Dihydroxyflavone gr-br	7,8		0.22	0.43	0.32	0.57
Chrysin br-g	5,7		0.48	0.57	0.52	0.33
Apigenin g	5,7,4'		0.29	0.46	0.18	0.48
Acacetin gr-g	5,7	4'	0.42	0.54	0.45	0.30
Apigenin-7,4'-dimethyl ether gr-g	5	7,4'	0.56	0.66	0.86	0.14
Luteolin y	5,7,3', 4'		0.22	0.38	0.06	0.53
Diosmetin gr	5,7,3'	4'	0.23	0.43	0.41	0.46
Tangeretin gr		5,6,7,8,4'	0.25	0.48	0.89	0.26
Flavonols:						
Galangin g	3,5,7		0.51	0.59	0.41	0.38
Kaempferol y-g	3,5,7,4'		0.32	0.48	0.14	0.57
Kaempferid b-g	3,5,7	4'	0.48	0.59	0.38	0.37
Kaempferol-7,4'-dimethyl ether b-g	3,5	7,4'	0.63	0.73	0.74	0.18
Kaempferol-3,4'-dimethyl ether ol	5,7	3,4'	0.43	0.55	0.61	0.35
Kaempferol-3,7,4'-trimethyl ether g	5	3,7,4'	0.61	0.69	0.87	0.19
Quercetin br-o	3,5,7,3', 4'		0.23	0.41	0.06	0.67
Tamarixetin g	3,5,7,3'	4'	0.31	0.48	0.19	0.53
Rhamnetin o	3,5,3', 4'	7	0.29	0.48	0.24	0.45
Quercetin-3,7-dimethylether o	5,3', 4'	3,7	0.27	0.46	0.47	0.46
Morin g	3,5,7,2', 4'		0.17	0.29	0.02	0.73
Fisetin o	3,7,3', 4'		0.21	0.36	0.08	0.70
Robinetin o	3,7,3', 4', 5'		0.08	0.22	0.01	0.82
Myricetin o-br	3,5,7,3', 4', 5'		0.13	0.32	0.02	0.76
Quercetagetin r-br	3,5,6,7,3', 4'		0.00	0.00	0.01	0.95

[*]Pigment colours are observed under UV-366 after spraying plates with Naturstoffreagenz A followed by polyethyleneglycol (PEG-4000). b = blue, br = brown, g = green, gr = gray, o = orange, ol = olive, r = red, y = yellow.
[†]Systems: 1 and 2, silica gel (60 F$_{254}$, 0.25 mm), benzene–ethyl acetate–formic acid (40:10:5) and toluene–ethyl formate–formic acid (50:40:10); 3, polyamid (DC-alufolien F$_{254}$, 0.15 mm), toluene–butan-2-one–methanol (60:25:15); 4, reversed-phase (C-18 F$_{254}$, 0.25 mm) methanol–formic acid–water (58:10:16).

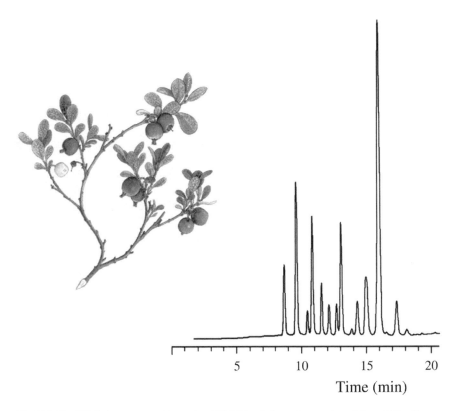

Fig. 10.2 The HPLC chromatogram (detected at 520 ± 20 nm) shows the complex mixture of 15 anthocyanidin 3-*O*-monoglycosides from fruits of *Vaccinium uliginosum*. It is achieved on reversed-phase C_{18} column (250×4.6 mm, 3 μm) with the solvents water–formic acid, 9:1 (A) and water–methanol–formic acid, 4:5:1 (B) using 10–80% B in A (linear gradient for 21 min).

array and mass spectrometric detector systems increases the amount of information obtained in an HPLC run. Spectra obtained during elution of the individual HPLC peaks combined with retention times are often used to characterise the separated flavonoids (Santos-Buelga *et al.*, 2003). Spectra obtained at several points during elution of each HPLC peak can also be used for determination of peak purity. The use of HPLC with colourimetric array detection for analysis of flavonoids has recently been reviewed (Manach, 2003).

10.2.4.3 Capillary electrophoresis
Capillary zone electrophoresis (CZE) is becoming increasingly recognised for analysis of flavonoids (Marchart & Kopp, 2003). It has the advantage of being fairly rapid, and may have separating power that exceeds HPLC dramatically (Steuer *et al.*, 1990). The separation mechanism in CZE is based on differences in the electrophoretic mobilities of the flavonoids. Flavonoids not carrying a charge

must thus be ionised by using a suitable buffer. The effects of several capillary electrophoresis (CE) factors including separation voltage and acidity and concentration of running buffer have recently been investigated to find optimum conditions for determination of flavonoids in *Hippophae rhamnoides* and its phytopharmaceuticals (Chu *et al.*, 2003).

10.2.4.4 Hydrolysis

Most flavonoid *O*-glycosides are cleaved in refluxing HCl solutions (50% MeOH-2N HCl) after 30–60 min, yielding sugar(s) and aglycones. Alkaline hydrolysis in 2M NaOH solutions at room temperature specifically removes acyl groups, while various enzymes (e.g. β-galactosidase) have been used to remove particular sugars (e.g. β-linked galactose) from particular flavonoid positions. For more information about hydrolysis and the following product analysis, see for instance Markham (1982).

10.2.4.5 UV spectroscopy

The different flavonoid classes have rather characteristic UV spectra including two maxima, whose absorption range and magnitude depend mainly on the nature

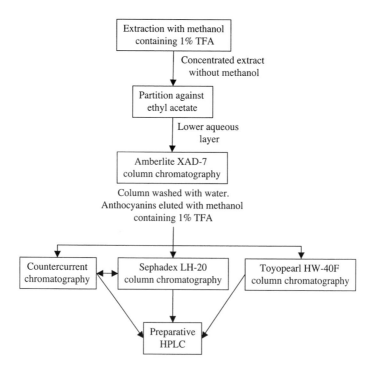

Fig. 10.3 Scheme for isolation of anthocyanins. The methods along the central vertical lines are sufficient for isolation of most anthocyanins.

of the C-ring and the substituents on the A- and B-rings. Spectral properties of the various flavonoid classes have been clearly set out by Markham and coauthors (Mabry *et al.*, 1970; Markham, 1982). They have compiled diagnostic spectra recorded after addition of shift reagents like NaOMe, NaOAc, AlCl$_3$ (+HCl), and H$_3$BO$_3$ (+HCl) to small amounts of purified flavonoids in UV cuvettes. The spectra reveal effects that are very useful for determination of flavonoid substitution patterns.

10.2.4.6   Mass spectrometry

The application of MS to the analysis of flavonoids has increased with the development of soft ionisation techniques. The combination of HPLC coupled simultaneously to a diode-array detector and to a mass spectrometer equipped with an electrospray (LC-ESI) or atmospheric pressure chemical ionisation source is at present the method of choice for determination of molecular masses (De Pascual-Teresa & Rivas-Gonzalo, 2003; De Rijke *et al.*, 2003). The mobile phase used in LC-ESI contains easily ionised components (e.g. trifluoroacetic acid), from which a charge may be transferred to the flavonoid, $[M + H]^+$. Both sodium $[M + 23]^+$ and potassium $[M + 39]^+$ adducts may be seen in these spectra. Depending on the energy of the ion source, glycosidic moieties may fragment off the flavonoid. Collision-induced dissociation in triple quadruple or ion trap mass spectrometers allows generation and analysis of accurate daughter fragments. Matrix-assisted laser desorption ionisation (MALDI) MS (Wang & Sporns, 2000) and capillary electrophoresis coupled to MS detection (Huck *et al.*, 2002) are alternative methods for characterisation of flavonoids.

10.2.4.7   Nuclear magnetic resonance spectroscopy

NMR is the single most important technique for structure elucidation of flavonoids. For ^{1}H NMR analysis at least 0.2 mg sample is needed, while ^{13}C NMR and 2-D NMR techniques (see Section 10.1.3) normally require ten times more material. Flavonoids may be dissolved in a range of deuterated solvents. We have found methanol-d_6 and DMSO-d_6 suitable for most flavonoids. Anthocyanins require the addition of an acid to ensure conversion to the flavylium form: 5% and 30% CF$_3$COOD in methanol-d_6 and DMSO-d_6, respectively. In some cases it has been necessary to dissolve apolar flavonoids in acetone-d_6 or CDCl$_3$. A protocol treating experimental details of using modern NMR techniques for anthocyanin analysis has recently been published (Andersen & Fossen, 2003). Several authors have compiled NMR data on individual flavonoids (Markham & Chari, 1976; Agrawal, 1989; Markham & Geiger, 1993; Andersen & Fossen, 2003). Figure 10.4 shows assignments of ^{1}H and ^{13}C NMR chemical shifts of some typical flavonoid classes. The one-bond ^{1}H-^{13}C correlations ($^1J_{CH}$) of the sugar observed in the HSQC spectrum of a flavonoid trisaccharide are shown in Fig. 10.10. Both one-bond and long-range ^{1}H-^{13}C correlations are revealed in the HSQC–TOCSY spectrum of the flavonol kaempferol 3-*O*-(2-*O*-β-glucosyl-β-glucoside)-7-*O*-β-glucosiduronic acid (Fig. 10.5). The use of liquid

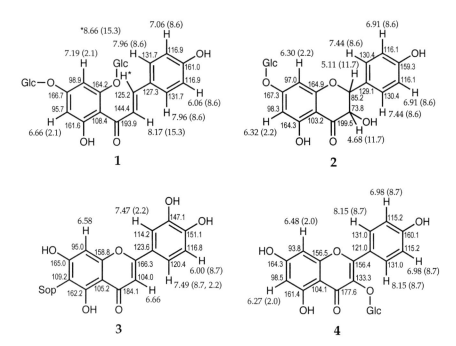

Fig. 10.4 ¹H and ¹³C NMR chemical shifts of representatives of various flavonoid classes. ¹H-¹H coupling constants are given in brackets. **1**, the chalcone chalcononaringenin 2′, 4′-di-*O*-glucoside; **2**, the dihydro-flavonol aromadendrin 7-*O*-glucoside; **3**, the flavone luteolin 6-*C*-(2″-glucosylglucoside); **4**, the flavonol kaempferol 3-*O*-glucoside. **1** dissolved in pyridine-d_5 is recorded by Iwashina and Kitajima (2000), while **2–4** dissolved in CD$_3$OD are recorded in our laboratories. Glc = glucoside, Sop = sophoroside.

chromatography coupled to on-line NMR (LC-NMR) for identification of di- and triglycosylated flavonoids is exciting (Andrade *et al.*, 2002). LC-NMR applied on polyphenols has recently been reviewed (Wolfender *et al.*, 2003).

10.2.4.8 Circular dichroism spectroscopy

Circular dichroism (CD) spectroscopy is used to determine absolute configuration at the stereocentres of chiral flavonoids (Li *et al.*, 2002), usually by comparison with similar data of compounds with known stereochemistry. CD spectra have recently been used to calculate the association constant between the flavonol quercetin and human serum albumin (Zsila *et al.*, 2003); they have also given insight into association properties of anthocyanins (Hoshino, 1992). The use of chiroptical spectroscopies for structure elucidation of flavonoids has been reviewed by Levai (1998).

10.2.5 Quantification

The most convenient method for the estimation of the concentration of pure flavonoids is by their UV-absorbance related to a standard curve based on a

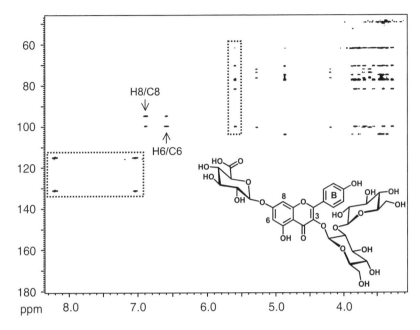

Fig. 10.5 The HSQC–TOCSY NMR spectrum of kaempferol 3-*O*-(2-*O*-β-glucosyl-β-glucoside)-7-*O*-β-glucosiduronic acid in CD_3OD. The crosspeaks within the left rectangle represent correlations between the proton and corresponding carbon atoms of the B-ring. The right rectangle shows the correlations between the anomeric proton of the glucose unit in the 3-position and the carbon atoms of this sugar. The spectrum is recorded by Ms Kjersti Kalberg and Dr Torgils Fossen, University of Bergen.

pure standard. However, HPLC techniques in combination with UV detection at specific wavelengths provide the means of very accurate determination of individual flavonoids in mixtures. A comparison of integration data of individual flavonoid peaks on the chromatogram, with the peak corresponding to known amounts of a standard flavonoid, is the basis for calculations. Even though the molar absorptivity coefficients are different for individual flavonoids, it is accepted practice to use the same value for all peaks in the chromatogram. However, the choice of detection wavelengths is very important since the flavonoids in the sample may have very different UV spectra. The quantification of flavonoids by CE using a flavonol as internal standard correlates with the results achieved by spectrophotometric determination (Marchart & Kopp, 2003).

10.3   Anthocyanins

10.3.1   Introduction

Improvements in methods and instrumentation used for separation and structure elucidation of anthocyanins have made it easier to use smaller quantities of material to achieve results at higher levels of precision. The number of different anthocyanidin and anthocyanin structures has thus increased dramatically in recent years, and according to our files there are in excess of 540 with complete structures. Each anthocyanin consists of an aglycone (anthocyanidin) (Fig. 10.6), sugar(s) and in many cases acyl group(s) (Andersen, 2001a). While more than 26 monomeric anthocyanidins have been properly identified, nearly all anthocyanins are based on just six anthocyanidins. The recent findings of different pyranoantho-cyanins (Fig. 10.6) in wine (Fulcrand *et al.*, 1998), roses (Fukui *et al.*, 2002), red onion (Fossen & Andersen, 2003) and strawberry (Andersen *et al.*, 2004) are included in this number. The 3-deoxyanthocyanidins are the only anthocyanidins stable enough to occur in their non-glycosylated form in plants. The various anthocyanin acyl moieties have been indicated in Fig. 10.7.

Analytical methods for the extraction, separation and characterisation of antho-cyanins have been treated in a number of reviews (Strack & Wray, 1989; DaCosta *et al.*, 2000; Delgado-Vargas *et al.*, 2000; Andersen, 2001b; Takeoka & Dao, 2002; Wrolstad *et al.*, 2002; Andersen & Fossen, 2003; Francis & Andersen, 2003; Rivas-Gonzalo, 2003). Mazza & Miniati (1993) have published an impressive compilation of the anthocyanin content of a variety of fruits, vegetables and grains. Since anthocyanins constitute one among the flavonoid classes, Section 10.2 should also be consulted when considering techniques discussed in Section 10.3. The space of this chapter allows just highlighting of the most useful and recent techniques used in analysis of anthocyanin. A general isolation scheme is presented in Fig. 10.3, and Table 10.2 contains a collection of ^{13}C NMR chemical shifts on anthocyanidins.

10.3.2 Sample preparation and extraction

The anthocyanin may be extracted from fresh, frozen or freeze-dried material. Even anthocyanins in herbarium material have been analysed in chemotaxonomic con-texts. However, most anthocyanins show instability during storage towards a

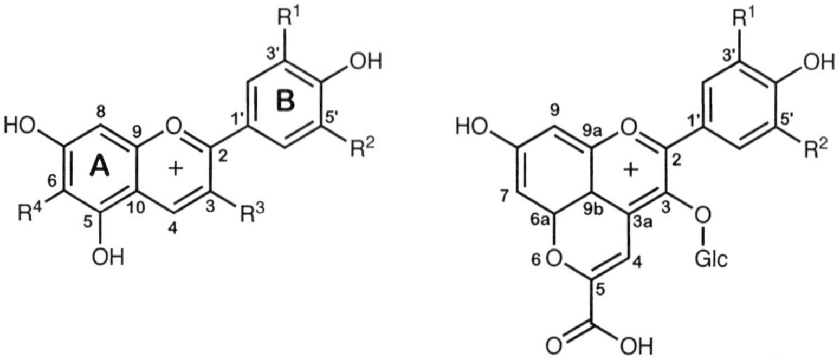

Fig. 10.6 *Left:* a general structure representing most anthocyanidins found in nature. The six common anthocyanidins: R^1 and R^2 = H, OH or OMe; R^3 = OH; R^4 = H. *Right:* pyranoanthocyanins.

Anthocyanin number

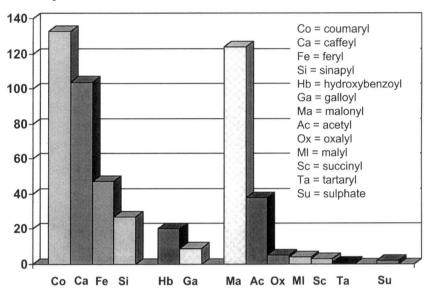

Fig. 10.7 The various acyl moieties identified in anthocyanins. The vertical axis represents number of anthocyanins.

variety of parameters, including oxygen and high temperatures (Andersen, 2001a). Most anthocyanins are especially unstable in weakly acidic and neutral aqueous solutions.

Due to their polar nature, anthocyanins are mainly extracted with alcohols (methanol) and water containing some acid. Solvent evaporation under acidic conditions may however result in partial or total anthocyanin hydrolysis (Revilla *et al.*, 1998). Weak organic acids (formic, acetic or trifluoracetic acid) in the solvents are preferable when the anthocyanins are acylated with aliphatic acids. We have experienced esterification of dicarboxylic anthocyanin moieties (e.g. malonic acid) in alcoholic solvents during storage (Fig. 10.8). Even esterification of the sugar hydroxyl groups in solvents containing formic acid has been reported, although this latter effect has not been any problem in our labs. The use of acetone for extraction of anthocyanins allows reproducibility, avoids problems with pectins and permits low temperatures for sample concentration (Garcia-Viguera *et al.*, 1998). Lu & Foo (2001) have, however, shown that anthocyanins may undergo facile reactions with acetone to give rise to pyranoanthocyanins. The use of enzymes may improve extraction (Pardo *et al.*, 1999). Solid-phase extraction has been used to improve the stability of unstable anthocyanidins by their deposition on C_{18} cartridges (Dao *et al.*, 1998).

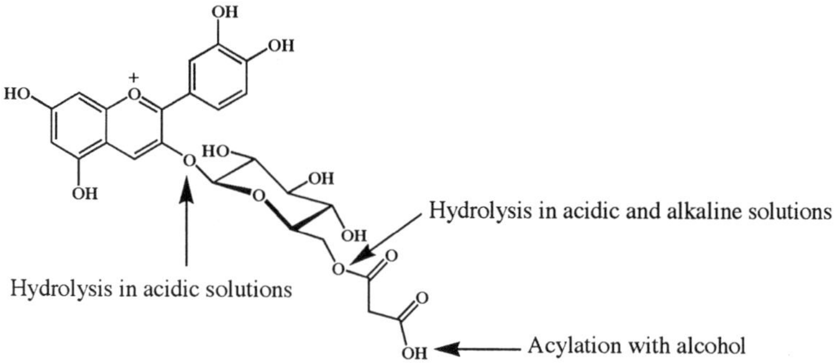

Fig. 10.8 Potential modifications of anthocyanins during extraction and storage.

10.3.3   Purification and separation

A typical isolation scheme used in our labs involves extraction of approximately 200 g plant material followed by purification by partition against ethyl acetate (or chloroform) (Fig. 10.3). Extraction of anthocyanins is non-selective and normally gives a pigment mixture in addition to large amounts of impurities.

10.3.3.1   Chromatography

Small disposable cartridges packed with C_{18} silica-based material provide a simple and convenient means for purification of anthocyanins (Hong & Wrolstad, 1990; Kraemer-Schafhalter et al., 1998). For higher isolation scales we routinely use Amberlite XAD-7 to eliminate polar, non-phenolic compounds (Andersen, 1988). After loading of the column (18×2.6 cm), it is washed with 2 l of H_2O followed by 300 ml MeOH containing 0.5% TFA for elution of the anthocyanins. Similarly, Nørbæk and Kondo (1998) have eluted anthocyanins extracted from *Crocus* flowers stepwise from an Amberlite XAD-7 with 4–20% aqueous CH_3CN containing 0.5% TFA. Sephadex LH-20 material has proved to be especially useful for purification of individual anthocyanins (Kim et al., 1989). In our labs Sephadex LH-20 columns (100×5 cm or 100×1 cm) are frequently used for separation and purification of individual anthocyanins using both isocratic (MeOH–H_2O–TFA ($30.5 : 70 : 0.5$)) and step elution (20–60% MeOH–H_2O (0.1% TFA)). This column material may be used repeatedly after washing with 0.2M NaOH. Toyopearl HW-40F gel gives chromatograms with higher resolution between the anthocyanidin 3-rutinoside and anthocyanidin 3-glucoside bands than Sephadex LH-20 gel (Frøytlog et al., 1998). Other column supports include polyvinylpyrrolidone, polyamide and perlon (Strack & Wray, 1989). For semi-preparative isolation of anthocyanins, we have used extensively a reversed-phase C_{18} HPLC column (250×22 mm, 10 μm) with formic acid in water–methanol mixtures as mobile phase.

Anthocyanin mixtures have recently been successfully fractionated using HSCCC and centrifugal partition chromatography (Renault et al., 1997;

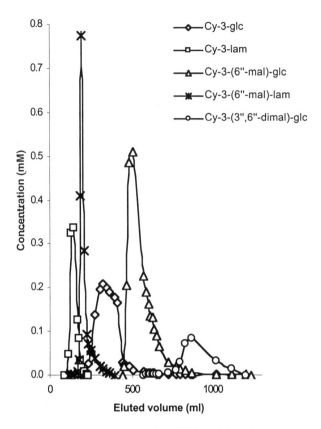

Fig. 10.9 High-speed countercurrent chromatography (HSCCC) of the anthocyanins in a partly purified extract (750 mg) from red onion (Torskangerpoll *et al.*, 2001). Cy = cyanidin; glc = glucoside; lam = laminariobioside; mal = malonyl.

Degenhardt *et al.*, 2000; Torskangerpoll *et al.*, 2001; Fig. 10.9). These all-liquid techniques have their major strength in the preparative scale.

10.3.4 Characterisation, identification and quantification

Recent advances in methodology and instrumentation have established 'soft' techniques in MS and 2-D NMR spectroscopy as essential tools for characterisation of novel anthocyanins. However, verification of known anthocyanins and determination of their concentration demand less sophisticated techniques. HPLC has revolutionised anthocyanin studies due to its high efficiency, sensitivity, speed and accurate quantitation (Takeoka & Dao, 2002, and references therein). More information about HPLC applied to anthocyanins is found in Section 10.2.4. A typical HPLC separation of a complex mixture of anthocyanidin 3-*O*-mono-glycosides from a crude fruit extract is shown in Fig. 10.2.

10.3.4.1   Chemical degradation

Chemical degradation of anthocyanins will partly or totally cleave each pigment into anthocyanidin, sugar and sometimes acid units. Individual units may be identified by chromatography with reference standards. Total acid hydrolysis is performed by mixing equal amounts of the aqueous anthocyanin solution and 3M HCl followed by heating at 90°C for 45–120 min. After cooling, the anthocyanidins are extracted with 1-pentanol. The stability of anthocyanidins in 1-pentanol is relatively high; however, the high boiling point of this solvent is very inconvenient. Takeoka *et al.* (1997) have suggested the use of ethyl acetate for extraction of the anthocyanidins from the hydrolysis mixture. Samples taken at suitable intervals during controlled acid hydrolysis may reveal intermediates, whose identities are useful for the determination of sugar sequences and linkage positions.

Acyl groups attached by ester linkages to anthocyanin sugar moieties may be removed by mild treatment with alkali. This may be accomplished by keeping the sample in a syringe for 2 h in 2M NaOH at room temperature under air-free conditions. Liberated acid is extracted with hot ether. Lack of retention time changes (TLC or HPLC) before and after alkaline treatment may indicate absence of acylation.

10.3.4.2   Thin-layer chromatography

The use of TLC for analysis of anthocyanins has been reviewed recently (Francis & Andersen, 2003). A TLC solvent, concentrated HCl–formic acid–water (25 : 24 : 51) (FHW), on cellulose layers has been used for the simultaneous separation of anthocyanidins and their mono-, di- and triglycosides. This system is particularly valuable for structure elucidation of anthocyanins using hydrolysis techniques. For assessment of anthocyanin structure, it is useful to develop the cellulose plate first in an aqueous solvent like FHW, and then in an alcoholic solvent such as BAW. With otherwise similar structures, the R_F values of the anthocyanins developed with both FHW and BAW increase normally with addition of acyl moieties, and increase (FHW) and decrease (BAW) with increasing number of sugar units. The chromatographic behaviour of anthocyanins on cellulose TLC and PC is comparable, and a large number of anthocyanins have been provided with R_F data from PC (Harborne, 1967). An HPTLC densitometric method for quantitative analysis of the anthocyanins of mallow flowers on cellulose plates has proved to be more sensitive than an HPLC-DAD system (detection at 530 nm) (Farina *et al.*, 1995).

Anthocyanins are visible at the concentration levels encountered on chromatograms. Examination under UV light is however worthwhile since the 3,5-diglycosides of pelargonidin, peonidin and malvidin are distinguished by their fluorescence from the corresponding 3-glycosides. After spraying the dried chromatograms with aluminum chloride (3% in methanol), anthocyanins with free adjacent hydroxyls on the B-ring of the aglycone (delphinidin-, cyanidin- and petunidin-derivatives) turn blue.

10.3.4.3   Electrophoresis

The use of CE has in recent years developed into an important tool for analysis of anthocyanins (DaCosta *et al.*, 2000). Compared with HPLC, CE may even entail increased peak resolution and improved detection limits. CE is carried out with low solvent consumption within relatively short analysis time. By using a running buffer (pH 2.1) containing a quaternary ammonium salt with a relatively long alkyl chain, it has been possible to separate pure anthocyanins (Bicard *et al.*, 1999). A method has recently been developed for the isolation and rapid identification of anthocyanins from floricultural crops based on high-voltage paper electrophoresis with bisulphite buffer (Asenstorfer *et al.*, 2003a). Paper electrophoresis has also been used over the pH range 1.2–10.4 to measure apparent pKa values for malvidin 3-*O*-glucoside to highlight the complex nature of its equilibrium forms (Asenstorfer *et al.*, 2003b). A CE method has been proposed for the quantitative determination of anthocyanins in wine as an alternative to HPLC (Saenz-Lopez, 2003).

10.3.4.4   UV-visible absorption spectroscopy and CIELAB colours

In contrast to other flavonoids, anthocyanins absorb light in the visible area around 520 nm under acidic conditions (Fig. 10.1). For characterisation purposes under controlled solvent conditions, the absorption maximum in this region is related to the nature of the aglycone, position of sugar substituents, presence of aromatic acyl groups, co-pigmentation, self-association, etc. (Strack & Wray, 1989). It is standard procedure to measure UV-Vis spectra of pure anthocyanins dissolved in MeOH containing 0.01% HCl. During the last decade it has been common to record anthocyanin spectra with a diode array detector during on-line HPLC (Santos-Buelga *et al.*, 2003).

Giusti and Wrolstad (2001) have described in detail a procedure for quantitation of monomeric anthocyanins, even in the presence of interfering compounds. This report contains also a good compilation of molar absorptivity (ε) coefficients.

Several authors have shown that the CIELAB system is useful to describe colours of flowers (Nørbæk *et al.*, 1998; Hashimoto *et al.*, 2002) as well as pure anthocyanins (Heredia *et al.*, 1998; Giusti *et al.*, 1999; Torskangerpoll & Andersen, 2004).

10.3.4.5   Mass spectrometry

Recent developments in MS have made it, together with NMR, the most important tool for research and quality control related to anthocyanins. Some applications of MS to analysis of flavonoids in general are presented in Section 10.2.4. DaCosta *et al.* (2000) have reviewed the specific use of LC-MS for anthocyanin analysis. In a recent study Oliveira *et al.* (2001) described the use of electrospray ionisation MS, in combination with collision-induced dissociation and tandem MS, for the structural characterisation of anthocyanidins and anthocyanins.

10.3.4.6   Nuclear magnetic resonance spectroscopy
Today it is possible to make complete assignments of all proton and carbon atoms in NMR spectra of most isolated anthocyanins. These assignments are normally based on chemical shifts (δ) and coupling constants (J) observed in 1-D ^{1}H and ^{13}C NMR spectra combined with correlations observed as crosspeaks in various homo- and heteronuclear 2-D NMR experiments (see Sections 10.1.3 and 10.2.4). Recently, a protocol including preparation of anthocyanin NMR samples, details of various NMR experiments applied on anthocyanins (^{1}H, CAPT, DQF-COSY, TOCSY, HSQC, HMBC, NOESY and ROESY), and NMR data on selected anthocyanins and their building blocks has been published (Andersen & Fossen, 2003). A ^{1}H-^{1}H HSQC spectrum of the sugar region of cyanidin 3-O-(2″-O-β-glucopyranosyl-6″-O-α-rhamnopyranosyl-β-glucopyranoside) is shown in Fig. 10.10. Table 10.2 shows ^{13}C NMR chemical shifts of the anthocyanidins and most anthocyanins.

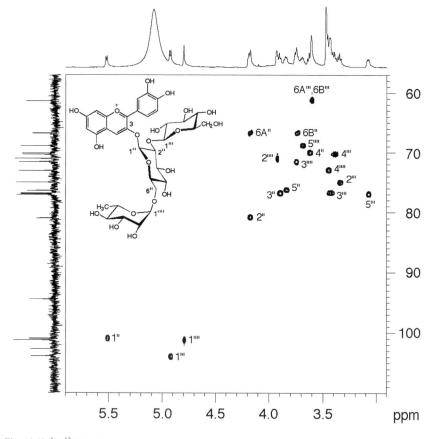

Fig. 10.10 ^{1}H-^{13}C HSQC spectrum of the sugar region of cyanidin 3-O-(2″-O-β-glucopyranosyl-6″-O-α-rhamnopyranosyl-β-glucopyranoside) dissolved in CD$_3$OD containing 5% CF$_3$COOD showing all the $^1J_{CH}$ correlations, and thus all the ^{1}H and ^{13}C chemical shifts of the three sugar units except the crosspeak of the methyl group of rhamnose (H6‴/C6‴) at 1.2/17.1 ppm.

Table 10.2 ^{13}C NMR chemical shifts of the aglycones of various anthocyanins representing the anthocyanidins of most anthocyanins in nature

Aglycone*	1	2	3	4	5	6	7†	8	9	10	11
2	163.85	164.36	164.19	164.49	164.14	163.98	173.8	162.28	161.62	166.46	165.01
3	145.31	145.64	145.49	145.96	145.41	144.74	112.0	145.85	145.94	135.95	135.25
4 (3a)	137.08	137.03	137.35	136.61	136.45	136.68	149.9	134.39	133.47	150.44	109.79
5 (6a)	157.48	159.55	159.29	159.03	159.22	159.16	158.6	141.95	141.64	154.48	153.07
6 (7)	103.49	103.50	103.42	103.29	103.43	102.39	105.5	135.62	135.53	101.77	101.30
7 (8)	170.59	170.56	170.70	170.38	170.59	170.76	172.1	159.37	159.78	169.33	168.40
8 (9)	95.28	95.19	95.24	95.03	95.20	95.23	98.5	95.24	95.07	101.53	101.05
9 (9a)	157.44	157.75	157.86	157.72	157.67	157.91	160.2	151.56	151.17	154.48	153.20
10 (9b)	113.39	113.45	113.63	113.29	113.34	113.59	114.5	114.04	113.87	111.00	109.67
1′	120.61	121.31	121.14	120.07	119.94	119.87	121.6	121.38	120.10	120.95	120.33
2′	135.67	118.56	115.19	112.62	108.68	110.83	134.1	118.11	112.06	135.04	118.5
3′	117.91	147.41	149.51	147.56	149.88	149.85	118.9	147.19	147.25	117.40	146.14
4′	166.57	155.78	156.37	144.71	147.32	146.39	168.3	154.97	143.69	165.75	153.89
5′	117.91	117.48	117.55	147.56	145.10	149.85	118.9	117.51	147.25	117.40	116.89
6′	135.67	128.22	128.84	112.62	114.75	110.83	134.1	127.68	143.69	135.04	126.52
OCH₃			56.91		57.20	57.34					
(4)										107.48	106.8
(5)										154.55	154.46
COOH										161.17	160.52

*The labels in the brackets within the table represent skeleton positions of the pyranoanthocyanins **10–11** (see Fig. 10.6)
†Swinny *et al.*, 2002. The other NMR results are obtained in our labs.
1 = pelargonidin 3-glucoside (H, H, O-, H)
2 = cyanidin 3-glucoside (OH, H, O-, H)
3 = peonidin 3-glucoside (OMe, H, O-, H)
4 = delphinidin 3-galactoside (OH, OH, O-, H)
5 = petunidin 3-sambubioside (OMe, OH, O-, H)
6 = malvidin 3-rutinoside (OMe, OMe, O-, H)
7 = apigeninidin 5-glucoside (H, H, H, H)
8 = 6-hydroxycyanidin 3-rutinoside (OH, H, O-, OH)
9 = 6-hydroxydelphinidin 3-rutinoside (OH, OH, O-, OH)
10 = 5-carboxpyranopelargonidin 3-glucoside (H, H)
11 = 5-carboxpyranocyanidin 3-glucoside (OH, H)
The four labels within the brackets behind **1–9** correspond to R^1, R^2, R^3 and R^4 in Fig. 10.6 (left), while the two labels within the brackets behind **10–11** correspond to R^1 and R^2 in Fig. 10.6 (right).
Pigments **1–10** are dissolved in CD$_3$OD containing 1–5% CF$_3$COOD, while **11** is dissolved in CF$_3$COOD-DMSO-d_6 (1:4).

10.3.4.7 Miscellaneous *in vivo* techniques

Strategies for production of anthocyanins from plant cell and tissue cultures have been in focus in recent years. A method using a non-destructive image processing system has been used to estimate individual cell characteristics such as the quantitative content of anthocyanins in each cell (Miyanaga *et al.*, 2000). With respect to other *in vivo* analysis, radiolabelling of anthocyanins (e.g. with L-[^{14}C]-phenylalanine) (Wharton & Nicholson, 2000), and the production of ^{13}C-labelled anthocyanins in cell cultures (Krisa *et al.*, 1999) are promising.

10.4   Chlorophylls

10.4.1   Introduction

The chlorophylls are omnipresent in photosynthetic tissue where they are responsible for photoreception (Formaggio *et al.*, 2001). These tetrapyrrole pigments contain a centrally placed chelated magnesium ion and have typical green hues, although the presence of other pigments can mask their presence in intact tissue (Schoefs, 2002). The structures of chlorophyll *a* and chlorophyll *b* are shown in Fig. 10.11. These pigments differ only in that one methyl group of chlorophyll *a* is replaced by a formyl group in chlorophyll *b*. These two pigments were for a time thought to be the only natural chlorophylls in higher plants, but it has since been shown that a variety of precursors and degradation products are also present in nature. The normal ratio of relative amounts of chlorophylls *a* and *b* found in higher plants is about 3 : 1, although the preponderance of chlorophyll *a* is greater in those grown in bright conditions and less in those grown in shady sites (Gross, 1991). The C-10 epimeric chlorophylls *a'* and *b'* are now believed to be of natural occurrence in the living tissue rather than to be mere artefacts. Pheophytins *a* and *b* that lack the magnesium ion of the chlorophylls, the pheophorbides *a* and *b* that lack both the magnesium ion and the phytyl side chain, and the chlorophyllides *a* and *b* that are the products of hydrolysis of the phytyl ester are all normally regarded as decomposition products formed during extraction.

10.4.2   Sample preparation and extraction

Chlorophylls are relatively polar and are thus normally extracted with methanol, acetone or other organic solvents miscible with water (Minguez-Mosquera *et al.*, 2002a). The sample should be finely divided to ensure adequate contact between the solvent and the material being extracted. In a typical extraction a sample (50 g) is macerated with acetone (100 ml) containing a small amount of sodium carbonate (2 g), and the resulting suspension allowed to stir for a few minutes in reduced light at room temperature. The presence of sodium carbonate takes into account the unusual sensitivity of the chlorophylls to acids, and avoiding heat and light will minimise other factors leading to structural modification. The mixture is filtered and extraction repeated with a fresh portion of acetone and sodium carbonate. Four or five repetitions will normally be needed to arrive at a point where the extracts are virtually colourless, although it should be noted that the resulting exhausted plant material may still be coloured due to the presence of other pigments. The extracts

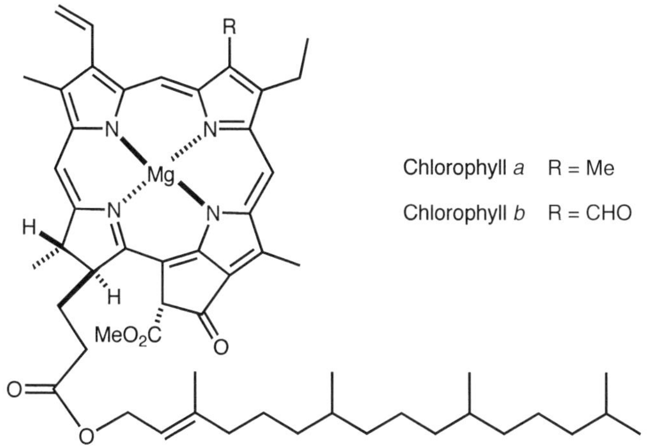

Fig. 10.11 Structures of chlorophyll *a* and chlorophyll *b*.

are then combined to give a single global extract (*c.* 500 ml). This extract is then concentrated under reduced pressure at temperatures under 40°C. Diethyl ether (100 ml) and water (100 ml) containing 10% sodium chloride are added to the residue, the mixture is shaken and the ether layer removed and retained. The water layer is re-extracted with similar amounts of ether twice, the ether extracts combined and washed with one portion of water. The resulting ether solution is then taken to dryness under reduced pressure. Any remaining traces of water may be removed by azeotropic distillation with benzene and small amounts of methanol. The pigment residues, usually dissolved in a small volume of acetone or diethyl ether, are stored at 4°C under a nitrogen atmosphere until required.

10.4.3   Purification and separation

10.4.3.1   Dioxane complexes

It may be desirable to purify chlorophylls prior to further analysis, and this can be done by means of their readily formed and moderately stable dioxane complexes – a procedure that provides for removal of nearly all of the non-chlorophyll impurities present in the crude extract (Iriyama *et al.*, 1974). Samples of finely ground plant material are extracted three or four times with small portions of isopropanol. Dioxane is added to the combined extracts (1:7), and thereafter water added drop by drop until turbidity occurs. A chlorophyll–dioxane complex forms if this mixture is allowed to stand in the refrigerator for about 2–3 h. The complex is finely divided and is best collected by centrifugation. The chlorophylls are readily recovered from the complex by dissolution in acetone or diethyl ether. This cycle may be repeated to provide increasingly pure complexes and thus chlorophyll samples. The complex itself may be dried under reduced pressure and stored under nitrogen at −30°C.

10.4.3.2 Thin-layer and high-performance liquid chromatography
The purified extracts can be separated by various chromatographic techniques, chief of which are TLC and HPLC. HPLC is particularly useful as the eluted pigment can be monitored continuously on-line and contamination between zones being separated minimised.

TLC of chlorophylls can be carried out in a wide range of systems but chief of these are separations on cellulose and silica layers using somewhat polar solvent systems (Francis & Andersen, 2003). Retention behaviour for illustrative systems may be found in Table 10.3. There is a clear relationship between structural changes and retention data, and it should be noted in particular that the additional carbonyl group found in the *b* series results in increased polarity and hence greater retention for each member of the *b* series as compared with the *a* series. Otherwise it is clear that the magnesium-free derivates are less retained than the corresponding magnesium-containing derivatives. The derivatives resulting from the hydrolysis of the phytyl group and thus having an additional hydroxyl group are more retained than the intact esters.

A wide variety of HPLC systems have been employed for the separation of chlorophylls (Minguez-Mosquera *et al.*, 2002a). Both normal and reversed phase, and isocratic and gradient systems have found application. While hydrocarbons containing polar organic modifiers are usually employed with silica normal phase systems, methanol–water mixtures are the general preference for reversed phase separations with RP-18 as the usual stationary phase. The various systems, and their advantages and disadvantages, have been recently reviewed (Minguez-Mosquero *et al.*, 2002a). The system of choice varies considerably depending on

Table 10.3 Thin-layer chromatography of chlorophylls and derivatives (R_F values)

Derivative	R	R′	Metal	Colour	System 1	System 2
Pheophytin *a*	Me	Phytyl	—	grey	0.93	0.40
Pheophytin *b*	CHO	Phytyl	—	yellow-brown	0.88	0.33
Chlorophyll *a′*	Me	Phytyl	Mg	blue-green	0.80	0.31
Chlorophyll *a*	Me	Phytyl	Mg	blue-green	0.76	0.27
Chlorophyll *b′*	CHO	Phytyl	Mg	yellow-green	0.60	0.25
Chlorophyll *b*	CHO	Phytyl	Mg	yellow-green	0.57	0.22
Pheophorbide *a*	Me	H	—	grey	0.36	—
Pheophorbide *b*	CHO	H	—	yellow-brown	0.18	—
Chlorophyllide *a*	Me	H	Mg	blue-green	0.08	—
Chlorophyllide *b*	CHO	H	Mg	yellow-green	0.05	—

Structures can be derived by referring to Fig. 10.11.
System 1: cellulose layer (0.1 mm, Merck Art. 5716), light petroleum (40–60°C)–acetone (80:20).
System 2: silica gel 60 (0.25 mm, Merck Art. 5721), diethyl ether–acetone–isooctane (20:20:60).
All compounds show red fluorescence under UV light.

the application, although the order of elution follows that for TLC for the normal phase systems, and is reversed for the reversed phase cases. Considerable efforts are still being made to find improved HPLC systems. Many of the systems have been designed to separate both chlorophylls and carotenoids in a single operation, and a number of these have been newly discussed (Schoefs, 2002, 2003).

10.4.4   Characterisation and identification

10.4.4.1   Visible absorption spectroscopy

All chlorophyll derivatives show a bright red fluorescence under UV light. This together with their chromatographic behaviour will often serve to provide a preliminary identity. Visible light absorption spectra of chlorophylls show two main maxima, one at about 440 nm and one at about 650 nm; e.g. see the spectra of chlorophylls a and b (Fig. 10.12). Taken together these maxima result in a variety of brown-grey-green shades for the various derivatives, and the resulting visual impressions can be found in Table 10.3. A list of visible absorption maxima in a variety of solvents has been recently provided (Minguez-Mosquera et al., 2002a).

10.4.4.2   Preparation of chlorophyll derivatives

Standard samples of chlorophyll a and chlorophyll b are readily obtained from parsley (Petroselinum crispum) as described above. The extract will contain smaller amounts of many modified chlorophyll derivatives, but it is usually more convenient to obtain the latter by appropriate chemical modification of the parent compounds. Heating a pyridine solution of the parent chlorophylls a and b for 1 h in a sealed tube at 60°C provides the epimers, chlorophyll a' and

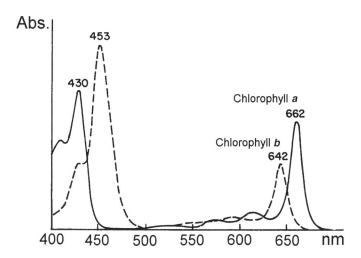

Fig. 10.12 Visible light absorption spectra of chlorophyll a and chlorophyll b in diethyl ether (adapted from Gross, 1991).

chlorophyll b'. Pheophytin a and b are produced when an ether solution of the parent chlorophylls is stirred for 2–5 min with 1 M HCl (2 : 1). Similarly, pheophorbides a and b are prepared from ethereal solutions by stirring for 2–5 min with more concentrated acid, 30% (9.5 M) HCl (1 : 2). The pigments prepared by acid treatment should be transferred to diethyl ether by adding water and washing the resultant ether layer with distilled water until neutrality is reached.

10.4.4.3   Nuclear magnetic resonance spectroscopy and mass spectrometry

^{13}C- and ^{1}H-NMR spectroscopy give immediate information on the type of chlorophyll being examined and will very quickly show if new or unknown types are present. A good source for NMR data on chlorophyll derivatives is available (Abraham & Rowan, 1991).

The low thermal stability and high molecular weights of chlorophylls have made it difficult to use MS, as this traditionally depends on providing molecules in the gas phase. However, modern techniques involving desorption methods and surface bombardment have slowly allowed advances in the use of MS for chlorophylls. The use of HPLC-MS methods has been particularly useful and electrospray ionisation combined with an ion-trap mass spectrometer has recently been used to great effect to show not only a pseudo-molecular $(M + H)$ ion but also series of decompositions from this ion for chlorophylls (Schoefs, 2001).

10.4.5   Quantification

Solutions of chlorophylls in a number of different solvents can be used to determine the amounts present. However, the most usual method is to use ether solutions and standard equations (Lichtenthaler & Wellburn, 1983) to allow individual determinations of chlorophyll a and chlorophyll b:

$$\text{Chlorophyll } a \text{ (mg/l)} = 10.05\text{A}(660.6) - 0.97\text{A}(642.2)$$
$$\text{Chlorophyll } b \text{ (mg/l)} = 16.36\text{A}(642.2) - 2.43\text{A}(660.6)$$

Densitometry can readily be applied to the measurement of thin-layer plates provided that suitable reference pigments are available. Chlorophyll a and chlorophyll b are commercially available, while other standards can be obtained as indicated above.

10.5   Carotenoids

10.5.1   Introduction

Carotenoids, named originally because of their isolation from carrots (*Dauca carota*), are yellow to red-coloured terpenoids containing eight isoprene units arranged in a symmetrical linear pattern. The chain itself contains a series of conjugated double bonds that are responsible for the typical visible light absorption of this class of compounds. The number of double bonds is normally nine or more, although shorter chromophores are occasionally found. There is little skeletal variation, and it is almost entirely restricted to cyclisation at the end(s) of the chain where simple functionalisation can also occur. As can be seen from the typical carotenoid structures illustrated in Fig. 10.13, only oxygen functions are found, e.g. as in ketones, alcohols, epoxides. These oxygenated compounds are sometimes known as xanthophylls to distinguish them from the hydrocarbon carotenes.

Carotenoids must be present in all photosynthetic tissue as, in addition to functioning as auxiliary pigments, they are responsible for protecting the chlorophylls from photodynamic destruction (Young & Frank, 1995). Carotenoids are also found in many flowers and fruits where they are valuable in terms of providing signals to other species on the availability of food sources, and in this way ensure pollination and the spread of seeds. Numerous reviews are available describing the distribution of carotenoids (Goodwin, 1980; Gross, 1991), and these are useful as a source of data on related species, and in finding good sources for reference compounds.

Carotenoids are almost always studied as the free C_{40}-compounds, but it is important to note that they often occur in nature in the form of esters and as lipid and protein complexes of various types. The interested reader will find considerably more detail on this aspect of carotenoid occurrence elsewhere, including in Chapter 3.

10.5.2   Sample preparation, extraction and saponification

The majority of carotenoids are relatively non-polar, and this allows their ready extraction with the organic solvents (Britton et al., 1995). Hydrocarbons, ethers,

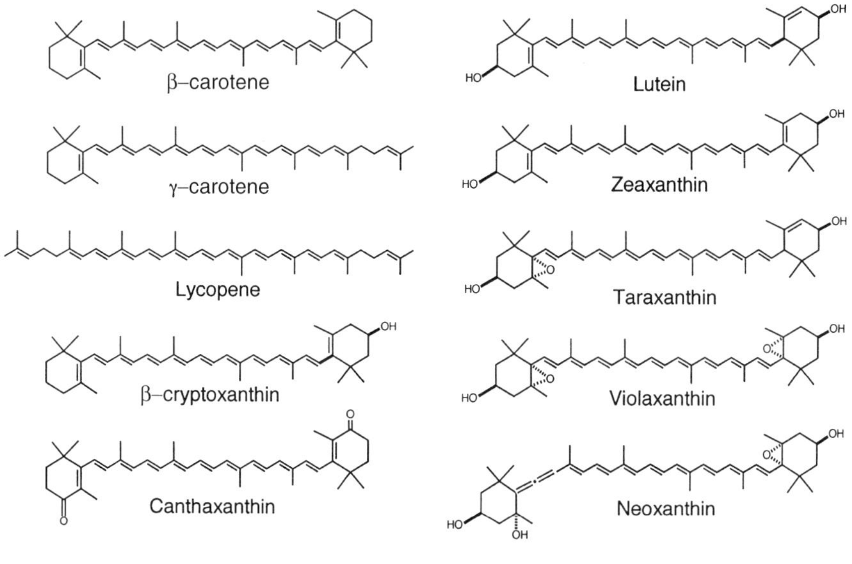

Fig. 10.13 Structures of some typical carotenoids.

acetone and the lower alcohols are mostly used for this purpose, although other solvents with low boiling points can be used. Chlorinated solvents should be carefully checked for possible contamination with hydrochloric acid, which will destroy most carotenoids. More recently extraction with liquid carbon dioxide or other compromised gasses has been used.

The ease with which raw materials may be extracted varies greatly, and while soft tissues can often be simply cut into pieces, it may be necessary to chop or grind skins, seed coats, etc. where the carotenoids are incorporated into woody material or within cell walls.

The material to be extracted is cut up or macerated, eventually ground if intractable, and extracted with acetone. This will dehydrate the material and make it more amenable to further extraction. After filtration the residue is re-extracted with successive portions of acetone or acetone containing increasing amounts of methanol until the extract is colourless. Should the residue still be coloured, it is advisable to carry out one or more extractions with petroleum ether or hexane, or with mixtures of these with acetone. Highly polar carotenoids are sometimes present and while these are difficult to extract, additional extraction steps with portions of methanol will normally lead to satisfactory results. The combined extract is concentrated under reduced pressure at not more than 40°C. The pigments are extracted with diethyl ether and washed with a portion of 10% aqueous sodium chloride and then with a portion of water. The diethyl ether layer is then taken to dryness under reduced pressure; any traces of water remaining are removed by azeotropic distillation with benzene/methanol. The isolated pigment mixture is dissolved in diethyl ether and should then be stored under a nitrogen atmosphere at low temperature (−30°C).

While successive extractions with solvents of increasing polarity normally provide the most successful strategy, some plant tissues may be highly waxy and in such cases it may be advantageous to de-wax such material with a quick hydrocarbon wash. Such pre-extracts can often be incorporated into the total extract at a later stage. The main extraction series should be continued as described above until the resulting solutions are colourless, at which time the extracts can be unified.

The global extract prepared as above will almost inevitably contain water, and in many cases chlorophylls. Water is easily removed by means of azeotropic distillation under reduced pressure, using small amounts of methanol and benzene after prior removal of the bulk of the solvent.

10.5.2.1 *Saponification*

Most procedures now include a saponification step at this point. Carotenoid alcohols often occur in flowers, fruits, etc. as long-chain fatty acid esters, and the mixture of compounds available can be greatly simplified by introducing a saponification step. This serves in addition to destroy glycerides present, thus allowing removal of many colourless lipids, and it causes destruction of the chlorophylls providing easily removed water-soluble by-products.

Saponification is executed by taking the pigment extract to dryness under reduced pressure, and adding a 1 : 1 mixture of diethyl ether and 10% methanolic potassium hydroxide or sodium hydroxide. The mixture is then allowed to stand for 6 h and thereafter one volume of diethyl ether and one volume of water containing 10% NaCl are added. The mixture is shaken, the phases allowed to separate and the ether phase removed. Two further diethyl ether extractions are carried out, and the extracts then mixed. The extract is then washed to neutrality with water containing 10% NaCl. The extract is concentrated, and the last traces of water are removed by azeotropic distillation with benzene and methanol under reduced pressure. The residue is then dissolved in acetone or diethyl ether and stored under nitrogen at reduced pressure. Colourless precipitates, usually sterols or proteins, may be removed by centrifugation or filtering through a wad of glass wool. Finally, the pigments are dissolved in acetone and stored at low temperature.

10.5.3   Purification and separation

Carotenoids are separated by column chromatography, TLC and HPLC (Britton *et al.*, 1995). Occasionally, it is possible to isolate the main compound(s) in relatively pure state by direct crystallisation and even student exercises for doing this have been described, e.g. lycopene from tomato fruits and capsanthin from saponified extracts of red bell peppers (Ikan, 1969).

10.5.3.1   Column chromatography
Open column chromatography was for many years the main method for the isolation of carotenoids, but it is now mainly used for preliminary separation. The number of systems developed is very large and many of them are rarely used in modern practice. Silica gel and alumina have been regularly used as stationary phases, but a wide variety of others have found more specialised applications, and magnesium oxide, calcium carbonate and sucrose have been extensively used. Developing solvents are almost always mixtures of aliphatic hydrocarbons with more polar modifiers such as diethyl ether, acetone and various alcohols, although systems containing chloroform and acetonitrile have also found use. Elution from the column is usually carried out using a hydrocarbon carrier with increasing amounts of acetone or methanol. Modifications employing flash chromatography have largely followed this same pattern. Indications as to what systems are likely to be successful in particular cases can be found in previous reviews (Britton *et al.*, 1995; Minguez-Mosquera *et al.*, 2002b).

10.5.3.2   Thin-layer chromatography
The availability of TLC for semi-preparative and preparative scale work led to this technique replacing open column methods. A very large number of thin-layer systems have been employed, but with experience and the availability of high-quality commercial thin-layers, the number of variations employed has decreased. Nevertheless, chromatography on silica gel layers remains a simple and useful way of carrying out separation of carotenoids. Separation of the more common pigments

may be readily achieved using a hydrocarbon developer, typically hexane or light petroleum (40–60°C), containing polar modifiers such as acetone or alcohols. Carotenes move readily in the unmodified hydrocarbons, and β-carotene moves close to the solvent front in all systems and may be used as a standard. It is given the value of 1.00 and relative retentions then quoted as R_β values. TLC systems of this type separate the pigments into groups mainly according to the oxygen-bearing functional groups. Carotenoid hydrocarbons are least retained, and are followed by mono-alcohols, diols, and thereafter by poly-oxygenated compounds. Epoxy groups have less effect on retention than hydroxyls. The same applies to carbonyl groups in acetone-based systems, but changes can be seen in these effects with other polar modifiers. Table 10.4 presents R_F values for systems where light petroleum (40–60°C) is modified with 20% acetone, 40% acetone and 20% *tert*-butanol, and the results there serve to indicate the general trends found. Carotenoids are also readily separated on reversed phase TLC systems, where the order of retention is largely speaking the opposite of that found in the normal phase systems. R_F values using a solvent containing light petroleum (40–60°C)–acetonitrile–methanol (20 : 40 : 40) on RP-18 layers are also given in Table 10.4.

All of the above systems can be used for the separation of carotenoids with varying degrees of success depending on the compounds that are present in a particular case. They may be freely combined to give more detailed results. Semi-preparative work is possible since commercial plates with high loading capacity are now available.

Table 10.4 Thin-layer chromatography of carotenoids from established sources. R_β values (β-carotene = 1.00) for systems 1,2 and 3, and R_F values for system 4

Carotenoid		Colour	System 1	System 2	System 3	System 4
β-carotene	a	yellow	1.00	1.00	1.00	0.13
γ-carotene	a	orange	0.99	1.00	1.00	0.17
Lycopene	b	red	0.97	1.00	1.00	0.23
β-cryptoxanthin	a	yellow	0.34	0.72	0.78	0.31
Canthaxanthin	c	orange	0.36	0.71	0.69	0.38
Lutein	d	yellow	0.07	0.44	0.56	0.55
Zeaxanthin	e	yellow	0.07	0.44	0.55	0.57
Taraxanthin	e	yellow	—	0.41	0.43	0.62
Violaxanthin	d	yellow	—	0.33	0.30	0.68
Neoxanthin	d	yellow	—	0.18	0.13	0.72

System 1: silica gel 60 (0.25 mm, Merck, Art. 5721) 20% acetone/p.e.
System 2: silica gel 60 (0.25 mm, Merck, Art. 5721) 40% acetone/p.e.
System 3: silica gel 60, 20% *tert*-butanol/p.e.
System 4: RP-18 F$_{254}$, p.e.–acetonitrile–methanol (20:40:40).
Solvent compositions by volume, p.e.=petroleum ether (40–60°C).
Sources: a, *Sorbus aucuparia* berries; b, *Solanum lycopersicum* fruit; c, commercial; d, *Petroselinum crispum*; e, *Taraxacum officinale* flowers.

Structural information can be obtained directly from TLC by further treatment of the plates. Exposing the dried developed plate to hydrochloric acid fumes produces rapid changes in the colour of carotenoid epoxides from yellow to blue or green colours. Compounds having α- or β-groups may be distinguished by spraying with silver nitrate (Isaksen & Francis, 1990).

10.5.3.3 High-performance liquid chromatography
The introduction of HPLC has led to a revolution in the isolation of carotenoids (Pfander et al., 1994). HPLC has been used both analytically and preparatively in the carotenoid field. The availability of good preparative HPLC systems, using commercially filled columns or those of the self-packing variety have allowed adaption of much of the earlier column chromatography work to find renewed applications under somewhat modified conditions. It should be noted that most of the present HPLC systems use reversed phase material and have been developed either from scratch or as a result of previous TLC systems. This technique allows the rapid separation of not only the compounds themselves but also their *cis/trans* isomers, and indeed enantiomers and diastereomers as readily prepared derivatives.

10.5.4 Characterisation and identification

Visible light absorption spectroscopy, infrared spectroscopy, NMR and MS have all proved invaluable in specific cases. While a certain technique may prove especially useful in one case, the best results are always obtained when they are used in combination. This is particularly well illustrated in relation to stereochemical aspects of carotenoids (Liaaen-Jensen, 1997). An excellent source of many earlier references to spectroscopic data is also very useful (Straub, 1987).

10.5.4.1 Visible and infrared spectroscopy
Much of the early work on the identification of carotenoids was based on their characteristic visible light absorption spectra, and this technique is still widely applied (Britton, 1995). The spectra of three carotenoids are shown in Fig. 10.14 and clearly illustrate some of the principle features. The main absolute max shows that each additional conjugated double bond in the acyclic chromophore results in some 15 nm shift, while conjugated double bonds within a ring contribute only some 5 nm to the maximum position. Another effect that is clearly seen is that the double bonds of the latter type change the appearance of the spectrum by reducing the fine structure, i.e. the depth of the observed minima, and indeed change the lowest wavelength maximum into a mere inflection. The effect of conjugated ketone functions is normally somewhat greater than that of an additional C–C double bond in the same position.

Infrared spectroscopy was historically the next technique introduced in carotenoid work. Infrared allowed immediate disclosure of non-conjugated functionalities, such as alcohol and ether functions, and carbonyls of whatever type (Bernhard & Grosjean, 1995). The original discovery of the presence

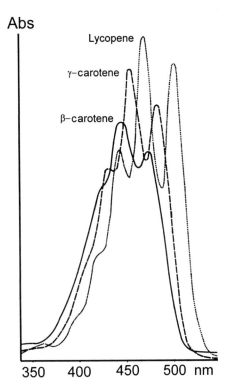

Abs

Lycopene

γ–carotene

β–carotene

350   400   450   500 nm

Fig. 10.14 Visible absorption spectra of β-carotene, γ-carotene and lycopene in petroleum ether (adapted from Vetter *et al.*, 1971).

of allene and acetylene functions was largely due to the application of this method.

10.5.4.2 Nuclear magnetic resonance spectroscopy

A very large number of carotenoids have been examined by NMR spectroscopy and a recent review discusses many applications in the carotenoid field (Englert, 1995). However, some salient facts should be mentioned: the normal situation is that changes in the chain are immediately seen, but it is particularly in the finding of unusual end-groups or structural units that NMR spectroscopy has come into its own. Examples of this are to be found in the discovery of such compounds as peridinin with an atypical chain, and in cucurbitaxanthin with an unexpectedly complicated end-group (Matsuno *et al.*, 1986).

10.5.4.3 Mass spectrometry

MS normally furnishes a weak, but distinct, molecular ion that is readily identified by losses of 92 and 106 mass units, corresponding to the elements of toluene and xylene respectively. The normal pattern of fragmentation shows ions due to losses of any functionalities present, e.g. 18 (water), 31 (OMe), 32 (MeOH), 43 (MeCO), 59 (MeCOO), 60 (MeCOOH), and those resulting from cleavage of the conjugated chain. The composite picture thus obtained is quite informative and when taken with other spectroscopic data, often proves conclusive (Enzell & Back, 1995). Unusual features within the conjugated chain are readily revealed by MS, e.g. loroxanthin (Nitsche, 1974). Recently, the unique advantages of LC/MS systems have been used to great effect and a good review of the techniques involved has been published (van Breemen, 1997).

10.5.4.4   Stereochemical investigations

The application of ORD/CD in particular requires the availability of pure compounds to provide good spectra. Improving chromatographic techniques have made this possible and this has led to the identification of previously unknown optical isomers and isomeric mixtures. A good example of this kind of application is to be found in lactucaxanthin (Siefermann-Harms, 1981), a lutein isomer found in lettuce.

Stereochemical investigations have proved the great advantages of modern techniques in combination, with NMR often playing a crucial role. The identification of the isomeric dinochromes used a variety of NMR experiments (DQF-COSY, NOESY, HSQC and HMBC) to uncover the niceties of the stereochemical relationships within each molecule. Some key NOESY data from this work are provided for dinochrome A in Fig. 10.15 (Maoka et al., 2002).

10.5.4.5   Microchemistry

Microchemistry is important in carotenoid chemistry, even though the compounds are highly sensitive and relatively few reactions are thus applicable. These have been discussed in a number of texts, but particularly important have been simple oxidation and reduction methods, and the derivatisation of alcohol functions to esters and ethers (Eugster, 1995).

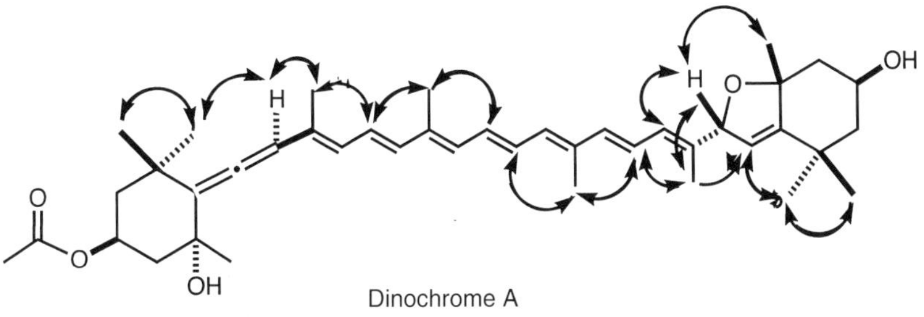

Dinochrome A

Fig. 10.15 Structure of dinochrome A, showing some key NOESY data (adapted from Maoka et al., 2002).

10.5.5   Quantitative aspects

The quantification of carotenoids is readily carried out by visible light absorption, either in solution or by densitometry on TLC plates. A requirement for accurate results is the availability of data for the pure compounds to be measured. Much data of this type are available and can be found in the original literature and in several monographs (Foppen, 1971; Britton, 1995). However, even when the identities of the compounds are not known, approximations can be made assuming a notional average carotenoid with an assumed specific extinction coefficient $E_{1\,\text{cm}}^{1\%}$ of about 2300 depending on the solvent used (Davis, 1976).

10.6   Betalains

10.6.1   Introduction

The betalains are divided into two groups: the red-violet betacyanins and yellow betaxanthins, isolated mainly from closely related families of the order Caryophyllales (covered in detail in Chapter 6). Betacyanins occur as *O*-glycosides of betanidin, isobetanidin (Fig. 10.16) and rarely 2-descarboxybetanidin (Fig. 10.17) with a sugar attached to one of the hydroxyl groups of the dihydroindole unit. Glucose is the most common glycosyl moiety, while apiose, rhamnose and glucuronic acid occur much less frequently. The sugar residues (mono-, di- or trisaccharides) are often acylated with hydroxycinnamic acids (*p*-coumaric, caffeic, ferulic or sinapic). Acylation may also include malonic, 3-hydroxy-3-methylglutaric, citric and sulphuric acid. The yellow betaxanthins (Fig. 10.16), which are conjugates of betalamic acid with amino acids or the corresponding amines, have hitherto not been reported to be glycosylated (or acylated).

Techniques used for betalain analysis have been methodically reviewed previously (Steglich & Strack, 1990; Strack *et al.*, 1993; Strack & Wray, 1994; Jackman & Smith, 1996; Delgado-Vargas *et al.*, 2000). The nature of newly identified betalains, as well as definitive structural elucidations carried out within the last ten years on compounds previously described in terms of putative structures, has recently been compiled (Strack *et al.*, 2003). The present chapter focuses on the most appropriate current methodologies used for extraction, isolation, structure elucidation and quantification of these highly polar nitrogen-containing pigments. The structure of several reported betalains has yet to be fully elucidated. There exist ^{13}C data for only one betalain (neobetain) (Alard *et al.*, 1985), no heteronuclear NMR data, and analysis may be limited by lack of commercially available standards.

10.6.2   Sample preparation and extraction

Betalain-containing materials (fresh or frozen) are generally macerated or ground before extraction with aqueous methanol or ethanol (20–80%), or pure water.

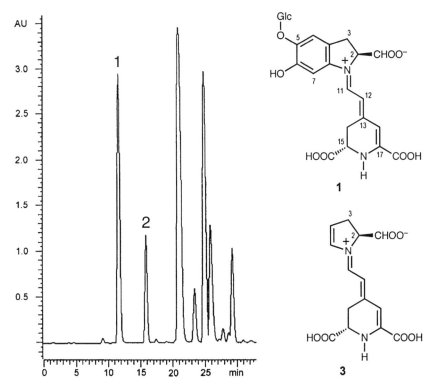

Fig. 10.16 HPLC separation of betacyanins from pitaya fruits detected at 538 nm on a reversed-phase C$_{18}$ column (250 × 3 mm, 5 μm) with solvent A (0.2% formic acid–water, v/v) and solvent B (acetonitril–water, 4:1, v/v) using gradient (linear) elution from 5–9% B in A during 33–min and 1 ml/min as flow. **1**= betanidin 5-*O*-β-glucoside (2*S*, 15*S*); **2**= isobetanidin 5-*O*-β-glucoside (2*S*,15*R*); **3**= indicaxanthin (a betaxanthin). The chromatogram is adapted from Stintzing *et al.* (2002b).

It is recommended that the extraction be carried out at low temperature and in darkness. The addition of ascorbic acid (*c.* 50 mM) in the extraction medium has been recommended to inhibit possible oxidation by polyphenoloxidases (Schliemann *et al.*, 1999). Degradative enzymes in aqueous extracts have also been inactivated by short heat treatment; this may however destroy some of the pigments.

10.6.3 *Purification and separation*

Betalains often occur as complex mixtures. They are easily decomposed during the purification steps, which render isolation of larger amounts difficult. Betacyanins have been successfully purified by precipitation during controlled acidification with HCl (Wyler *et al.*, 1959). However, betacyanins involving aliphatic acids contain ester linkages labile towards mineral acids like HCl.

10.6.3.1   Column chromatography

Various types of column support have been used for purification and separation of betalains. In recent years a procedure involving Sephadex G-25 (Adams & von Elbe, 1977), followed by Sephadex LH-20 chromatography prior to preparative HPLC (Schliemann et al., 1996) seems to be very useful. In a recent paper, solid-phase extraction with C_{18} sorbent has been used for semi-purification and concentration of betalain samples, including separation of betaxanthins from betacyanins using pH-dependent retention characteristics (Stintzing et al., 2002a). For preparative purifications, the use of ion exchange (e.g. Dowex 50W-X2 or Dowex 1X8), followed by column chromatography using combinations of polyamide, Polyclar-AT, polyvinylpyrrolidone, Sephadex G-15 or G-25, are common procedures.

10.6.3.2   High-speed countercurrent chromatography

HSCCC has recently been applied to isolate betalains on a preparative scale (Degenhardt & Winterhalter, 2001). Due to the hydrophilic character of betalains, a very polar solvent system consisting of ethanol–acetonitrile–ammonium sulphate solution (saturated)–water $(1.0 : 0.5 : 1.2 : 1.0)$ was used to separate the pigments in a commercial red beet juice concentrate without further work-up of the sample. The major peak consisted of a mixture of betanin and its C_{15} epimer isobetanin.

10.6.4   Characterisation, identification and quantification

Known betalains may be identified by comparison of their chromatographic, spectroscopic and electrophoretic properties with those of standards (Mabry & Dreiding, 1968; Strack & Wray, 1994; Stintzing et al., 2002a). Additional structure information may be achieved after appropriate hydrolysis by standard identification of amino acids, sugars and acyl moieties (Strack et al., 1993; Stintzing et al., 2002a). Betaxanthin standards may be synthesised by the reaction of betalamic acid with amino compounds (Trezzini & Zrÿd, 1991; Stintzing et al., 2002a).

10.6.4.1   Thin-layer chromatography

TLC on cellulose (Macherey-Nagel, Avicel), with the solvent ethyl acetate–formic acid–water $(33 : 7 : 10)$, has been used for separation of a complex mixture of betacyanins from bracts of *Bougainvillea glabra* (Heuer et al., 1994). High sample load (200 mg pigment from beetroot) has been administrated by pre-run in the polar solvent, 2-propanol–ethanol–water–acetic acid $(6 : 7 : 6 : 1)$ (Bilyk, 1981). After development of over 10 cm, followed by extensive drying of the plate, betanin was finally separated from the betaxanthins using two successive developments (15 cm) with the same solvent components in other proportions $(10 : 4 : 4 : 1)$. Separation of betaxanthins has been carried out on DEAE cellulose (Strack et al., 1987). Two developments with 2-propanol–water–acetic acid $(75 : 20 : 5)$ were required; however, this system failed to separate the major pigments extracted from beetroot.

10.6.4.2 High-performance liquid chromatography

HPLC with reverse-phase supports (C_{18} or C_8) has been widely used for separation of betalain mixtures (Fig. 10.16). Solvents consisting of water–methanol or water–acetonitrile mixtures acidified with acetic, formic or phosphoric acid have been favoured for both analytical (Strack *et al.*, 1993) and (semi-)preparative (Forni *et al.*, 1992) work.

10.6.4.3 Capillary electrophoresis

Valuable electrophoretic applications have been performed both on paper (e.g. Impellizzeri *et al.*, 1973) and cellulose thin-layer plates (e.g. Alard *et al.*, 1985). Betacyanins migrate as anions at pH values between 2 and 7, while they are immobile zwitterions at pH 2.0. Later CZE with multi-wavelength detection has been used for separation of the major betacyanins and the corresponding agly-cones from extracts of beetroot (Stuppner & Egger, 1996). Recently, investigations have been conducted into the use of low-level electric currents and voltages to extract, transport and collect secondary metabolites like betanin from beetroot cells (Yang *et al.*, 2003).

10.6.4.4 UV and infrared spectroscopy and quantification

Betacyanins and betaxanthins exhibit intense absorption around 540 nm and 480 nm, respectively. When betacyanins are acylated with aromatic acids, a second absorption maximum is shown in the region 300–330 nm, where the absorption of non-acylated betacyanins are weak. The molar absorptivities (ε) of betalains vary considerably with structure, solvent and even concentration. For quantification purposes the ε-values 61 000 and 48 000 may be used for pure isolated betacyanins and betaxanthins recorded at their visible absorption maxima, respectively (Wyler & Meuer, 1979). In mixtures, and when degradation products and other interfering substances are present, HPLC is the method of choice for quantitative analysis of individual and total betalains (Schwartz *et al.*, 1981). Absorption bands in Fourier transform infrared (FTIR) spectra may be used to characterise typical functional groups of betalains (Cai *et al.*, 1998).

10.6.4.5 Mass spectrometry

The molecular masses of betalains ($[M+H]^+$) have in recent years been determined mainly by electrospray MS in the positive ionisation mode (Strack *et al.*, 2003). A coupling with HPLC allows screening of individual betalains in mixtures without prior isolation. Most betalains will also produce daughter ions, recorded for instance at m/z 389 for betacyanins, corresponding to $[(iso)betanidin + H]^+$. The number of betacyanins in crude mixtures may also by estimated by electrospray tandem MS by parent ion scans selecting the $[(iso)betanidin + H]^+$ fragment ion at m/z 389 as daughter ion (Heuer *et al.*, 1994). A micro-HPLC system using electrospray ionisation and time-of-flight (TOF) mass detection has been used to obtain accurate masses of betalains from different *Celosia* species (Schliemann *et al.*, 2001).

10.6.4.6   Nuclear magnetic resonance spectroscopy

Proton NMR data on betalains have been recorded in DMSO-d_6 and CD$_3$OD containing CF$_3$COOD or traces of DCl. However, solvents containing acid, especially mineral acids like DCl, may hydrolyse linkages between, in particular, sugars and aliphatic acyl moieties. Some carbon-linked protons may be difficult to observe in spectra due to exchange with deuterium from the solvent. Proton NMR shifts on betalains are difficult to assign based on 1-D ^{1}H NMR spectra alone. In recent years 2-D homonuclear techniques (^{1}H-^{1}H DQF-COSY and ^{1}H-^{1}H TOCSY) have provided valuable information for assignments. A useful compilation of ^{1}H NMR data has been presented by Strack *et al.* (1993), and ^{1}H NMR data on the unusual betacyanidin, 2-descarboxybetanidin, are shown in Fig. 10.17.

Betacyanin aglycones occur either as betanidin, isobetanidin or 2-descarboxybetanidin (Figs 10.16 and 10.17). A betacyanin (2S,15S) seems to have H-11, H-12 and H-14A at slightly higher fields (0.02–0.04 ppm) than the corresponding iso-form (2S,15R) (Strack *et al.*, 1993). The double doublet of H-2 around 5.6 ppm of betanidin and isobetanidin is replaced by a two-proton triplet around 4.4 ppm in 2-descarboxybetanidin (Kobayashi *et al.*, 2001). Betacyanins are known to exist as E/Z isomers at the partial C-12/C-13 double bond. Signals for H-11 and H-12 of the 12 Z-isomer are normally detected at lower and higher field, respectively, compared to those of the 12 E-isomer.

The linkage position of the glycosyl moiety to either C-5 or C-6 of betacyanidins should clearly be revealed by a long-range coupling between anomeric sugar proton and the aglycone C-5 or C-6 in heteronuclear HMBC NMR spectra.

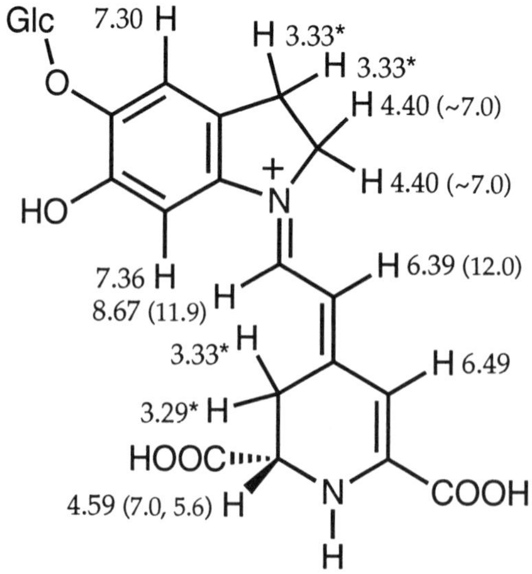

Fig. 10.17 ^{1}H NMR shifts (coupling constants in brackets) of 2-descarboxybetanin in CD$_3$OD/DCl (Kobayashi *et al.*, 2001). *from 2-D COSY.

However, heteronuclear NMR spectra have not yet been published for betalains. A small chemical shift difference between H-4 and H-7 (~ 0.1 ppm) is nevertheless characteristic for glycosyl substitution at the C-5 hydroxyl group of betanidin, as opposed to that at C-6 (~ 0.8 ppm). For the betaxanthins the chemical shifts and coupling constants in the 1,7-diazaheptamethinium system closely resemble those found in the betacyanins (Strack *et al.*, 1993).

10.6.5   Differentiation between betalains and anthocyanins

The most reliable method of distinguishing between betacyanins and anthocyanins is electrophoresis using acid buffer (pH 2–4), when betacyanins migrate as anions and anthocyanins as cations. However, anthocyanins acylated with dicarboxylic acid groups will also migrate as anions at pH around 4. Betaxanthins are similarly recognised by their anionic mobility, whereas water-soluble yellow flavonoids are immobile under these conditions. When an unknown sample is applied on a cellulose TLC plate and developed in 1-butanol–acetic acid–water (6 : 1 : 2) for 2 h, betalains will move rather slowly compared to anthocyanins. These mobilities will be reversed using aqueous solvent containing some acid as mobile phase.

10.7   Quinones

10.7.1   Introduction

The natural quinone pigments belong to a wide selection of structural types (Thomson, 1957, 1971, 1987, 1997), although the naphthoquinones and the anthraquinones are perhaps the best known (see also Chapter 7). Natural pigments containing more extensive chromophores are well documented, including dimeric structures such as hypericin, and some representative structures can be found in Fig. 10.18. Unlike the other common pigment classes the quinones are biosynthesised by way of a variety of pathways, and this may go some way towards explaining the fact that the functions of the quinone pigments in the plant are not well understood. It should be noted that they can, and frequently do, occur in tissues that are not directly visible to the naked eye, e.g. in heartwood and in roots. They can provide almost any colour of compound and have found historic uses as dyes for this reason. Their medical applications cover human history from the use of the emetic properties of rhubarb by early civilisations to current developments in the field of antibiotics. The naphthoquinone shikonin is one of the first compounds produced by modern industrial biotechnology.

10.7.2   Extraction and hydrolysis

The great variation of structure found among the natural quinone pigments makes it impossible to recommend general methods for all of these pigments. Methods of extraction and chromatography necessarily reflect polarity, and thus a plethora of

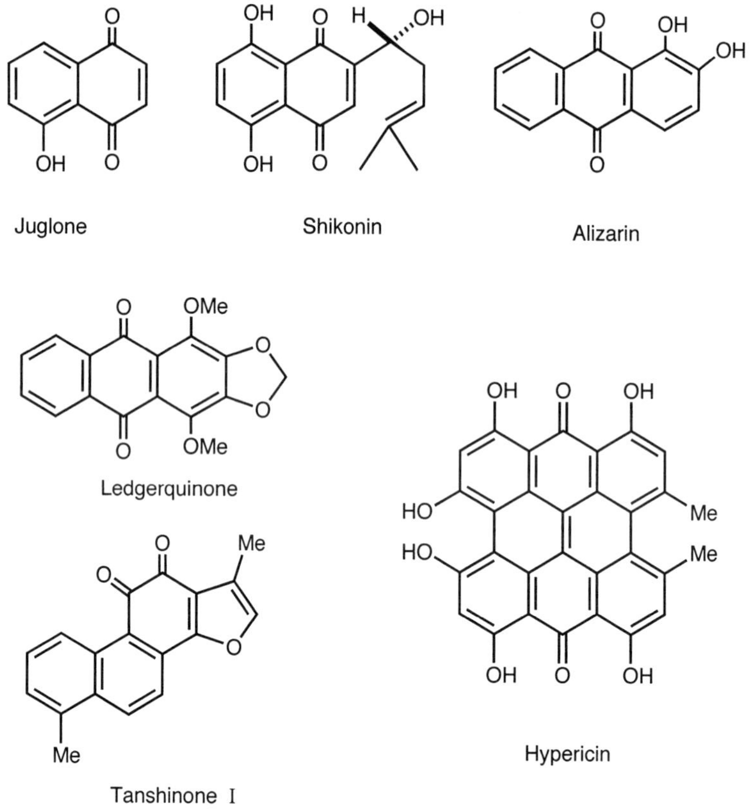

Juglone   Shikonin   Alizarin

Ledgerquinone

Tanshinone I   Hypericin

Fig. 10.18 Some representative natural pigments with quinonoid structures.

solvents have been used in work with quinones. The normal procedure is to use extraction with a polar organic solvent, methanol being a case in point, at either room temperature or using a Soxhlet apparatus. It is advisable to carry out serial extractions with several solvents of increasing polarity to ensure complete extraction.

Hydrolysis of glycosides is readily performed by taking the initial extract to dryness and heating it under reflux for 25 min with 7.5% hydrochloric acid. After cooling, the aglycones are extracted with portions of diethyl ether. A method involving direct hydrolysis of the glycosides on the TLC plate has been developed (Ebel & Kaal, 1980).

10.7.3 Purification and separation

Column chromatography, TLC and more recently HPLC have all found applications in the separation of quinones on both analytical and preparative scales. The variety of methods is endless and reflects the huge differences in structure and polarity present in these compounds. There are a number of reviews available in the literature (Thomson, 1976; Rauwald, 1990; Kocjan, 2000), although in most cases it is advisable to examine reports describing extraction of similar plant species, or referring to compounds of the same type.

10.7.3.1 Thin-layer chromatography

Some data from a TLC investigation of the anthraquinone glycosides of rhubarb and buckthorn are to be found in Table 10.5. The structures of the compounds can be derived from Fig. 10.19. The findings shown are quite typical and illustrate the problems often met. The systems chosen utilise commercial silica gel plates, although they require somewhat different developing solvents for the intact glycosides (ethyl acetate–methanol–water in the proportions 100 : 13.5 : 10) and aglycones (light petroleum 40–60°C–ethyl acetate–formic acid in the proportions 75 : 25 : 1). The retention of the glycosides is largely determined by the sugar moieties, with retention increasing along the series apiosyl, rhamnosyl to glucosyl, thus mirroring the deoxypentosyl, deoxyhexosyl to hexosyl structures. The free aglycones follow the polarities expected on the basis of the oxygen functionalities being carried out.

10.7.4 Characterisation and identification

Widely varying chemical sensitivity restricts not only the methods of separation that can be applied at the initial stages of investigation but also the number of chemical derivatives that can be usefully prepared during characterisation. However, a great deal of spectroscopic data is available for individual compounds and

Table 10.5 Thin-layer chromatography of anthraquinone glycosides and aglycones obtained by hydrolysis (R_F values)

Compound	Intact glycoside System 1	Hydrolysate System 2
Emodin-6-*O*-apioside (frangulin B)	0.59	0.36
Emodin-6-*O*-rhamnoside (frangulin A)	0.50	0.36
Emodin-8-*O*-glucoside	0.41	0.36
Emodin-6-*O*-apiosyl-8-*O*-glucoside (glucofrangulin B)	0.23	0.36
Emodin-6-*O*-rhamnosyl-8-*O*-glucoside (glucofrangulin A)	0.17	0.36
Chrysophanol-8-*O*-glucoside	0.41	0.65
Physcion-8-*O*-glucoside	0.41	0.59
Aloe-emodin-8-*O*-glucoside	0.36	0.13

System 1: intact glycosides using as solvent ethyl acetate–methanol–water (100:13.5:10) on silica gel 60 layers (0.25 mm, Merck Art. 5715).
System 2: hydrolysates (aglycones) using as solvent light petroleum (40–60°C)–ethyl acetate–formic acid (75:25:1) on silica gel 60 layers (0.25 mm, Merck Art. 5715).

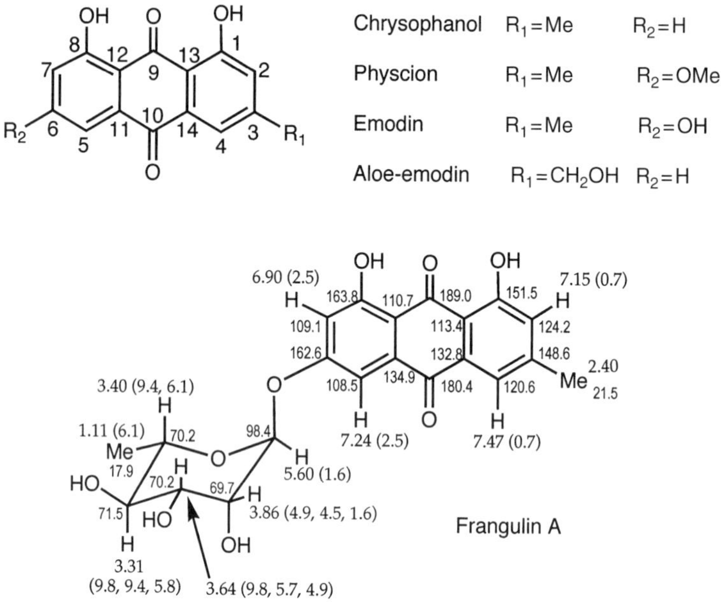

Fig. 10.19 Some anthraquinone pigments, and ^{13}C and ^{1}H-NMR data for frangulin A.

can be readily accessed in collected form through a valuable series of books on these compounds (Thomson, 1957, 1971, 1987, 1997).

10.7.4.1 Visible spectrometry and mass spectrometry

The use of visible light absorption spectroscopy allows the classification of the pigments into groups in the most general way, but gives little further information on unknown compounds. Similarly, infrared spectroscopy is of little use beyond providing confirmation of the type of compound being examined and indicating some of the groups present. Mass spectrometry has proved to give somewhat mixed results, depending on the type of quinone being investigated, but can often be used to provide a good indication of the molecular weight. Quinones often disproportionate in the system and 'molecular ions' should be regarded with caution as they may in fact often occur either two units above or below the actual molecular weight due to hydrogen transfer processes. While losses of 28 mass units, corresponding to the elements of carbon monoxide, are common in the various quinones and can confirm molecular weights, these losses often dominate the spectrum to an extent where other information is largely suppressed (Zeller & Mueller, 1988).

10.7.4.2 Nuclear magnetic resonance spectroscopy

NMR spectroscopy is capable of giving a great deal of information on quinones, in particular once a partial structure has been established, but distance effects are often seen in chemical shifts and these must be interpreted with care. An example of this type is provided by Frangulin A, the NMRs of which are presented in Fig. 10.19 (Francis *et al.*, 1998).

References

Abraham, R. J. & Rowan, A. E. (1991) Nuclear magnetic resonance spectroscopy of chlorophyll, in *Chlorophylls* (ed. H. Scheer), CRC Press, Boca Raton, Florida, USA, pp. 797–834.

Adams, J. P. & von Elbe, J. H. (1977) Betanine separation and quantification by chromatography on gels. *J. Food Sci.*, **42**, 410–414.

Agrawal, P. K. (1989) *Carbon-13 NMR of Flavonoids*, Elsevier, Amsterdam, The Netherlands.

Alard, D., Wray, V., Grotjahn, L., Reznik, H. & Strack, D. (1985) Neobetanin: isolation and identification from *Beta vulgaris*. *Phytochemistry*, **24**, 2383–2385.

Amaya, I. & Valpuesta, V. (2002) Improving natural pigments by genetic modification of crop plants. *Colour in Food*, (ed. D. B. MacDougall), Woodhead Publishing, Cambridge, UK, pp. 283–296.

Andersen, Ø. M. (1988) Semipreparative isolation and structure determination of pelargonidin 3-*O*-α-L-rhamnopyranosyl-(1→2)-β-D-glucopyranoside and other anthocyanins from the tree *Dacrycarpus dacrydioides*. *Acta Chem. Scand.*, **42**, 462–468.

Andersen, Ø. M. (2001a) Anthocyanins, in *Encyclopedia of Life Sciences*. www.els.net, Macmillan Publishers, New York, USA.

Andersen, Ø. M. (2001b) How easy is it nowadays to analyse anthocyanins, in *Polyphenols 2000* (eds S. Martens, D. Treutter & G. Forkmann), Technische Universitat Munchen, Freising, Germany, pp. 49–59.

Andersen, Ø. M. & Fossen, T. (2003) Characterization of anthocyanins by NMR, Unit F.1.4, in *Current Protocols in Food Analytical Chemistry*, (ed. R. Wrolstad), John Wiley, New York, USA, pp. 1–24.

Andersen, Ø. M., Fossen, T., Torskangerpoll, K., Hauge, U. & Fossen, A. (2004) Anthocyanin with novel aglycone, 5-carboxypyranopelargonidin, from strawberries, *Fragaria ananassa*. *Phytochemistry*, **65**, 405–410.

Andrade, F. D. P., Santos L. C., Datchler, M., Albert, K. & Vilegas, W. (2002) Use of on-line liquid chromatography-nuclear magnetic resonance spectroscopy for the rapid investigation of flavonoids from Sorocea bomplandii. *J. Chromatogr. A*, **953**, 287–291.

Asenstorfer, R. E., Iland, P. G., Tate, M. E. & Jones, G. P. (2003b) Charge equilibria and pKa of malvidin-3-glucoside by electrophoresis. *Anal. Biochem.*, **318**, 291–299.

Asenstorfer, R. E., Morgan, A. L., Hayasaka, Y., Sedgley, M. & Jones, G. P. (2003a) Purification of anthocyanins from species of *Banksia* and *Acacia* using high-voltage paper electrophoresis. *Phytochem. Anal.*, **14**, 150–154.

Bernhard, K. & Grosjean, M. (1995) in *Carotenoids*,Vol. 1B: *Spectroscopy* (eds Britton, G. Liaaen-Jensen, S & Pfander, H.), Birkhäuser, Basel, Switzerland, pp. 117–134.

Bicard, V., Fougerousse, A. & Brouillard, R. (1999) Analysis of natural anthocyanins by capillary zone electrophoresis in acidic media. *J. Liq. Chrom. Rel. Technol.*, **22**, 541–550.

Bilyk, A. (1981) Thin-layer chromatography separation of beet pigments. *J. Food Sci.*, **46**, 298–299.

Bjoern, L. O. (2002) Spectral Tuning in biology, in *Photobiology* (ed. L. O. Bjoern), Kluwer Academic Publishers, Dordrecht, The Netherlands, pp. 115–151.

Britton, G. (1995) UV/Visible Spectroscopy, in *Carotenoids*, Vol. 1B: *Spectroscopy* (eds G. Britton, S. Liaaen-Jensen & H. Pfander), Birkhäuser, Basel, Switzerland, pp. 13–62.

Britton, G., Liaaen-Jensen, S. & Pfander, H. (eds) (1995) *Carotenoids*, Vol 1A: *Isolation and Analysis*, Birkhäuser, Basel, Switzerland, pp. 81–108.

Bohm, B. A. (1998) *Introduction to Flavonoids*, Harwood Academic Publisher, Singapore.

Cai, Y., Sun, M., Wu, H., Huang, R. & Corke, H. (1998) Characterization and Quantification of betacyanin pigments from diverse *Amaranthus* species. *J. Agric. Food Chem.*, **46**, 2063–2070.

Chen, L.-J., Games, D. E. & Jones, J. (2003) Isolation and identification of four flavonoid constituents from the seeds of *Oroxylum indicum* by high-speed counter-current chromatography. *J. Chromatogr. A*, **988**, 95–105.

Chu, Q. C., Qu, W. Q., Peng, Y. Y., Cao, Q. H. & Ye, J. N. (2003) Determination of flavonoids in *Hippophae rhamnoides* L. and its phytopharmaceuticals by capillary electrophoresis with electrochemical detection. *Chromatographia*, **58**, 67–71.

DaCosta, C. T. Horton, D. & Margolis, S. A. (2000) Analysis of anthocyanins in foods by liquid chromatography, liquid chromatography-mass spectrometry and capillary electrophoresis. *J. Chromatogr. A*, **881**, 403–410.

Dao, L. T., Takeoka, G. R., Edwards, R. H. & Berrios, J. D. (1998) Improved method for the stabilization of anthocyanidins. *J. Agric. Food Chem.*, **46**, 3564–3569.

Davis, B. H. (1976) Carotenoids, in *Chemistry and Biochemistry of Plant Pigments*,Vol. 2, 2nd edn (ed. T. W. Goodwin), Academic Press, London, UK, pp. 38–165.

De Jong, M. (1997) Natural pigments: red from green. *Chemisch Magazine (Rijswijk, Netherlands)* pp. 28–31.

Degenhardt, A. & Winterhalter, P. (2001) Isolation of natural pigments by high speed CCC. *J. Liq. Chrom. Rel. Technol.*, **24**, 1745–1764.

Delgado-Vargas, F., Jiménez, A. R. & Paredes-López, O. (2000) Natural pigments: carotenoids, anthocyanins, and betalains – characteristics, biosynthesis, processing and stability. *Cri. Rev. Food Sci. Nutr.*, **40**, 173–289.

De Pascual-Teresa, S. & Rivas-Gonzalo, J. C. (2003) Application of LC-MS for the identification of polyphenols, in *Methods in Polyphenol Analysis* (eds C. Santos-Buelga & G. Williamson), Royal Society of Chemistry, Cambridge, UK, pp. 48–62.

De Rijke, E., Zappey, H., Ariese, F., Gooijer, C. & Brinkman, U. A. T. (2003) Liquid chromatography with atmospheric pressure chemical ionization and electrospray ionization mass spectrometry of flavonoids with triple-quadrupole and ion-trap instruments. *J. Chromatogr. A*, **984**, 45–58.

Ebel, S. & Kaal, M. (1980) Analysis of anthraquinone drugs. I. Hydrolysis of glycosides directly on TLC plates. *Planta Med.*, **40**, 271–277.

Englert, G. (1995) NMR spectroscopy, in *Carotenoids*,Vol. 1B: *Spectroscopy* (eds Britton, G., Liaaen-Jensen, S. & Pfander, H.), Birkhauser, Basel, pp. 147–260.

Enzell, C. R. & Back, S. (1995). Mass spectroscopy, in *Carotenoids*,Vol. 1B: *Spectroscopy* (eds G. Britton, S. Liaaen-Jensen & H. Pfander), Birkhäuser, Basel, Switzerland, pp. 261–320.

Escribano-Bailon, M. T. & Santos-Buelga, C. (2003) Polyphenol extraction from foods, in *Methods in Polyphenol Analysis* (eds C. Santos-Buelga & G. Williamson), Royal Society of Chemistry, Cambridge, UK, pp. 1–16.

Eugster, C. H. (1995) Chemical derivatization: Microscale tests for the presence of common functional groups in carotenoids, in *Carotenoids*, Vol. 1A: *Isolation and Analysis* (eds G. Britton, S. Liaaen-Jensen & H. Pfander), Birkhäuser, Basel, Switzerland, pp. 71–80.

Farina, A., Doldo, A., Cotichini, V., Rajevic, M., Quaglia, M. G. Mulinacci, N. & Vincieri, F. F. (1995) HPTLC and reflectance mode densitometry of anthocyanins in *Malva silvestris* L.: a comparison with gradient-elution reversed-phase HPLC. *J. Pharm. Biomed Anal.*, **14**, 203–211.

Foppen, F. H. (1971) Tables for the identification of carotenoid pigments. *Chromatogr. Reviews*, **14**, 133–298.

Formaggio, E., Cinque, G. & Bassi, R. (2001) Functional architecture of the major light-harvesting complex from higher plants. *J. Mol. Biol.*, **314**, 1157–1166.

Forni, E., Polesello, A., Montefiori, D. & Maestrelli, A. (1992) High-performance liquid chromatographic analysis of the pigments of blood-red prickly pear (*Opuntia ficus indica*). *J. Chromatogr. A*, **593**, 177–183.

Fossen, T. & Andersen, Ø. M. (2003) Anthocyanins from red onion, *Allium cepa*, with novel aglycon. *Phytochemistry*, **62**, 1217–1220.

Fossen, T., Slimestad, R., Øvstedal, D. O. & Andersen, Ø. M. (2000) Covalent anthocyanin-flavonol complexes from flowers of chive, *Allium schoenoprasum. Phytochemistry*, **54**, 317–323.

Francis, G. W. & Andersen, Ø. M. (2003) Natural pigments, in *Handbook of Thin-Layer Chromatography*, 3rd edn (eds J. Sherma & B. Fried), Marcel Dekker, New York, USA, pp. 697–732.

Francis, G. W. & Isaksen, M. (1988) Thin-layer chromatography of carotenoids with tertiary alcohol-petroleum ether solutions as developing solvents. *J. Food Sci.*, **53**, 979–980.

Francis, G. W., Aksnes, D. W. & Holt, Ø. (1998) Assignment of the 1H and 13C NMR spectra of anthraquinone glycosides from *Rhamnus frangula. Magn. Res. Chem.*, **36**, 769–772.

Frøytlog, C., Slimestad, R. & Andersen, Ø. M. (1998) Combination of chromatographic techniques for preparative isolation of anthocyanins – applied on blackcurrant *(Ribes nigrum)* fruits. *J. Chromatogr. A*, **825**, 89–95.

Fukui, Y., Takaaki, K., Katsuyoshi, M., Takashi, I. & Kyosuke, N. (2002) Structure of rosacyanin B, a novel pigment from the petals of *Rosa hybrida. Tetrahedron Lett.*, **43**, 2637–2639.

Fulcrand, H., Benabdeljalil, C., Rigaud, J., Cheynier, V. & Moutounet, M. (1998) A new class of wine pigments generated by reaction between pyruvic acid and grape anthocyanins. *Phytochemistry*, **47**, 1401–1407.

Gakh, E. G., Dougall, D. K. & Baker, D. C. (1998) Proton nuclear magnetic resonance studies of mono-acylated anthocyanins from the wild carrot: Part 1. Inter- and intra-molecular interactions in solution. *Phytochem. Anal.*, **9**, 28–34.

Garcia-Viguera, C., Zafrilla, P. Tomas-Barberan, F. A. (1998) Use of acetone as an extraction solvent for anthocyanins from strawberry fruit. *Phytochem. Anal.*, **9**, 274–277.

Giusti, M. M., Rodriguez-Saona, L. E. & Wrolstad, R. E. (1999) Molar absorptivity and color characteristics of acylated and non-acylated pelargonidin-based anthocyanins. *J. Agric. Food Chem.*, **47**, 4631–4637.

Giusti, M. M. & Wrolstad R. E. (2001) Characterization and measurement of anthocyanins by UV-visible spectroscopy, in *Current Protocols in Food Analytical Chemistry*, John Wiley, New York, USA.

Goodwin, T. W. (ed.) (1976) *Chemistry and Biochemistry of Plant Pigments*, Vols 1 and 2, 2nd edn, Academic Press, London, UK.

Goodwin, T. W. (1980) *The Biochemistry of the Carotenoids*, Vol. I: *Plants*, Chapman & Hall, London, UK.

Gonnet, J. F. (1998) Colour effects of co-pigmentation of anthocyanins revisited – 1. A colorimetric definition using the CIELAB scale. *Food Chem.*, **63**, 409–415.

Gross, J. (1991) *Pigments in Vegetables*, Van Nostrand Reinhold, New York, USA.

Harborne, J. B. (1967) *Comparative Biochemistry of the Flavonoids,* Academic Press, London, UK.

Harborne, J. B. (ed.) (1988) *The Flavonoids: Advances in Research since 1980*, Chapman & Hall, London, UK.

Harborne, J. B. (ed.) (1994) *The Flavonoids: Advances in Research since 1986*, Chapman & Hall, London, UK.

Harborne, J. B. & Mabry, T. J. (eds) (1982) *The Flavonoids: Advances in Research*, Chapman & Hall, London, UK.

Harborne, J. B. & Turner, B. L. (1984) *Plant Chemosystematics*, Academic Press, London, UK.

Harborne, J. B. & Baxter, H. (1999) *The Handbook of Natural Flavonoids*, John Wiley, Chichester, UK.

Harborne, J. B., Mabry, T. J. & Mabry, H. (eds) (1975) *The Flavonoids*, Academic Press, New York, USA.

Harborne, J. B. & Williams, C. A. (1998) Anthocyanins and other flavonoids. *Nat. Prod. Rep.*, **15**, 631–652.

Harborne, J. B. & Williams, C. A. (2001) Anthocyanins and other flavonoids. *Nat. Prod. Rep.*, **18**, 310–333.

Hashimoto, F., Tanaka, M., Maeda, H., Fukuda, S., Shimizu, K. & Sakata, Y. (2002) Changes in flower coloration and sepal anthocyanins of cyanic *Delphinium* cultivars during flowering. *Biosci. Biotech. Biochem.*, **66**, 1652–1659.

Heredia, F. J., Francia-Aricha, E. M., Rivas-Gonzalo, J. C., Vicario, I. M. & Santos-Buelga, C. (1998) Chromatic characterization of anthocyanins from red grapes – I. pH effect. *Food Chem.*, **63**, 491–498.

Heuer, S., Richter, S., Metzger, J. W., Wray, V., Nimtz, M. & Strack, D. (1994) Betacyanins from bracts of *Bougainvillea glabra. Phytochemistry*, **37**, 761–767.

Hong, V. & Wrolstad, R. E. (1990) Use of HPLC separation/photodiode array detection for characterization of anthocyanins. *J. Agric. Food Chem.*, **38**, 708–715.

Hoshino, T. (1992) Self-association of flavylium cations of anthocyanidin 3,5-diglucosides studied by circular dichroism and proton NMR. *Phytochemistry*, **31**, 647–654.

Huck, C. W., Stecher, G., Ahrer, W., Stoggi, W. M., Buchberger, W. & Bonn, G. K. (2002) Analysis of three flavonoids by CE-UV and CE-ESI-MS. Determination of naringenin from a phytomedicine. *J. Sep. Sci.*, **25**, 904–908.

Ikan, R. (1969) *Natural Products, A Laboratory Guide*, Israel University Press, Jerusalem.

Impellizzeri, G., Piattelli, M. & Sciuto, S. (1973) Acylated betacyanins from *Drosanthemum floribundum*. *Phytochemistry*, **12**, 2295–2296.

Iriyama, K., Ogura, N. & Takamiya, A. (1974) A simple method for extraction and partial purification of chlorophyll from plant material, using dioxane. *J. Biochem.*, **76**, 901–904.

Isaksen, M. & Francis, G. W. (1986) Reversed-phase thin-layer chromatography of carotenoids. *J. Chromatogr.*, **355**, 358–362.

Isaksen, M. & Francis, G. W. (1990) Silver ion spray reagent for the discrimination of β- and ε-end groups in carotenoids on thin-layer chromatograms. *Chromatographia*, **29**, 363–365.

Iwashina, T. & Kitajima, J. (2000) Chalcone and flavonol glycosides from *Asarum canadense* (Aristolochiaceae). *Phytochemistry*, **55**, 971–974.

Jackman, R. L. & Smith, J. L. (1996) Anthocyanins and betalains, in *Natural Food Colourants* (eds G. A. F. Hendry & J. D. Houghton), Chapman & Hall, London, UK, pp. 244–309.

Kim, J. H., Nonaka, G., Fujieda, K. & Uemoto, S. (1989) Anthocyanidin malonylglucosides in flowers of *Hibiscus syriacus*. *Phytochemistry*, **28**, 1503–1506.

Kobayashi, N. Wray, V. & Schliemann, W. (2001) Formation and occurrence of dopamine-derived betacyanins. *Phytochemistry*, **56**, 429–436.

Kocjan, B. (2000) Detection and identification of quinones in TLC. *J. Planar Chromatography-Modern TLC*, **13**, 396–397.

Kraemer-Schafhalter, A., Fuchs, T. H. & Pfannhauser, W. (1998) Solid-phase extraction (SPE) – a comparison of 16 materials for the purification of anthocyanins from *Aronia melanocarpa* var Nero. *J. Sci. Food Agric.*, **78**, 435–440.

Krisa, S., Teguo, P. W., Decendit, A., Deffieux, G., Vercauteren, J. & Merillon, J.-M. (1999) Production of 13C-labelled anthocyanins by *Vitis vinifera* cell suspension cultures. *Phytochemistry*, **51**, 651–656.

Levai, A. (1998) Utilization of the chiroptical spectroscopies for the structure elucidation of flavonoids and related benzopyran derivatives. *Acta Chim. Slovenica*, **45**, 267–284.

Li, X.-C., Joshi, A. S., Tan, B., ElSohly, H. N., Walker, L. A., Zjawiony, J. K. & Ferreira, D. (2002) Absolute configuration, conformation, and chiral properties of flavanone-(3→8'')-flavone biflavonoids from *Rheedia acuminata*. *Tetrahedron*, **58**, 8709–8717.

Liaaen-Jensen, S. (1997) Stereochemical aspects of carotenoids. *Pure Appl. Chem.*, **69**, 2027–2038.

Lichtenthaler, H. K. & Wellburn, A. R. (1983) Determinations of total carotenoids and chlorophylls a and b of leaf extracts in different solvents. *Biochem. Soc. Trans.*, **603**, 591–592.

Lu, Y. R. & Foo, L. Y. (2001) Unusual anthocyanin reaction with acetone leading to pyranoanthocyanin formation. *Tetrahedron Lett.*, **42**, 1371–1373.

Mabry, T. J. & Dreiding, A. S. (1968) The betalains, in *Recent Advances in Phytochemistry* (eds T. J. Mabry, R. E. Alston & V. C. Runeckles), Appleton-Century-Crofts, New York, USA, pp. 145–69.

Mabry, T. J., Markham, K. R. & Thomas, M. B. (1970) *The Systematic Identification of Flavonoids*, Springer-Verlag, New York, USA.

Manach, C. (2003) The use of HPLC with coulometric array detection in the analysis of flavonoids in complex matrixes, in *Methods in Polyphenol Analysis* (eds C. Santos-Buelga & G. Williamson), Royal Society of Chemistry, Cambridge, UK, pp. 63–91.

Maoka, T., Tsushima, M. & Nishino, H. (2002) Isolation and characterisation of Dinochrome A and B, anticarcinogenic active carotenoids from the fresh water red tide *Peridinium bipes*, *Chem. Pharm. Bull.*, **50**, 1630–1633.

Marchart, E. & Kopp, B. (2003) Capillary electrophoretic separation and quantification of flavone-*O*- and *C*-glycosides in *Achillea setacea* W. et K. *J. Chromatogr. B*, **792**, 363–368.

Markham, K. R. (1982) *Techniques of Flavonoid Identification*, Academic Press, London, UK.

Markham, K. R. & Bloor, S. J. (1998) Analysis and identification of flavonoids in practise, in *Flavonoids in Health and Disease* (eds C. A. Rice-Evans & L. Packer), Marcel Dekker, New York, USA, pp. 1–33.

Markham, K. R. & Chari, V.M. (1982) Carbon-13 NMR spectroscopy of flavonoids, in *The Flavonoids: Advances in Research* (eds J. B. Harborne & T. J. Mabry), Chapman & Hall, London, UK, pp. 19–134.

Markham, K. R. & Geiger, H. (1993) ^{1}H nuclear magnetic resonance spectroscopy of flavonoids and their glycosides in hexadeuterodimethylsulfoxide, in *The Flavonoids: Advances in Research Since 1986* (eds J. B. Harborne), Chapman & Hall, London, UK, pp. 441–497.

Markham, K. R. & Wilson, R. D. (1988) Paper chromatographic mobilities of a range of flavone and flavonol-*O*-glycosides. *Phytochem. Bull.*, **20**, 8–12.

Matsuno, T., Tani, Y., Maoka, T., Matsuo, K. & Komori, T. (1986) Isolation and structural elucidation of cucurbitaxanthin A and B from pumpkin *Cucurbita maxima. Phytochemistry*, **25**, 2837–2840.

Mazza, G. & Miniati, E. (1993) *Anthocyanins in fruits, vegetables and grains*, CRC Press, Boca Raton, Florida, USA.

Merken, H. M. & Beecher, G. R. (2000) Measurement of food flavonoids by high-performance liquid chromatography. *J. Agric. Food Chem.*, **48**, 577–599.

Minguez-Mosquera, M. I., Gandul-Rojas, B., Gallardo-Guerrero, L. & Jaren-Galan, M. (2002a) Chlorophylls, in *Methods of Analysis for Functional Foods and Nutraceuticals* (ed. W. J. Hurst), CRC Press, Boca Raton, Florida, USA, pp. 159–218.

Minguez-Mosquera, M. I., Hornero-Mendez, D & Perez-Galvez, A. (2002b) Carotenoids and provitamin A in functional foods, in *Methods of Analysis for Functional Foods and Nutraceuticals* (ed. W. J. Hurst), CRC Press, Boca Raton, Florida, USA, pp. 101–157.

Miyanaga, K., Seki, M. & Furusaki, S. (2000) Quantitative determination of cultured strawberry-cell heterogeneity by image analysis: effects of medium modification on anthocyanin accumulation. *Biochem. Eng. J.*, **5**, 201–207.

Nerdal, W. & Andersen, Ø. M. (1992) Intermolecular aromatic acid association of an anthocyanin (petanin) evidenced by 2-dimensional nuclear Overhauser enhancement nuclear-magnetic-resonance experiments and distance geometry calculations. *Phytochem. Anal.*, **3**, 182–189.

Nitsche, H. (1974) Identity of loroxanthin with pyrenoxanthin, trollein, and trihydroxy-α-carotene. *Arch. Microbiol.* **95**, 79–90.

Nørbæk, R., Christensen, L. P. & Brandt, K. (1998) An HPLC investigation of flower colour and breeding of anthocyanins in species and hybrids of Alstroemeria. *Plant Breeding*, **117**, 63–67.

Nørbæk, R. & Kondo, T. (1998) Anthocyanins from flowers of *Crocus* (Iridaceae). *Phytochemistry*, **47**, 861–864.

Oliveira, M. C., Esperanca, P., Ferreira, M. A. A. (2001) Characterisation of anthocyanidins by electrospray ionisation and collision-induced dissociation tandem mass spectrometry. *Rap. Commun. Mass Spectrom.*, **15**, 1525–1532.

Pardo, F., Salinas, M. R., Alonso, G. L., Navarro, G. & Huerta, M. D. (1999) Effect of diverse enzyme preparations on the extraction and evolution of phenolic compounds in red wines. *Food Chem.*, **67**, 135–142.

Pfander, H., Riesen, R. & Niggli, U. (1994) HPLC and SFC of carotenoids – scope and limitations. *Pure Appl. Chem.*, **66**, 947–954.

Rauwald, H. W. (1990) Naturally occurring quinones and their related reduction forms: analysis and analytical methods. *PZ Wissenschaft*, **3**, 169–181.

Revilla, E., Ryan, J. M. & Martin-Ortega, G. (1998) Comparison of several procedures used for the extraction of anthocyanins from red grapes. *J. Agric. Food Chem.*, **46**, 4592–97.

Rivas-Gonzalo, J. C. (2003) Analysis of anthocyanins, in *Methods in Polyphenol Analysis* (eds C. Santos-Buelga & G. Williamson), Royal Society of Chemistry, Cambridge, UK, pp. 338–358.

Saenz-Lopez, R., Fernandez-Zurbano, P. & Tena, M. T. (2003) Development and validation of a capillary zone electrophoresis method for the quantitative determination of anthocyanins in wine. *J. Chromatogr. A*, **990**, 247–258.

Santos-Buelga, C. Garcia-Viguera, C. & Tomas-Barberan, F. A. (2003) On-line identification of flavonoids by HPLC coupled to diode array detection, in *Methods in Polyphenol Analysis* (eds C. Santos-Buelga & G. Williamson), Royal Society of Chemistry, Cambridge, UK, pp. 92–127.

Santos-Buelga, C. & Williamson, G. (eds) (2003) *Methods in Polyphenol Analysis*. Royal Society of Chemistry, Cambridge, UK.

Schliemann, W., Cai, Y., Degenkolb, T., Schmidt, J. & Corke, H. (2001) Betalains of *Celosia argentea*. *Phytochemistry*, **58**, 159–165.

Schliemann, W., Joy, R. W. IV, Komamine, A., Metzger, J. W., Nimtz, M., Wray, V. & Strack, D. (1996) Betacyanins from plants and cell cultures of *Phytolacca americana*. *Phytochemistry*, **42**, 1039–1046.

Schliemann, W., Kobayashi, N. & Strack, D. (1999) The decisive step in betaxanthin biosynthesis is a spontaneous reaction. *Plant Physiol.*, **119**, 1217–1232.

Schwartz, S. J., Hildenbrand, B. E. & von Elbe, J. H. (1981) Comparison of spectrophotometric and HPLC methods of quantify betacyanins. *J. Food Sci.*, **46**, 286–297.

Schoefs, B. (2001) Determination of tetrapyrrole pigments in foodstuffs. *Int. Lab.*, **33**, 34–39.

Schoefs, B. (2002) Chlorophyll and carotenoid analysis in food products. Properties of the pigments and methods of analysis. *Trends Food Sci. Tech.*, **13**, 361–71.

Schoefs, B. (2003) Chlorophyll and carotenoid analysis in food products. A practical case-by-case view. *Trends Anal. Chem.*, **22**, 335–339.

Siefermann-Harms, D., Hertzberg, S., Borch, G. & Liaaen-Jensen, S. (1981) Carotenoids of higher plants. Part 13. Lactucaxanthin, an ε,ε-carotene-3,3'-diol from *Lactuca sativa*. *Phytochemistry*, **20**, 85–88.

Sivam, G. (2002) Analysis of flavonoids, in *Methods of Analysis for Functional Foods and Nutraceuticals* (ed. W. J. Hurst), CRC Press, Boca Raton, Florida, USA, pp. 363–384.

Stahl, E. (1969) *Thin Layer Chromatography*, 2nd edn, Springer-Verlag, New York, USA.

Steglich, W. & Strack, D. (1990) Betalains, in *The Alkaloids* (ed. A. Brossi), Academic Press, Orlando, Florida, USA, pp. 1–62.

Steuer, W., Grant, I. & Erni, F. (1990) Chromatography and capillary zone electrophoresis in drug analysis. *J. Chromatogr. A*, **507**, 125–140.

Stintzing, F. C., Schieber, A. & Carle, R. (2002a) Identification of betalains from yellow beet (*Beta vulgaris* L.) and cactus pear [*Opuntia ficus-indica* (L.) Mill.] by high-performance liquid chromatography electrospray ionization mass spectrometry. *J. Agric. Food Chem.*, **50**, 2302–2307.

Stintzing, F. C., Schieber, A. & Carle, R. (2002b) Betacyanins in fruits of red-purple pitaya, *Hylocerus polyrhizus* (Weber) Britton and Rose. *Food Chem.*, **77**, 101–106.

Strack, D., Schmitt, D., Reznik, H., Boland, W., Grotjahn, L. & Wray, V. (1987) Humilixanthin, a new betaxanthin from *Rivina humilis*. *Phytochemistry*, **26**, 2285–2287.

Strack, D., Steglich, W. & Wray, V. (1993) Betalains, in *Methods in Plant Biochemistry* (eds P. M. Dey & J. B. Harborne), Vol. 8: *Alkaloids and Sulphur Compounds* (ed. P. G. Waterman), Academic Press, London, UK, pp. 421–450.

Strack, D., Vogt, T. & Schliemann, W. (2003) Recent advances in betalain research. *Phytochemistry*, **62**, 247–269.

Strack, D. & Wray, V. (1989) Anthocyanins, in *Methods in Plant Biochemistry* (eds P. M. Dey & J. B. Harborne), Academic Press, London, UK, pp. 325–355.

Strack, D. & Wray, V. (1994) Recent advances in betalain analysis, in *Caryophyllales, Evolution and Systematics* (eds H.-D. Behnke & T. J. Mabry), Springer-Verlag, Berlin, Germany, pp. 263–277.

Straub.O. (1987) *Key to Carotenoids*, Birkhäuser, Basel, Switzerland.

Stuppner, H. & Egger, R. (1996) Application of capillary zone electrophoresis to the analysis of betalains from *Beta vulgaris*. *J. Chromatogr. A*, **735**, 409–413.

Swinny, E. E., Bloor, S. J. & Wong, H. (2000) 1H and 13C NMR assignments for the 3-deoxyanthocyanins, luteolinidin-5-glucoside and apigeninidin-5-glucoside. *Magn. Reson. Chem.*, **38**, 1031–1033.

Takeoka, G. & Dao, L. (2002) Anthocyanins, in *Methods of Analysis for Functional Foods and Nutraceuticals* (eds W. J. Hurst), CRC Press, Boca Raton, Florida, USA, pp. 219–241.

Takeoka, G. R., Dao. L. T., Full, G. H., Wong, R. Y., Harden, L. A., Edwards, R. H. & Berrios, J. D. J. (1997) Characterization of black bean (*Phaseolus vulgaris* L.) anthocyanins. *J. Agric. Food Chem.*, **45**, 3395–3400.

Torskangerpoll, K. & Andersen, Ø. M. (2004) Colour stability of anthocyanins in aqueous solutions at various pH values. *Food Chem.*, in press.

Torskangerpoll, K., Chou, E. & Andersen, Ø. M. (2001) Separation of acylated anthocyanin pigments by high speed counter-current chromatography *J. Liq. Chrom. Rel. Technol.*, **24**, 1791–1799.

Thomson, R. H. (1957) *Naturally Occurring Quinones*, Academic Press, New York, USA.

Thomson, R. H. (1971) *Naturally Occurring Quinones*, 2nd edn, Academic Press, London, UK.

Thomson, R. H. (1976) Isolation and identification of quinines, in *Chemistry and Biochemistry of Plant Pigments*, 2nd edn (ed. T. W. Goodwin), Academic Press, London, UK, pp. 207–232.

Thomson, R. H. (1987) *Naturally Occurring Quinones III: Recent Advances.* Chapman & Hall, New York, USA.

Thomson, R. H. (ed.) (1997) *Naturally Occurring Quinones IV: Recent Advances*, 4th edn, Blackie, London, UK.

Trezzini, G. F. & Zrÿd, J.-P. (1991) Characterization of some natural and semi-synthetic betaxanthins. *Phytochemistry*, **30**, 1901–1903.

Van Breemen, R. B. (1997) Liquid chromatography/mass spectrometry of carotenoids. *Pure Appl. Chem.*, **69**, 2061–2066.

Vetter, W., Englert, G., Rigassi, N. & Schwieter, U. (1971) Spectroscopic methods, in *Carotenoids* (ed. O. Isler), Birkhäuser, Basel, Switzerland, pp. 189–266.

Wagner, H., Bladt, S. & Zgainski, E. M. (1984) *Plant Drug Analysis,* Springer-Verlag, Berlin, Germany.

Wang, J. & Sporns, P. (2000) MALDI-TOF MS analysis of food flavonol glycosides. *J. Agric. Food Chem.*, **48**, 1657–1662.

Waterman, P. G. & Mole, S. (1994) *Analysis of Phenolic Plant Metabolites*, Blackwell, Oxford, UK.

Wharton, P. S. & Nicholson, R. L. (2000) Temporal synthesis and radiolabelling of the sorghum 3-deoxyanthocyanidin phytoalexins and the anthocyanin, cyanidin 3-dimalonyl glucoside. *New Phytologist*, **145**, 457–469.

Wolfender, J.-L., Ndjoko, K. & Hostettmann, K. (2003) Application of LC-NMR in the structure elucidation of polyphenols, in *Methods in Polyphenol Analysis* (eds C. Santos-Buelga & G. Williamson), Royal Society of Chemistry, Cambridge, UK, pp. 128–156.

Wollenweber, E. (1982) Thin-layer chromatography of flavonoids on polyamide layers. *GIT Fachzeitschrift fuer das Laboratorium (Suppl. Chromatogr.)*, 50–54.

Wrolstad, R. E., Durst, R. W. Giusti, M. M. & Rodriguez-Saona, L. E. (2002) Analysis of anthocyanins in nutraceuticals. *ACS Symposium Series*, **803**, 42–62.

Wyler, H. & Meuer, U. (1979) Biogenesis of betacyanins: experiments with [2-14C]dopaxanthine. *Helv. Chim. Acta*, **62**, 1330–1339.

Wyler, H., Vincenti, G., Mercier, M., Sassu, G. & Dreiding, A. S. (1959) Zur konstitution des randenfarbstoffes betanin. *Helv. Chim. Acta*, **42**, 1696–1698.

Yang, R. Y. K., Bayraktar, O. & Pu, H. T. (2003) Plant-cell bioreactors with simultaneous electropermeabilization and electrophoresis. *J. Biotech.*, **100**, 13–22.

Young, A. J. & Frank, H. A. (1996) Energy transfer reactions involving carotenoids: quenching of chlorophyll fluorescence. *J. Photochem. Photobiol., B: Biology*, **36**, 3–15.

Zeller, K. P. & Mueller, R. (1988) Mass spectra of quinines, in *Chem. Quinonoid Compounds 2*, Part 1(eds S. Patai & Z. Rappoport) John Wiley, Chichester, UK, pp. 87–109.

Zsila, F., Bikadi, Z. & Simonyi, M. (2003) Probing the binding of the flavonoid, quercetin to human serum albumin by circular dichroism, electronic absorption spectroscopy and molecular modelling methods. *Biochem. Pharmacol.*, **65**, 447–456.

Index